PARTIAL DIFFERENTIAL EQUATIONS

PARTIAL DIFFERENTIAL EQUATIONS

Theory and Completely Solved Problems

T. HILLEN
I. E. LEONARD
H. VAN ROESSEL
Department of Mathematical and Statistical Sciences
University of Alberta

WILEY

A JOHN WILEY & SONS, INC., PUBLICATION

Cover art: Water Wave courtesy of Brocken Inaglory: Coronal Mass Ejection courtesy of NASA/SDO and AIA, EVE, and HMI science teams.

Library of Congress Cataloging-in-Publication Data:

Hillen, Thomas, 1966–
Partial differential equations : theory and completely solved problems / Thomas Hillen, I. Ed Leonard, Henry van Roessel.
p. cm.
Includes bibliographical references and index.
ISBN 978-1-118-06330-9 (hardback)
1. Differential equations, Partial. I. Leonard, I. Ed., 1938– II. Van Roessel, Henry, 1956 III. Title.
QA377.H55 2012
515′.353–dc23 2012017382

Printed in the United States of America.

CONTENTS

Preface xi

PART I THEORY

1 Introduction **3**

1.1 Partial Differential Equations 4
1.2 Classification of Second-Order Linear PDEs 7
1.3 Side Conditions 10
1.3.1 Boundary Conditions on an Interval 12
1.4 Linear PDEs 12
1.4.1 Principle of Superposition 14
1.5 Steady-State and Equilibrium Solutions 16
1.6 First Example for Separation of Variables 19
1.7 Derivation of the Diffusion Equation 24
1.7.1 Boundary Conditions 25
1.8 Derivation of the Heat Equation 26
1.9 Derivation of the Wave Equation 29
1.10 Examples of Laplace's Equation 33
1.11 Summary 37
1.11.1 Problems and Notes 38

2 Fourier Series **39**

2.1 Piecewise Continuous Functions 39
2.2 Even, Odd, and Periodic Functions 41
2.3 Orthogonal Functions 43
2.4 Fourier Series 48
2.4.1 Fourier Sine and Cosine Series 53
2.5 Convergence of Fourier Series 56
2.5.1 Gibbs' Phenomenon 60
2.6 Operations on Fourier Series 63
2.7 Mean Square Error 74
2.8 Complex Fourier Series 78
2.9 Summary 81
2.9.1 Problems and Notes 82

3 Separation of Variables **83**

3.1 Homogeneous Equations 83
3.1.1 General Linear Homogeneous Equations 89
3.1.2 Limitations of the Method of Separation of Variables 93
3.2 Nonhomogeneous Equations 95
3.2.1 Method of Eigenfunction Expansions 100
3.3 Summary 111
3.3.1 Problems and Notes 113

4 Sturm-Liouville Theory **115**

4.1 Formulation 115
4.2 Properties of Sturm-Liouville Problems 119
4.3 Eigenfunction Expansions 127
4.4 Rayleigh Quotient 134
4.5 Summary 141
4.5.1 Problems and Notes 143

5 Heat, Wave, and Laplace Equations **145**

5.1 One-Dimensional Heat Equation 145
5.2 Two-Dimensional Heat Equation 150
5.3 One-Dimensional Wave Equation 153
5.3.1 d'Alembert's Solution 157
5.4 Laplace's Equation 163
5.4.1 Potential in a Rectangle 163
5.5 Maximum Principle 167

5.6 Two-Dimensional Wave Equation 168
5.7 Eigenfunctions in Two Dimensions 172
5.8 Summary 176
5.8.1 Problems and Notes 177

6 Polar Coordinates 179

6.1 Interior Dirichlet Problem for a Disk 179
6.1.1 Poisson Integral Formula 186
6.2 Vibrating Circular Membrane 188
6.3 Bessel's Equation 191
6.3.1 Series Solutions of ODEs 191
6.4 Bessel Functions 195
6.4.1 Properties of Bessel Functions 201
6.4.2 Integral Representation of Bessel Functions 204
6.5 Fourier-Bessel Series 210
6.6 Solution to the Vibrating Membrane Problem 214
6.7 Summary 218
6.7.1 Problems and Notes 220

7 Spherical Coordinates 221

7.1 Spherical Coordinates 221
7.1.1 Derivation of the Laplacian 222
7.2 Legendre's Equation 224
7.3 Legendre Functions 227
7.3.1 Legendre Polynomials 228
7.3.2 Fourier-Legendre Series 245
7.3.3 Legendre Functions of the Second Kind 248
7.3.4 Associated Legendre Functions 249
7.4 Spherical Bessel Functions 252
7.5 Interior Dirichlet Problem for a Sphere 253
7.6 Summary 257
7.6.1 Problems and Notes 259

8 Fourier Transforms 261

8.1 Fourier Integrals 261
8.1.1 Fourier Integral Representation 261
8.1.2 Examples 264
8.1.3 Fourier Sine and Cosine Integral Representations 268
8.1.4 Proof of Fourier's Theorem 271

8.2 Fourier Transforms 277
8.2.1 Operational Properties of the Fourier Transform 281
8.2.2 Fourier Sine and Cosine Transforms 284
8.2.3 Operational Properties of the Fourier Sine and Cosine Transforms 288
8.2.4 Fourier Transforms and Convolutions 289
8.2.5 Fourier Transform of a Gaussian Function 294
8.3 Summary 297
8.3.1 Problems and Notes 298

9 Fourier Transform Methods in PDEs 299

9.1 The Wave Equation 300
9.1.1 d'Alembert's Solution to the One-Dimensional Wave Equation 300
9.2 The Heat Equation 305
9.2.1 Heat Flow in an Infinite Rod 305
9.2.2 Fundamental Solution to the Heat Equation 306
9.2.3 Error Function 308
9.2.4 Heat Flow in a Semi-infinite Rod: Dirichlet Condition 311
9.2.5 Heat Flow in a Semi-infinite Rod: Neumann Condition 317
9.3 Laplace's Equation 319
9.3.1 Laplace's Equation in a Half-Plane 319
9.3.2 Laplace's Equation in a Semi-infinite Strip 324
9.4 Summary 328
9.4.1 Problems and Notes 329

10 Method of Characteristics 331

10.1 Introduction to the Method of Characteristics 331
10.2 Geometric Interpretation 335
10.3 d'Alembert's Solution 344
10.4 Extension to Quasilinear Equations 348
10.5 Summary 350
10.5.1 Problems and Notes 351

PART II EXPLICITLY SOLVED PROBLEMS

11 Fourier Series Problems 355

12 Sturm-Liouville Problems 387

13 Heat Equation Problems 425

14 Wave Equation Problems 481

15 Laplace Equation Problems 533

16 Fourier Transform Problems 567

17 Method of Characteristics Problems 597

18 Four Sample Midterm Examinations 615

18.1 Midterm Exam 1 615
18.2 Midterm Exam 2 619
18.3 Midterm Exam 3 623
18.4 Midterm Exam 4 626

19 Four Sample Final Examinations 631

19.1 Final Exam 1 631
19.2 Final Exam 2 640
19.3 Final Exam 3 646
19.4 Final Exam 4 653

Appendix A: Gamma Function 661

Bibliography 667

Index 671

PREFACE

This textbook on linear partial differential equations (PDEs) consists of two parts. In Part I we present the theory, with an emphasis on completely solved examples and intuition. In Part II we present a collection of exercises containing over 150 explicitly solved problems for linear PDEs and boundary value problems. These problems are based on more than 30 years of collective experience in teaching introductory PDE courses at several North American universities.

Many excellent introductory textbooks on PDEs are available, and over the years we have used the monographs by Asmar [5], Brown and Churchill [11], Haberman [25], Keane [31], and Powers [41], to name a few. These books give a concise, detailed, and easily accessible introduction to linear PDEs, and provide a number of solved examples. However, students always ask for additional problems with detailed solutions, and they tend to benefit from a drill-like repetition of problems and solutions. Here we address exactly this need. While Part I presents the theory behind linear PDEs and introduces methods and techniques for solving them, the problems in Part II allow students to learn and repeat arguments in hands-on exercises. The problems in Part II are all completely solved and explained in great detail.

The final two chapters contain four sample midterm examinations and four sample final examinations. These sample examinations are actual exams given between 2004 and 2009 at the University of Alberta. They provide students with a useful guideline as to what to expect as well as an opportunity to test their abilities.

To help students use the text, we have incorporated two special features. First, we rank the problems according to their difficulty; of course, this is a subjective task, but it gives a good indication of the level of difficulty anticipated. We use

rank 1	✗	for very simple problems,
rank 2	✗✗	for simple problems,
rank 3	✗✗✗	for more involved problems, and
rank 4	✗✗✗✗	for difficult problems.

The second feature of the text is a detailed summary at the end of each chapter, with cross-references to the solved problems in Part II.

Most colleges and universities now teach undergraduate courses in boundary value problems, Fourier series, Laplace transforms, or Fourier transforms, and then give applications to PDEs. The audience usually consists of third-year students of mathematics, engineering, and other sciences. The requirements are a solid grounding in calculus, linear algebra, and elementary differential equations. Students need to be familiar with the methods and techniques for solving linear ordinary differential equations (ODEs). Many ODE courses do not cover advanced topics such as Bessel's equation or Legendre's equation, and hence we include a full treatment in the chapters that deal with PDE problems in polar and spherical coordinates. In the past, physics and engineering have been a major source of interesting PDE problems; nowadays, problems come from other areas as well, such as mathematical biology. These problems address such topics as the spread of epidemics, survival or extinction of populations, or the invasion of healthy tissue by cancer cells; see, for example, [2] or [19].

Although not all PDEs can be solved by separation of variables or transform methods, most of this text focuses on these two methods. This is not surprising, since they form the backbone of any study of PDEs. Furthermore, the method of characteristics is also covered.

A choice of topics had to be made, and we chose to focus on the above three methods: *separation of variables*, *Fourier transforms*, and the *method of characteristics*, and to illustrate them in great detail. Thus, topics not included are *Green's functions* or *numerical methods*.

To prepare a syllabus for a one-semester course using this text, we recommend the following:

Chapters 1, 2, 3, 4, 5
Sections 6.1, 6.2, 6.6, 6.7, 7.1, 7.2, 7.4, 7.5
Chapters 8, 9, 10

Acknowledgments

We would like to thank our families for their support, without which this book could not have been written. T. H. would like to thank Lisa for her help in organizing and indexing the solved problems from Part II. I. E. L. would like to thank Amanda for reading the entire manuscript numerous times. T. H. and I. E. L. would especially like to thank Michael Keane for visiting Edmonton to discuss an earlier version of the book.

We will maintain a website at

`http://www.math.ualberta.ca/~thillen/pde-book-page.html`

with supplemental material and a list of errors and typos that readers find in the text.

We hope that this book will provide useful assistance to all those interested in learning to solve linear PDEs and provide a glimpse of the beautiful theory behind them.

THOMAS, ED, AND HENRY

Edmonton, Alberta, Canada
March 2012

PART I

THEORY

CHAPTER 1

INTRODUCTION

Many physical, biological, and engineering problems can be expressed mathematically by means of partial differential equations (PDEs) together with initial and/or boundary conditions. Partial differential equations are used in basically all scientific areas: for example, Schrödinger's equation in quantum mechanics, Maxwell's equations in electrodynamics, reaction-diffusion equations in chemistry and mathematical biology, models for spatial spread of populations and heat conduction problems, and the Black-Scholes formula for financial markets. Defining PDEs mathematically is quite simple, since a PDE is an equation that involves partial derivatives. The fascinating aspect of PDEs is that most of them can be classified into three classes: **elliptic, parabolic**, and **hyperbolic**. For illustrative purposes, we call these three classes the three "kingdoms." Each of these kingdoms has a "monarch," that is, a simple equation that exhibits most of the properties of the group. Elliptic equations are represented by **Laplace's equation**; parabolic equations are represented by the **heat equation**; and hyperbolic equations are represented by the **wave equation**. These classifications are defined for second-order linear equations in Section 1.2. The properties of these types, however, carry much further, and some higher-order equations exhibit "wave-like" or "diffusion-like" behavior. The study of these three basic equations, which represent the three subgroups, is the content of this book. If

Partial Differential Equations: Theory and Completely Solved Problems, First Edition. By T. Hillen, I. E. Leonard, H. van Roessel

you learn the methods for solving the wave equation, you will be able to study fluid flow in a pipeline, and through Schrödinger's equation you will gain an understanding of quantum mechanics. Laplace's equation is a prototype for Maxwell's equations in electrostatics, two-dimensional steady-state incompressible fluid flow, and the statics of buildings and bridges. The theory of the heat equation prepares you for the study of reaction-diffusion equations in population biology and for heat flow problems in conducting materials.

In this book, we deal almost exclusively with **linear** PDEs (the simplest type), and we use primarily one technique to solve them. This technique, called **separation of variables**, involves reducing (i.e., simplifying) the PDEs to ordinary differential equations (ODEs), which can then be solved using ODE methods. Generally, a PDE will have infinitely many solutions. To isolate a unique solution, we introduce side conditions (auxiliary conditions) that typically appear as **initial conditions** and **boundary conditions**. Before we delve into the theory, we recall some basic facts about functions and their partial derivatives. We then properly define the concepts of elliptic, parabolic, and hyperbolic PDEs.

1.1 PARTIAL DIFFERENTIAL EQUATIONS

Let $f : \Omega \to \mathbb{R}$ be a function defined on an open set $\Omega \subset \mathbb{R}^2$. The **partial derivatives** of $f(x, y)$ are defined as

$$\frac{\partial}{\partial x} f(x,y) = \lim_{h \to 0} \frac{f(x+h,y) - f(x,y)}{h},$$
$$\frac{\partial}{\partial y} f(x,y) = \lim_{h \to 0} \frac{f(x,y+h) - f(x,y)}{h},$$

provided that the limits exist. Thus, we differentiate with respect to one of the variables while holding the other variable fixed. Alternative notations, which we use in this text, include

$$\frac{\partial f}{\partial x}(x,y) = \frac{\partial f(x,y)}{\partial x} = f_x(x,y), \quad \frac{\partial f}{\partial y}(x,y) = \frac{\partial f(x,y)}{\partial y} = f_y(x,y)$$

or, even simpler,

$$\frac{\partial f}{\partial x} = f_x, \qquad \frac{\partial f}{\partial y} = f_y.$$

Further notations, which we do not use here but which can be found in other books, include

$$f_x = f_1 = \partial_x f = D_1 f, \quad f_y = f_2 = \partial_y f = D_2 f.$$

As a rule of thumb, the more important a concept is in mathematics, the more notations it has.

For example, the **gradient** of a function $f(x, y)$ is defined as

$$\nabla f(x, y) = (f_x(x, y), f_y(x, y))$$

and may be denoted by

$$\nabla f(x, y) = \frac{\partial f}{\partial x}(x, y)\,\widehat{\mathbf{i}} + \frac{\partial f}{\partial y}(x, y)\,\widehat{\mathbf{j}}$$

or, simply,

$$\nabla f = f_x\,\widehat{\mathbf{i}} + f_y\,\widehat{\mathbf{j}}$$

or in many other ways.

Similarly, the **Laplacian** of a function $f(x, y, z)$ is defined as

$$\triangle f(x, y, z) = \text{div (grad } f) = \nabla \cdot \nabla f(x, y, z) = \nabla^2 f(x, y, z)$$

and may also be expressed in many different ways, for example,

$$\triangle f(x, y, z) = \frac{\partial^2 f}{\partial x^2}(x, y, z) + \frac{\partial^2 f}{\partial y^2}(x, y, z) + \frac{\partial^2 f}{\partial z^2}(x, y, z)$$

or, simply,

$$\triangle f = \frac{\partial^2 f}{\partial x^2} + \frac{\partial^2 f}{\partial y^2} + \frac{\partial^2 f}{\partial z^2}.$$

Hence, partial derivatives are quite important! In general, a **partial differential equation** for an unknown function $u(x, y)$, $u(x, y, z)$, or $u(x, y, z, t)$, $\ldots$ can be written as a function:

$$F(x, y, u, u_x, u_y, u_{xx}, u_{xy}, u_{yy}, u_{xxx}, \ldots) = 0.$$

We now define some terminology. In the equations above, u (the unknown function) is referred to as a **dependent** variable, with all remaining variables x, y, z, t termed **independent** variables.

A **linear differential operator** is an operator containing partial derivatives such that

$$L(v + w) = Lv + Lw \qquad \text{and} \qquad L(\lambda v) = \lambda Lv$$

for all functions v and w in the domain of L and all scalars λ.

A PDE in some unknown function u is said to be **linear** if the equation can be written in the form

$$Lu = f,$$

where L is a linear differential operator and the function f does not depend on u or any of its derivatives. If $f = 0$, the equation is said to be **homogeneous**, while if $f \neq 0$, the equation is said to be **nonhomogeneous**.

The **order** of a PDE refers to the order of the highest-order derivative that appears in the equation. Finally, the **dimension** of a PDE refers to the number of independent variables present. If there is a clear distinction between time and space variables, dimension is also used for the spatial part alone.

Example 1.1. Find the dimension and order of the following PDEs. Which are linear, and which are homogeneous?

Heat equation: $$u_t = Du_{xx} + f(x) \tag{1.1}$$

Wave equation: $$u_{tt} - c^2 u_{xx} = 0 \tag{1.2}$$

Laplace's equation: $$u_{xx} + u_{yy} = 0 \tag{1.3}$$

Advection equation: $$\frac{\partial u}{\partial x} + \frac{\partial u}{\partial y} = 0 \tag{1.4}$$

$$\frac{\partial^2 u}{\partial x^2} + e^y \sin z \frac{\partial^2 u}{\partial x\, \partial z} = u \tag{1.5}$$

$$\frac{\partial^2 u}{\partial x\, \partial y} = \sin u \tag{1.6}$$

KdV equation: $$u_t + uu_{xx} + u_{xxx} = 1 \tag{1.7}$$

Solution.

- Equation (1.1) is a two-dimensional second-order linear nonhomogeneous (for $f \neq 0$) PDE. It is sometimes called the one-dimensional heat equation since the space variable x is one-dimensional. It can be written as $Lu = f$, with the differential operator

$$L = \frac{\partial}{\partial t} - D\frac{\partial^2}{\partial x^2}.$$

- Equation (1.2) is a two-dimensional second-order linear homogeneous PDE. It is also sometimes called the one-dimensional wave equation, since, again, the space variable x is one-dimensional. It can be written as $Lu = 0$, with

$$L = \frac{\partial^2}{\partial t^2} - c^2\frac{\partial^2}{\partial x^2}.$$

- Equation (1.3) is a two-dimensional second-order linear homogeneous PDE, with

$$L = \frac{\partial^2}{\partial x^2} + \frac{\partial^2}{\partial y^2}.$$

- Equation (1.4) is a two-dimensional first-order linear homogeneous PDE, with

$$L = \frac{\partial}{\partial x} + \frac{\partial}{\partial y}.$$

- Equation (1.5) is a three-dimensional second-order linear homogeneous PDE, with

$$L = \frac{\partial^2}{\partial x^2} + e^y \sin z\frac{\partial^2}{\partial x\, \partial z} - 1.$$

- Equation (1.6) is a two-dimensional second-order nonlinear nonhomogeneous PDE.
- Equation (1.7) is a two-dimensional third-order nonlinear nonhomogeneous PDE.

∎

1.2 CLASSIFICATION OF SECOND-ORDER LINEAR PDES

Definitions of the kingdoms of PDEs, elliptic, parabolic, and hyperbolic, can be obtained from a study of second-order linear PDEs. A general second-order linear constant-coefficient homogeneous PDE can be written as

$$au_{xx} + 2bu_{xy} + cu_{yy} + du_x + eu_y + fu = 0, \tag{1.8}$$

with real constant coefficients a, b, c, d, e, f (placing the factor 2 in front of the b-term is just a convention to make the PDE look nicer in the end). The type of this equation is defined by its **principal part**, which consists of the highest-order terms,

$$au_{xx} + 2bu_{xy} + cu_{yy}.$$

This expression can be written in abstract matrix notation as

$$\begin{pmatrix} \dfrac{\partial}{\partial x} & \dfrac{\partial}{\partial y} \end{pmatrix} \begin{pmatrix} a & b \\ b & c \end{pmatrix} \begin{pmatrix} \dfrac{\partial}{\partial x} \\ \dfrac{\partial}{\partial y} \end{pmatrix} u.$$

Here, we just pretend that $\partial/\partial x$ and $\partial/\partial y$ are symbols that can be entered as components of a vector. The interpretation is that this vector is applied as the derivative operators on the function $u(x, y)$:

$$\begin{aligned}
\begin{pmatrix} \dfrac{\partial}{\partial x} & \dfrac{\partial}{\partial y} \end{pmatrix} \begin{pmatrix} a & b \\ b & c \end{pmatrix} \begin{pmatrix} \dfrac{\partial}{\partial x} \\ \dfrac{\partial}{\partial y} \end{pmatrix} u &= \begin{pmatrix} \dfrac{\partial}{\partial x} & \dfrac{\partial}{\partial y} \end{pmatrix} \begin{pmatrix} a & b \\ b & c \end{pmatrix} \begin{pmatrix} \dfrac{\partial u}{\partial x} \\ \dfrac{\partial u}{\partial y} \end{pmatrix} \\
&= \begin{pmatrix} \dfrac{\partial}{\partial x} & \dfrac{\partial}{\partial y} \end{pmatrix} \begin{pmatrix} au_x + bu_y \\ bu_x + cu_y \end{pmatrix} \\
&= au_{xx} + bu_{yx} + bu_{xy} + cu_{yy} \\
&= au_{xx} + 2bu_{xy} + cu_{yy}.
\end{aligned}$$

For the last equality, we use the assumption that $u(x, y)$ is twice continuously differentiable in order that the mixed partial derivatives are identical, that is, $u_{xy} = u_{yx}$.

The matrix $\begin{pmatrix} a & b \\ b & c \end{pmatrix}$ is said to be the **coefficient matrix** of the PDE, or the **symbol** of the PDE. The classification of PDEs is based on the relative sign of the eigenvalues of the symbol. Notice that the symbol is symmetric; hence, the eigenvalues are real. As in linear algebra, the determinant of this matrix gives the product of the eigenvalues:

$$\lambda_1\lambda_2 = \det A = ac - b^2,$$

which we use to define the **type** of the equation.

Definition 1.1. *(Type of PDE)*
The PDE

$$au_{xx} + 2bu_{xy} + cu_{yy} + du_x + eu_y + fu = 0$$

is said to be

- **elliptic** if and only if $ac - b^2 > 0$; that is, the eigenvalues of A have the same sign and are not zero, that is, both are positive or both are negative.
- **parabolic** if and only if $ac - b^2 = 0$; that is, at least one eigenvalue is zero.
- **hyperbolic** if and only if $ac - b^2 < 0$; that is, the eigenvalues have opposite signs and are nonzero.

Now that we have defined our PDE kingdoms, we introduce their corresponding rulers:

- **Laplace's equation** in two dimensions is given by $u_{xx} + u_{yy} = 0$, and its symbol is $A = \begin{pmatrix} 1 & 0 \\ 0 & 1 \end{pmatrix}$ with determinant $\det A = 1 > 0$. Hence, Laplace's equation is elliptic.
- The **heat equation** in one (spatial) dimension is given by $u_t = ku_{xx}$, and its symbol is $A = \begin{pmatrix} k & 0 \\ 0 & 0 \end{pmatrix}$ with determinant $\det A = 0$. Hence, the heat equation is parabolic.
- The **wave equation** in one (spatial) dimension is given by $u_{tt} - c^2u_{xx} = 0$, and its symbol is $A = \begin{pmatrix} -c^2 & 0 \\ 0 & 1 \end{pmatrix}$ with determinant $\det A = -c^2 < 0$. Hence, the wave equation is hyperbolic.

Example 1.2. Classify the following second-order linear PDEs.

1. $u_t + 2u_{tt} + 3u_{xx} = 0$
2. $17u_{yy} + 3u_x + u = 0$
3. $4u_{xy} + 2u_{xx} + u_{yy} = 0$
4. $u_{yy} - u_{xx} - 2u_{xy} = 0$

Solution.

1. Symbol $A = \begin{pmatrix} 2 & 0 \\ 0 & 3 \end{pmatrix}$, $\det A = 6 > 0$, elliptic

2. Symbol $A = \begin{pmatrix} 0 & 0 \\ 0 & 17 \end{pmatrix}$, $\det A = 0$, parabolic

3. Symbol $A = \begin{pmatrix} 2 & 2 \\ 2 & 1 \end{pmatrix}$, $\det A = -2 < 0$, hyperbolic

4. Symbol $A = \begin{pmatrix} -1 & -1 \\ -1 & 1 \end{pmatrix}$, $\det A = -2 < 0$, hyperbolic

■

One would expect that a classification scheme should not depend on the coordinate system in which the PDE is expressed. To see that this is indeed the case, consider a change of independent variable:

$$\xi = \xi(x, y), \qquad \eta = \eta(x, y).$$

The transformation is **nonsingular** if the Jacobian determinant of the transformation is nonzero; that is,

$$J = \left| \frac{\partial(\xi, \eta)}{\partial(x, y)} \right| \neq 0.$$

Let us denote the transformed dependent variable as $w(\xi, \eta) = u(x, y)$. Then the PDE (1.8) becomes

$$\alpha w_{\xi\xi} + 2\beta w_{\xi\eta} + \gamma w_{\eta\eta} + \delta w_{\xi} + \epsilon w_{\eta} + f w = 0,$$

where

$$\begin{aligned}
\alpha &= a\xi_x^2 + 2b\xi_x\xi_y + c\xi_y^2, \\
\beta &= a\xi_x\eta_x + b(\xi_x\eta_y + \xi_y\eta_x) + c\xi_y\eta_y, \\
\gamma &= a\eta_x^2 + 2b\eta_x\eta_y + c\eta_y^2, \\
\delta &= d\xi_x + e\xi_y, \\
\epsilon &= d\eta_x + e\eta_y.
\end{aligned}$$

Computing the symbol A for the transformed equation and its determinant gives

$$\det A = \alpha\gamma - \beta^2 = (ac - b^2)\, J^2.$$

The sign of $\alpha\gamma - \beta^2$ is the same as that of $ac - b^2$. Hence, the classification of PDEs is invariant under a change of coordinates.

1.3 SIDE CONDITIONS

Remember from the methods for ODEs that when solving linear ODEs, one usually finds a "general" solution that involves a number of undetermined constants and that to find these constants, some side conditions are needed. Quite often, initial conditions are used to identify a unique solution. This idea is similar for PDEs: Here the PDE alone does not give rise to uniqueness, as can be seen from Laplace's equation in two dimensions.

Example 1.3. Consider the two-dimensional Laplace equation

$$u_{xx} + u_{yy} = 0.$$

This equation is a second-order linear homogeneous PDE. Solutions to this equation include:

$$\begin{aligned}
u(x,y) &= cxy, & u(x,y) &= c\sin nx \cosh ny,\\
u(x,y) &= c(x^2 - y^2), & u(x,y) &= c\,e^{-y}\cos x,\\
u(x,y) &= c(x^3 - 3xy^2), & u(x,y) &= c\log(x^2 + y^2),\\
u(x,y) &= c(x^4 - 6x^2y^2 + y^4), & u(x,y) &= c\tan^{-1}(y/x),\\
u(x,y) &= c(x^5 - 10x^3y^2 + 5xy^4), & u(x,y) &= c\,e^{\sin x \cosh y}\sin(\cos x \sinh y),
\end{aligned}$$

where, in each case, the constant c is arbitrary. The list goes on. Polynomial solutions of any order exist, as do solutions involving various combinations of exponential and trigonometric functions, as well as many others. Linear combinations of solutions are again solutions. ∎

We have just seen that PDEs alone often have infinitely many solutions. To get a unique solution to a particular problem, additional conditions must be applied. These auxiliary conditions are most often of two types: **initial conditions** and **boundary conditions**. Initial conditions are typically given at a chosen start time, usually $t = 0$. Boundary conditions are used to describe what the system does on the domain boundaries.

As an example, we introduce classical boundary conditions for the heat equation: Let $\Omega \subset \mathbb{R}^n$ be a given domain whose boundary is smooth (here, **smooth** means that at each point there exists a unique normal vector to the boundary $\partial\Omega$). The heat equation on Ω reads

$$u_t = k\Delta u.$$

1. **Initial conditions**: We prescribe the initial temperature distribution in Ω at some specific time (usually, $t = 0$). In three dimensions, these take the form

$$u(x,y,z,0) = u_0(x,y,z).$$

2. **Boundary conditions**: We give conditions on the boundary $\partial\Omega$ for all times. The boundary conditions we consider in this book are divided into three categories:

(a) **Dirichlet conditions**: We prescribe u on $\partial\Omega$. These take the form

$$u(x, y, z, t) = g(x, y, z, t)$$

for $(x, y, z) \in \partial\Omega$. Common examples are **homogeneous Dirichlet boundary conditions**:

$$u(x, y, z, t) = 0$$

on $\partial\Omega$.

(b) **Neumann conditions**: We prescribe the heat flow through the boundary $\partial\Omega$. These take the form

$$\frac{\partial u}{\partial n} = g$$

for $(x, y, z) \in \partial\Omega$, where $\partial u/\partial n = \nabla u \cdot \mathbf{n}$ is a directional derivative ($\mathbf{n}$ being the outward-pointing unit normal to $\partial\Omega$). Common examples occur when there is no heat flow through the boundary (representing perfect insulation) and are called **homogeneous Neumann boundary conditions**:

$$\frac{\partial u}{\partial n} = 0$$

on $\partial\Omega$.

(c) **Robin conditions**: We prescribe a mixture of Dirichlet and Neumann boundary conditions. These take the form

$$\alpha u + \beta \frac{\partial u}{\partial n} = g$$

for $(x, y, z) \in \partial\Omega$ and $t \geqslant 0$. This type of boundary condition occurs when, for example, Newton's law of cooling is applied. Newton's law of cooling states that the rate at which heat is transferred across a boundary is proportional to the temperature difference across the boundary. If we denote the temperature outside the region Ω by T, Newton's law of cooling can be written as

$$\kappa u + \nu \frac{\partial u}{\partial n} = \kappa T.$$

A complete problem for a PDE consists of the PDE plus an appropriate number of side conditions. For example, a complete problem for a general heat equation is given by

$$\frac{\partial u}{\partial t} = \nabla \cdot (K(x)\nabla u) + Q(x), \qquad (x, y, z) \in \Omega, \quad t \geqslant 0, \tag{1.9}$$

$$u(x, y, z, 0) = f(x, y, z), \qquad (x, y, z) \in \Omega, \tag{1.10}$$

$$\alpha u(x, y, z, t) + \beta \frac{\partial u}{\partial n}(x, y, z, t) = g(x, y, z, t), \quad (x, y, z) \in \partial\Omega, \quad t \geqslant 0. \tag{1.11}$$

We see that equation (1.11) corresponds to

(i) Dirichlet conditions, if $\alpha \neq 0,\ \beta = 0$;

(ii) Neumann conditions, if $\alpha = 0,\ \beta \neq 0$;

(iii) Robin conditions, if $\alpha \neq 0,\ \beta \neq 0$.

If a PDE is studied on an infinite domain, appropriate decay conditions are typically used. For example, we might require that

$$\lim_{x\to\infty} |u(x,t)| < c_1 e^{-c_2 x^2},$$

with appropriate constants c_1, c_2.

1.3.1 Boundary Conditions on an Interval

Most of this book deals with PDEs on n-dimensional intervals:

$$I = [a_1, b_1] \times \cdots \times [a_n, b_n].$$

Hence, the following rule of thumb can be applied:

> **Rule of Thumb for Side Conditions:**
> To formulate a complete problem for a PDE on an interval, the number of side conditions for each of the variables $t, x, y, \ldots$ corresponds to the maximum order of the derivative with respect to that variable.

For example, the heat equation $u_t = k(u_{xx} + u_{yy})$ on $I = [0,1] \times [0,1]$ needs one initial condition (order of time derivative is 1), two boundary conditions for x, and two boundary conditions for y. These could be of Dirichlet form: for example,

$$\begin{aligned} &u(0,y,t) = f_1(y,t), \quad u(1,y,t) = f_2(y,t),\\ &u(x,0,t) = f_3(x,t), \quad u(x,1,t) = f_4(x,t),\\ &u(x,y,0) = g(x,y). \end{aligned}$$

The one-dimensional wave equation $u_{tt} - c^2 u_{xx} = 0$ on $[0,1]$ needs two initial conditions, usually for location $u(x,0) = f(x)$ and initial velocity $u_t(x,0) = g(x)$, and two boundary conditions, typically for $u(0,t)$ and $u(1,t)$. The rule of thumb above is a nice tool to check if the correct number of side conditions is given. The rule can be extended to more general domains (e.g., circular domains), but one has to be careful to gain the right intuition. We give several additional examples later.

1.4 LINEAR PDEs

In this section we explore linear PDEs a bit more and present a very important tool called the **superposition principle**. This principle is the very foundation of our solution theory.

Recall that every linear PDE can be written in one of two forms:

$$Lu = 0 \qquad \text{(homogeneous)}$$

or

$$Lu = f \qquad \text{(nonhomogeneous)}$$

for some linear differential operator L. What exactly do we mean by a linear differential operator? The definition is analogous to that in linear algebra.

Definition 1.2. (*Linear Operator*)
An operator L with domain of definition $D(L)$ is said to be **linear** if it satisfies the following two properties:

(i) $L(u_1 + u_2) = Lu_1 + Lu_2$, for two functions $u_1, u_2 \in D(L)$; and

(ii) $L(cu) = cLu$, for any constant $c \in \mathbb{R}$ and $u \in D(L)$.

For differential operators, we usually take the domain as the set of functions that are continuously differentiable on the underlying set Ω.

Example 1.4. If

$$L = \frac{\partial}{\partial x} + \frac{\partial}{\partial y},$$

the equation $Lu = 0$ is equivalent to

$$u_x + u_y = 0.$$

The domain of L is

$$D(L) = \{\text{set of continuously differentiable functions}\},$$

and it is easily seen that the advection equation is linear. ∎

Example 1.5. If

$$L = \frac{\partial^2}{\partial x^2} + e^y \sin x \frac{\partial}{\partial y} - 1,$$

the equation $Lu = 0$ is equivalent to

$$u_{xx} + e^y \sin x\, u_y = u.$$

The domain of L is

$$D(L) = \{\text{set of twice continuously differentiable functions}\},$$

and again, checking that L is linear is straightforward. ∎

1.4.1 Principle of Superposition

The principle of superposition makes it much easier to deal with linear PDEs than nonlinear PDEs.

Theorem 1.3. *(Principle of Superposition)*
If u_1 and u_2 are solutions to a linear homogeneous PDE $Lu = 0$, then $c_1u_1 + c_2u_2$ is also a solution for arbitrary constants c_1 and c_2.

Proof. We have $Lu_1 = 0$ and $Lu_2 = 0$, since u_1 and u_2 are solutions. Therefore,

$$L(c_1u_1 + c_2u_2) = L(c_1u_1) + L(c_2u_2) = c_1Lu_1 + c_2Lu_2 = c_1(0) + c_2(0) = 0.$$

□

Example 1.6. Consider Laplace's equation in two dimensions, $Lu = 0$, where

$$L = \Delta = \frac{\partial^2}{\partial x^2} + \frac{\partial^2}{\partial y^2}.$$

Let

$$u_n(x, y) = \cos nx \sinh ny$$

for $n = 1, 2, 3, \ldots$. For each n, u_n is a solution to Laplace's equation, since

$$\begin{aligned} Lu_n &= \frac{\partial^2}{\partial x^2}(\cos nx \sinh ny) + \frac{\partial^2}{\partial y^2}(\cos nx \sinh ny) \\ &= -n^2 \cos nx \sinh ny + n^2 \cos nx \sinh ny = 0. \end{aligned}$$

Hence, by the principle of superposition,

$$u(x, y) = \sum_{n=1}^{N} a_n u_n(x, y) = \sum_{n=1}^{N} a_n \cos nx \sinh ny$$

is also a solution for any integer N and any constants a_n. ■

Thus, the principle of superposition gives us a means of constructing new solutions if a few solutions are already known. In general, this does *not* hold for nonlinear equations, as the following example illustrates.

Example 1.7. *(Burger's Equation)*
Consider the following two-dimensional first-order *nonlinear* PDE:

$$u_x + uu_y = 0.$$

The functions

$$u_1(x, y) = 1 \qquad \text{and} \qquad u_2(x, y) = \frac{y}{1 + x}$$

are solutions to the PDE, since

$$\frac{\partial u_1}{\partial x} + u_1 \frac{\partial u_1}{\partial y} = 0 + (1)(0) = 0$$

and

$$\frac{\partial u_2}{\partial x} + u_2 \frac{\partial u_2}{\partial y} = -\frac{y}{(1+x)^2} + \frac{y}{1+x} \cdot \frac{1}{1+x} = 0.$$

However, the sum of these two solutions,

$$u(x,y) = u_1(x,y) + u_2(x,y) = 1 + \frac{y}{1+x},$$

is not a solution, since

$$\frac{\partial u}{\partial x} + u \frac{\partial u}{\partial y} = -\frac{y}{(1+x)^2} + \left(1 + \frac{y}{1+x}\right) \frac{1}{1+x} = \frac{1}{1+x} \neq 0.$$

■

In general, the principle of superposition holds only for linear homogeneous PDEs. For nonhomogeneous PDEs, we have the following result:

Theorem 1.4. If

(a) u_p is a solution to the nonhomogeneous PDE $Lu = f$ (i.e., $Lu_p = f$); and

(b) u_h is a solution to the homogeneous PDE $Lu = 0$ (i.e., $Lu_h = 0$);

then $u = cu_h + u_p$ is a solution to $Lu = f$ for any constant c.

Proof. $Lu = L(cu_h + u_p) = cLu_h + Lu_p = c(0) + f = f$. □

This result means that if a particular solution to a linear nonhomogeneous PDE is known and a solution to the homogeneous counterpart is also known, then a new solution to the nonhomogeneous PDE may be constructed. In fact, any solution to a linear nonhomogeneous PDE will be of this form. Suppose that $\widetilde{u}$ is any solution to the nonhomogeneous PDE $Lu = f$ (i.e., $L\widetilde{u} = f$). Define $u_h = \widetilde{u} - u_p$. Then

$$Lu_h = L(\widetilde{u} - u_p) = L\widetilde{u} - Lu_p = f - f = 0;$$

thus, u_h is a solution to the homogeneous PDE $Lu = 0$. Hence, *any solution* $\widetilde{u}$ of the nonhomogeneous PDE $Lu = f$ can be written as the sum of a solution to the homogeneous PDE and any particular solution to the nonhomogeneous PDE:

$$\widetilde{u} = u_h + u_p.$$

1.5 STEADY-STATE AND EQUILIBRIUM SOLUTIONS

Definition 1.5. A **steady-state** or **equilibrium solution** to a boundary value–initial value problem for a PDE is a solution that does not depend on time; that is,

$$u(x,t) = \bar{u}(x).$$

Example 1.8. (*Diffusion Through a Cell Membrane*)
Suppose that we want to compute the concentration $u(x,t)$ of a nutrient, for example, oxygen, through a cell membrane of thickness ℓ. We assume that the oxygen concentration inside and outside the cell are constant with values c_1 (inside) and c_2 (outside). The corresponding boundary value–initial value problem employs the diffusion equation:

$$\begin{aligned}
u_t &= Du_{xx},\\
u(0,t) &= c_1,\\
u(\ell,t) &= c_2,\\
u(x,0) &= f(x).
\end{aligned}$$

This is a nonhomogeneous Dirichlet problem for the diffusion equation, where D denotes the diffusion coefficient of oxygen in the cell membrane and $f(x)$ denotes the initial distribution of oxygen. Later we learn how to solve this model completely. Here, we are only interested in a steady-state distribution. Since a steady-state solution does not depend on time,

$$u_t(x,t) = 0,$$

so that

$$u(x,t) = \overline{u}(x).$$

Thus, u depends only on the spatial variable x, and

$$\frac{d^2\overline{u}}{dx^2} = u_{xx} = 0.$$

Integrating twice, we get

$$\overline{u}'(x) = A \qquad \text{and} \qquad \overline{u}(x) = Ax + B,$$

where A and B are two unknown constants, which can be determined from the boundary conditions. At $x = 0$, we have

$$\overline{u}(0) = c_1 = B;$$

hence, $B = c_1$. At $x = \ell$, we have

$$\overline{u}(\ell) = c_2 = A\ell + c_1;$$

hence, $A = (c_2 - c_1)/\ell$, and we find the steady-state solution

$$\overline{u}(x) = \frac{c_2 - c_1}{\ell} x + c_1.$$

To make sure that we have found the right solution, we have to check that it is a solution to the original problem. It is a linear function; therefore, $\overline{u}'' = 0$. At $x = 0$, we have $\overline{u}(0) = c_1$, and at $x = \ell$, we find $\overline{u}(\ell) = c_2$.

The concentration profile through a membrane is a linear function that interpolates between the two concentration levels c_1 and c_2. ∎

Example 1.9. Next, we want to find the steady-state solution to a homogeneous Neumann problem for the heat equation on $[0, \ell]$:

$$\begin{aligned}
u_t(x,t) &= k u_{xx}(x,t), \\
u_x(0,t) &= 0, \\
u_x(\ell,t) &= 0, \\
u(x,0) &= f(x).
\end{aligned}$$

As before, we find that at steady state we have $u(x,t) = \overline{u}(x)$ and $\overline{u}''(x) = 0$. Hence, $\overline{u}(x)$ is a linear function $\overline{u}(x) = Ax + B$ with two unknown constants A and B. Using the boundary conditions, we find that

$$\overline{u}'(0) = A = 0 \qquad \text{and} \qquad \overline{u}'(\ell) = A = 0,$$

and both boundary conditions require that $A = 0$; hence, $\overline{u}(x) = B$ is a constant function.

Since we have already used both boundary conditions, how can we find the missing constant B? To answer this question, we need a deeper understanding of the conservation of mass property of the diffusion equation or the conservation of energy property of the heat equation with homogeneous Neumann boundary conditions. The integral

$$M(t) = \int_0^\ell u(x,t)\, dx$$

can be understood as the total mass (in the case of the diffusion equation) or as the total heat energy (in the case of the heat equation) in the system. Using the fundamental theorem of calculus, we find that

$$\begin{aligned}\frac{d}{dt}M(t) &= \frac{d}{dt}\int_0^\ell u(x,t)\,dx \\ &= \int_0^\ell u_t(x,t)\,dx \\ &= \int_0^\ell k u_{xx}(x,t)\,dx \\ &= k\left[u_x(\ell,t) - u_x(0,t)\right] \\ &= 0.\end{aligned}$$

Hence, $M(t)$ = constant, and the total mass (heat energy) is conserved. Thus, we expect that the steady-state system has the same total mass (heat energy) as the initial system. The initial mass (heat energy) is

$$M_0 = M(0) = \int_0^\ell f(x)\,dx,$$

and we require that

$$M_0 = \int_0^\ell \overline{u}(x)\,dx = \int_0^\ell B\,dx = B\ell.$$

Hence,

$$B = \frac{M_0}{\ell},$$

and the steady-state solution to the problem above is

$$\overline{u}(x) = \frac{M_0}{\ell} = \frac{1}{\ell}\int_0^\ell f(x)\,dx.$$

∎

In the next example, we add a source term to the heat equation.

Example 1.10. Find the steady-state solution to the following PDE on $[0, 2\pi]$:

$$\begin{aligned}u_t &= 3u_{xx} + 9\sin x, \\ u(x,0) &= 9\sin x, \\ u(0,t) &= 9, \\ u_x(2\pi,t) &= 0.\end{aligned}$$

Solution. The steady-state solution $\overline{u}(x)$ satisfies

$$3\overline{u}'' + 9\sin x = 0.$$

Hence, $\overline{u}'' = -3\sin x$, which has the general solution

$$\overline{u}(x) = 3\sin x + c_1 x + c_2.$$

The boundary condition at $x = 0$ gives $\overline{u}(0) = 9 = c_2$. For the boundary at $x = 2\pi$, we need the derivative $\overline{u}'(x) = 3\cos x + c_1$. The corresponding boundary condition gives

$$\overline{u}'(2\pi) = 0 = 3 + c_1;$$

hence, $c_1 = -3$, and the steady-state solution is

$$\overline{u}(x) = 3\sin x - 3x + 9.$$

∎

1.6 FIRST EXAMPLE FOR SEPARATION OF VARIABLES

Separation of variables is a method that tries to separate the dependence of the corresponding variables. Suppose that we have a linear PDE for some unknown function $u(x, y)$. First, we look for a solution (*separated solution*) of the form

$$u(x, y) = X(x) \cdot Y(y).$$

So far, we do not know if this idea works. Nevertheless, we substitute this assumed form of the solution into the PDE, perform some manipulations, and, we hope, end up with ODEs for the unknown functions X and Y that we can solve.

Example 1.11. Consider the one-dimensional heat equation with Dirichlet boundary conditions on the interval $[0, \ell]$:

$$\frac{\partial u}{\partial t} = k\frac{\partial^2 u}{\partial x^2}, \qquad x \in \Omega = [0, \ell], \quad t > 0, \tag{1.12}$$

$$u(0, t) = 0, \tag{1.13}$$

$$u(\ell, t) = 0, \tag{1.14}$$

$$u(x, 0) = f(x) = 3\sin\frac{2\pi x}{\ell}. \tag{1.15}$$

Here, $u(x, t)$ represents the temperature in a bar at time t and position x. Trying the method of separation of variables, we assume a solution of the form

$$u(x, t) = X(x) \cdot T(t). \tag{1.16}$$

Then

$$\frac{\partial u}{\partial t}(x, t) = X(x) \cdot T'(t) \qquad \text{and} \qquad \frac{\partial^2 u}{\partial x^2}(x, t) = X''(x) \cdot T(t).$$

Substituting these expressions into equation (1.12) gives

$$X \cdot T' = kX'' \cdot T,$$

which can be rewritten as

$$\frac{T'}{kT} = \frac{X''}{X}.$$

Now, the left-hand side is a function of the time t only, whereas the right-hand side is a function of the position x only; since x and t are *independent* of each other, both sides must equal a constant, which we call $-\lambda$. Hence, we get two equations:

$$\frac{T'}{kT} = -\lambda \qquad \text{and} \qquad \frac{X''}{X} = -\lambda,$$

and the value of this separation constant λ is, at this point, unknown. Thus, we get two ODEs, one for T and one for X:

$$T' + \lambda kT = 0 \qquad \text{and} \qquad X'' + \lambda X = 0.$$

Now insert equation (1.16) into the boundary conditions (1.13) and (1.14):

$$u(0,t) = X(0) \cdot T(t) = 0,$$
$$u(\ell,t) = X(\ell) \cdot T(t) = 0.$$

If $X(0) \neq 0$ or $X(\ell) \neq 0$, then $T(t) = 0$ for all t; however, this would imply that $u \equiv 0$, which would mean that the initial condition (1.15) is not satisfied. Thus, the only conclusion is that

$$X(0) = 0 \qquad \text{and} \qquad X(\ell) = 0.$$

The problems for T and X become

$$T' + \lambda kT = 0, \quad t > 0, \qquad X'' + \lambda X = 0, \quad x \in [0, \ell], \tag{1.17}$$
$$X(0) = X(\ell) = 0. \tag{1.18}$$

The equation for T is easy to solve:

$$T(t) = ce^{-\lambda kt},$$

where c is an arbitrary constant, with no restriction as yet on λ.
Now we turn to the problems (1.17) and (1.18) for X. Note that $X(x) \equiv 0$ is a solution (called the **trivial solution**) to the problem. However, the trivial solution leads to

$$u(x,t) \equiv 0,$$

which does not satisfy the initial condition (1.15). Therefore, we seek nontrivial solutions to (1.17) and (1.18) for X. As we shall see, nontrivial solutions exist only for certain values of λ, called the **eigenvalues**. The corresponding nontrivial solutions X are called the **eigenfunctions**. More complicated problems will result in more complicated eigenvalue problems, known as Sturm-Liouville problems. We discuss these later.

The solutions to (1.17) and (1.18) for X will depend on the sign of λ. There are three cases to consider: λ negative, positive, or zero.

- Case (i): $\lambda < 0$. If $\lambda = -\mu^2 \neq 0$, the X equation becomes

  $$X'' - \mu^2 X = 0,$$

  with general solution

  $$X(x) = A\cosh \mu x + B\sinh \mu x.$$

  The boundary condition on the left-hand side gives us

  $$X(0) = 0,$$

  which implies that $A = 0$. The solution is now

  $$X(x) = B\sinh \mu x,$$

  and the boundary condition on the right-hand side now gives us

  $$X(\ell) = 0,$$

  which implies that

  $$B\sinh \mu\ell = 0,$$

  which in turn implies that

  $$B = 0 \qquad \text{or} \qquad \sinh \mu\ell = 0.$$

  However, $\mu\ell \neq 0$; thus, we must have $B = 0$, which implies that $X(x) \equiv 0$. Therefore, there are no nontrivial solutions for the case $\lambda < 0$.

- Case (ii): $\lambda = 0$. If $\lambda = 0$, the X equation becomes $X'' = 0$ with general solution

  $$X(x) = Ax + B.$$

  However, the boundary conditions imply that $A = B = 0$, which again leads to the trivial solution. Therefore, there are no nontrivial solutions for the case $\lambda = 0$.

- Case (iii): $\lambda > 0$. If $\lambda = \mu^2 \neq 0$, then the X equation becomes

  $$X'' + \mu^2 X = 0,$$

  with general solution

  $$X(x) = A\cos \mu x + B\sin \mu x.$$

The boundary condition on the left-hand side gives us

$$X(0) = 0,$$

which implies that $A = 0$, so that

$$X(x) = B \sin \mu x.$$

The boundary condition on the right-hand side now gives us

$$X(\ell) = 0,$$

which implies that

$$B \sin \mu\ell = 0,$$

so that

$$B = 0 \qquad \text{or} \qquad \sin \mu\ell = 0.$$

Since a nontrivial solution results only when $B \neq 0$, we must have $\sin \mu\ell = 0$, which occurs for $\mu\ell = n\pi, n = \pm 1, \pm 2, \ldots$. Thus, we get nontrivial solutions only for

$$\lambda = \lambda_n = \mu_n^2 = \frac{n^2\pi^2}{\ell^2},$$

where $n = 1, 2, 3, \ldots$, and these are the eigenvalues. The corresponding nontrivial solutions

$$X = X_n(x) = \sin \mu_n x = \sin \frac{n\pi x}{\ell},$$

for $n = 1, 2, 3, \ldots$, are the eigenfunctions.

If we now substitute these eigenvalues into the solution to the T equation, we get

$$T = T_n(t) = e^{-\lambda_n kt},$$

and for each $n = 1, 2, 3, \ldots$ we have a solution,

$$u_n(x,t) = X_n(x)T_n(t),$$

that satisfies the PDE and the boundary conditions. Since the PDE is linear, a multiple of a solution is also a solution. Thus, for each $n = 1, 2, 3, \ldots$,

$$u_n(x,t) = b_n e^{-\lambda_n kt} \sin \frac{n\pi x}{\ell}$$

is also a solution.

All that remains is to satisfy the initial condition (1.15):

$$u(x,0) = f(x).$$

If we take $n = 2$ and $b_2 = 3$, then

$$u_2(x,0) = 3\sin\frac{2\pi x}{\ell} = f(x).$$

Therefore, the solution to the problem is

$$u(x,t) = 3e^{-4\pi^2 kt/\ell^2}\sin\frac{2\pi x}{\ell}.$$

■

In the previous example, the initial condition was of a very specific form. In fact, the initial condition $f(x)$ is one of the eigenfunctions, $\sin(n\pi x/\ell)$ for $n = 2$. Now, suppose that the initial condition had been $f(x) = x(\ell - x)$. In this case, the initial condition is not an eigenfunction and hence cannot be satisfied for any n. However, our method can still work for this case. Since the PDE and the boundary conditions are both linear and homogeneous, we can apply the principle of superposition to obtain

$$u(x,t) = \sum_{n=1}^{N} b_n u_n(x,t)$$

or, more generally,

$$u(x,t) = \sum_{n=1}^{\infty} b_n u_n(x,t) = \sum_{n=1}^{\infty} b_n X_n(x)\cdot T_n(t) = \sum_{n=1}^{\infty} b_n e^{-\lambda_n kt}\sin\frac{n\pi x}{\ell}.$$

Now, applying the initial condition, we obtain

$$f(x) = \sum_{n=1}^{\infty} b_n X_n(x);$$

that is,

$$x(\ell - x) = \sum_{n=1}^{\infty} b_n \sin\frac{n\pi x}{\ell}.$$

At this point, a few questions come to mind:

1. Does the infinite series above converge?
2. What class of functions can be represented by an infinite series of the type given above?
3. If a function f can be represented by such a series, how does one find the constants b_n?

We return to these questions in the next chapter.

1.7 DERIVATION OF THE DIFFUSION EQUATION

In this section we derive the diffusion equation (or heat equation) from the point of view of population dynamics.

Assume that bacteria swim in a cylindrical tank. Let $u(x,t)$ denote the cell density at time t at location $x \in \Omega$, and let $j(x,t)$ denote the bacterial flux at (x,t). The domain Ω is an interval in the real line $\mathbb{R}$ (Fig. 1.1) We model the change in bacterial density by a word equation:

$$\begin{pmatrix}\text{change of } u\\ \text{in } [a,b]\end{pmatrix} = \begin{pmatrix}\text{net influx or outflux}\\ \text{through } a \text{ or } b\end{pmatrix},$$

where we assume that the bacteria neither die nor divide on the time scale of the experiment. Putting this word equation into a formula, we obtain

$$\frac{\partial}{\partial t}\int_a^b u(x,t)\,dx = j(a,t) - j(b,t).$$

Now we employ **Fick's first law**, which states that the flux is proportional to the negative of the gradient of u; that is,

$$j(x,t) = -k\frac{\partial u}{\partial x}(x,t),$$

so that

$$\begin{aligned}\frac{\partial}{\partial t}\int_a^b u(x,t)\,dx &= -k\frac{\partial u}{\partial x}(a,t) + k\frac{\partial u}{\partial x}(b,t)\\ &= k\int_a^b \frac{\partial^2 u}{\partial x^2}(x,t)\,dx.\end{aligned}$$

We assume that $\partial u/\partial t$ is continuous and apply Leibniz's formula,

$$\frac{\partial}{\partial t}\int_a^b u(x,t)\,dx = \int_a^b \frac{\partial u}{\partial t}(x,t)\,dx;$$

hence,

$$\int_a^b \left(\frac{\partial u}{\partial t}(x,t) - k\frac{\partial^2 u}{\partial x^2}(x,t)\right)\,dx = 0,$$

which must be true for all choices of $a,\, b \in \Omega$ with $a \leqslant b$.

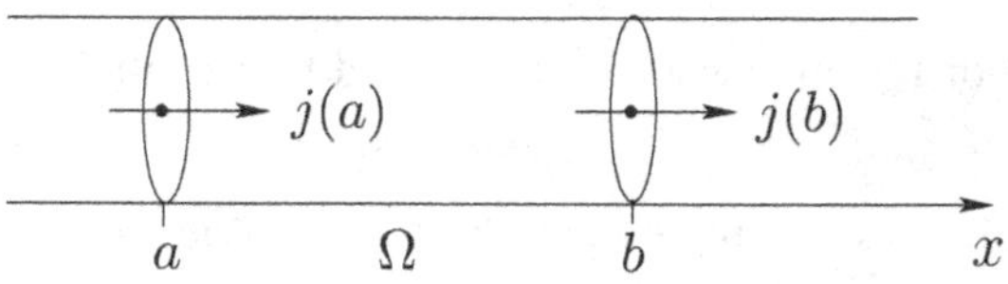

Figure 1.1 Diffusion in a pipe.

Hence, the integrand must be equal to zero pointwise, and we get

$$\frac{\partial u}{\partial t}(x,t) = k\frac{\partial^2 u}{\partial x^2}(x,t)$$

for all $x \in \Omega$ and $t \geqslant 0$. This equation is called the **one-dimensional diffusion equation**, or **Fick's second law**. In the next section we will see that the heat equation has the same form (mathematically), but that the physical interpretation is different.

1.7.1 Boundary Conditions

Let us discuss the boundary conditions in the context of the diffusion of bacteria. According to the rule of thumb for side conditions, we need one condition for the initial distribution of bacteria,

$$u(x,0) = f(x)$$

and two boundary conditions. We consider only boundary conditions of the following forms:

(a) Dirichlet boundary conditions on $[0, \ell]$,

$$u(0,t) = g_1(t) \qquad \text{and} \qquad u(\ell,t) = g_2(t),$$

correspond to a pipe attached to a tank on each side, where the concentration of bacteria in the tanks, $g_1(t)$ and $g_2(t)$, can be controlled (Fig. 1.2). The homogeneous Dirichlet boundary conditions

$$u(0,t) = 0 \qquad \text{and} \qquad u(\ell,t) = 0$$

describe a pipe with open ends.

(b) Neumann boundary conditions on $[0, \ell]$,

$$\frac{\partial u}{\partial x}(0,t) = \psi_1(t) \qquad \text{and} \qquad \frac{\partial u}{\partial x}(\ell,t) = \psi_2(t),$$

correspond to valves or pumps at the boundaries where the influx or outflux can be controlled (Fig. 1.3).

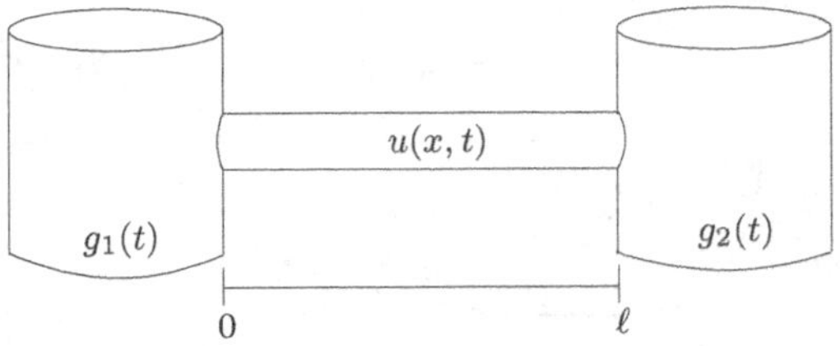

Figure 1.2 Dirichlet boundary conditions in a pipe.

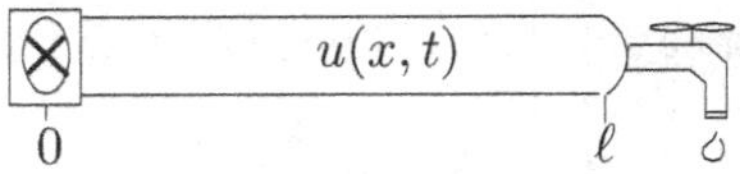

Figure 1.3 Neumann boundary conditions in a pipe.

The homogeneous Neumann boundary conditions

$$\frac{\partial u}{\partial x}(0, t) = 0 \qquad \text{and} \qquad \frac{\partial u}{\partial x}(\ell, t) = 0$$

describe a closed pipe.

(c) Robin boundary conditions describe a mixture of the conditions described above, that is, where the concentration and the influx and outflux can be controlled.

Example 1.12. Find a model for particles that diffuse in a cylinder of length $\ell = 10$, with initial distribution $f(x) = e^{-5x}$, and where the cylinder has one closed end ($x = 0$) and one open end ($x = \ell$).
Solution. The boundary value–initial value problem satisfied in this case is given by

$$\begin{aligned} &u_t = Du_{xx}, \qquad 0 \leqslant x \leqslant \ell, \quad t \geqslant 0, \\ &u(x, 0) = e^{-5x}, \\ &u_x(0, t) = 0, \\ &u(\ell, t) = 0. \end{aligned}$$

■

1.8 DERIVATION OF THE HEAT EQUATION

Consider a region $\Omega \subset \mathbb{R}^3$ (Fig. 1.4), consisting of some substance with boundary given by $\partial\Omega = S$. Let us introduce the following notation:

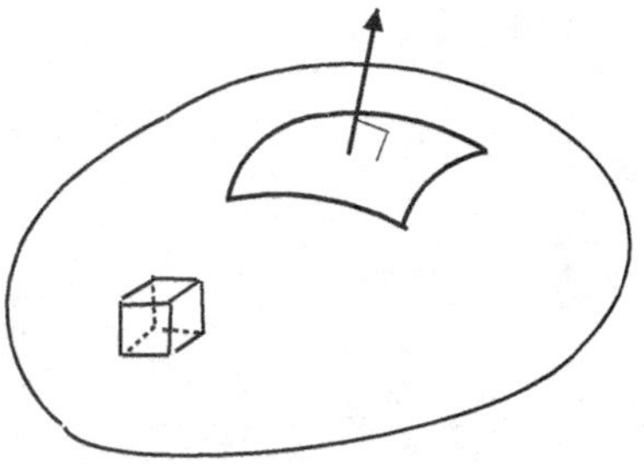

Figure 1.4 Region Ω.

$$
\begin{aligned}
u &\equiv \text{temperature at the point } (x, y, z) \in \Omega \text{ at time } t; \\
K &\equiv \text{thermal conductivity of substance at } (x, y, z); \\
\rho &\equiv \text{density of substance at } (x, y, z); \\
c &\equiv \text{heat capacity per unit mass at } (x, y, z); \\
\mathbf{J} &\equiv \text{heat flux vector (gives magnitude and direction of heat flow)}; \\
h &\equiv \text{rate of internal heat generation per unit volume at } (x, y, z).
\end{aligned}
$$

The quantities above have the following physical dimensions:

$$
\begin{aligned}
[u] &\equiv T; & T &\equiv \text{temperature;} \\
[K] &\equiv \frac{H}{LTt}; & H &\equiv \text{heat energy;} \\
[\rho] &\equiv \frac{m}{L^3}; & L &\equiv \text{length;} \\
[c] &\equiv \frac{H}{mT}; & t &\equiv \text{time;} \\
[h] &\equiv \frac{H}{L^3 t}; & m &\equiv \text{mass.}
\end{aligned}
$$

Suppose that we let

$$
\begin{aligned}
dV &\equiv \text{an element of volume;} \\
dS &\equiv \text{an element of surface area; and} \\
\mathbf{n} &\equiv \text{outward-pointing unit normal vector at } dS.
\end{aligned}
$$

The vector ∇u points in the direction of the most rapid increase in u; that is, ∇u is orthogonal to the surface $u \equiv$ constant. We use **Fourier's law**, which states that heat flows from regions of high temperature to those of low temperature, where the flux $\mathbf{J}$ satisfies

$$
\mathbf{J} = -K \nabla u \tag{1.19}
$$

(see Fig. 1.5).

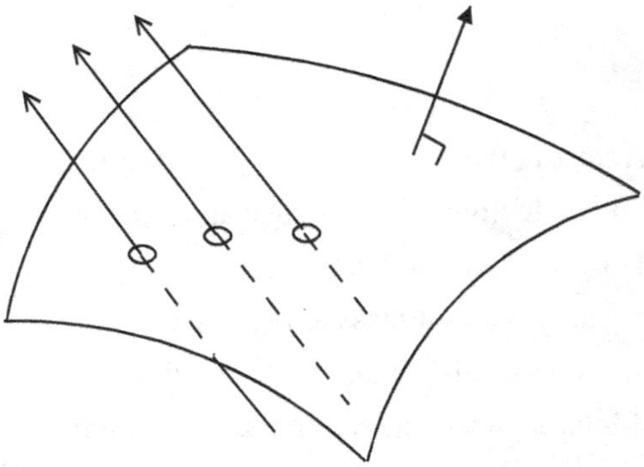

Figure 1.5 Flux **J** through a surface element dS.

We have

$$\left\{\begin{matrix}\text{amount of heat energy}\\ \text{in volume element } dV\end{matrix}\right\} = \text{mass} \times \text{heat capacity} \times \text{temperature}$$
$$= \rho c u \, dV;$$

$$\left\{\begin{matrix}\text{total amount of}\\ \text{heat energy in } \Omega\end{matrix}\right\} = \iiint\limits_{\Omega} \rho c u \, dV;$$

$$\left\{\begin{matrix}\text{rate at which heat}\\ \text{energy enters through } dS\end{matrix}\right\} = -\mathbf{J} \cdot \mathbf{n} \, dS;$$

$$\left\{\begin{matrix}\text{total rate at which heat energy}\\ \text{enters through the boundary } S\end{matrix}\right\} = -\iint\limits_{S} \mathbf{J} \cdot \mathbf{n} \, dS;$$

$$\left\{\begin{matrix}\text{total rate at which heat}\\ \text{energy is generated internally}\end{matrix}\right\} = \iiint\limits_{\Omega} h \, dV.$$

The conservation of energy implies that

$$\left\{\begin{matrix}\text{rate of change of}\\ \text{heat energy in } \Omega\end{matrix}\right\} = \left\{\begin{matrix}\text{rate at which heat energy}\\ \text{enters through boundary } S\end{matrix}\right\} + \left\{\begin{matrix}\text{rate at which heat energy}\\ \text{is generated internally}\end{matrix}\right\}.$$

In other words,

$$\frac{\partial}{\partial t}\iiint\limits_{\Omega} \rho c u \, dV = -\iint\limits_{S} \mathbf{J}\cdot\mathbf{n}\, dS + \iiint\limits_{\Omega} h \, dV.$$

Using the divergence theorem from advanced calculus, we can rewrite this as

$$\iiint\limits_{\Omega} \rho c \frac{\partial u}{\partial t}\, dV = -\iiint\limits_{\Omega} \nabla\cdot\mathbf{J}\, dV + \iiint\limits_{\Omega} h\, dV$$

or

$$\iiint\limits_{\Omega} \left\{\rho c\frac{\partial u}{\partial t} + \nabla\cdot\mathbf{J} - h\right\} dV = 0.$$

The equation above must hold for an arbitrary region Ω; thus, the integrand must be identically zero. Using the definition for the flux vector in equation (1.19), we get the nonhomogeneous heat equation or diffusion equation

$$\frac{\partial u}{\partial t} = \frac{1}{\rho c}\nabla\cdot(K\nabla u) + \frac{h}{\rho c}.$$

Often, ρ, c, and K are constant. In that case, we define

$$k = \frac{K}{\rho c} \qquad \text{and} \qquad F = \frac{h}{\rho c},$$

where the constant k is called the **thermal diffusivity** and F, which is not necessarily constant, is called the **forcing term**. Hence, we get the heat equation in standard form:

$$\frac{\partial u}{\partial t} = k\nabla^2 u + F. \tag{1.20}$$

Written explicitly in terms of Cartesian coordinates yields

$$\frac{\partial u}{\partial t} = k\left(\frac{\partial^2 u}{\partial x^2} + \frac{\partial^2 u}{\partial y^2} + \frac{\partial^2 u}{\partial z^2}\right) + F, \quad \text{(three-dimensional heat equation)} \tag{1.21}$$

$$\frac{\partial u}{\partial t} = k\left(\frac{\partial^2 u}{\partial x^2} + \frac{\partial^2 u}{\partial y^2}\right) + F, \quad \text{(two-dimensional heat equation)} \tag{1.22}$$

$$\frac{\partial u}{\partial t} = k\frac{\partial^2 u}{\partial x^2} + F. \quad \text{(one-dimensional heat equation)} \tag{1.23}$$

1.9 DERIVATION OF THE WAVE EQUATION

Here we derive the one-dimensional wave equation for small vertical displacements of a perfectly elastic string of length ℓ in a resisting medium with restoring forces (like Hooke's law). We also assume that there are external forces (such as gravity) acting. We assume that the string is stretched under tension and fastened at two points A and B, and we let x denote the distance from the end A toward the end B, and let t denote the time.

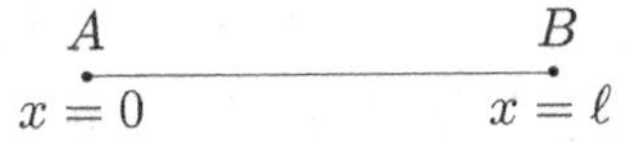

At time $t = 0$, the string is set in motion, and we let $u(x, t)$ denote the vertical displacement of the string at position x, at time t. We assume that the string is flexible, that is, that there is no resistance to bending, so the tension $T(x, t)$ in the string at position x, at time t acts tangentially (Fig. 1.6). Applying Newton's second law to the small portion of the string between x and $x + \Delta x$, if ρ is the mass per unit length, we have

$$\begin{aligned}\frac{\partial}{\partial t}\left(\rho \Delta x \frac{\partial u}{\partial t}\right) &\approx T(x+\Delta x, t)\sin\theta(x+\Delta x, t) - T(x,t)\sin\theta(x,t) \\ &\quad - \alpha\frac{\partial u}{\partial t}\Delta x - \beta u \Delta x + Q(x,t)\Delta x,\end{aligned}$$

where $\alpha > 0$ and $\beta > 0$.

The term $-\alpha\partial u/\partial t$ represents any resisting force per unit length, the term $-\beta u$ represents any restoring force per unit length, and the term $Q(x, t)$ represents any external forces (such as gravity) per unit length. Dividing by Δx and letting $\Delta x \to 0$, we get equality in the limit, so that u satisfies

$$\frac{\partial}{\partial t}\left(\rho\frac{\partial u}{\partial t}\right) = \frac{\partial}{\partial x}\left(T(x,t)\sin\theta(x,t)\right) - \alpha\frac{\partial u}{\partial t} - \beta u(x,t) + Q(x,t) \tag{1.24}$$

for $t \geqslant 0,\ 0 \leqslant x \leqslant \ell$.

Now we make some simplifying assumptions.

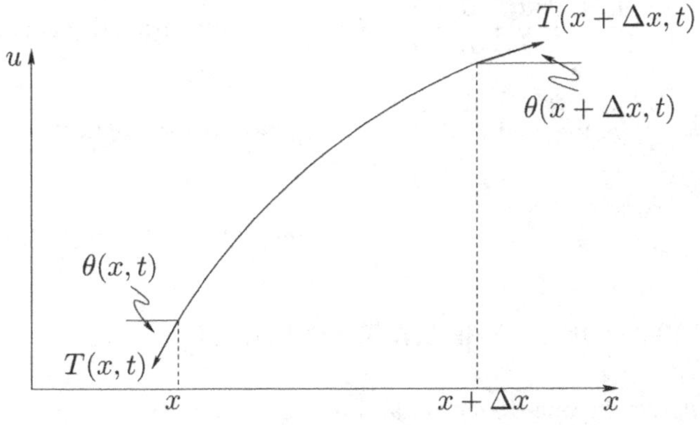

Figure 1.6

- For *small vertical displacements*, we have the approximation

$$\sin\theta(x,t) \approx \tan\theta(x,t) = \frac{\partial u}{\partial x}$$

and the PDE (1.24) becomes

$$\frac{\partial}{\partial t}\left(\rho\frac{\partial u}{\partial t}\right) = \frac{\partial}{\partial x}\left(T(x,t)\frac{\partial u}{\partial x}\right) - \alpha\frac{\partial u}{\partial t} - \beta u + Q(x,t) \qquad (1.25)$$

for $t \geqslant 0,\ 0 \leqslant x \leqslant \ell$.

- If the string is *perfectly elastic*, the tension is constant, and $T \approx T_0$, the initial tension; if the string is made from a *uniform material*, the density $\rho(x)$ is constant, and the PDE (1.25) becomes

$$\rho\frac{\partial^2 u}{\partial t^2} = T_0\frac{\partial^2 u}{\partial x^2} - \alpha\frac{\partial u}{\partial t} - \beta u + Q(x,t) \qquad (1.26)$$

for $t \geqslant 0,\ 0 \leqslant x \leqslant \ell$.

- If the *tension T is large compared to $Q(x,t)$*, we may neglect $Q(x,t)$ and the PDE (1.26) becomes

$$\rho\frac{\partial^2 u}{\partial t^2} = T_0\frac{\partial^2 u}{\partial x^2} - \alpha\frac{\partial u}{\partial t} - \beta u \qquad (1.27)$$

for $t \geqslant 0,\ 0 \leqslant x \leqslant \ell$. This equation is called the **telegrapher's equation**, and models, among other things, electromagnetic wave transmission in a wire. In this context, it usually is written as

$$\frac{\partial^2 u}{\partial x^2} = LC\frac{\partial^2 u}{\partial t^2} + (RC + LG)\frac{\partial u}{\partial t} + RGu,$$

where u is either the magnitude v of the voltage at any point in the wire or the current i at any point in the wire. Here, R is the series resistance per unit length, L is the inductance per unit length, C is the capacitance per unit length, and G is the conductance per unit length.

- If there are *no frictional forces* and *no restoring forces*, then the PDE (1.27) becomes

$$\rho\frac{\partial^2 u}{\partial t^2} = T_0\frac{\partial^2 u}{\partial x^2},$$

that is,

$$\frac{\partial^2 u}{\partial t^2} = c^2\frac{\partial^2 u}{\partial x^2} \qquad (1.28)$$

for $t \geqslant 0,\ 0 \leqslant x \leqslant \ell$, where $c^2 = T_0/\rho$, and c is the velocity of wave propagation along the string. This is the **one-dimensional wave equation** and models sound waves, water waves, vibrations in solids, and longitudinal or torsional vibrations in a rod, among other things.

From our rule of thumb for side conditions, we need two boundary conditions and two initial conditions. The initial conditions usually take the following form:

1. **Initial displacement**: $u(x,0) = f(x),\ 0 \leqslant x \leqslant \ell$, and
2. **Initial velocity**: $v(x,0) = \dfrac{\partial u}{\partial t}(x,0) = g(x),\ 0 \leqslant x \leqslant \ell.$

Typical boundary conditions are of the same form as those given in the discussion of the one-dimensional heat equation.

(a) **Dirichlet conditions**:

$$u(0,t) = g_1(t),$$
$$u(\ell,t) = g_2(t)$$

for $t \geqslant 0$. Here the ends move with time in vertical motion only. For homogeneous Dirichlet conditions,

$$u(0,t) = 0,$$
$$u(\ell,t) = 0$$

for $t \geqslant 0$, and the ends of the string are tied down.

(b) **Neumann conditions**:

Here the tensile force $T\partial u/\partial x$ is specified at the ends:

$$T(0,t)\frac{\partial u}{\partial x}(0,t) = g_1(t),$$
$$T(\ell,t)\frac{\partial u}{\partial x}(\ell,t) = g_2(t)$$

for $t \geqslant 0$. For constant tensile force, we can assume that we have homogeneous Neumann conditions,

$$\frac{\partial u}{\partial x}(0,t) = 0,$$
$$\frac{\partial u}{\partial x}(\ell,t) = 0$$

for $t \geqslant 0$. These conditions can be achieved, for example, by attaching the ends of the string to a frictionless sleeve that moves vertically.

(c) **Robin conditions**:

$$T(0,t)\frac{\partial u}{\partial x}(0,t) = k_1 u(0,t),$$
$$T(\ell,t)\frac{\partial u}{\partial x}(\ell,t) = -k_2 u(\ell,t),$$

where the spring constants are $k_1 > 0$ and $k_2 > 0$, and both springs have the other end fixed. Or, the other ends of the springs can move vertically,

$$T(0,t)\frac{\partial u}{\partial x}(0,t) = k_1[u(0,t) - d_1(t)],$$
$$T(\ell,t)\frac{\partial u}{\partial x}(\ell,t) = -k_2[u(\ell,t) - d_2(t)]$$

for $t \geqslant 0$.

The higher-dimensional wave equations are given by

$$\frac{\partial^2 u}{\partial t^2} = c^2\left(\frac{\partial^2 u}{\partial x^2} + \frac{\partial^2 u}{\partial y^2}\right) + F, \qquad \text{(two-dimensional wave equation)} \tag{1.29}$$

$$\frac{\partial^2 u}{\partial t^2} = c^2\left(\frac{\partial^2 u}{\partial x^2} + \frac{\partial^2 u}{\partial y^2} + \frac{\partial^2 u}{\partial z^2}\right) + F. \qquad \text{(three-dimensional wave equation)} \tag{1.30}$$

In general, we have

$$\frac{\partial^2 u}{\partial t^2} = c^2\nabla^2 u + F.$$

A complete problem for the wave equation consists of the PDE itself plus two initial conditions plus boundary conditions:

$$\frac{\partial^2 u}{\partial t^2} = c^2\nabla^2 u + F \qquad (x,y,z) \in \Omega,\ t \geqslant 0, \tag{1.31}$$

$$u(x,y,z,0) = f(x,y,z) \qquad (x,y,z) \in \Omega, \tag{1.32}$$

$$\frac{\partial u}{\partial t}(x,y,z,0) = g(x,y,z) \qquad (x,y,z) \in \Omega, \tag{1.33}$$

$$\alpha u(x,y,z,t) + \beta\frac{\partial u}{\partial n}(x,y,z,t) = h(x,y,z,t) \quad (x,y,z) \in \partial\Omega,\ t \geqslant 0. \tag{1.34}$$

1.10 EXAMPLES OF LAPLACE'S EQUATION

From the preceding two sections we have

$$u_t = k\Delta u + F, \qquad \text{(heat equation)}$$
$$u_{tt} = c^2\Delta u + F. \qquad \text{(wave equation)}$$

If we look for a steady-state, that is, time-independent, solution to either the heat equation or the wave equation, we get an equation of the form

$$\Delta u = F. \qquad \text{(Poisson's equation)}$$

If $F \equiv 0$, we have

$$\Delta u = 0. \qquad \text{(Laplace's equation)}$$

Laplace's equation, also called the **potential equation**, occurs in many areas of physics, such as hydrodynamics, elasticity, and electromagnetic field theory. A complete problem for Laplace's equation consists of the PDE itself plus boundary conditions:

$$\Delta u = 0, \qquad (x, y, z) \in \Omega, \tag{1.35}$$

$$\alpha u(x, y, z) + \beta \frac{\partial u}{\partial n}(x, y, z) = f(x, y, z), \quad (x, y, z) \in \partial\Omega. \tag{1.36}$$

As an example of a problem involving Laplace's equation in two dimensions, we present the following.

Example 1.13. (*Laplace's Equation in a Rectangle*)
Formulate a boundary value problem to determine the steady-state temperature distribution of a thin conducting plate that occupies the rectangular region

$$\{(x, y) : 0 \leqslant x \leqslant a,\ 0 \leqslant y \leqslant b\}$$

in the plane. We assume that the top and bottom of the plate are insulated and that the temperature on the boundary is specified. The temperature $u = u(x, y)$ at a point (x, y) in the plate must satisfy Laplace's equation in two dimensions,

$$\frac{\partial^2 u}{\partial x^2} + \frac{\partial^2 u}{\partial y^2} = 0$$

for $0 \leqslant x \leqslant a,\ 0 \leqslant y \leqslant b.$

Laplace's equation above is second-order in x and second-order in y; hence, according to our rule of thumb for side conditions, we need to specify four side conditions. The domain of our problem has four boundaries (the four sides of the rectangle $[0, a] \times [0, b]$). Hence, boundary conditions can be defined quite naturally; for example, we might require that

$$u(x, 0) = x(a - x), \qquad u(x, b) = 0, \quad 0 \leqslant x \leqslant a,$$
$$u(0, y) = y(b - y), \qquad u(a, y) = 0, \quad 0 \leqslant y \leqslant b.$$

In this case, the complete problem for Laplace's equation has the form

$$\frac{\partial^2 u}{\partial x^2} + \frac{\partial^2 u}{\partial y^2} = 0, \qquad 0 \leqslant x \leqslant a,\ 0 \leqslant y \leqslant b,$$

$$u(x,0) = x(a-x),$$

$$u(x,b) = 0,$$

$$u(0,y) = y(b-y),$$

$$u(a,y) = 0.$$

∎

An interesting modification arises if we consider the same problem for a thin conducting plate in the form of a circular disk.

Example 1.14. (*Dirichlet Problem for Laplace's Equation in a Disk*)
We look for the steady-state temperature distribution of a plate of radius $a > 0$. As before, we assume that the top and bottom of the plate are insulated and that the temperature on the outer boundary is specified. This is a classic problem known as the **interior Dirichlet problem for Laplace's equation in a disk.**

Thus, we want to formulate a boundary value problem to solve the Dirichlet problem for Laplace's equation

$$\Delta u = 0$$

in the disk

$$D(a) = \{\, (x,y) \in \mathbb{R}^2 \mid x^2 + y^2 \leqslant a^2 \,\}.$$

Again, we need four boundary conditions; however, the disk does not seem to have enough boundaries. Where do the other three boundary conditions come from?

To find them, we have to use the appropriate coordinate system. If we reformulate the disk problem in planar polar coordinates (r, θ), where $x = r\cos\theta$ and $y = r\sin\theta$, the disk above can then be described as

$$D(a) = \{(r,\theta) \mid 0 \leqslant r \leqslant a,\ -\pi \leqslant \theta \leqslant \pi\}.$$

With this parametrization, there are clearly four boundaries:

$$\{(r,\theta)\ :\ r = 0,\ -\pi \leqslant \theta \leqslant \pi\},$$
$$\{(r,\theta)\ :\ r = a,\ -\pi \leqslant \theta \leqslant \pi\},$$
$$\{(r,\theta)\ :\ 0 \leqslant r \leqslant a,\ \theta = \pi\},$$
$$\{(r,\theta)\ :\ 0 \leqslant r \leqslant a,\ \theta = -\pi\}.$$

Now we can specify our four boundary conditions; however, these need to be physically meaningful.

(i) The condition on the outer boundary $r = a$ is straightforward: If $f(\theta)$ denotes the given temperature distribution on the rim of the disk, then we assume that

$$u(a, \theta) = f(\theta)$$

for $-\pi \leqslant \theta \leqslant \pi$.

(ii) The boundaries at $\theta = -\pi$ and $\theta = \pi$ are artificial since the disk has not been cut or damaged in the transition from $-\pi$ to π. These boundaries are nonphysical; they arise only because of the specific parametrization we have chosen. Hence, we need to ensure that these artificial boundaries do not introduce errors in our solution. Thus, we assume that the solution u and the normal derivative $\partial u/\partial\theta$ are continuous across this boundary:

$$u(r, -\pi) = u(r, \pi),$$

$$\frac{\partial u}{\partial \theta}(r, -\pi) = \frac{\partial u}{\partial \theta}(r, \pi)$$

for $0 \leqslant r \leqslant a$. These conditions arise frequently and are called **periodic boundary conditions** or **periodicity conditions**.

(iii) The final boundary arises for $r = 0$, that is, at the origin. This boundary is also artificial, and arises from the singularity in the coordinate transformation at $r = 0$. Here, we assume that the solution remains finite:

$$\lim_{r \to 0+} u(r, \theta) = u(0, \theta) \quad \text{(finite)}$$

for $-\pi \leqslant \theta \leqslant \pi$. Later, when we solve this problem completely, we will see that there are solutions which are not bounded at 0, and we need to exclude them as nonphysical.

Now, from (i), (ii), (iii), we have our four boundary conditions, as required.

To obtain a complete problem for Laplace's equation, we need to express the Laplacian Δ in polar coordinates. This will be done in detail in Chapter 15; Exercise 15.2, here we just give the result. The interior Dirichlet problem for Laplace's equation on the disk $D(a)$ consists of the boundary value problem

$$\begin{aligned}
&\frac{\partial^2 u}{\partial r^2} + \frac{1}{r}\frac{\partial u}{\partial r} + \frac{1}{r^2}\frac{\partial^2 u}{\partial \theta^2} = 0, && 0 < r < a,\ -\pi < \theta < \pi,\\
&u(r, -\pi) = u(r, \pi), && 0 < r < a,\\
&\frac{\partial u}{\partial \theta}(r, -\pi) = \frac{\partial u}{\partial \theta}(r, \pi), && 0 < r < a,\\
&\lim_{r \to 0+} u(r, \theta) = u(0, \theta), && -\pi < \theta < \pi, \quad \text{(finite)}\\
&u(a, \theta) = f(\theta), && -\pi < \theta < \pi.
\end{aligned}$$

∎

Both of the examples above can be solved using separation of variables, and we do this later in Section 5.4.1 and Example 6.1.

1.11 SUMMARY

To learn a new language, we must first learn some basic vocabulary. In mathematics, the language of science, we must do the same thing. In this introductory chapter we defined the elementary notions of **partial derivatives**, **gradient**, and **Laplacian**, and saw that **partial differential equations** (PDEs) can be **homogeneous**, **nonhomogeneous**, **linear**, or **nonlinear**. Some linear homogeneous PDEs can be solved completely, using **separation of variables** and the **superposition principle**. Example 1.11 applies this method to the heat equation, and we recommend that it be studied thoroughly.

We classified second-order linear homogeneous PDEs into three kingdoms: **elliptic**, **parabolic**, and **hyperbolic**. Elliptic PDEs are represented by Laplace's equation $\nabla^2 u = 0$, which appears in the study of steady-state temperature distributions and in potential theory in physics. The parabolic kingdom is represented by the heat equation $u_t = D\nabla^2 u$, also called the diffusion equation. This model applies to heat conduction, diffusion, or population dynamics. The hyperbolic kingdom is represented by the wave equation $u_{tt} = c^2\nabla^2 u$, which models oscillatory phenomena, as for example, vibrations of a guitar string.

We saw that just having a PDE is not enough; to guarantee a unique solution we needed to impose **side conditions**. These usually arise as **initial conditions** or as **boundary conditions**. However, as we see later, they can have other forms, such as boundedness conditions or integral conditions. We provided a **rule of thumb for side conditions**, where the number of side conditions corresponds to the order of the highest-order derivative for each of the variables appearing in the PDE. The most commonly used boundary conditions are the **Dirichlet**, **Neumann**, and **Robin boundary conditions**, and we gave physical interpretations of those.

Finally, we discussed **steady states**. Recall that to find the equations governing the steady state, if it exists, we set all the time derivatives equal to zero: $u_t = 0, u_{tt} = 0$, etc. In most cases, the PDE then becomes so simple that it can be solved very easily.

1.11.1 Problems and Notes

Problems from Part II:

Exercise	11.1	11.14	11.17	13.1	14.11	15.1
Notes						

Final exam questions:

Exercise	19.2	19.3	19.12	19.15
Notes				

CHAPTER 2

FOURIER SERIES

2.1 PIECEWISE CONTINUOUS FUNCTIONS

In most physical problems, we may expect to use "nice" smooth functions. However, it will be convenient to allow for certain types of discontinuous functions. Consider the behavior of a function f near a point x_0. In particular, consider the following one-sided limits:

$$f(x_0^+) = \lim_{x \to x_0^+} f(x), \qquad f(x_0^-) = \lim_{x \to x_0^-} f(x).$$

We make the following definition:

Definition 2.1. Let the function f be defined on an open interval containing the point x_0.

(i) If $f(x_0^+) = f(x_0^-) = f(x_0)$, f is **continuous** at x_0; otherwise, f is **discontinuous** at x_0.

(ii) If f is discontinuous at x_0 and if both $f(x_0^+)$ and $f(x_0^-)$ exist, f is said to have a **discontinuity of the first kind** or a **simple discontinuity** at x_0.

Partial Differential Equations: Theory and Completely Solved Problems, First Edition. By T. Hillen, I. E. Leonard, H. van Roessel

(iii) A simple discontinuity of f of the first kind at x_0 is said to be

(a) a **removable discontinuity** if $f(x_0^+) = f(x_0^-) \neq f(x_0)$ and

(b) a **jump discontinuity** if $f(x_0^+) \neq f(x_0^-)$, regardless of the value $f(x_0)$.

(iv) Any discontinuity of f at x_0 not of the first kind is said to be a **discontinuity of the second kind** at x_0.

Example 2.1. The function f defined by

$$f(x) = \begin{cases} e^x, & \text{for } x \neq 1, \\ 2, & \text{for } x = 1, \end{cases}$$

has a removable discontinuity at $x_0 = 1$, while the function g defined by

$$g(x) = \begin{cases} \sin(1/x), & \text{if } x \neq 0, \\ 0, & \text{if } x = 0, \end{cases}$$

has a discontinuity of the second kind at $x_0 = 0$. ∎

Definition 2.2. A function f is **piecewise continuous** (PWC), sometimes called **sectionally continuous**, on an interval (a, b) if

(i) f is continuous for $x \in (a, b)$ except possibly at a finite number of points;

(ii) $f(x^+)$ exists for all $x \in [a, b)$;

(iii) $f(x^-)$ exists for all $x \in (a, b]$.

Thus, a function f is piecewise continuous on an interval (a, b) if and only if f has at most a finite number of discontinuities, none of which is worse than a jump discontinuity.

Example 2.2. The function

$$f(x) = \begin{cases} x, & 0 < x \leqslant 1, \\ -1, & 1 < x \leqslant 2, \\ 1, & 2 < x < 3, \end{cases}$$

is piecewise continuous on the interval $(0, 3)$. ∎

Example 2.3. The function $f(x) = 1/x$ is not piecewise continuous on $(0, 1)$, since

$$f(0^+) = \lim_{x \to 0+} f(x) = \lim_{x \to 0+} \frac{1}{x}$$

does not exist. ∎

It is convenient to introduce the following notation:

Notation. $PWC(a, b)$ denotes the set of all PWC functions on (a, b).

The following properties of piecewise continuous functions are easily verified.

Theorem 2.3. [*Properies of* $PWC(a, b)$]

(i) If $f, g \in PWC(a, b)$, then $\alpha f + \beta g \in PWC(a, b)$ for all $\alpha, \beta \in \mathbb{R}$.

(ii) If $f, g \in PWC(a, b)$, then $f \cdot g \in PWC(a, b)$.

(iii) If $f \in PWC(a, b)$, then $\int_a^b |f(x)|\, dx$ exists.

From property (i) it follows that $PWC(a, b)$ is a vector space.

Definition 2.4. A function f is **piecewise smooth** (abbreviated PWS) on (a, b) if

(i) $f \in PWC(a, b)$ and

(ii) $f' \in PWC(a, b)$.

Again, we introduce the following notation.

Notation. $PWS(a, b)$ denotes the set of all PWS functions on (a, b). It is clear that $PWS(a, b) \subset PWC(a, b)$.

2.2 EVEN, ODD, AND PERIODIC FUNCTIONS

We begin with a definition.

Definition 2.5. Let f be a function whose domain $D(f)$ is symmetric, that is, $-x \in D(f)$ whenever $x \in D(f)$; then we say that

(i) f is **even** if $f(-x) = f(x)$ for all $x \in D(f)$.

(ii) f is **odd** if $f(-x) = -f(x)$ for all $x \in D(f)$.

(iii) f is **periodic with period** p if $x + p \in D(f)$ whenever $x \in D(f)$, and $f(x + p) = f(x)$ for all $x \in D(f)$.

Example 2.4. The following functions are defined for all real numbers, that is, $D(f) = \mathbb{R}$, and

(i) $f(x) = x^n$ is even if n is an even integer.

(ii) $f(x) = x^n$ is odd if n is an odd integer.

(iii) $f(x) = \cos x$ is even and 2π-periodic.

(iv) $f(x) = \sin x$ is odd and 2π-periodic.

(v) $f(x) = e^x$ is not even, odd, or periodic.

(vi) $f(x) = \cosh x$ is even.

(vii) $f(x) = \sinh x$ is odd.

∎

Note: If f is integrable and periodic with period p, the integral of f over any full period is a constant; that is,

$$\int_a^{a+p} f(x)\,dx = \int_0^p f(x)\,dx \tag{2.1}$$

for any real number a.

Example 2.5. If a function f is defined on an interval (a, b), it is sometimes useful to extend the definition of f to the entire real line.

The **periodic extension** of f, denoted $\bar{f}$, is defined as

$$\bar{f}(x) = f(x + np) \quad \text{for} \quad a - np < x < b - np, \quad n \in \mathbb{Z},$$

where $p = b - a$. ∎

Frequently, in applications, we will encounter a function f defined on an interval $(0, \ell)$, and we will need to construct periodic extensions of f.

Definition 2.6. If the function f is defined on the interval $(0, \ell)$:

(i) The **odd extension** of f on $(-\ell, \ell)$, denoted f_{odd}, is defined by

$$f_{\text{odd}}(x) = \begin{cases} f(x), & \text{for} \quad 0 < x < \ell, \\ -f(-x), & \text{for} \ -\ell < x < 0, \end{cases}$$

and

(ii) The **even extension** of f on $(-\ell, \ell)$, denoted f_{even}, is defined by

$$f_{\text{even}}(x) = \begin{cases} f(x), & \text{for} \quad 0 < x < \ell, \\ f(-x), & \text{for} \ -\ell < x < 0. \end{cases}$$

Note: If we then extend f_{even} and f_{odd} to the entire real line, then we get the periodic extensions $\bar{f}_{\text{even}}$ and $\bar{f}_{\text{odd}}$, which are periodic on $\mathbb{R}$ with period $p = 2\ell$.

2.3 ORTHOGONAL FUNCTIONS

We introduce orthogonality of functions in a function space similar to the way in which orthogonality of vectors was defined in $\mathbb{R}^3$; before doing this, however, we review some facts from linear algebra.

Vectors in the three-dimensional vector space $\mathbb{R}^3$ will be denoted by boldface roman letters, such as $\mathbf{a}$ or $\mathbf{A}$. Recall that the standard Euclidean basis for $\mathbb{R}^3$ is given by

$$\mathbf{i} = (1, 0, 0), \qquad \mathbf{j} = (0, 1, 0), \qquad \mathbf{k} = (0, 0, 1).$$

For convenience, we rewrite this basis as

$$\{\mathbf{e}_1, \mathbf{e}_2, \mathbf{e}_3\} = \{\mathbf{i}, \mathbf{j}, \mathbf{k}\}.$$

Given two vectors in $\mathbb{R}^3$,

$$\mathbf{v} = (a_1, a_2, a_3) = a_1\mathbf{e}_1 + a_2\mathbf{e}_2 + a_3\mathbf{e}_3 = \sum_{i=1}^{3} a_i\mathbf{e}_i,$$

$$\mathbf{w} = (b_1, b_2, b_3) = b_1\mathbf{e}_1 + b_2\mathbf{e}_2 + b_3\mathbf{e}_3 = \sum_{i=1}^{3} b_i\mathbf{e}_i;$$

the **dot product** of these two vectors is given by

$$\mathbf{v} \cdot \mathbf{w} = a_1b_1 + a_2b_2 + a_3b_3 = \sum_{i=1}^{3} a_ib_i.$$

To see this, we include all the details using the familiar properties of the dot product:

$$\mathbf{v} \cdot \mathbf{w} = \left(\sum_{i=1}^{3} a_i\mathbf{e}_i\right) \cdot \left(\sum_{j=1}^{3} b_j\mathbf{e}_j\right) = \sum_{i=1}^{3}\sum_{j=1}^{3} a_ib_j\mathbf{e}_i \cdot \mathbf{e}_j = \sum_{i=1}^{3}\sum_{j=1}^{3} a_ib_j\delta_{ij},$$

where

$$\delta_{ij} = \begin{cases} 1, & \text{if } i = j, \\ 0, & \text{if } i \neq j, \end{cases}$$

is the **Kronecker delta**. Thus,

$$\mathbf{v} \cdot \mathbf{w} = \sum_{i=1}^{3} a_ib_i.$$

In certain cases, representing a vector as a linear combination of the basis vectors, or the calculation of the dot product of two vectors, becomes very simple, as the following example shows.

Example 2.6. Consider the two sets of basis vectors in $\mathbb{R}^2$,

$$\mathbf{u}_1 = (1,0),\ \mathbf{u}_2 = (1,1) \qquad \text{and} \qquad \mathbf{w}_1 = (1,1),\ \mathbf{w}_2 = (1,-1).$$

Suppose that we want to write the vector $\mathbf{v} = (1,2)$ in terms of $\{\mathbf{u}_1, \mathbf{u}_2\}$ and $\{\mathbf{w}_1, \mathbf{w}_2\}$.
If $\mathbf{v} = a_1\mathbf{u}_1 + a_2\mathbf{u}_2$, then

$$\begin{aligned} \mathbf{v}\cdot\mathbf{u}_1 &= a_1\mathbf{u}_1\cdot\mathbf{u}_1 + a_2\mathbf{u}_2\cdot\mathbf{u}_1, \\ \mathbf{v}\cdot\mathbf{u}_2 &= a_1\mathbf{u}_1\cdot\mathbf{u}_2 + a_2\mathbf{u}_2\cdot\mathbf{u}_2, \end{aligned}$$

which becomes

$$\begin{aligned} 1 &= a_1 + a_2, \\ 3 &= a_1 + 2a_2. \end{aligned}$$

Therefore, $a_1 = -1$ and $a_2 = 2$, so that

$$\mathbf{v} = -\mathbf{u}_1 + 2\mathbf{u}_2.$$

On the other hand, if $\mathbf{v} = b_1\mathbf{w}_1 + b_2\mathbf{w}_2$, then

$$\begin{aligned} \mathbf{v}\cdot\mathbf{w}_1 &= b_1\mathbf{w}_1\cdot\mathbf{w}_1 + b_2\mathbf{w}_2\cdot\mathbf{w}_1, \\ \mathbf{v}\cdot\mathbf{w}_2 &= b_1\mathbf{w}_1\cdot\mathbf{w}_2 + b_2\mathbf{w}_2\cdot\mathbf{w}_2, \end{aligned}$$

which becomes

$$\begin{aligned} 3 &= 2b_1, \\ -1 &= 2b_2. \end{aligned}$$

Therefore, $b_1 = 3/2$ and $b_2 = -1/2$, so that

$$\mathbf{v} = \frac{3}{2}\mathbf{w}_1 - \frac{1}{2}\mathbf{w}_2.$$

∎

While this is a simple two-dimensional example, it is clear that the second calculation is much simpler. The reason for this simplicity is that the basis vectors $\{\mathbf{w}_1, \mathbf{w}_2\}$ form an orthogonal set. The significance of this becomes more evident in higher-dimensional spaces. For example, if in $\mathbb{R}^3$ we have

$$\begin{aligned} \mathbf{v} &= a_1\mathbf{u}_1 + a_2\mathbf{u}_2 + a_3\mathbf{u}_3, \\ \mathbf{v} &= b_1\mathbf{w}_1 + b_2\mathbf{w}_2 + b_3\mathbf{w}_3, \end{aligned}$$

where $\{\mathbf{u}_1, \mathbf{u}_2, \mathbf{u}_3\}$ and $\{\mathbf{w}_1, \mathbf{w}_2, \mathbf{w}_3\}$ are bases for $\mathbb{R}^3$, a little manipulation yields

$$\mathbf{v}\cdot\mathbf{u}_i = a_1\mathbf{u}_1\cdot\mathbf{u}_i + a_2\mathbf{u}_2\cdot\mathbf{u}_i + a_3\mathbf{u}_3\cdot\mathbf{u}_i, \qquad i = 1,2,3,$$

and

$$\mathbf{v} \cdot \mathbf{w}_i = b_1 \mathbf{w}_1 \cdot \mathbf{w}_i + b_2 \mathbf{w}_2 \cdot \mathbf{w}_i + b_3 \mathbf{w}_3 \cdot \mathbf{w}_i, \qquad i = 1, 2, 3.$$

In matrix form, the first set of equations becomes

$$\begin{pmatrix} \mathbf{u}_1 \cdot \mathbf{u}_1 & \mathbf{u}_2 \cdot \mathbf{u}_1 & \mathbf{u}_3 \cdot \mathbf{u}_1 \\ \mathbf{u}_1 \cdot \mathbf{u}_2 & \mathbf{u}_2 \cdot \mathbf{u}_2 & \mathbf{u}_3 \cdot \mathbf{u}_2 \\ \mathbf{u}_1 \cdot \mathbf{u}_3 & \mathbf{u}_2 \cdot \mathbf{u}_3 & \mathbf{u}_3 \cdot \mathbf{u}_3 \end{pmatrix} \begin{pmatrix} a_1 \\ a_2 \\ a_3 \end{pmatrix} = \begin{pmatrix} \mathbf{v} \cdot \mathbf{u}_1 \\ \mathbf{v} \cdot \mathbf{u}_2 \\ \mathbf{v} \cdot \mathbf{u}_3 \end{pmatrix},$$

and solving this equation requires the inversion of a 3×3 matrix. However, if the second set of basis vectors is orthogonal, that is, $\mathbf{w}_i \cdot \mathbf{w}_j = 0$ for $i \neq j$, then in matrix form, the second set of equations becomes

$$\begin{pmatrix} \mathbf{w}_1 \cdot \mathbf{w}_1 & 0 & 0 \\ 0 & \mathbf{w}_2 \cdot \mathbf{w}_2 & 0 \\ 0 & 0 & \mathbf{w}_3 \cdot \mathbf{w}_3 \end{pmatrix} \begin{pmatrix} b_1 \\ b_2 \\ b_3 \end{pmatrix} = \begin{pmatrix} \mathbf{v} \cdot \mathbf{w}_1 \\ \mathbf{v} \cdot \mathbf{w}_2 \\ \mathbf{v} \cdot \mathbf{w}_3 \end{pmatrix},$$

which is easily solved to get

$$b_1 = \frac{\mathbf{v} \cdot \mathbf{w}_1}{\mathbf{w}_1 \cdot \mathbf{w}_1}, \qquad b_2 = \frac{\mathbf{v} \cdot \mathbf{w}_2}{\mathbf{w}_2 \cdot \mathbf{w}_2}, \qquad b_3 = \frac{\mathbf{v} \cdot \mathbf{w}_3}{\mathbf{w}_3 \cdot \mathbf{w}_3}.$$

Using summation notation, we have

$$\mathbf{v} = \sum_{i=1}^{3} b_i \mathbf{w}_i,$$

and for $j = 1, 2, 3$, we have

$$\mathbf{v} \cdot \mathbf{w}_j = \left(\sum_{i=1}^{3} b_i \mathbf{w}_i \right) \cdot \mathbf{w}_j = \sum_{i=1}^{3} b_i \mathbf{w}_i \cdot \mathbf{w}_j = b_j \mathbf{w}_j \cdot \mathbf{w}_j,$$

since $\{\mathbf{w}_1, \mathbf{w}_2, \mathbf{w}_3\}$ is an orthogonal basis for $\mathbb{R}^3$. Therefore, again we find that

$$b_j = \frac{\mathbf{v} \cdot \mathbf{w}_j}{\mathbf{w}_j \cdot \mathbf{w}_j}$$

for $j = 1, 2, 3$.

Recall that the **norm** of a vector, or the **length** of a vector, is given by

$$\|\mathbf{v}\| = \sqrt{\mathbf{v} \cdot \mathbf{v}}.$$

If $\{\mathbf{e}_1, \mathbf{e}_2, \mathbf{e}_3\}$ is an orthogonal basis for $\mathbb{R}^3$ and $\mathbf{v} \in \mathbb{R}^3$ has components a_1, a_2, and a_3 with respect to this basis, that is,

$$\mathbf{v} = \sum_{i=1}^{3} a_i \mathbf{e}_i,$$

then

$$\|\mathbf{v}\|^2 = \mathbf{v}\cdot\mathbf{v} = \left(\sum_{i=1}^{3} a_i\mathbf{e}_i\right)\cdot\left(\sum_{j=1}^{3} a_j\mathbf{e}_j\right)$$

$$= \sum_{i=1}^{3}\sum_{j=1}^{3} a_i a_j \mathbf{e}_i\cdot\mathbf{e}_j = \sum_{i=1}^{3} a_i^2\mathbf{e}_i\cdot\mathbf{e}_i = \sum_{i=1}^{3} a_i^2\|\mathbf{e}_i\|^2.$$

If the $\mathbf{e}_i$ are unit vectors, that is, $\|\mathbf{e}_i\| = 1$, for $i = 1, 2, 3$, then

$$\|\mathbf{v}\| = \sqrt{a_1^2 + a_2^2 + a_3^2}.$$

Now we are ready to introduce an inner product and orthogonality for functions.

Definition 2.7. An **inner product** on a vector space X is any function $\langle \mathbf{u}, \mathbf{v}\rangle$ that acts on pairs of vectors $\mathbf{u}$ and $\mathbf{v}$ in X and satisfies the following properties: For any $\mathbf{u}$, $\mathbf{v}$, $\mathbf{w} \in X$ and $\lambda \in \mathbb{R}$:

(i) $\langle \mathbf{u}, \mathbf{u}\rangle \geqslant 0$ and $\langle \mathbf{u}, \mathbf{u}\rangle = 0$ if and only if $\mathbf{u} = \mathbf{0}$,

(ii) $\langle \mathbf{u}, \mathbf{v}\rangle = \langle \mathbf{v}, \mathbf{u}\rangle$,

(iii) $\langle \mathbf{u}, \mathbf{v}+\mathbf{w}\rangle = \langle \mathbf{u}, \mathbf{v}\rangle + \langle \mathbf{u}, \mathbf{w}\rangle$, and

(iv) $\langle \lambda\mathbf{u}, \mathbf{v}\rangle = \lambda\langle \mathbf{u}, \mathbf{v}\rangle$.

The dot product is but one example of an inner product. We want to generalize these concepts of inner product and norms to PWC functions.

Here, we give a heuristic motivation for the definition of an inner product for PWC functions. Let

$$f, g, w \in PWC(a, b) \quad \text{with} \quad w \geqslant 0,$$

and recall from elementary calculus the definition of the integral as the limit of a Riemann sum:

$$\int_a^b f(x)g(x)w(x)\,dx = \lim_{N\to\infty}\sum_{i=1}^{N} f(x_i)g(x_i)w(x_i)\Delta x_i,$$

where $a = x_0 < x_1 < \cdots < x_N = b$ is a partition of the interval (a, b) and $\Delta x_i = x_i - x_{i-1}$ for $i = 1, 2, \ldots, N$.

If we define

$$a_i = f(x_i)\sqrt{w(x_i)\Delta x_i}, \qquad b_i = g(x_i)\sqrt{w(x_i)\Delta x_i},$$

for $i = 1, 2, \ldots, N$, then

$$\int_a^b f(x)g(x)w(x)\,dx = \lim_{N\to\infty} \sum_{i=1}^{N} a_i b_i,$$

and the sum on the right looks like an inner product in N dimensions. This leads us to make the following definition.

Definition 2.8. Let $f, g, w \in PWC(a, b)$ with $w(x) \geqslant 0$. The **inner product** of f and g with **weight** w is defined as

$$\langle f, g\rangle = \int_a^b f(x)g(x)w(x)\,dx.$$

The function $w(x)$ is called the **weight function**. Quite often, we use $w(x) = 1$.

Note: It is easily verified that this definition satisfies all the properties given in the definition of an inner product except for one. Namely, it is possible that f is in $PWC(a, b)$ and

$$\langle f, f\rangle = \int_a^b f(x)^2 w(x)\,dx = 0,$$

but $f \neq 0$; that is, it is not true that $f(x) = 0$ for all $x \in (a, b)$. In this case, f is called a **null function**. This does not present a problem, since any $f \in PWC(a, b)$ has only finitely many simple discontinuities, and if we regard any two functions $f, g \in PWC$ as identical whenever they differ at only finitely many points, the difficulty above disappears (the value of their integrals is the same, since the integral is not affected by changing the value of the function at finitely many points).

Recall that we saw earlier that $PWC(a, b)$ is a vector space. We can think of Definition 2.8 as an inner product defined on the "infinite-dimensional" vector space $PWC(a, b)$. We can extend many concepts from linear algebra in $\mathbf{R}^n$ to linear algebra in $PWC(a, b)$.

Definition 2.9. The **norm** of $f \in PWC(a, b)$ with weight w is $\|f\| = \sqrt{\langle f, f\rangle}$.

Definition 2.10. If $f, g, w \in PWC(a, b)$ with $w(x) \geqslant 0$, f and g are said to be **orthogonal** on (a, b) relative to the weight w if $\langle f, g\rangle = 0$.

We illustrate these concepts with the following important example.

Example 2.7. Define an inner product on the vector space $PWC(a, b)$ by

$$\langle f, g\rangle = \int_a^b f(x)g(x)\,dx,$$

with weight function $w(x) \equiv 1$. For each nonnegative integer n, the functions

$$c_n(x) = \cos\frac{n\pi x}{\ell},$$

$$s_n(x) = \sin\frac{n\pi x}{\ell},$$

where $\ell = (b-a)/2$, are in $PWC(a,b)$, are defined for all real numbers x, and are periodic with period 2ℓ. For example,

$$c_n(x+2\ell) = \cos\frac{n\pi(x+2\ell)}{\ell} = \cos\left(\frac{n\pi x}{\ell} + 2n\pi\right) = \cos\frac{n\pi x}{\ell} = c_n(x),$$

and similarly for s_n. Thus, for $n = 0$, we have

$$\langle c_0, c_0\rangle = \int_a^b c_0^2(x)\,dx = \int_a^b dx = b - a = 2\ell.$$

For $n \neq 0$, we have

$$\langle c_n, c_m\rangle = \int_a^b c_n(x)c_m(x)\,dx = \int_0^{2\ell} c_n(x)c_m(x)\,dx$$

from (2.1), since the product $c_n c_m$ is periodic with period 2ℓ, and

$$\int_0^{2\ell} \cos\frac{n\pi x}{\ell}\cos\frac{m\pi x}{\ell}\,dx = \frac{\ell}{\pi}\int_0^{2\pi} \cos n\xi \cos m\xi\,d\xi,$$

after making the substitution $\xi = \pi x/\ell$. Therefore,

$$\langle c_n, c_m\rangle = \frac{\ell}{\pi}\,\delta_{nm}\pi = \delta_{nm}\,\ell.$$

A similar calculation yields

$$\langle s_n, s_m\rangle = \delta_{nm}\,\ell$$

and

$$\langle s_n, c_m\rangle = 0.$$

Therefore, the set

$$\{\,c_0,\,c_1,\,s_1,\,c_2,\,s_2,\,\dots\} = \left\{1,\,\cos\frac{\pi x}{\ell},\,\sin\frac{\pi x}{\ell},\,\cos\frac{2\pi x}{\ell},\,\sin\frac{2\pi x}{\ell},\,\dots\right\}$$

is an orthogonal set of functions on (a,b) with respect to the inner product above. ∎

2.4 FOURIER SERIES

If f is periodic with period p, one might attempt to represent f with an infinite series of p-periodic functions. Notice that the set

$$\left\{1,\,\cos\frac{\pi x}{\ell},\,\sin\frac{\pi x}{\ell},\,\cos\frac{2\pi x}{\ell},\,\sin\frac{2\pi x}{\ell},\,\dots\right\} \tag{2.2}$$

consists of functions having a common period $p = 2\ell$. Suppose that f is 2ℓ-periodic; we could try to represent f as follows:

$$f(x) = a_0 + \sum_{n=1}^{\infty} \left(a_n \cos \frac{n\pi x}{\ell} + b_n \sin \frac{n\pi x}{\ell} \right).$$

Since each term in the series has period 2ℓ, we would expect the sum, if it exists, to be 2ℓ-periodic also. Several questions come to mind:

1. How does one find the constants a_0, a_n, b_n?

2. Once the constants are found, does the series converge?

3. If the series converges, does it converge to f?

In Example 2.7 we saw that the elements of the set (2.2) are orthogonal, which makes the calculation of the constants in the series fairly straightforward. Writing f in the more compact notation

$$f(x) = a_0 c_0 + \sum_{n=1}^{\infty} [a_n c_n(x) + b_n s_n(x)]$$

and using the inner product of Example 2.7, we get

$$\begin{aligned}
\langle f, c_0 \rangle &= \left\langle a_0 c_0 + \sum_{n=1}^{\infty} (a_n c_n + b_n s_n), c_0 \right\rangle \\
&= a_0 \langle c_0, c_0 \rangle + \sum_{n=1}^{\infty} (a_n \langle c_n, c_0 \rangle + b_n \langle s_n, c_0 \rangle) \\
&= 2\ell\, a_0 + \sum_{n=1}^{\infty} (a_n \ell\, \delta_{n0} + 0) = 2\ell\, a_0.
\end{aligned}$$

Therefore,

$$a_0 = \frac{1}{2\ell} \langle f, c_0 \rangle .$$

For $m \neq 0$, we have

$$\begin{aligned}
\langle f, c_m \rangle &= \left\langle a_0 c_0 + \sum_{n=1}^{\infty} (a_n c_n + b_n s_n), c_m \right\rangle \\
&= a_0 \langle c_0, c_m \rangle + \sum_{n=1}^{\infty} (a_n \langle c_n, c_m \rangle + b_n \langle s_n, c_m \rangle) \\
&= 0 + \sum_{n=1}^{\infty} (a_n \ell\, \delta_{nm} + 0) = \ell\, a_m.
\end{aligned}$$

Therefore,

$$a_m = \frac{1}{\ell} \langle f, c_m \rangle,$$

and similarly,

$$b_m = \frac{1}{\ell} \langle f, s_m \rangle$$

for $m \geqslant 1$. This prompts the following definition:

Definition 2.11. (*Fourier Series*)
The series

$$a_0 + \sum_{n=1}^{\infty} \left(a_n \cos \frac{n\pi x}{\ell} + b_n \sin \frac{n\pi x}{\ell} \right),$$

where $\ell = (b - a)/2$, and

$$a_0 = \frac{1}{2\ell} \langle f, c_0 \rangle = \frac{1}{2\ell} \int_a^b f(x)\, dx,$$

$$a_n = \frac{1}{\ell} \langle f, c_n \rangle = \frac{1}{\ell} \int_a^b f(x) \cos \frac{n\pi x}{\ell}\, dx,$$

$$b_n = \frac{1}{\ell} \langle f, s_n \rangle = \frac{1}{\ell} \int_a^b f(x) \sin \frac{n\pi x}{\ell}\, dx,$$

for $n = 1, 2, \ldots$, is called the **Fourier series** for f, and a_0, a_n, and b_n are called the **Fourier coefficients** of f.

Note that if $f \in PWC(a, b)$, then all of the integrals above exist. We have answered question 1 above: We have simple formulas for the Fourier coefficients. Again, we reiterate that the simplicity of these formulas is due to the orthogonality of the sine and cosine functions on the given interval with respect to the given inner product. It remains to determine whether this infinite series converges, and if so, to what function it converges. Until convergence can be determined, we write

$$f(x) \sim a_0 + \sum_{n=1}^{\infty} \left(a_n \cos \frac{n\pi x}{\ell} + b_n \sin \frac{n\pi x}{\ell} \right)$$

and say that the function f is **represented by its Fourier series on the interval** (a, b).

Example 2.8. (*Sawtooth Function*)
Find the Fourier series for the 2π-periodic function f defined by

$$f(x) = \begin{cases} x, & \text{for} \quad 0 < x < \pi, \\ 0, & \text{for} \ -\pi < x < 0, \end{cases}$$

and $f(x+2\pi) = f(x)$ otherwise.

Solution. We take the interval to be $(a, b) = (-\pi, \pi)$, so that $\ell = \pi$, and

$$c_n(x) = \cos nx \qquad \text{and} \qquad s_n(x) = \sin nx.$$

The Fourier coefficients of f are

$$a_0 = \frac{1}{2\ell}\langle f, c_0\rangle = \frac{1}{2\pi}\int_{-\pi}^{\pi} f(x)\,dx = \frac{1}{2\pi}\int_0^{\pi} x\,dx = \frac{\pi}{4},$$

$$a_n = \frac{1}{\ell}\langle f, c_n\rangle = \frac{1}{\pi}\int_0^{\pi} x\cos nx\,dx = \frac{(-1)^n - 1}{n^2\pi},$$

$$b_n = \frac{1}{\ell}\langle f, s_n\rangle = \frac{1}{\pi}\int_0^{\pi} x\sin nx\,dx = \frac{(-1)^{n+1}}{n},$$

where we use integration by parts to evaluate the integrals for a_n and b_n. Therefore, the Fourier series for f is given by

$$\begin{aligned} f(x) &\sim a_0 + \sum_{n=1}^{\infty}(a_n\cos nx + b_n\sin nx) \\ &= \frac{\pi}{4} + \sum_{n=1}^{\infty}\left[\frac{(-1)^n - 1}{n^2\pi}\cos nx + \frac{(-1)^{n+1}}{n}\sin nx\right]; \end{aligned}$$

that is,

$$f(x) \sim \frac{\pi}{4} - \frac{2}{\pi}\cos x + \sin x - \frac{1}{2}\sin 2x - \frac{2}{9\pi}\cos 3x + \frac{1}{3}\sin 3x - \cdots.$$

Note that the function is left undefined for $x = n\pi$, where n is an integer, but the series itself is defined (although it may not converge) at these points.

∎

Another example is given below.

Example 2.9. (*Linear Function*)
Find the Fourier series for the 2ℓ-periodic function f defined by

$$f(x) = x, \quad -\ell < x < \ell,$$

and $f(x+2\ell) = f(x)$.

Solution. In this case,

$$c_n(x) = \cos\frac{n\pi x}{\ell} \qquad \text{and} \qquad s_n(x) = \sin\frac{n\pi x}{\ell}$$

for $-\ell < x < \ell$, and the Fourier coefficients of f are given by

$$a_0 = \frac{1}{2\ell}\langle f, c_0\rangle = \frac{1}{2\ell}\int_{-\ell}^{\ell} x\,dx = 0,$$

$$a_n = \frac{1}{\ell}\langle f, c_n\rangle = \frac{1}{\ell}\int_{-\ell}^{\ell} x\cos\frac{n\pi x}{\ell}\,dx = 0, \quad n \geqslant 1,$$

$$b_n = \frac{1}{\ell}\langle f, s_n\rangle = \frac{1}{\ell}\int_{-\ell}^{\ell} x\sin\frac{n\pi x}{\ell}\,dx = \frac{2\ell}{n\pi}(-1)^{n+1}, \quad n \geqslant 1.$$

Therefore, the Fourier series for f is given by

$$x \sim \frac{2\ell}{\pi}\sum_{n=1}^{\infty}\frac{(-1)^{n+1}}{n}\sin\frac{n\pi x}{\ell} \tag{2.3}$$

for $-\ell < x < \ell$.

Again, note that the function f is left undefined for $x = (2n+1)\ell$, where n is an integer, but that the series itself is defined (although it may not converge) at these points. In fact, at $x = \ell$, the left-hand side of (2.3) equals ℓ, whereas the right-hand side equals 0. This indicates that at least at one point, the Fourier series does not converge to the function.

∎

Another observation concerning this last example: The function f is an odd function, and the Fourier series is a sum of odd terms, which should not be surprising.

In general, we have the following theorem (the proof is left as an exercise):

Theorem 2.12. For $f \in PWC(-\ell, \ell)$, the following are true:

(a) If f is an odd function,

$$f(x) \sim \sum_{n=1}^{\infty} b_n \sin\frac{n\pi x}{\ell};$$

that is, the Fourier series for f contains only sine terms.

(b) If f is an even function,

$$f(x) \sim a_0 + \sum_{n=1}^{\infty} a_n \cos\frac{n\pi x}{\ell};$$

that is, the Fourier series for f contains only cosine terms.

2.4.1 Fourier Sine and Cosine Series

If the function f is defined on the interval $[-\ell, \ell]$, the Fourier coefficients are uniquely determined, and the Fourier series representation for f on the interval may contain both sine and cosine terms. However, the PDEs we solve using separation of variables typically have boundary conditions specified on the subinterval $[0, \ell]$, and usually lead to series containing only sine terms or to series containing only cosine terms. If the function f is defined only on the interval $[0, \ell]$, we may extend it to the interval $[-\ell, \ell]$ as either an odd function f_{odd} or as an even function f_{even}. From Theorem 2.12, the Fourier series for f_{odd} on the interval $[-\ell, \ell]$ contains only sine terms, while the Fourier series for f_{even} on the interval $[-\ell, \ell]$ contains only cosine terms. Both f_{odd} and f_{even} agree with f on the interval $[0, \ell]$; hence, f has two different Fourier series representations on the interval $[0, \ell]$.

(i) The **Fourier sine series** for f on $[0, \ell]$,

$$f(x) \sim \sum_{n=1}^{\infty} b_n \sin \frac{n\pi x}{\ell},$$

where

$$b_n = \frac{1}{\ell} \int_{-\ell}^{\ell} f_{\text{odd}}(x) \sin \frac{n\pi x}{\ell}\, dx = \frac{2}{\ell} \int_0^{\ell} f(x) \sin \frac{n\pi x}{\ell}\, dx$$

for $n \geqslant 1$.

(ii) The **Fourier cosine series** for f on $[0, \ell]$,

$$f(x) \sim a_0 + \sum_{n=1}^{\infty} a_n \cos \frac{n\pi x}{\ell},$$

where

$$a_0 = \frac{1}{2\ell} \int_{-\ell}^{\ell} f_{\text{even}}(x)\, dx = \frac{1}{\ell} \int_0^{\ell} f(x)\, dx$$

and

$$a_n = \frac{1}{\ell} \int_{-\ell}^{\ell} f_{\text{even}}(x) \cos \frac{n\pi x}{\ell}\, dx = \frac{2}{\ell} \int_0^{\ell} f(x) \cos \frac{n\pi x}{\ell}\, dx$$

for $n \geqslant 1$.

Example 2.10. Find the Fourier sine series and the Fourier cosine series, both of which represent the function

$$f(x) = \begin{cases} 2x, & \text{for } \ 0 < x < 1, \\ 2, & \text{for } \ 1 < x < 2, \end{cases}$$

on the interval $0 < x < 2$. The graph of this function is shown in Figure 2.1.

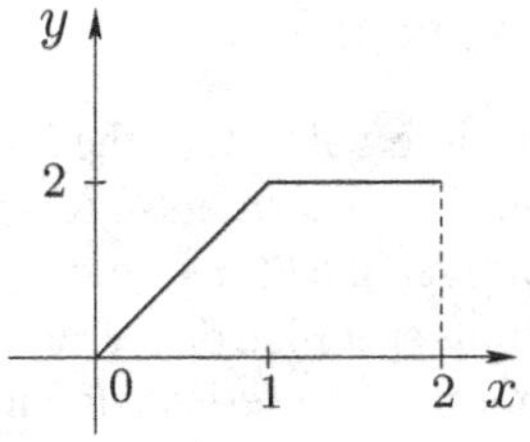

Figure 2.1

• Fourier Sine Series

To find the Fourier sine series for $f(x)$ on the interval $[0, 2]$, we consider the odd extension f_{odd} to the interval $[-2, 2]$,

$$f_{\text{odd}}(x) = \begin{cases} -2, & \text{for} \quad -2 < x < -1, \\ 2x, & \text{for} \quad -1 < x < 0, \\ 2x, & \text{for} \quad 0 < x < 1, \\ 2, & \text{for} \quad 1 < x < 2, \end{cases}$$

whose graph is shown in Figure 2.2. Since f_{odd} is an odd function on $[-2, 2]$, the Fourier coefficients satisfy $a_0 = 0$ and $a_n = 0$ for $n \geqslant 1$, while

$$b_n = \frac{1}{2}\int_{-2}^{2} f_{\text{odd}}(x)\sin\frac{n\pi x}{2}\,dx = \int_0^2 f(x)\sin\frac{n\pi x}{2}\,dx$$

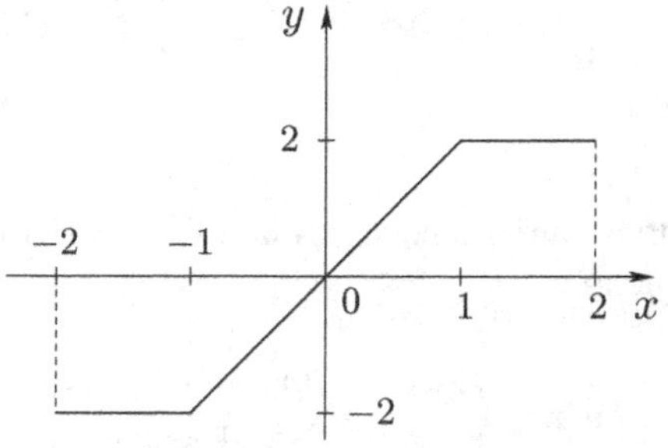

Figure 2.2

since $f_{\text{odd}} = f$ on the interval $[0, 2]$. Therefore, for $n \geqslant 1$,

$$\begin{aligned} b_n &= \int_0^1 2x \sin \frac{n\pi x}{2}\, dx + \int_1^2 2 \sin \frac{n\pi x}{2}\, dx \\ &= -\frac{4}{n\pi} x \cos \frac{n\pi x}{2} \bigg|_0^1 + \frac{4}{n\pi} \int_0^1 \cos \frac{n\pi x}{2}\, dx - \frac{4}{n\pi} \cos \frac{n\pi x}{2} \bigg|_1^2 \\ &= -\frac{4}{n\pi} \cos \frac{n\pi}{2} + \frac{8}{n^2\pi^2} \sin \frac{n\pi x}{2} \bigg|_0^1 - \frac{4}{n\pi} \left(\cos n\pi - \cos \frac{n\pi}{2} \right) \\ &= \frac{8}{n^2\pi^2} \sin \frac{n\pi}{2} - \frac{4}{n\pi} (-1)^n, \end{aligned}$$

and the Fourier sine series for $f(x)$ on the interval $[0, 2]$ is given by

$$f(x) = \sum_{n=1}^{\infty} b_n \sin \frac{n\pi x}{2} = \sum_{n=1}^{\infty} \left(\frac{8}{n^2\pi^2} \sin \frac{n\pi}{2} - \frac{4}{n\pi} (-1)^n \right) \sin \frac{n\pi x}{2}$$

for $0 < x < 2$.

• Fourier Cosine Series

To find the Fourier cosine series for $f(x)$ on the interval $[0, 2]$, we consider the even extension f_{even} to the interval $[-2, 2]$,

$$f_{\text{even}}(x) = \begin{cases} 2, & \text{for} \quad -2 < x < -1, \\ -2x, & \text{for} \quad -1 < x < 0, \\ 2x, & \text{for} \quad 0 < x < 1, \\ 2, & \text{for} \quad 1 < x < 2, \end{cases}$$

whose graph is shown in Figure 2.3. Since f_{even} is an even function on $[-2, 2]$, the

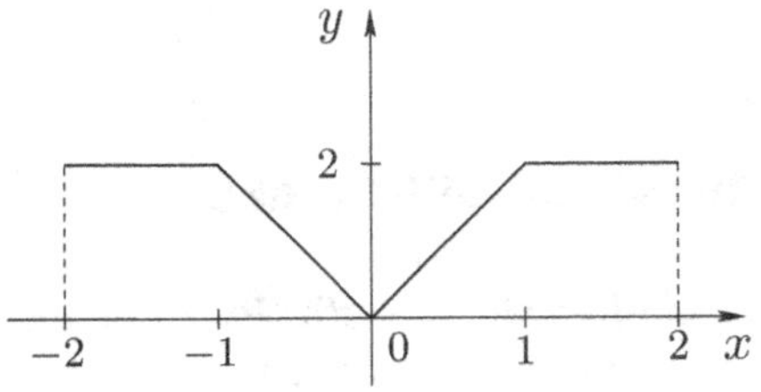

Figure 2.3

Fourier coefficients satisfy $b_n = 0$ for $n \geqslant 1$, while

$$\begin{aligned}
a_0 &= \frac{1}{4}\int_{-2}^{2} f_{\text{even}}(x)\,dx \\
&= \frac{1}{2}\int_{0}^{2} f_{\text{even}}(x)\,dx = \frac{1}{2}\int_{0}^{2} f(x)\,dx \\
&= \frac{1}{2}\int_{0}^{1} 2x\,dx + \frac{1}{2}\int_{1}^{2} 2\,dx \\
&= \frac{1}{2} + 1 = \frac{3}{2}.
\end{aligned}$$

Also, for $n \geqslant 1$,

$$\begin{aligned}
a_n &= \frac{1}{2}\int_{-2}^{2} f_{\text{even}}(x)\cos\frac{n\pi x}{2}\,dx = \int_{0}^{2} f(x)\cos\frac{n\pi x}{2}\,dx \\
&= \int_{0}^{1} 2x\cos\frac{n\pi x}{2}\,dx + \int_{1}^{2} 2\cos\frac{n\pi x}{2}\,dx \\
&= \frac{4}{n\pi}x\sin\frac{n\pi x}{2}\bigg|_{0}^{1} - \frac{4}{n\pi}\int_{0}^{1}\sin\frac{n\pi x}{2}\,dx + \frac{4}{n\pi}\sin\frac{n\pi x}{2}\bigg|_{1}^{2} \\
&= \frac{4}{n\pi}\sin\frac{n\pi}{2} + \frac{8}{n^2\pi^2}\cos\frac{n\pi x}{2}\bigg|_{0}^{1} + \frac{4}{n\pi}\left(\sin n\pi - \sin\frac{n\pi}{2}\right) \\
&= \frac{8}{n^2\pi^2}\left(\cos\frac{n\pi}{2} - 1\right).
\end{aligned}$$

The Fourier cosine series for $f(x)$ on the interval $[0, 2]$ is given by

$$f(x) = a_0 + \sum_{n=1}^{\infty} a_n \cos\frac{n\pi x}{2} = \frac{3}{2} + \frac{8}{\pi^2}\sum_{n=1}^{\infty}\frac{1}{n^2}\left(\cos\frac{n\pi}{2} - 1\right)\cos\frac{n\pi x}{2}$$

for $0 < x < 2$. ∎

2.5 CONVERGENCE OF FOURIER SERIES

Recall that a function f is **piecewise smooth** (PWS) on the interval (a, b) if both f and f' are piecewise continuous on (a, b). That is, both f and f' are continuous on (a, b) except at a finite number of points where they have a **simple** discontinuity (**removable** discontinuity or **jump** discontinuity).

Example 2.11. If

$$f(x) = \begin{cases} x^2, & \text{for} \quad 0 < x < 1, \\ x, & \text{for} \ -1 < x < 0, \end{cases}$$

then

$$f'(x) = \begin{cases} 2x, & \text{for} \quad 0 < x < 1, \\ 1, & \text{for} \ -1 < x < 0, \end{cases}$$

so that $f \in PWS(-1, 1)$. ∎

Example 2.12. If

$$f(x) = x^{2/3}$$

for $-1 < x < 1$, then

$$f'(x) = \frac{2}{3x^{1/3}},$$

and $f \in PWC(-1, 1)$. However, $f' \notin PWC(-1, 1)$, since $x_0 = 0$ is a discontinuity of the second kind for f'; hence, $f \notin PWS(-1, 1)$. ∎

The theorem that follows, **Dirichlet's theorem** or **Fourier's theorem**, describes under what conditions the Fourier series of a function f converges and to what function the series converges.

Theorem 2.13. (*Dirichlet's Theorem*)
Let $f(x)$ be piecewise smooth on the interval $(-\ell, \ell)$. The Fourier series

$$a_0 + \sum_{n=1}^{\infty} \left(a_n \cos \frac{n\pi x}{\ell} + b_n \sin \frac{n\pi x}{\ell} \right),$$

where

$$a_0 = \frac{1}{2\ell} \int_{-\ell}^{\ell} f(t)\, dt,$$

$$a_n = \frac{1}{\ell} \int_{-\ell}^{\ell} f(t) \cos \frac{n\pi t}{\ell}\, dt, \quad n \geqslant 1,$$

$$b_n = \frac{1}{\ell} \int_{-\ell}^{\ell} f(t) \sin \frac{n\pi t}{\ell}\, dt, \quad n \geqslant 1,$$

has the following properties:

(i) If $f(x)$ is continuous at x_0, where $-\ell < x_0 < \ell$, then

$$f(x_0) = a_0 + \sum_{n=1}^{\infty} \left(a_n \cos \frac{n\pi x_0}{\ell} + b_n \sin \frac{n\pi x_0}{\ell} \right);$$

that is, the Fourier series converges to $f(x_0)$.

(ii) If $f(x)$ has a jump discontinuity at x_0, where $-\ell < x_0 < \ell$, then

$$\frac{f(x_0^+) + f(x_0^-)}{2} = a_0 + \sum_{n=1}^{\infty} \left(a_n \cos \frac{n\pi x_0}{\ell} + b_n \sin \frac{n\pi x_0}{\ell} \right);$$

that is, the Fourier series converges to the **average** or **mean** of the jump.

(iii) At the endpoints $x_0 = \pm\ell$, the Fourier series converges to

$$\frac{f(-\ell^+) + f(\ell^-)}{2}.$$

As usual, we write

$$f(x_0) \sim a_0 + \sum_{n=1}^{\infty} \left(a_n \cos \frac{n\pi x_0}{\ell} + b_n \sin \frac{n\pi x_0}{\ell} \right)$$

and say that $f(x)$ is **represented by its Fourier series** on the interval $(-\ell, \ell)$.

Note: The Fourier series

$$a_0 + \sum_{n=1}^{\infty} \left(a_n \cos \frac{n\pi x}{\ell} + b_n \sin \frac{n\pi x}{\ell} \right)$$

is periodic with period 2ℓ and converges for **all** real numbers x. For each real number x_0 it converges to the 2ℓ-periodic extension $\bar{f}(x_0)$ of f whenever $\bar{f}$ is continuous at x_0, and to

$$\frac{\bar{f}(x_0^+) + \bar{f}(x_0^-)}{2}$$

whenever $\bar{f}$ has a jump discontinuity at x_0. This allows us to sketch the graph of the Fourier series once we know what the graph of the function $f(x)$ is, without computing the Fourier series coefficients a_0, a_n, and b_n for $n \geqslant 1$.

Example 2.13. Sketch the graph of the Fourier series of the function

$$f(x) = e^{-x}$$

for $-a < x < a$. Note that the function is undefined for $x = -a$ and $x = a$, which does not affect the integrals when we compute the Fourier coefficients. We can sketch the graph of the Fourier series using Dirichlet's theorem, and then go back and redefine the function at the points of discontinuity so that the Fourier series converges to the function for all $x \in [-a, a]$.

Solution. From Dirichlet's theorem, since the $2a$-periodic extension $\bar{f}$ of the function

$$f(x) = e^{-x}, \quad -a < x < a,$$

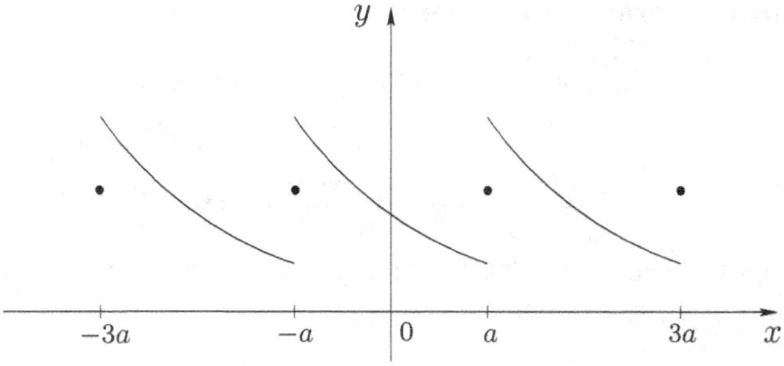

Figure 2.4 Fourier series of f.

is piecewise smooth; then the Fourier series of f converges to $\bar{f}(x)$ at all points where $\bar{f}$ is continuous and to

$$\frac{\bar{f}(x_n^-) + \bar{f}(x_n^+)}{2} = \cosh a$$

at all points $x_n = (2n-1)a$ where $\bar{f}$ is discontinuous. The graph of the Fourier series of f is sketched in Figure 2.4.

∎

Example 2.14. (*Linear Function*)
In Example 2.9 we found the Fourier series for the 2ℓ-periodic function

$$f(x) = x, \quad -\ell < x < \ell,$$

and $f(x+2\ell) = f(x)$ to be

$$f(x) \sim \frac{2\ell}{\pi} \sum_{n=1}^{\infty} \frac{(-1)^{n+1}}{n} \sin \frac{n\pi x}{\ell}.$$

Using Dirichlet's theorem, the graph of the Fourier series is as shown in Figure 2.5. If $\bar{f}$ is the 2ℓ-periodic extension of f, then

$$\bar{f}(\ell^+) = -\ell \qquad \text{and} \qquad \bar{f}(\ell^-) = +\ell.$$

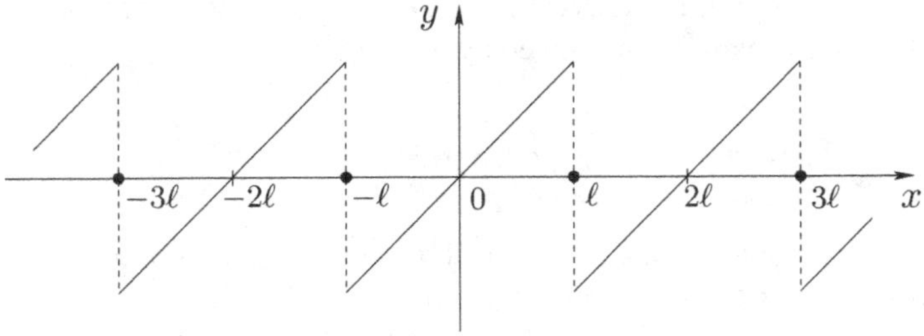

Figure 2.5 Fourier series of the linear function.

If $\bar{f}$ is the 2ℓ-periodic extension of f, then

$$\bar{f}(\ell^+) = -\ell \qquad \text{and} \qquad \bar{f}(\ell^-) = +\ell.$$

Therefore,

$$\frac{\bar{f}(\ell^+) + \bar{f}(\ell^-)}{2} = \frac{-\ell + \ell}{2} = 0.$$

From Dirichlet's theorem, since the function f is piecewise smooth on the interval $(-\ell, \ell)$, the Fourier series evaluated at $x = \ell$ converges to 0; that is,

$$\frac{2\ell}{\pi} \sum_{n=1}^{\infty} \frac{(-1)^{n+1}}{n} \sin n\pi = 0.$$

If we evaluate the Fourier series at $x = \ell/2$, again from Dirichlet's theorem, since f is continuous at $x = \ell/2$, the series converges to $f(\ell/2)$; that is,

$$\begin{aligned} \frac{\ell}{2} &= \frac{2\ell}{\pi} \sum_{n=1}^{\infty} \frac{(-1)^{n+1}}{n} \sin \frac{n\pi}{2} \\ &= \frac{2\ell}{\pi} \sum_{n=1}^{\infty} \left\{ \frac{(-1)^{2n}}{2n-1} \sin \frac{(2n-1)\pi}{2} + \frac{(-1)^{2n+1}}{2n} \sin n\pi \right\}. \end{aligned}$$

Therefore,

$$\frac{\pi}{4} = \sum_{n=1}^{\infty} \frac{1}{2n-1} \sin \frac{(2n-1)\pi}{2} = \sum_{n=0}^{\infty} \frac{(-1)^n}{2n+1};$$

that is,

$$\frac{\pi}{4} = 1 - \frac{1}{3} + \frac{1}{5} - \frac{1}{7} + \frac{1}{9} - \frac{1}{11} + \cdots.$$

This is *Leibniz's formula* for $\pi/4$, discovered by Leibniz in 1673. ∎

2.5.1 Gibbs' Phenomenon

We saw from Dirichlet's theorem that if f is a piecewise smooth function on the interval $[-\ell, \ell]$ and f is continuous at $x_0 \in (-\ell, \ell)$, the Fourier series for f converges to $f(x_0)$; that is, the partial sums, given by

$$S_N(x_0) = a_0 + \sum_{n=1}^{N} \left(a_n \cos \frac{n\pi x_0}{\ell} + b_n \sin \frac{n\pi x_0}{\ell} \right)$$

for $N \geqslant 1$, converge to $f(x_0)$. However, if f is discontinuous at $x_0 \in (-\ell, \ell)$, the partial sums converge to the average of the jump:

$$\frac{f(x_0^+) + f(x_0^-)}{2},$$

and as noticed in 1899 by the mathematical physicist J. W. Gibbs, the partial sums tend to overshoot $f(x_0^+)$ and $f(x_0^-)$ in the direction opposite the jump, even for very large values of N. The nature of this deviation of the partial sums of f from the values of f near x_0 is commonly referred to as *Gibbs' phenomenon*, even though it was discovered 60 years earlier by the mathematician Wilbraham. Gibbs' phenomenon is illustrated in the example below for the linear function discussed in Example 2.9.

Example 2.15. (*Gibbs' Phenomenon*)
For the 2ℓ-periodic function f defined by

$$f(x) = x, \quad -\ell < x < \ell,$$

and $f(x + 2\ell) = f(x)$, we found the Fourier series for f to be

$$x \sim \frac{2\ell}{\pi} \sum_{n=1}^{\infty} \frac{(-1)^{n+1}}{n} \sin \frac{n\pi x}{\ell} \tag{2.4}$$

for $-\ell < x < \ell$.

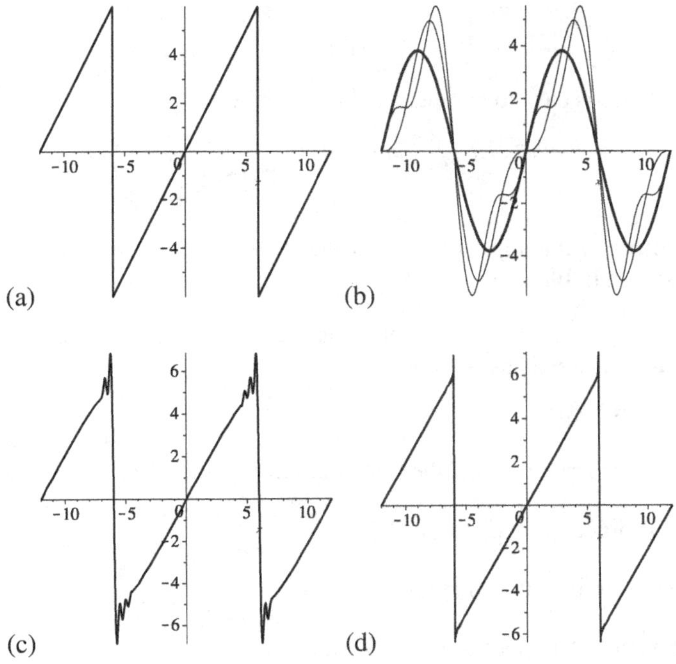

Figure 2.6 Gibbs' phenomenon for the linear function f.

Figure 2.6 shows:

(a) The graph of the Fourier series of Example 2.9 for $\ell = 2\pi$, on the interval $[-4\pi, 4\pi]$.

(b) The graphs of the first three partial sums $S_1(x), S_2(x), S_3(x)$, where the first partial sum $S_1(x)$ is in boldface type.

(c) The 25 th-order approximation $S_{25}(x)$.

(d) The 500 th-order approximation $S_{500}(x)$.

We observe a clear overshooting at the jump discontinuity, even for the 500 th-order approximation. In Figure 2.6 we show how the partial sums of the Fourier series approximate the Fourier series in Figure 2.5.

∎

Gibbs' phenomenon shows up even more dramatically in the following example.

Example 2.16. (*Square Wave*)
Consider the piecewise smooth function g defined on $[-2\pi, 2\pi]$:

$$g(x) = \begin{cases} -1, & \text{for } -2\pi \le x < -\pi, \\ 0, & \text{for } \quad -\pi \le x < \pi, \\ 1, & \text{for } \quad \pi \leqslant x \leqslant 2\pi. \end{cases}$$

The Fourier series is easy to compute and is given by

$$\sum_{n=1}^{\infty} \frac{2}{n\pi} \left(\cos \frac{n\pi}{2} - \cos n\pi \right) \sin \frac{nx}{2}.$$

The function g and the approximation to the Fourier series for g by partial sums $S_N(x)$ are shown in Figure 2.7.

Again, we clearly see strong oscillations near the jump discontinuities, even for very high approximations such as $S_{500}(x)$.

Figure 2.7 shows:

(a) The Fourier series of g and the first three partial sums $S_1(x)$, $S_2(x)$, and $S_3(x)$.

(b) The 5 th-order approximation $S_5(x)$.

(c) The 25 th-order approximation $S_{25}(x)$.

(d) The 500 th-order approximation $S_{500}(x)$.

∎

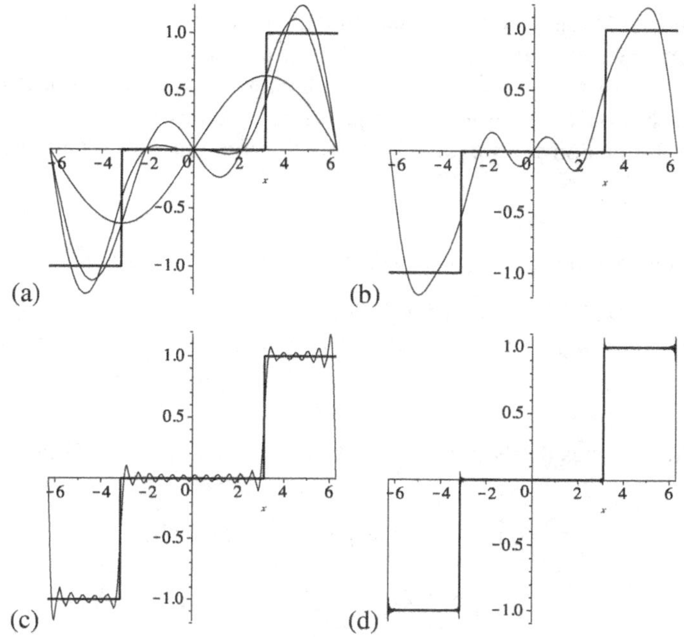

Figure 2.7 Gibbs' phenomenon for the square wave g.

2.6 OPERATIONS ON FOURIER SERIES

Here we examine what manipulations may legitimately be carried out on Fourier series.

Theorem 2.14. (*Uniqueness of Fourier Series*)
If f is 2ℓ-periodic and piecewise smooth on the interval $(-\ell, \ell)$, its Fourier series is unique.

Proof. The Fourier coefficients of f are determined uniquely by

$$a_0 = \frac{1}{2\ell} \int_{-\ell}^{\ell} f(t)\, dt,$$

$$a_n = \frac{1}{\ell} \int_{-\ell}^{\ell} f(t) \cos \frac{n\pi t}{\ell}\, dt, \quad n \geqslant 1,$$

$$b_n = \frac{1}{\ell} \int_{-\ell}^{\ell} f(t) \sin \frac{n\pi t}{\ell}\, dt, \quad n \geqslant 1.$$

If we assume that f and g are piecewise smooth on the interval $(-\ell, \ell)$ and have the same Fourier series, then from Dirichlet's theorem, the Fourier series of the function $f - g$ converges to $f(x_0) - g(x_0) = 0$ at each point x_0 where $f - g$ is continuous. Since $f - g$ has only finitely many simple discontinuities on the interval $(-\ell, \ell)$, $f - g$ is a null function and we identify it with the zero function. Therefore, $f = g$. □

Next, we show that the operation of finding the Fourier series is linear.

Theorem 2.15. (*Linearity of Fourier Series*)
If f and g are piecewise continuous on $(-\ell, \ell)$ and c_1 and c_2 are scalars, the Fourier series of

$$c_1 f(x) + c_2 g(x)$$

is the sum of c_1 times the Fourier series of $f(x)$ and c_2 times the Fourier series of $g(x)$.

Proof. Suppose that the Fourier series of f and g are given by

$$f(x) \sim A_0 + \sum_{n=1}^{\infty} \left(A_n \cos \frac{n\pi x}{\ell} + B_n \sin \frac{n\pi x}{\ell} \right)$$

and

$$g(x) \sim C_0 + \sum_{n=1}^{\infty} \left(C_n \cos \frac{n\pi x}{\ell} + D_n \sin \frac{n\pi x}{\ell} \right),$$

where

$$A_0 = \frac{1}{2\ell} \int_{-\ell}^{\ell} f(x)\, dx,$$

$$A_n = \frac{1}{\ell} \int_{-\ell}^{\ell} f(x) \cos \frac{n\pi x}{\ell}\, dx,$$

$$B_n = \frac{1}{\ell} \int_{-\ell}^{\ell} f(x) \sin \frac{n\pi x}{\ell}\, dx$$

and

$$C_0 = \frac{1}{2\ell} \int_{-\ell}^{\ell} g(x)\, dx,$$

$$C_n = \frac{1}{\ell} \int_{-\ell}^{\ell} g(x) \cos \frac{n\pi x}{\ell}\, dx,$$

$$D_n = \frac{1}{\ell} \int_{-\ell}^{\ell} g(x) \sin \frac{n\pi x}{\ell}\, dx.$$

If c_1 and c_2 are scalars and the Fourier series of $c_1 f + c_2 g$ is

$$c_1 f(x) + c_2 g(x) \sim E_0 + \sum_{n=1}^{\infty} \left(E_n \cos \frac{n\pi x}{\ell} + F_n \sin \frac{n\pi x}{\ell} \right),$$

then

$$\begin{aligned} E_0 &= \frac{1}{2\ell} \int_{-\ell}^{\ell} [c_1 f(x) + c_2 g(x)]\, dx \\ &= \frac{c_1}{2\ell} \int_{-\ell}^{\ell} f(x)\, dx + \frac{c_2}{2\ell} \int_{-\ell}^{\ell} g(x)\, dx = c_1 A_0 + c_2 C_0. \end{aligned}$$

Also,

$$\begin{aligned} E_n &= \frac{1}{\ell} \int_{-\ell}^{\ell} [c_1 f(x) + c_2 g(x)] \cos \frac{n\pi x}{\ell}\, dx \\ &= \frac{c_1}{\ell} \int_{-\ell}^{\ell} f(x) \cos \frac{n\pi x}{\ell}\, dx + \frac{c_2}{\ell} \int_{-\ell}^{\ell} g(x) \cos \frac{n\pi x}{\ell}\, dx \\ &= c_1 A_n + c_2 C_n. \end{aligned}$$

Similarly,

$$\begin{aligned} F_n &= \frac{1}{\ell} \int_{-\ell}^{\ell} [c_1 f(x) + c_2 g(x)] \sin \frac{n\pi x}{\ell}\, dx \\ &= \frac{c_1}{\ell} \int_{-\ell}^{\ell} f(x) \sin \frac{n\pi x}{\ell}\, dx + \frac{c_2}{\ell} \int_{-\ell}^{\ell} g(x) \sin \frac{n\pi x}{\ell}\, dx \\ &= c_1 B_n + c_2 D_n \end{aligned}$$

for $n \geqslant 1$. Therefore, the Fourier series for $c_1 f + c_2 g$ is

$$\begin{aligned} c_1 f(x) + c_2 g(x) &\sim c_1 A_0 + c_2 C_0 + \sum_{n=1}^{\infty} \Big[(c_1 A_n + c_2 C_n) \cos \frac{n\pi x}{\ell} \\ &\qquad + (c_1 B_n + c_2 D_n) \sin \frac{n\pi x}{\ell} \Big] \\ &= c_1 \left[A_0 + \sum_{n=1}^{\infty} \left(A_n \cos \frac{n\pi x}{\ell} + B_n \sin \frac{n\pi x}{\ell} \right) \right] \\ &\qquad + c_2 \left[C_0 + \sum_{n=1}^{\infty} \left(C_n \cos \frac{n\pi x}{\ell} + D_n \sin \frac{n\pi x}{\ell} \right) \right] \\ &\sim c_1 f(x) + c_2 g(x). \end{aligned}$$

□

Theorem 2.16. (*Bessel's Inequality*)
Let $f(x)$ be piecewise smooth on the interval $[-\ell, \ell]$. If the Fourier series of f is

$$f(x) \sim a_0 + \sum_{n=1}^{\infty} \left(a_n \cos \frac{n\pi x}{\ell} + b_n \sin \frac{n\pi x}{\ell} \right),$$

where

$$a_0 = \frac{1}{2\ell} \int_{-\ell}^{\ell} f(t)\, dt,$$

$$a_n = \frac{1}{\ell} \int_{-\ell}^{\ell} f(t) \cos \frac{n\pi t}{\ell}\, dt, \quad n \geqslant 1,$$

$$b_n = \frac{1}{\ell} \int_{-\ell}^{\ell} f(t) \sin \frac{n\pi t}{\ell}\, dt, \quad n \geqslant 1,$$

then

$$a_0^2 + \frac{1}{2} \sum_{k=1}^{\infty} (a_k^2 + b_k^2) \leqslant \frac{1}{2\ell} \int_{-\ell}^{\ell} f(t)^2\, dt.$$

Proof. Let $S_n(x)$ be the nth partial sum of the Fourier series; that is,

$$S_n(x) = a_0 + \sum_{k=1}^{n} \left(a_k \cos \frac{k\pi x}{\ell} + b_k \sin \frac{k\pi x}{\ell} \right).$$

Using the orthogonality, we have

$$\int_{-\ell}^{\ell} a_0\, S_n(t)\, dt = 2\ell\, a_0^2 = a_0 \int_{-\ell}^{\ell} f(t)\, dt,$$

$$\int_{-\ell}^{\ell} a_k \cos \frac{k\pi t}{\ell}\, S_n(t)\, dt = \ell\, a_k^2 = a_k \int_{-\ell}^{\ell} f(t) \cos \frac{k\pi t}{\ell}\, dt,$$

$$\int_{-\ell}^{\ell} b_k \sin \frac{k\pi t}{\ell}\, S_n(t)\, dt = \ell\, b_k^2 = b_k \int_{-\ell}^{\ell} f(t) \sin \frac{k\pi t}{\ell}\, dt$$

for $k = 1, 2, \ldots, n$. Adding the equations on the left-hand side, we have

$$\int_{-\ell}^{\ell} S_n(t)^2\, dt = 2\ell\, a_0^2 + \ell \sum_{k=1}^{n} (a_k^2 + b_k^2), \tag{2.5}$$

and adding the equations on the right-hand side, we have

$$2\ell\, a_0^2 + \ell \sum_{k=1}^{n} (a_k^2 + b_k^2) = \int_{-\ell}^{\ell} f(t)\, S_n(t)\, dt \tag{2.6}$$

for $n \geqslant 1$. Therefore, from (2.5) and (2.6) we have

$$\int_{-\ell}^{\ell} S_n(t)^2\,dt = 2\ell\,a_0^2 + \ell\sum_{k=1}^{n}(a_k^2 + b_k^2) = \int_{-\ell}^{\ell} f(t)\,S_n(t)\,dt. \tag{2.7}$$

Now

$$\begin{aligned}
0 &\leqslant \int_{-\ell}^{\ell} [f(t) - S_n(t)]^2\,dt \\
&= \int_{-\ell}^{\ell} f(t)^2\,dt - 2\int_{-\ell}^{\ell} f(t)\,S_n(t)\,dt + \int_{-\ell}^{\ell} S_n(t)^2\,dt \\
&= \int_{-\ell}^{\ell} f(t)^2\,dt - \int_{-\ell}^{\ell} S_n(t)^2\,dt \\
&= \int_{-\ell}^{\ell} f(t)^2\,dt - \left[2\ell\,a_0^2 + \ell\sum_{k=1}^{n}(a_k^2 + b_k^2)\right],
\end{aligned}$$

so that

$$a_0^2 + \frac{1}{2}\sum_{k=1}^{n}(a_k^2 + b_k^2) \leqslant \frac{1}{2\ell}\int_{-\ell}^{\ell} f(t)^2\,dt$$

for all $n \geqslant 1$. Letting $n \to \infty$, we have

$$a_0^2 + \frac{1}{2}\sum_{k=1}^{\infty}(a_k^2 + b_k^2) \leqslant \frac{1}{2\ell}\int_{-\ell}^{\ell} f(t)^2\,dt,$$

and the series

$$a_0^2 + \frac{1}{2}\sum_{k=1}^{\infty}(a_k^2 + b_k^2)$$

converges. □

Note: For the result above, we only need the function f to be piecewise continuous on the interval $[-\ell, \ell]$ so that the integral

$$\int_{-\ell}^{\ell} f(t)^2\,dt$$

exists and is finite.

Also, if we assume that f is piecewise smooth on the interval $[-\ell, \ell]$ and that $f(\ell) = f(-\ell)$, we can use Bessel's inequality to show that the Fourier series for f converges at each point of the interval. It is a little more difficult to show that it converges to f at the points of continuity of f; however, we do have the following result concerning uniform convergence; first the definition.

Definition 2.17. (*Uniform Convergence*)
We say that a sequence of functions $\{f_n(x)\}_{n\geqslant 1}$, defined on an interval I, **converges uniformly** to the function $f(x)$ on the interval I if and only if given any $\epsilon > 0$ there exists an N (depending only on ϵ) such that whenever $n > N$,

$$|f_n(x) - f(x)| < \epsilon$$

for all $x \in I$.

Theorem 2.18. (*Uniform Convergence of Fourier Series*)
If f is piecewise smooth on the interval $[-\ell, \ell]$ and $f(\ell) = f(-\ell)$, the Fourier series

$$a_0 + \sum_{n=1}^{\infty} \left(a_n \cos \frac{n\pi x}{\ell} + b_n \sin \frac{n\pi x}{\ell} \right)$$

converges absolutely and uniformly on $[-\ell, \ell]$.

Proof. Since f' is piecewise continuous on the interval $[-\ell, \ell]$, the Fourier coefficients of f' exist. Thus, Fourier series of f' can be written as

$$f'(x) \sim \alpha_0 + \sum_{n=1}^{\infty} \left(\alpha_n \cos \frac{n\pi x}{\ell} + \beta_n \sin \frac{n\pi x}{\ell} \right),$$

where

$$\alpha_0 = \frac{1}{2\ell} \int_{-\ell}^{\ell} f'(t)\, dt,$$

$$\alpha_n = \frac{1}{\ell} \int_{-\ell}^{\ell} f'(t) \cos \frac{n\pi t}{\ell}\, dt,$$

$$\beta_n = \frac{1}{\ell} \int_{-\ell}^{\ell} f'(t) \sin \frac{n\pi t}{\ell}\, dt$$

for $n \geqslant 1$. Now

$$\alpha_0 = \frac{1}{2\ell} \int_{-\ell}^{\ell} f'(t)\, dt = \frac{1}{2\ell} [f(\ell) - f(-\ell)] = 0,$$

and integrating by parts, we have

$$\alpha_n = n\, b_n \qquad \text{and} \qquad \beta_n = -n\, a_n$$

for $n \geqslant 1$.

From the Cauchy-Schwarz inequality and Bessel's inequality applied to f', we have

$$\begin{aligned}\sum_{n=1}^{N}\sqrt{a_n^2+b_n^2} &= \sum_{n=1}^{N}\frac{1}{n}\sqrt{\alpha_n^2+\beta_n^2}\\ &\leqslant \left(\sum_{n=1}^{N}\frac{1}{n^2}\right)^{1/2}\left(\sum_{n=1}^{N}(\alpha_n^2+\beta_n^2)\right)^{1/2}\\ &\leqslant \frac{\pi}{\sqrt{6}}\left(\sum_{n=1}^{N}(\alpha_n^2+\beta_n^2)\right)^{1/2}\\ &\leqslant \frac{\pi}{\sqrt{6\,\ell}}\left(\int_{-\ell}^{\ell} f'(t)^2\,dt\right)^{1/2}\end{aligned}$$

for all $N \geqslant 1$, and so

$$\sum_{n=1}^{\infty}\sqrt{a_n^2+b_n^2}$$

converges. For each $n \geqslant 1$, we have

$$|a_n| \leqslant \sqrt{a_n^2+b_n^2} \qquad \text{and} \qquad |b_n| \leqslant \sqrt{a_n^2+b_n^2},$$

so that

$$\begin{aligned}\sum_{n=1}^{\infty}\left|a_n\cos\frac{n\pi t}{\ell}+b_n\sin\frac{n\pi t}{\ell}\right| &\leqslant \sum_{n=1}^{\infty}\left(|a_n|+|b_n|\right)\\ &\leqslant 2\sum_{n=1}^{\infty}\sqrt{a_n^2+b_n^2}\\ &\leqslant \frac{2\pi}{\sqrt{6\,\ell}}\left(\int_{-\ell}^{\ell} f'(t)^2\,dt\right)^{1/2} < \infty.\end{aligned}$$

By the Weierstrass M-test, the Fourier series for f converges absolutely and uniformly on the interval $[-\ell, \ell]$.

□

Note: Since $f(-\ell) = f(\ell)$, if f is continuous on $[-\ell, \ell]$, the 2ℓ-periodic extension $\bar{f}$ of f is continuous at each real number x, and by Dirichlet's theorem, the Fourier series of f converges for all real numbers x to $\bar{f}(x)$. Therefore, the Fourier series for f converges absolutely and uniformly to f on $[-\ell, \ell]$.

The next few theorems, which summarize the results on term-by-term differentiation of Fourier series, Fourier cosine series, and Fourier sine series, all follow from Dirichlet's theorem and will be stated without proof.

Theorem 2.19. *(Term-by-Term Differentiation of Fourier Series)*
Let f be a function such that

(i) f is continuous on the interval $-\pi \leqslant x \leqslant \pi$;

(ii) $f(-\pi) = f(\pi)$; and

(iii) f' is piecewise smooth on the interval $-\pi < x < \pi$.

The Fourier series representation

$$f(x) = a_0 + \sum_{n=1}^{\infty} (a_n \cos nx + b_n \sin nx),$$

where

$$a_0 = \frac{1}{2\pi} \int_{-\pi}^{\pi} f(x)\, dx,$$

$$a_n = \frac{1}{\pi} \int_{-\pi}^{\pi} f(x) \cos nx\, dx,$$

$$b_n = \frac{1}{\pi} \int_{-\pi}^{\pi} f(x) \sin nx\, dx$$

for $n \geqslant 1$, is differentiable at each point x_0 with $-\pi < x_0 < \pi$ at which $f''(x_0)$ exists, and

$$f'(x_0) = \sum_{n=1}^{\infty} n(-a_n \sin nx_0 + b_n \cos nx_0). \tag{2.8}$$

At a point x_0 with $-\pi < x_0 < \pi$ at which $f''(x_0)$ does not exist but where f' has one-sided derivatives, the series in (2.8) converges to

$$\frac{f'(x_0^+) + f'(x_0^-)}{2}.$$

Note: Conditions (i), (ii), and (iii) are needed to ensure that the periodic extension of f is continuous everywhere and that the Fourier series of f converges uniformly to this extension at each point. Condition (iii) is also needed to ensure that the Fourier series for f' converges. If we assume that f' is piecewise smooth, f' and f'' are both piecewise continuous, and Dirichlet's theorem applied to f' gives us the result.
If f or its periodic extension $\bar{f}$ fails to be continuous, the differentiated series may fail to converge at some points, as the next example shows.

Example 2.17. Going back to the linear function of Example 2.9, according to Dirichlet's theorem, we have

$$x = f(x) = \frac{2\ell}{\pi} \sum_{n=1}^{\infty} \frac{(-1)^{n+1}}{n} \sin \frac{n\pi x}{\ell}$$

for $-\ell < x < \ell$.

We have $f'(x) = 1$ for all $x \in (-\ell, \ell)$, and the differentiated Fourier series is

$$\frac{2\ell}{\pi}\sum_{n=1}^{\infty}\frac{(-1)^{n+1}}{n}\frac{n\pi}{\ell}\cos\frac{n\pi x}{\ell} = 2\sum_{n=1}^{\infty}(-1)^{n+1}\cos\frac{n\pi x}{\ell}.$$

However, the differentiated series does not converge for any $x \in (-\ell, \ell)$, since

$$\lim_{n\to\infty}\cos(n\pi x/\ell) \neq 0$$

for any $x \in (-\ell, \ell)$.

This does not contradict Theorem 2.19 on term-by-term differentiation of Fourier series, since f is not continuous on $[-\ell, \ell]$, so the Fourier series of f may not converge uniformly on $[-\ell, \ell]$. ∎

For term-by-term differentiation of Fourier cosine and sine series, we have the following results:

Theorem 2.20. (*Term-by-Term Differentiation of Fourier Cosine Series*)
Let f be a function such that

(i) f is continuous on the interval $0 \leqslant x \leqslant \pi$;

(ii) f' is piecewise continuous on the interval $0 < x < \pi$.

The Fourier cosine series representation

$$f(x) = a_0 + \sum_{n=1}^{\infty} a_n \cos nx,$$

where

$$a_0 = \frac{1}{\pi}\int_0^{\pi} f(x)\,dx, \quad a_n = \frac{2}{\pi}\int_0^{\pi} f(x)\cos nx\,dx$$

for $n \geqslant 1$ is differentiable at each point x_0 with $0 < x_0 < \pi$ at which $f''(x_0)$ exists, and

$$f'(x_0) = -\sum_{n=1}^{\infty} n a_n \sin n x_0. \tag{2.9}$$

At a point x_0 with $0 < x_0 < \pi$ at which $f''(x_0)$ does not exist but where f' has one-sided derivatives, the series in (2.9) converges to

$$\frac{f'(x_0^+) + f'(x_0^-)}{2}.$$

Theorem 2.21. (*Term-by-Term Differentiation of Fourier Sine Series*)
Let f be a function such that

(i) f is continuous on the interval $0 \leqslant x \leqslant \pi$;

(ii) $f(0) = f(\pi) = 0$; and

(iii) f' is piecewise continuous on the interval $0 < x < \pi$.

The Fourier sine series representation

$$f(x) = \sum_{n=1}^{\infty} b_n \sin nx,$$

where

$$b_n = \frac{2}{\pi} \int_0^{\pi} f(x) \sin nx \, dx$$

for $n \geqslant 1$ is differentiable at each point x_0 with $0 < x_0 < \pi$ at which $f''(x_0)$ exists, and

$$f'(x_0) = \sum_{n=1}^{\infty} n b_n \cos nx_0. \tag{2.10}$$

At a point x_0 with $0 < x_0 < \pi$ at which $f''(x_0)$ does not exist but where f' has one-sided derivatives, the series in (2.10) converges to

$$\frac{f'(x_0^+) + f'(x_0^-)}{2}.$$

Note: Conditions (i), (ii), and (iii) are needed to ensure that the periodic extension of f_{odd} is continuous everywhere and that the Fourier series of f_{odd} converges to this extension at each point. Condition (iii) is also needed to ensure that the Fourier cosine series for f' converges.

The following result on term-by-term integration of Fourier series does not require as many assumptions as the result for term-by-term differentiation. In fact, the series does not even need to converge for the integrated series to converge to the integral of the function.

Theorem 2.22. (*Term-by-Term Integration of Fourier Series*)
Let f be piecewise continuous on the interval $-\pi < x < \pi$, and suppose that

$$f(x) \sim a_0 + \sum_{n=1}^{\infty} (a_n \cos nx + b_n \sin nx)$$

on $(-\pi, \pi)$, where

$$a_0 = \frac{1}{2\pi} \int_{-\pi}^{\pi} f(t) \, dt, \quad a_n = \frac{1}{\pi} \int_{-\pi}^{\pi} f(t) \cos nt \, dt, \quad b_n = \frac{1}{\pi} \int_{-\pi}^{\pi} f(t) \sin nt \, dt$$

for $n \geqslant 1$; then

$$\int_{-\pi}^{x} f(t)\,dt = a_0(x+\pi) + \sum_{n=1}^{\infty} \frac{1}{n}\left\{a_n \sin nx - b_n\left[\cos nx + (-1)^{n+1}\right]\right\}$$

for $-\pi \leqslant x \leqslant \pi$.

Note that if $a_0 \neq 0$, the integrated series is **not** a Fourier series because of the presence of the term $a_0\,x$. Note also that the convergence results are stated only for functions defined on the interval $(-\pi, \pi)$, but are easily extended to functions defined on an arbitrary interval $(-\ell, \ell)$. For example, since the product of two piecewise continuous functions is piecewise continuous, we have

Theorem 2.23. Let f be piecewise continuous on $\mathbb{R}$ and 2ℓ-periodic with Fourier series representation

$$f(x) \sim a_0 + \sum_{n=1}^{\infty} a_n \cos\frac{n\pi x}{\ell} + b_n \sin\frac{n\pi x}{\ell},$$

where

$$a_0 = \frac{1}{2\ell}\int_{-\ell}^{\ell} f(t)\,dt,$$

$$a_n = \frac{1}{\ell}\int_{-\ell}^{\ell} f(t)\cos\frac{n\pi t}{\ell}\,dt,$$

$$b_n = \frac{1}{\ell}\int_{-\ell}^{\ell} f(t)\sin\frac{n\pi t}{\ell}\,dt$$

for $n \geqslant 1$. If g is piecewise continuous on the interval (α, β), then

$$\int_{\alpha}^{\beta} f(x)\,g(x)\,dx = \int_{\alpha}^{\beta} a_0\,g(x)\,dx + \sum_{n=1}^{\infty}\int_{\alpha}^{\beta}\left(a_n\cos\frac{n\pi x}{\ell} + b_n\sin\frac{n\pi x}{\ell}\right) g(x)\,dx.$$

Example 2.18. Going back again to the linear function of Example 2.9,

$$f(x) = x, \quad -\ell < x < \ell,$$

and $f(x+2\ell) = f(x)$, with Fourier series

$$x = \frac{2\ell}{\pi}\sum_{n=1}^{\infty}\frac{(-1)^{n+1}}{n}\sin\frac{n\pi x}{\ell}$$

for $0 < x < \ell$.

Using $g(x) = 1$, we get

$$\begin{aligned}\frac{\ell^2}{2} = \int_0^\ell x\,dx &= \frac{2\ell}{\pi}\sum_{n=1}^{\infty}\frac{(-1)^{n+1}}{n}\int_0^\ell \sin\frac{n\pi x}{\ell}\,dx \\ &= \frac{2\ell^2}{\pi^2}\sum_{n=1}^{\infty}\frac{(-1)^n}{n^2}[(-1)^n - 1].\end{aligned}$$

After a little manipulation,

$$\frac{\pi^2}{8} = \sum_{n=1}^{\infty}\frac{1}{(2n-1)^2} = 1 + \frac{1}{3^2} + \frac{1}{5^2} + \frac{1}{7^2} + \cdots.$$

∎

If we know the Fourier series of a function, we can determine differentiability properties of the function from its Fourier coefficients, as our final result shows.

Theorem 2.24. Let f be defined on $\mathbb{R}$, such that

(i) f is periodic, with Fourier coefficients a_n, b_n, and

(ii) there exists a positive integer k such that

$$\sum_{n=1}^{\infty} n^k\left(|a_n| + |b_n|\right)$$

converges,

then f has continuous derivatives f', f'', ..., $f^{(k)}$ whose Fourier series are the differentiated series of f.

2.7 MEAN SQUARE ERROR

We have seen that some functions can be represented by an infinite Fourier series; however, in practice, we can only evaluate **finite** sums. If we want to approximate a function $f(x)$ on the interval $(-\ell, \ell)$ by a **trigonometric polynomial** of the form

$$T_N(x) = \sum_{n=0}^{N}\left(A_n\cos\frac{n\pi x}{\ell} + B_n\sin\frac{n\pi x}{\ell}\right), \quad x \in (-\ell, \ell),$$

how should we choose the coefficients A_n and B_n in this finite sum or truncated Fourier series to get a good approximation?

We will use the L^2 norm on $PWC(-\ell, \ell)$,

$$\|f\|_{L^2} = \langle f, f\rangle^{1/2} = \left(\int_{-\ell}^{\ell} f(x)^2\,dx\right)^{1/2},$$

to measure the distance between two functions, so that if $f(x)$ and $g(x)$ are in $PWC(-\ell, \ell)$; then the distance between $f(x)$ and $g(x)$ is given by

$$\|f - g\|_{L^2} = \left(\int_{-\ell}^{\ell} |f(x) - g(x)|^2 \, dx \right)^{1/2}.$$

Now we return to the original question: How good an approximation is a truncated Fourier series?

Suppose that $f \in PWC(-\ell, \ell)$ has the Fourier series

$$f(x) \sim a_0 + \sum_{n=1}^{\infty} \left(a_n \cos \frac{n\pi x}{\ell} + b_n \sin \frac{n\pi x}{\ell} \right),$$

and let $S_N(x)$ be the truncated Fourier series of f; that is,

$$S_N(x) = a_0 + \sum_{n=1}^{N} \left(a_n \cos \frac{n\pi x}{\ell} + b_n \sin \frac{n\pi x}{\ell} \right).$$

Let $g_N(x)$ be another trigonometric polynomial,

$$g_N(x) = A_0 + \sum_{n=1}^{N} \left(A_n \cos \frac{n\pi x}{\ell} + B_n \sin \frac{n\pi x}{\ell} \right).$$

We want to determine the constants A_n and B_n so that g_N is the best approximation to f relative to the L^2–norm.

To this end, we define the **mean square error** E_N made in approximating f by g_N to be

$$E_N = \|f - g_N\| = \left(\int_{-\ell}^{\ell} |f(t) - g_N(t)|^2 \, dt \right)^{1/2}$$

and find A_n and B_n that minimize E_N. We have

$$E_N^2 = \|f - g_N\|^2 = \langle f - g_N, f - g_N \rangle = \langle f, f \rangle - 2 \langle f, g_N \rangle + \langle g_N, g_N \rangle, \quad (2.11)$$

and we examine each term in (2.11) individually. The first term is easy: $\langle f, f \rangle = \|f\|^2$.

Using the orthogonality, the second term becomes

$$\begin{aligned}
\langle f, g_N \rangle &= \left\langle f, A_0 + \sum_{n=1}^{N} \left(A_n \cos \frac{n\pi x}{\ell} + B_n \sin \frac{n\pi x}{\ell} \right) \right\rangle \\
&= A_0 \langle f, 1 \rangle + \sum_{n=1}^{N} \left(A_n \left\langle f, \cos \frac{n\pi x}{\ell} \right\rangle + B_n \left\langle f, \sin \frac{n\pi x}{\ell} \right\rangle \right) \\
&= 2\ell A_0 a_0 + \sum_{n=1}^{N} \ell \Big(A_n a_n + B_n b_n \Big) \\
&= 2\ell \left(a_0 A_0 + \frac{1}{2} \sum_{n=1}^{N} \Big(a_n A_n + b_n B_n \Big) \right),
\end{aligned}$$

while the third term becomes

$$\begin{aligned}
\langle g_N, g_N \rangle = &\left\langle A_0 + \sum_{n=1}^{N} \left(A_n \cos \frac{n\pi x}{\ell} + B_n \sin \frac{n\pi x}{\ell} \right), \right. \\
&\qquad \left. A_0 + \sum_{n=1}^{N} \left(A_n \cos \frac{n\pi x}{\ell} + B_n \sin \frac{n\pi x}{\ell} \right) \right\rangle \\
= &\, A_0^2 \langle 1, 1 \rangle + 2 \sum_{n=1}^{N} \left(A_0 A_n \left\langle 1, \cos \frac{n\pi x}{\ell} \right\rangle + A_0 B_n \left\langle 1, \sin \frac{n\pi x}{\ell} \right\rangle \right) \\
&+ \sum_{n=1}^{N} \sum_{m=1}^{N} \left(A_n A_m \left\langle \cos \frac{n\pi x}{\ell}, \cos \frac{m\pi x}{\ell} \right\rangle \right. \\
&\qquad + A_n B_m \left\langle \cos \frac{n\pi x}{\ell}, \sin \frac{m\pi x}{\ell} \right\rangle \\
&\qquad + A_m B_n \left\langle \cos \frac{m\pi x}{\ell}, \sin \frac{n\pi x}{\ell} \right\rangle \\
&\qquad \left. + B_n B_m \left\langle \sin \frac{n\pi x}{\ell}, \sin \frac{m\pi x}{\ell} \right\rangle \right).
\end{aligned}$$

Thus,

$$\begin{aligned}
\langle g_N, g_N \rangle &= 2\ell A_0^2 + \sum_{n=1}^{N} \sum_{m=1}^{N} \ell \Big(A_n A_m + B_n B_m \Big) \delta_{nm} \\
&= 2\ell \left(A_0^2 + \frac{1}{2} \sum_{n=1}^{N} (A_n^2 + B_n^2) \right).
\end{aligned}$$

Substituting these expressions into (2.11), we get

$$E_N^2 = \|f\|^2 - 2\langle f, g_N\rangle + \langle g_N, g_N\rangle$$

$$= \|f\|^2 - 4\ell\Big(a_0 A_0 + \frac{1}{2}\sum_{n=1}^{N}(a_n A_n + b_n B_n)\Big)$$

$$+ 2\ell\Big(A_0^2 + \frac{1}{2}\sum_{n=1}^{N}(A_n^2 + B_n^2)\Big);$$

that is,

$$E_N^2 = \|f\|^2 + 2\ell\left(A_0^2 - 2a_0A_0\right) + \ell\sum_{n=1}^{N}\left(A_n^2 - 2a_nA_n + B_n^2 - 2b_nB_n\right)$$

$$= \|f\|^2 + 2\ell\left((A_0 - a_0)^2 - a_0^2\right)$$

$$+ \ell\sum_{n=1}^{N}\left((A_n - a_n)^2 - a_n^2 + (B_n - b_n)^2 - b_n^2\right)$$

$$= \|f\|^2 + 2\ell\left[(A_0 - a_0)^2 + \frac{1}{2}\sum_{n=1}^{N}\left((A_n - a_n)^2 + (B_n - b_n)^2\right)\right]$$

$$- 2\ell\Big(a_0^2 + \frac{1}{2}\sum_{n=1}^{N}(a_n^2 + b_n^2)\Big).$$

The first and last terms are fixed, while only the middle term contains instances of A_n and B_n. The middle term, being the sum of squares, is nonnegative, so E_N^2 (and hence E_N) is minimized by setting $A_n = a_n$ and $B_n = b_n$ for all n, with $0 \leqslant n \leqslant N$.

Thus, the best approximation to f by trigonometric polynomials on the interval $(-\ell, \ell)$ in the L_2-norm is $g_N = S_N$, the truncated Fourier series. In this case, the minimum error is

$$E_N^2 = \|f\|^2 - 2\ell\left(a_0^2 + \frac{1}{2}\sum_{n=1}^{N}(a_n^2 + b_n^2)\right),$$

and since $E_N^2 \geqslant 0$, we get Bessel's inequality:

$$a_0^2 + \frac{1}{2}\sum_{n=1}^{N}(a_n^2 + b_n^2) \leqslant \frac{1}{2\ell}\|f\|^2 = \frac{1}{2\ell}\int_{-\ell}^{\ell} f(x)^2\,dx.$$

Since the inequality holds for any integer $N \geqslant 1$, letting $N \to \infty$ yields

$$a_0^2 + \frac{1}{2}\sum_{n=1}^{\infty}(a_n^2 + b_n^2) \leqslant \frac{1}{2\ell}\|f\|^2 = \frac{1}{2\ell}\int_{-\ell}^{\ell} f(x)^2\,dx.$$

Using both the fact that the Fourier series converges and Theorem 2.22 on term-by-term integration, this condition can be strengthened to what is called **Parseval's identity**,

$$a_0^2 + \frac{1}{2}\sum_{n=1}^{\infty}(a_n^2 + b_n^2) = \frac{1}{2\ell}\|f\|^2 = \frac{1}{2\ell}\int_{-\ell}^{\ell} f(x)^2\,dx.$$

Therefore, we have

$$E_N^2 = \ell \sum_{n=N+1}^{\infty} (a_n^2 + b_n^2)$$

for all $N \geqslant 1$, and from Bessel's inequality the series on the right converges, so that

$$\lim_{N\to\infty} E_N^2 = 0.$$

Thus, the mean square error tends to zero:

$$\lim_{N\to\infty} \|f - S_N\| = \lim_{N\to\infty} \left(\int_{-\ell}^{\ell} |f(t) - S_N(t)|^2\,dt\right)^{1/2} = 0,$$

and the Fourier series converges in L^2 to the function f.

2.8 COMPLEX FOURIER SERIES

A well-known result in complex variables is **Euler's formula**,

$$e^{i\theta} = \cos\theta + i\sin\theta,$$

from which we can deduce that

$$\cos\theta = \frac{e^{i\theta} + e^{-i\theta}}{2} = \cosh i\theta \qquad \text{and} \qquad \sin\theta = \frac{e^{i\theta} - e^{-i\theta}}{2i} = -i\sinh i\theta.$$

We make use of these formulas to write a Fourier series in complex form. Suppose that $f \in PWC(-\ell,\ell)$ has the Fourier series representation

$$f(x) \sim a_0 + \sum_{n=1}^{\infty}\left(a_n \cos\frac{n\pi x}{\ell} + b_n \sin\frac{n\pi x}{\ell}\right);$$

then

$$\begin{aligned} f(x) &\sim a_0 + \frac{1}{2}\sum_{n=1}^{\infty}\left\{a_n\left(e^{in\pi x/\ell} + e^{-in\pi x/\ell}\right) - ib_n\left(e^{in\pi x/\ell} - e^{-in\pi x/\ell}\right)\right\} \\ &= a_0 + \frac{1}{2}\sum_{n=1}^{\infty}\left\{\left(a_n - ib_n\right)e^{in\pi x/\ell} + \left(a_n + ib_n\right)e^{-in\pi x/\ell}\right\} \\ &= c_0 + \sum_{n=1}^{\infty}\left(c_n e^{in\pi x/\ell} + c_{-n} e^{-in\pi x/\ell}\right) \end{aligned}$$

so that

$$f(x) \sim c_0 + \sum_{n=1}^{\infty} \left(c_n e^{in\pi x/\ell} + c_{-n} e^{-in\pi x/\ell} \right),$$

where

$$c_0 = a_0, \qquad c_n = \frac{a_n - ib_n}{2}, \qquad c_{-n} = \frac{a_n + ib_n}{2}$$

for $n = 1, 2, 3, \ldots$. The Fourier series may be rewritten conveniently as

$$f(x) \sim \sum_{n=-\infty}^{\infty} c_n e^{in\pi x/\ell}.$$

The formulas for the Fourier coefficients:

$$a_0 = \frac{1}{2\ell} \int_{-\ell}^{\ell} f(t)\, dt,$$

$$a_n = \frac{1}{\ell} \int_{-\ell}^{\ell} f(t) \cos \frac{n\pi t}{\ell}\, dt,$$

$$b_n = \frac{1}{\ell} \int_{-\ell}^{\ell} f(t) \sin \frac{n\pi t}{\ell}\, dt$$

lead to

$$c_0 = a_0 = \frac{1}{2\ell} \int_{-\ell}^{\ell} f(t)\, dt,$$

$$c_n = \frac{a_n - ib_n}{2} = \frac{1}{2\ell} \int_{-\ell}^{\ell} f(t) \left(\cos \frac{n\pi t}{\ell} - i \sin \frac{n\pi t}{\ell} \right) dt$$

$$= \frac{1}{2\ell} \int_{-\ell}^{\ell} f(t) e^{-in\pi t/\ell}\, dt,$$

$$c_{-n} = \frac{a_n + ib_n}{2} = \frac{1}{2\ell} \int_{-\ell}^{\ell} f(t) \left(\cos \frac{n\pi t}{\ell} + i \sin \frac{n\pi t}{\ell} \right) dt$$

$$= \frac{1}{2\ell} \int_{-\ell}^{\ell} f(t) e^{in\pi t/\ell}\, dt.$$

Summarizing, we have the following:

Theorem 2.25. The complex Fourier series for $f \in PWC(-\ell, \ell)$ is

$$f(x) \sim \sum_{n=-\infty}^{\infty} c_n e^{in\pi x/\ell},$$

where

$$c_n = \frac{1}{2\ell}\int_{-\ell}^{\ell} f(t)e^{-in\pi t/\ell}\,dt$$

for $n \in \mathbf{Z}$.

We finish this section with an example.

Example 2.19. Calculate the complex Fourier series for

$$f(x) = x, \quad -\pi < x < \pi,$$

and $f(x + 2\pi) = f(x)$ otherwise.

Solution. The complex Fourier coefficients are

$$c_n = \frac{1}{2\pi}\int_{-\pi}^{\pi} f(t)e^{-int}\,dt$$

for $n \in \mathbf{Z}$, so that

$$c_0 = \frac{1}{2\pi}\int_{-\pi}^{\pi} f(t)\,dt = \frac{1}{2\pi}\int_{-\pi}^{\pi} t\,dt = 0$$

and

$$\begin{aligned} c_n &= \frac{1}{2\pi}\int_{-\pi}^{\pi} te^{-int}\,dt \\ &= \frac{1}{2\pi}\left(\frac{2\pi i}{n}\cos n\pi + \frac{2}{n^2}\sin n\pi\right) \\ &= \frac{i}{n}(-1)^n \end{aligned}$$

for $n \neq 0$. Therefore, since f is continuous for $-\pi < x < \pi$, from Dirichlet's theorem we have

$$x = \sum_{\substack{n=-\infty \\ n\neq 0}}^{\infty} \frac{i(-1)^n}{n} e^{inx}$$

for $-\pi < x < \pi$. ∎

2.9 SUMMARY

This chapter should be read with your linear algebra text close at hand. Many vector space concepts from linear algebra, such as linear independence, basis, inner product, and orthogonality, are generalized to function spaces. Instead of the vector space $\mathbb{R}^n$, we now use a function space of **piecewise continuous functions** on (a, b) : $PWC(a, b)$. In this book, we develop the theory for a general interval (a, b), but we restrict this summary to the cases of $(-\ell, \ell)$ and $(0, \ell)$, since these are the two most common examples. These functions are continuous except for a finite number of jump discontinuities, and $PWC(-\ell, \ell)$ is equipped with the **inner product**:

$$\langle f, g\rangle = \int_{-\ell}^{\ell} f(x)g(x)\,dx,$$

and thus orthogonality in $PWC(-\ell, \ell)$, is again denoted by $\langle f, g\rangle = 0$. There is one problem with this definition; namely, it is possible that $\langle f, f\rangle = 0$, without the function f being identically zero. We get around this problem by identifying two functions that differ on a finite set of points.

There is a special set of orthogonal functions in $PWC(-\ell, \ell)$:

$$\left\{\cos\frac{n\pi x}{\ell}, \sin\frac{n\pi x}{\ell}\right\}_{n=0,1,2,\ldots},$$

which forms an orthogonal basis of $PWC(-\ell, \ell)$. Here the notion of basis is different from that in $\mathbb{R}^n$: Each function $f \in PWC(-\ell, \ell)$ has a unique representation in terms of the given family of functions, which we call a **Fourier series**:

$$f(x) \sim a_0 + \sum_{n=1}^{\infty}\left(a_n \cos\frac{n\pi x}{\ell} + b_n \sin\frac{n\pi x}{\ell}\right),$$

with **Fourier coefficients**

$$a_0 = \frac{1}{2\ell}\int_{-\ell}^{\ell} f(x)\,dx,$$

$$a_n = \frac{1}{\ell}\int_{-\ell}^{\ell} f(x)\cos\frac{n\pi x}{\ell}\,dx,$$

$$b_n = \frac{1}{\ell}\int_{-\ell}^{\ell} f(x)\sin\frac{n\pi x}{\ell}\,dx.$$

Convergence of this infinite series is established by **Dirichlet's theorem**, which says that at each point in the interval, the series converges to the average of the left- and right-hand limits. Near a jump discontinuity, the Fourier series is oscillatory, even though it converges at the point of discontinuity; this is called **Gibbs' phenomenon.**

An **odd** function satisfies $f(-x) = -f(x)$ and has only sine terms in its Fourier series, while an **even** function satisfies $f(-x) = f(x)$ and has only cosine terms in its Fourier series. This observation allowed us to define **Fourier sine series** and **Fourier cosine series** for $f \in PWC(0,\ell)$. If for $f \in PWC(0,\ell)$ we use the odd extension f_{odd}, its Fourier series will become the sine series for f. If we use the even extension f_{even}, the Fourier series of f_{even} becomes the cosine series of f.

It should be noted that in higher-level PDE courses, and in many applications such as physics and engineering, the Fourier series is usually represented in **complex form**,

$$f(x) \sim \sum_{n=-\infty}^{\infty} c_n\, e^{in\pi x/\ell},$$

where

$$c_n = \frac{1}{2\ell}\int_{-\ell}^{\ell} f(t)e^{-in\pi t/\ell}\, dt.$$

This presentation has the advantage that only one formula for the coefficients c_n needs to be remembered. Also, this form leads directly to the **Fourier transform**, which we study in Chapter 8.

2.9.1 Problems and Notes

Problems from Part II:

Exercise	11.2	11.3	11.4	11.5	11.6	11.7
Notes						
Exercise	11.8	11.9	11.10	11.11	11.12	11.13
Notes						
Exercise	11.15	11.16	11.18	11.19		
Notes						

Midterm exam questions:

Exercise	18.2	18.5
Notes		

Final exam questions:

Exercise	19.6
Notes	

CHAPTER 3

SEPARATION OF VARIABLES

3.1 HOMOGENEOUS EQUATIONS

In Example 1.11 we applied separation of variables successfully to a linear homogeneous PDE: the one-dimensional heat equation with Dirichlet boundary conditions on the interval $(0, \ell)$:

$$\begin{aligned}
&\frac{\partial u}{\partial t} = k\frac{\partial^2 u}{\partial x^2}, \qquad x \in \Omega = (0, \ell), \quad t > 0, \\
&u(0, t) = 0, \\
&u(\ell, t) = 0, \\
&u(x, 0) = f(x) = 3\sin\frac{2\pi x}{\ell},
\end{aligned}$$

where k is a positive constant. We looked for a separated solution, that is, one of the form $u(x, t) = X(x) \cdot T(t)$, and substituting this into the heat equation, we obtained

$$\frac{T'(t)}{kT(t)} = \frac{X''(x)}{X(x)}.$$

Partial Differential Equations: Theory and Completely Solved Problems, First Edition. By T. Hillen, I. E. Leonard, H. van Roessel

The left-hand side of this equation is a function of t only, whereas the right-hand side is a function of x only; since the variables x and t are independent, this means that both sides must equal a constant, say $-\lambda$, so that

$$\frac{T'}{kT} = \frac{X''}{X} = -\lambda,$$

where we omit the dependence on x and t. This leads to two ordinary differential equations for X and T :

$$\begin{aligned} &X'' + \lambda X = 0, \quad 0 < x < \ell, \qquad T' + \lambda k T = 0, \quad t > 0,\\ &X(0) = 0,\\ &X(\ell) = 0. \end{aligned}$$

Here, we try the same method for the one-dimensional wave equation with homogeneous Neumann boundary conditions.

Example 3.1. Consider the motion of a vibrating string governed by the Neumann problem for the wave equation:

$$\frac{\partial^2 u}{\partial t^2} = c^2 \frac{\partial^2 u}{\partial x^2}, \quad 0 < x < \ell, \quad t > 0,$$

with boundary conditions

$$\frac{\partial u}{\partial x}(0,t) = 0 \qquad \text{and} \qquad \frac{\partial u}{\partial x}(\ell,t) = 0,$$

and initial conditions

$$u(x,0) = f(x) \qquad \text{and} \qquad \frac{\partial u}{\partial t}(x,0) = g(x).$$

(a) Use the method of separation of variables to solve this problem.

(b) Show that the initial conditions,

$$u(x,0) = f(x) \quad \text{and} \quad \frac{\partial u}{\partial t}(x,0) = g(x),$$

lead to the Fourier cosine series:

$$f(x) = a_0 + \sum_{n=1}^{\infty} a_n \cos\frac{n\pi x}{\ell} \quad \text{and} \quad g(x) = \sum_{n=1}^{\infty} b_n \frac{cn\pi}{\ell} \cos\frac{n\pi x}{\ell}.$$

(c) Solve for a_n, b_n by using the orthogonality of the eigenfunctions

$$\left\langle \cos\frac{n\pi x}{\ell}, \cos\frac{m\pi x}{\ell} \right\rangle = \begin{cases} \dfrac{\ell}{2}, & \text{if } n = m \neq 0,\\ \ell, & \text{if } n = m = 0,\\ 0, & \text{if } n \neq m \end{cases}$$

for n, $m \geqslant 0$.

(d) Determine the frequencies of oscillation ν_n of this string and describe situations that lead to high or low frequencies.

Solution.

(a) Assuming a solution of the form

$$u(x,t) = \phi(x) \cdot G(t),$$

and separating variables, we get

$$\frac{\phi''(x)}{\phi(x)} = \frac{G''(t)}{c^2\, G(t)} = -\lambda,$$

where λ is a constant. The partial differential equation is reduced to two ordinary differential equations.

• **Spatial equation**:

$$\begin{aligned} &\phi''(x) + \lambda\phi(x) = 0, \quad 0 < x < \ell, \\ &\phi'(0) = 0, \\ &\phi'(\ell) = 0. \end{aligned}$$

• **Time equation**:

$$G''(t) + \lambda c^2 G(t) = 0, \quad t > 0.$$

We consider the spatial equation first since it has a complete set of boundary conditions (see the rule of thumb for side conditions).

Case 1: $\lambda < 0$. Here we can write $\lambda = -\mu^2$, where $\mu \neq 0$. The general solution to $\phi'' - \mu^2\phi = 0$ in this case is given by

$$\phi(x) = A\cosh\mu x + B\sinh\mu x,$$

with

$$\phi'(x) = \mu A\sinh\mu x + \mu B\cosh\mu x.$$

Applying the first boundary condition, we get

$$\phi'(0) = \mu B = 0,$$

so that $B = 0$, and the solution is now

$$\phi(x) = A\cosh\mu x$$

with

$$\phi'(x) = \mu A\sinh\mu x.$$

Applying the second boundary condition, we get

$$\phi'(\ell) = \mu A\sinh\mu\ell = 0,$$

so that $A = 0$. Thus, the only solution is the trivial solution $\phi(x) \equiv 0$.

Case 2: $\lambda > 0$. Here we can write $\lambda = \mu^2$, where $\mu \neq 0$. The general solution to $\phi'' + \mu^2\phi = 0$ in this case is given by

$$\phi(x) = A\cos\mu x + B\sin\mu x$$

with

$$\phi'(x) = -\mu A\sin\mu x + \mu B\cos\mu x.$$

Applying the first boundary condition, we get

$$\phi'(0) = \mu B \cdot 1 = 0,$$

so that $B = 0$, and the solution is now

$$\phi(x) = A\cos\mu x$$

with

$$\phi'(x) = -\mu A\sin\mu x.$$

Applying the second boundary condition, we get

$$\phi'(\ell) = -\mu A\sin\mu\ell = 0,$$

and if $A = 0$, we get only the trivial solution. Therefore, we get nontrivial solutions if and only if $\sin\mu\ell = 0$, that is, if and only if

$$\mu = \frac{n\pi}{\ell},$$

where $n \geqslant 1$. The eigenvalues for this problem are

$$\lambda_n = \mu_n^2 = \frac{n^2\pi^2}{\ell^2},$$

with corresponding eigenfunctions

$$\phi_n(x) = \cos\frac{n\pi x}{\ell}$$

for $n = 1, 2, 3, \ldots$.

Case 3: $\lambda = 0$. The general solution to $\phi'' = 0$ in this case is given by

$$\phi(x) = Ax + B.$$

Applying the boundary conditions, we get

$$\phi'(0) = A = 0 \qquad \text{and} \qquad \phi'(\ell) = A = 0,$$

and the solution is constant, $\phi(x) = B$. The eigenvalue in this case is $\lambda_0 = 0$, with eigenfunction $\phi_0(x) = 1$.

The solutions to the time equation corresponding to these nontrivial solutions are

$$G_n(t) = a_n \cos\frac{n\pi ct}{\ell} + b_n \sin\frac{n\pi ct}{\ell}$$

for $n = 0, 1, 2, 3, \ldots$. For $n \geqslant 0$, each of the functions

$$u_n(x,t) = \phi_n(x) \cdot G_n(t) = \cos\frac{n\pi x}{\ell}\left(a_n \cos\frac{n\pi ct}{\ell} + b_n \sin\frac{n\pi ct}{\ell}\right)$$

is a solution to the partial differential equation and satisfies the boundary conditions. Since the partial differential equation is linear and homogeneous, by the superposition principle, we can write the solution as

$$u(x,t) = a_0 + \sum_{n=1}^{\infty}\left(a_n \cos\frac{n\pi ct}{\ell} + b_n \sin\frac{n\pi ct}{\ell}\right)\cos\frac{n\pi x}{\ell}.$$

Note that the constant a_0 corresponds to the $n = 0$ term.

(b) The initial displacement is

$$u(x,0) = f(x),$$

and we need

$$f(x) = a_0 + \sum_{n=1}^{\infty} a_n \cos\frac{n\pi x}{\ell}$$

for $0 < x < \ell$ in order to satisfy the initial condition for u. The initial velocity is

$$\frac{\partial u}{\partial t}(x,0) = g(x),$$

and we need

$$g(x) = \sum_{n=1}^{\infty} b_n \frac{n\pi c}{\ell} \cos\frac{n\pi x}{\ell}$$

for $0 < x < \ell$ in order to satisfy the initial condition for $\partial u/\partial t$.

(c) To determine the coefficients a_n and b_n, we use the fact that the eigenfunctions are orthogonal on the interval $[0, \ell]$; in fact, we have

$$\int_0^{\ell} \cos\frac{n\pi x}{\ell}\cos\frac{m\pi x}{\ell}\,dx = \begin{cases} 0, & \text{for } n, m \geqslant 0,\ n \neq m, \\ \dfrac{\ell}{2}, & \text{for } n = m \neq 0, \\ \ell, & \text{for } n = m = 0. \end{cases}$$

Starting with the initial displacement

$$f(x) = a_0 + \sum_{n=1}^{\infty} a_n \cos\frac{n\pi x}{\ell} = \sum_{n=0}^{\infty} a_n \cos\frac{n\pi x}{\ell}$$

for $0 < x < \ell$, we multiply both sides of this equation by $\cos(m\pi x/\ell)$ and integrate over the interval $[0, \ell]$ to get

$$\left\langle f(x), \cos\frac{m\pi x}{\ell}\right\rangle = \sum_{n=0}^{\infty} a_n \left\langle \cos\frac{n\pi x}{\ell}, \cos\frac{m\pi x}{\ell}\right\rangle.$$

Using the orthogonality of the eigenfunctions on the interval $[0, \ell]$, we have

$$\left\langle f(x), \cos\frac{m\pi x}{\ell}\right\rangle = \begin{cases} \ell a_0, & \text{for} \quad m = 0, \\ \dfrac{\ell a_m}{2}, & \text{otherwise.} \end{cases}$$

Therefore,

$$a_0 = \frac{1}{\ell}\langle f(x), 1\rangle \qquad \text{and} \qquad a_n = \frac{2}{\ell}\left\langle f(x), \cos\frac{n\pi x}{\ell}\right\rangle$$

for $n \geqslant 1$. Similarly, for the initial velocity, we obtain

$$b_n \frac{n\pi c}{\ell} = \frac{2}{\ell}\left\langle g(x), \cos\frac{n\pi x}{\ell}\right\rangle,$$

and therefore,

$$b_n = \frac{2}{n\pi c}\left\langle g(x), \cos\frac{n\pi x}{\ell}\right\rangle$$

for $n \geqslant 1$. The complete solution is

$$u(x,t) = a_0 + \sum_{n=1}^{\infty}\left(a_n \cos\frac{n\pi ct}{\ell} + b_n \sin\frac{n\pi ct}{\ell}\right)\cos\frac{n\pi x}{\ell},$$

with the coefficients a_0, a_n, b_n as above.

(d) The frequencies of oscillation are given by

$$\nu_n = \frac{1}{T_n},$$

where T_n is the length of one period. Since the sine and cosine are periodic with period 2π, we have

$$\frac{n\pi c}{\ell} T_n = 2\pi.$$

Hence,

$$T_n = \frac{2\ell}{nc},$$

and the frequencies are

$$\nu_n = \frac{nc}{2\ell}$$

for $n \geqslant 1$.

We see clearly that longer strings give lower frequencies than shorter strings. Also, larger values of n correspond to higher frequencies. The tension of the string is proportional to the parameter c. A larger tension c leads to higher frequencies, which, of course, is well known by anyone who plays a guitar or violin.

∎

3.1.1 General Linear Homogeneous Equations

Suppose that we consider a slightly more general linear homogeneous PDE:

$$a_1 u_{tt} + a_2 u_t = k_1 u_{xx} + k_2 u_x + k_3 u,$$

and try the preceding method of separation of variables. We look for a separated solution of the form

$$u(x,t) = X(x) \cdot T(t),$$

and substituting this into the PDE, we get

$$a_1 \frac{T''}{T} + a_2 \frac{T'}{T} = k_1 \frac{X''}{X} + k_2 \frac{X'}{X} + k_3.$$

- If the coefficients are all constant, the left-hand side of the equation is a function of time t only, and the right-hand side of the equation is a function of the spatial variable x only, and both sides must be constant:

  $$a_1 \frac{T''}{T} + a_2 \frac{T'}{T} = k_1 \frac{X''}{X} + k_2 \frac{X'}{X} + k_3 = -\lambda.$$

  Therefore, we obtain ODEs for X and T:

  $$k_1 X'' + k_2 X' + (k_3 + \lambda) X = 0, \qquad a_1 T'' + a_2 T' + \lambda T = 0,$$

  and once again, separation of variables works.

For example:

Example 3.2. The one-dimensional wave equation in the presence of a damping term, where, in the simplest case, resistance can be assumed to be proportional to velocity, is called the **damped one-dimensional wave equation**:

$$\frac{\partial^2 u}{\partial t^2} + 2k \frac{\partial u}{\partial t} = c^2 \frac{\partial^2 u}{\partial x^2}, \quad 0 < x < \ell, \quad t > 0.$$

Solve this equation subject to the boundary conditions

$$u(0,t) = 0 \qquad \text{and} \qquad u(\ell,t) = 0$$

for $t > 0$, and the initial conditions

$$u(x,0) = f(x) \qquad \text{and} \qquad \frac{\partial u}{\partial t}(x,0) = g(x)$$

for $0 < x < \ell$.

Solution.

(a) Assuming a product solution of the form $u(x,t) = X(x) \cdot T(t)$ and substituting this into the equation, we have

$$X \cdot T'' + 2kX \cdot T' = c^2 X'' \cdot T.$$

Dividing by $c^2 X \cdot T$ and separating variables, we obtain

$$\frac{T''}{c^2 T} + \frac{2kT'}{c^2 T} = \frac{X''}{X}.$$

Since x and t are independent variables and the left-hand side depends only on t while the right-hand side depends only on x, both sides must be constant; hence,

$$\frac{X''}{X} = -\lambda \qquad \text{and} \qquad \frac{T''}{c^2 T} + \frac{2kT'}{c^2 T} = -\lambda,$$

where λ is the separation constant. Thus, the functions $X(x)$ and $T(t)$ satisfy the following ordinary differential equations:

$$X'' + \lambda X = 0,$$
$$T'' + 2kT' + \lambda c^2 T = 0.$$

Now, we can satisfy the boundary conditions by requiring that

$$X(0) = X(\ell) = 0,$$

so X satisfies the boundary value problem:

$$X'' + \lambda X = 0, \quad 0 < x < \ell,$$
$$X(0) = 0,$$
$$X(\ell) = 0.$$

As in the previous problems, we get a nontrivial solution only if the separation constant λ is positive, say, $\lambda = \mu^2$, where $\mu \neq 0$. In this case, the equations for X and T become

$$X'' + \mu^2 X = 0, \quad 0 < x < \ell, \qquad T'' + 2kT' + \mu^2 c^2 T = 0, \quad t > 0,$$

$$X(0) = 0,$$

$$X(\ell) = 0,$$

where $\mu \neq 0$.

(b) The general solution to the equation

$$X'' + \mu^2 X = 0$$

is given by

$$X(x) = A\cos\mu x + B\sin\mu x,$$

where the constants are determined from the boundary conditions. Applying the boundary conditions, since $X(0) = 0$, we must have $A = 0$, and since $X(\ell) = 0$, nontrivial solutions arise only when $\sin\mu\ell = 0$, which happens if and only if $\mu\ell = n\pi$, where n is an integer. Therefore, nontrivial solutions to the boundary value problem for X occur only for

$$\mu_n = \frac{n\pi}{\ell},$$

and the solutions are

$$X_n(x) = \sin\frac{n\pi x}{\ell}$$

for $n = 1, 2, \ldots$.

(c) For each integer $n \geqslant 1$, the corresponding equation for T is

$$T'' + 2kT' + \left(\frac{n\pi c}{\ell}\right)^2 T = 0,$$

a second-order linear homogeneous constant-coefficient equation, and we know how to solve this by assuming a solution of the form $T(t) = e^{\alpha t}$ and substituting into the differential equation to get the characteristic equation.

Alternatively, we illustrate another method by making a simple transformation of the dependent variable $y(t) = e^{kt}T(t)$, so that

$$y'' = e^{kt}T'' + 2ke^{kt}T' + k^2e^{kt}T = e^{kt}\underbrace{\left[T'' + 2kT' + \mu_n^2c^2T\right]}_{0} + (k^2 - \mu_n^2c^2)y$$

and y satisfies the differential equation

$$y'' + (\mu_n^2c^2 - k^2)y = 0.$$

To find the corresponding solutions $y_n = e^{kt}T_n(t)$, we need to consider three cases, according to whether

$$\mu_n^2c^2 - k^2 = \frac{n^2\pi^2c^2}{\ell^2} - k^2$$

is zero, positive, or negative.

Case 1: $\mu_n^2 c^2 - k^2 = 0$. In this case, we have equal real roots, and the solution is

$$y_n = e^{kt} T_n(t) = a_n + b_n t,$$

where $k = n\pi c/\ell > 0$.

Case 2: $\mu_n^2 c^2 - k^2 < 0$.
In this case we have two distinct real roots, and the solution is

$$y_n = e^{kt} T_n(t) = a_n \cosh s_n t + b_n \sinh s_n t,$$

where $s_n = \sqrt{k^2 - \mu_n^2 c^2}$.

Case 3: $\mu_n^2 c^2 - k^2 > 0$. In this case we have two distinct imaginary roots, and the solution is

$$y_n = e^{kt} T_n(t) = a_n \cos s_n t + b_n \sin s_n t,$$

where $s_n = \sqrt{\mu_n^2 c^2 - k^2}$.

(d) Since both the partial differential equation and the boundary conditions are linear and homogeneous, we use the superposition principle to write the solution as a linear combination of the solutions that we found in (c):

$$u(x,t) = \sum_{n=1}^{\infty} u_n(x,t) = \sum_{n=1}^{\infty} X_n(x) \cdot T_n(t).$$

If $k\ell/\pi c$ is not a positive integer,

$$\mu_n^2 c^2 - k^2 \neq 0,$$

and either

$$1 \leqslant n < \frac{k\ell}{\pi c} \qquad \text{or} \qquad n > \frac{k\ell}{\pi c}.$$

Thus, we are in case 2 or case 3, and the solution is

$$u(x,t) = e^{-kt} \sum_{1 \leqslant n < k\ell/\pi c} \sin \frac{n\pi x}{\ell} \left(a_n \cosh s_n t + b_n \sinh s_n t\right)$$

$$+ e^{-kt} \sum_{k\ell/\pi c < n < \infty} \sin \frac{n\pi x}{\ell} \left(a_n \cos s_n t + b_n \sin s_n t\right),$$

where these sums run over nonnegative integers only, and

$$s_n = \sqrt{\left| k^2 - \left(\frac{n\pi c}{\ell}\right)^2 \right|}.$$

Also, to satisfy the initial conditions, a_n are the Fourier sine coefficients for the odd periodic extension of $f(x)$; that is,

$$a_n = \frac{2}{\ell}\left\langle f(x), \sin\frac{n\pi x}{\ell}\right\rangle = \frac{2}{\ell}\int_0^\ell f(x)\sin\frac{n\pi x}{\ell}\,dx$$

for $n = 1, 2, \ldots$. If we differentiate this expression for $u(x,t)$ with respect to t and set $t = 0$, we see that $-ka_n + s_n b_n$ are just the Fourier sine coefficients of the odd periodic extension of $g(x)$; that is,

$$-ka_n + s_n b_n = \frac{2}{\ell}\left\langle g(x), \sin\frac{n\pi x}{\ell}\right\rangle = \frac{2}{\ell}\int_0^\ell g(x)\sin\frac{n\pi x}{\ell}\,dx$$

for $n = 1, 2, \ldots$.

(e) If $k\ell/\pi c$ is a positive integer, we have to add the corresponding term in the sum when $n = k\ell/\pi c$. In this case, if $n_0 = k\ell/\pi c$, the solution is as in (d), with the additional term

$$\left(a_{k\ell/\pi c}e^{-kt} + b_{k\ell/\pi c}te^{-kt}\right)\sin\frac{kx}{c},$$

where $a_{k\ell/\pi c}$ and $b_{k\ell/\pi c}$ are determined from the equations

$$a_{k\ell/\pi c} = \frac{2}{\ell}\left\langle f(x), \sin\frac{kx}{c}\right\rangle = \frac{2}{\ell}\int_0^\ell f(x)\sin\frac{kx}{c}\,dx$$

and

$$-ka_{k\ell/\pi c} + b_{k\ell/\pi c} = \frac{2}{\ell}\left\langle g(x), \sin\frac{kx}{c}\right\rangle = \frac{2}{\ell}\int_0^\ell g(x)\sin\frac{kx}{c}\,dx.$$

■

3.1.2 Limitations of the Method of Separation of Variables

Although quite powerful, the method of separation of variables does have limitations. Hence, before we apply this method to more complicated problems, such as nonhomogeneous problems or problems in higher spatial dimensions, we discuss a few cases where it is not immediately clear whether or not the method is applicable.

- If the coefficients are not constant, for example, if the PDE is

$$u_t = xtu_{xx},$$

separation of variables leads to

$$\frac{T'}{tT} = \frac{xX''}{X},$$

which leads to the following ODEs for X and T :

$$xX'' + \lambda X = 0, \qquad T' + \lambda t T = 0,$$

where λ is the separation constant, and again, separation of variables seems to work.

- However, for the PDE

$$u_t = (x+t)u_{xx},$$

separation of variables leads to

$$\frac{T'}{T} = (x+t)\frac{X''}{X},$$

and separation of variables does not work.

The success of the method of separation of variables in Examples 1.11, 3.1, and 3.2 depended not only on the fact that the PDEs were linear and homogeneous, but also on the fact that the boundary conditions were linear and homogeneous. Inserting

$$u(x,t) = X(x) \cdot T(t)$$

into homogeneous Dirichlet or Neumann boundary conditions such as

$$u(0,t) = u(\ell,t) = 0 \qquad \text{or} \qquad \frac{\partial u}{\partial x}(0,t) = \frac{\partial u}{\partial x}(\ell,t) = 0$$

leads to

$$X(0) = X(\ell) = 0 \qquad \text{or} \qquad X'(0) = X'(\ell) = 0.$$

More generally, inserting the separated solution $u(x,t) = X(x) \cdot T(t)$ into homogeneous Robin boundary conditions such as

$$\alpha u(0,t) + \beta \frac{\partial u}{\partial x}(0,t) = 0$$

leads to

$$\alpha X(0) + \beta X'(0) = 0.$$

What happens when either the PDE or the boundary conditions are no longer homogeneous?

- For a nonhomogeneous PDE such as

$$u_t = k u_{xx} + \gamma,$$

substituting

$$u(x,t) = X(x) \cdot T(t)$$

into the PDE leads to

$$\frac{T'}{kT} = \frac{X''}{X} + \frac{\gamma}{kXT},$$

and the method of separation of variables does not work.

- For the nonhomogeneous boundary conditions

$$u(0,t) = A \qquad \text{and} \qquad u(\ell,t) = B,$$

we get the inconsistency

$$T(t) = \frac{A}{X(0)} = \frac{B}{X(\ell)},$$

and again, separation of variables does not work.

In summary, we conclude that if we have a linear homogeneous PDE with linear homogeneous boundary conditions, the method of separation of variables *may or may not* work. Its success depends on the particular problem. For nonhomogeneous problems, the method does not work. However, all is not lost, as we shall see in the next section.

3.2 NONHOMOGENEOUS EQUATIONS

We saw in Section 3.1 that the method of separation of variables cannot be applied directly to problems where either the PDE or the boundary conditions are nonhomogeneous. To handle problems of this type, we split the problem into two parts:

(i) An ODE, which takes care of the nonhomogeneities, and which can be solved by ODE methods.

(ii) A homogeneous PDE, which we can solve by the method of separation of variables.

After we solve (i) and (ii), we combine these solutions to obtain the solution to the original problem. This is best illustrated by means of an example.

Example 3.3. Consider the one-dimensional nonhomogeneous heat equation

$$\frac{\partial u}{\partial t} = \frac{\partial^2 u}{\partial x^2} + \gamma, \quad 0 < x < 1, \quad t > 0,$$

$$u(0,t) = A,$$

$$u(1,t) = B,$$

$$u(x,0) = 0,$$

where γ, A, and B are positive constants. This problem governs the temperature inside a one-dimensional rod of unit length in which heat is generated internally. Notice that neither the PDE nor the boundary conditions are homogeneous.

If we look for a separated solution $u(x,t) = X(x) \cdot T(t)$, substituting into the PDE, we get

$$\frac{T'}{kT} = \frac{X''}{X} + \frac{\gamma}{kXT},$$

which is not separable; substituting into the boundary conditions, we get

$$T(t) = \frac{A}{X(0)} = \frac{B}{X(\ell)},$$

which is inconsistent.

To solve the problem we split the solution into two parts,

$$u(x,t) = v(x) + w(x,t), \tag{3.1}$$

where v is a function of x only. The goal is to let $v(x)$ cover the nonhomogeneous parts. Inserting this into the original PDE yields

$$w_t = w_{xx} + v'' + \gamma.$$

Now we have a single PDE in two unknown functions, v and w. We impose the condition

$$v'' + \gamma = 0,$$

on v, and get a linear, homogeneous PDE $w_t = w_{xx}$ for w. Inserting (3.1) into the boundary conditions, we get

$$\begin{aligned} u(0,t) &= v(0) + w(0,t) = A, \\ u(1,t) &= v(1) + w(1,t) = B, \end{aligned}$$

and if we impose the conditions $v(0) = A$ and $v(1) = B$ on the function v, then w also satisfies homogeneous boundary conditions: $w(0,t) = 0$ and $w(1,t) = 0$.
Now, v satisfies the boundary value problem

$$\begin{aligned} &v'' + \gamma = 0, \quad 0 < x < 1, \\ &v(0) = A, \\ &v(1) = B. \end{aligned}$$

The initial condition becomes

$$u(x,0) = v(x) + w(x,0) = 0,$$

that is, $w(x,0) = -v(x)$, so that w satsfies the boundary value–initial value problem

$$\frac{\partial w}{\partial t} = \frac{\partial^2 w}{\partial x^2}, \quad 0 < x < 1, \quad t > 0,$$

$$w(0,t) = 0,$$

$$w(1,t) = 0,$$

$$w(x,0) = -v(x).$$

Note: The nonhomogeneities of the original problem have been shifted over to the ODE for v, while the problem for w is a linear homogeneous PDE with linear homogeneous boundary conditions. We can solve the equation for v by direct integration, and we obtain

$$v(x) = -\tfrac{1}{2}\gamma x^2 + (\tfrac{1}{2}\gamma - A + B)x + A.$$

To solve the problem for w, we apply separation of variables:

$$w(x,t) = X(x) \cdot T(t),$$

and X and T satisfy the problems

$$\begin{aligned} &X'' + \lambda X = 0, \quad 0 < x < 1, \qquad && T' + \lambda T = 0, \quad t > 0, \\ &X(0) = 0, && \\ &X(1) = 0, && \end{aligned}$$

where λ is the separation constant.

We have solved this boundary value problem for X with Dirichlet conditions previously (twice); hence, from now on we just cite the eigenvalues and corresponding eigenfunctions:

$$\lambda_n = n^2\pi^2 \qquad \text{and} \qquad X_n(x) = \sin n\pi x$$

for $n = 1, 2, \ldots$. The corresponding solutions to the problem for T are

$$T_n(t) = e^{-n^2\pi^2 t}$$

and the product solutions for the equation for w are

$$w_n(x,t) = X_n(x) \cdot T_n(t) = e^{-n^2\pi^2 t} \sin n\pi x$$

for $n = 1, 2, \ldots$.

Now w_n is a solution to the PDE for w that also satisfies the homogeneous boundary conditions for each $n \geqslant 1$, and in order to satisfy the initial condition we need

$$w(x,0) = -v(x), \quad 0 < x < 1.$$

Applying the superposition principle, we write

$$w(x,t) = \sum_{n=1}^{\infty} b_n X_n(x) \cdot T_n(t) = \sum_{n=1}^{\infty} b_n e^{-n^2\pi^2 t} \sin n\pi x,$$

and applying the initial condition $w(x,0) = -v(x)$, we get

$$-v(x) = \sum_{n=1}^{\infty} b_n \sin n\pi x,$$

a Fourier sine series for $-v(x)$.

From the orthogonality of the eigenfunctions on the interval $(0,1)$, we have

$$\begin{aligned} b_n &= -2 \int_0^1 v(x) \sin n\pi x \, dx \\ &= -2 \int_0^1 [-\tfrac{1}{2}\gamma x^2 + (\tfrac{1}{2}\gamma - A + B)x + A] \sin n\pi x \, dx \\ &= \frac{2\gamma}{n^3\pi^3}[(-1)^n - 1] - \frac{2}{n\pi}[A - (-1)^n B]. \end{aligned} \tag{3.2}$$

The final solution is

$$u(x,t) = v(x) + w(x,t),$$

that is,

$$\begin{aligned} u(x,t) &= v(x) + \sum_{n=1}^{\infty} b_n e^{-n^2\pi^2 t} \sin n\pi x \\ &= -\tfrac{1}{2}\gamma x^2 + \left(\tfrac{1}{2}\gamma - A + B\right) x + A \\ &\quad + \sum_{n=1}^{\infty} \left\{ \frac{2\gamma}{n^3\pi^3}[(-1)^n - 1] - \frac{2}{n\pi}[A - (-1)^n B] \right\} e^{-n^2\pi^2 t} \sin n\pi x \end{aligned}$$

for $0 < x < 1,\ t > 0$.

Note that

$$\lim_{t\to\infty} u(x,t) = \lim_{t\to\infty} \{v(x) + w(x,t)\} = v(x),$$

and $v(x)$ is called the **equilibrium solution** or **steady-state solution**, while $w(x,t)$ is called the **transient solution**.

For the special case $\gamma = 0$, the solution reduces to

$$u(x,t) = (B - A)x + A - \frac{2}{\pi}\sum_{n=1}^{\infty} \frac{1}{n}[A - (-1)^n B]e^{-n^2\pi^2 t} \sin n\pi x$$

for $0 < x < 1,\ t > 0$.

Here we plot some results with the following parameter values:

$$\gamma = 15, \qquad A = 1, \qquad B = 2.$$

- Figure 3.1 contains a plot of the steady-state solution $v(x)$.
- Figure 3.2 shows approximations to $u(x,t)$ by using a truncated series with $N = 5$ and $N = 50$, at early times.
- In Figure 3.3 we show the evolution of $u(x,t)$ as time $t \to 0.6$. Notice that convergence to the limiting solution takes place very quickly.

∎

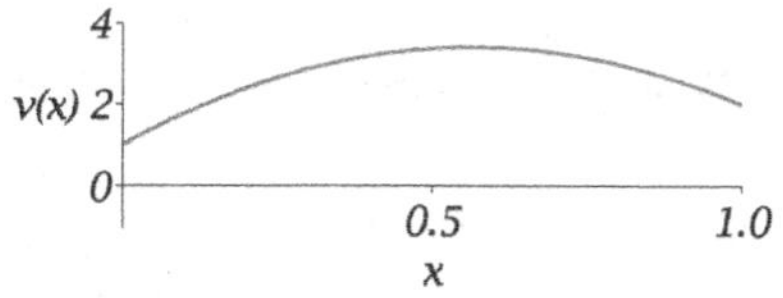

Figure 3.1 Steady-state solution $v(x)$

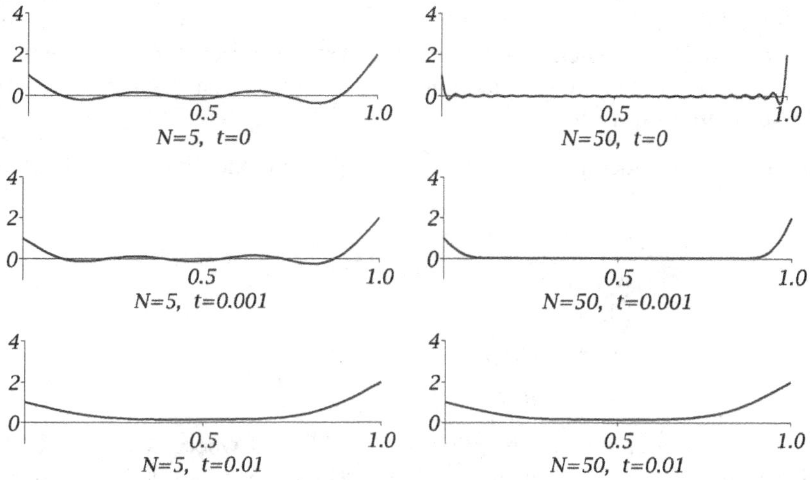

Figure 3.2 Various approximations for $u(x,t)$ with $N = 5$ and $N = 50$.

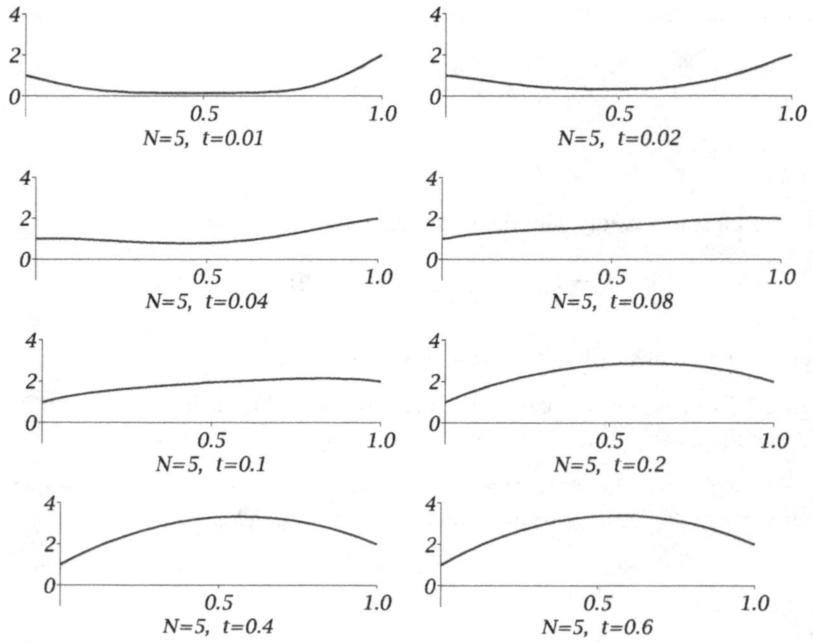

Figure 3.3 Time evolution of $u(x,t)$ with $N = 5$.

3.2.1 Method of Eigenfunction Expansions

Example 3.3 dealt with a nonhomogeneous problem in which the nonhomogeneities, in both the PDE and boundary conditions, were constant. In the next example we consider a problem with nonhomogeneities that are not constant. We will see that we need to extend our method of separation of variables into what is called the **method of eigenfunction expansions.**

Example 3.4. Consider the following nonhomogeneous one-dimensional heat equation:

$$\begin{aligned}
&\frac{\partial u}{\partial t} = \frac{\partial^2 u}{\partial x^2} + \gamma(x,t), \quad 0 < x < 1, \quad t > 0,\\
&u(0,t) = A(t),\\
&u(1,t) = B(t),\\
&u(x,0) = g(x),
\end{aligned}$$

where $\gamma(x,t)$ is a continuous function of x and t, and where $A(t)$ and $B(t)$ are continuously differentiable functions of t. The technique employed in Example 3.3, namely, writing

$$u(x,t) = v(x) + w(x,t),$$

will not work in this case, since the nonhomogeneous terms depend on t. To solve this problem, we proceed in four steps.

Step 1: First we split the solution into two parts,

$$u(x,t) = v(x,t) + w(x,t),$$

where now $v(x,t)$ depends on both x and t. Still, we use $v(x,t)$ to carry the burden of the nonhomogeneous boundary conditions, so we assume that $v(x,t)$ satisfies the boundary conditions

$$\begin{aligned} v(0,t) &= A(t), \\ v(1,t) &= B(t) \end{aligned}$$

and nothing else at this point. This has the effect of moving the nonhomogeneous boundary conditions up into the PDE itself. We use the simplest such function,

$$v(x,t) = A(t) + x\left[B(t) - A(t)\right],$$

so that $v(0,t) = A(t)$ and $v(1,t) = B(t)$, and

$$u(x,t) = A(t) + x\left[B(t) - A(t)\right] + w(x,t). \tag{3.3}$$

The boundary conditions are now

$$\begin{aligned} u(0,t) &= v(0,t) + w(0,t) = A(t) + w(0,t), \\ u(1,t) &= v(1,t) + w(1,t) = B(t) + w(1,t), \end{aligned}$$

and w satisfies the boundary conditions

$$w(0,t) = 0 \qquad \text{and} \qquad w(1,t) = 0.$$

Substituting expression (3.3) into the PDE, we get

$$A'(t) + x\left[B'(t) - A'(t)\right] + \frac{\partial w}{\partial t} = \frac{\partial^2 w}{\partial x^2} + \gamma(x,t)$$

and

$$A(0) + x\left[B(0) - A(0)\right] + w(x,0) = g(x)$$

for $0 < x < 1$, and $t > 0$. Therefore, $w(x,t)$ satisfies the nonhomogeneous PDE with homogeneous boundary conditions:

$$\frac{\partial w}{\partial t} = \frac{\partial^2 w}{\partial x^2} + F(x,t), \quad 0 < x < 1, \quad t > 0, \tag{3.4}$$

$$w(0,t) = 0,$$

$$w(1,t) = 0,$$

$$w(x,0) = f(x),$$

where

$$F(x,t) = \gamma(x,t) - A'(t) - x\left[B'(t) - A'(t)\right],$$
$$f(x) = g(x) - A(0) - x\left[B(0) - A(0)\right].$$

Step 2: As is usual with nonhomogeneous equations, we first find the solution to the associated homogeneous problem

$$\frac{\partial w}{\partial t} = \frac{\partial^2 w}{\partial x^2}, \quad 0 < x < 1,\ t > 0, \tag{3.5}$$
$$w(0,t) = 0,$$
$$w(1,t) = 0,$$

using separation of variables. Assuming a solution of the form $w(x,t) = X(x) \cdot T(t)$, we get two ordinary differential equations:

$$X''(x) + \lambda X(x) = 0, \quad 0 < x < 1 \qquad T'(t) + \lambda k\, T(t) = 0, \quad t > 0$$
$$X(0) = 0,$$
$$X(1) = 0.$$

The eigenvalues are

$$\lambda_n = n^2\pi^2$$

with corresponding eigenfunctions

$$X_n(x) = \sin n\pi x$$

for $n \geqslant 1$.

Step 3: Now, we are not solving the T equation and finding the general solution to the homogeneous problem as we did in Example 1.11. Instead, we use the method of eigenfunction expansions to write the solution to the nonhomogeneous problem (3.4) in terms of the eigenfunctions of the homogeneous problem (3.5):

$$w(x,t) = \sum_{n=1}^{\infty} w_n(t) X_n(x) = \sum_{n=1}^{\infty} w_n(t) \sin n\pi x, \tag{3.6}$$

where (similar to the method of variation of parameters) the coefficients $w_n(t)$ depend on t. Assuming that the functions $f(x)$ and $F(x,t)$ are piecewise smooth on the interval $[0,1]$ (as functions of x), from Dirichlet's convergence theorem we can expand these functions in terms of the eigenfunctions, and we have

$$f(x) = \sum_{n=1}^{\infty} f_n \sin n\pi x$$

and

$$F(x,t) = \sum_{n=1}^{\infty} F_n(t) \sin n\pi x,$$

where

$$f_n = 2\langle f(x), \sin n\pi x\rangle = 2\int_0^1 f(x)\sin n\pi x\, dx$$

and

$$F_n(t) = 2\langle F(x,t), \sin n\pi x\rangle = 2\int_0^1 F(x,t)\sin n\pi x\, dx$$

for $n \geqslant 1$.

Step 4: Next, we force expression (3.6) to be a solution to the nonhomogeneous equation (3.4) by requiring that each $w_n(t)$ satisfies a first-order ordinary differential equation together with an initial condition. We look at the initial condition first; when $t = 0$ we want

$$w(x,0) = f(x) = \sum_{n=1}^{\infty} w_n(0)\sin n\pi x,$$

and from the orthogonality of the eigenfunctions on the interval $[0,1]$, we need

$$w_n(0) = 2\int_0^1 f(x)\sin n\pi x\, dx = f_n$$

for $n \geqslant 1$. Now, from (3.6) we have

$$\frac{\partial w}{\partial t} = \sum_{n=1}^{\infty} \frac{dw_n(t)}{dt}\sin n\pi x \qquad \text{and} \qquad \frac{\partial^2 w}{\partial x^2} = -\sum_{n=1}^{\infty} n^2\pi^2 w_n(t)\sin n\pi x,$$

and substituting these expressions into (3.3), after some simplification, we obtain

$$\sum_{n=1}^{\infty}\left[\frac{dw_n(t)}{dt} + n^2\pi^2 w_n(t) - F_n(t)\right]\sin n\pi x = 0.$$

Again, from the orthogonality of the eigenfunctions on $[0,1]$, we have

$$\frac{dw_n(t)}{dt} + n^2\pi^2 w_n(t) = F_n(t),$$

and the w_n's satisfy the following initial value problem:

$$\frac{dw_n(t)}{dt} + n^2\pi^2 w_n(t) = F_n(t), \quad t > 0,$$

$$w_n(0) = f_n$$

for $n \geqslant 1$.

This is a first-order linear nonhomogeneous ODE, and as such, can always be solved by means of an integrating factor. Thus, the solution is

$$w_n(t) = e^{-n^2\pi^2 t}\left[f_n + \int_0^t e^{n^2\pi^2 s} F_n(s)\,ds\right],$$

and the solution to the original problem is

$$u(x,t) = v(x,t) + w(x,t);$$

that is,

$$u(x,t) = A(t) + x\,[B(t) - A(t)] + \sum_{n=1}^{\infty} w_n(t)\sin n\pi x$$

for $0 < x < 1,\ t > 0$, where

$$\begin{aligned}
w_n(t) &= e^{-n^2\pi^2 t}\left[f_n + \int_0^t e^{n^2\pi^2 s} F_n(s)\,ds\right] \\
&= f_n e^{-n^2\pi^2 t} + \int_0^t F_n(s) e^{-n^2\pi^2 (t-s)}\,ds, \\
F_n(t) &= 2\int_0^1 F(x,t)\sin n\pi x\,dx, \\
f_n &= 2\int_0^1 f(x)\sin n\pi x\,dx, \\
F(x,t) &= \gamma(x,t) - A'(t) - x\,[B'(t) - A'(t)], \\
f(x) &= g(x) - A(0) - x\,[B(0) - A(0)]
\end{aligned}$$

for $0 < x < 1,\ t > 0$.

Now we consider some **special cases**:

(a) $\gamma \equiv 0,\ A \equiv 0$, and $B \equiv 0$ (i.e., the homogeneous case). Here the boundary value–initial value problem is

$$\begin{aligned}
&\frac{\partial u}{\partial t} = \frac{\partial^2 u}{\partial x^2}, \quad 0 < x < 1, \quad t > 0, \\
&u(0,t) = 0, \\
&u(\ell,t) = 0, \\
&u(x,0) = g(x),
\end{aligned}$$

so that

$$F(x,t) \equiv 0, \qquad f(x) = g(x), \qquad \text{and} \qquad w_n(t) = g_n e^{-n^2\pi^2 t}.$$

The solution is

$$u(x,t) = \sum_{n=1}^{\infty} g_n e^{-n^2\pi^2 t} \sin n\pi x,$$

where $g_n = \langle g(x), \sin n\pi x\rangle$ for $n \geqslant 1$.

(b) γ, A and B constant, $g(x) \equiv 0$ (i.e., Example 3.3). Here, the boundary value–initial value problem is

$$\begin{aligned}
&\frac{\partial u}{\partial t} = \frac{\partial^2 u}{\partial x^2} + \gamma, \quad 0 < x < 1, \quad t > 0,\\
&u(0,t) = A,\\
&u(1,t) = B,\\
&u(x,0) = 0,
\end{aligned}$$

so that

$$A' \equiv 0, \quad B' \equiv 0, \quad F(x,t) = \gamma, \quad \text{and} \quad f(x) = -A - (B-A)x$$

and

$$f_n = 2\int_0^1 f(x)\sin n\pi x\,dx = \frac{2}{n\pi}[A - (-1)^n(B-A)],$$

$$F_n(t) = 2\int_0^1 \gamma \sin n\pi x\,dx = \frac{2\gamma}{n\pi}[1-(-1)^n].$$

Therefore,

$$\begin{aligned}
w_n(t) &= e^{-n^2\pi^2 t}\left[f_n + F_n\int_0^t e^{n^2\pi^2 s}\,ds\right]\\
&= e^{-n^2\pi^2 t}\left[f_n + \frac{F_n}{n^2\pi^2}\left(e^{n^2\pi^2 t} - 1\right)\right]\\
&= \left(f_n - \frac{F_n}{n^2\pi^2}\right)e^{-n^2\pi^2 t} + \frac{F_n}{n^2\pi^2},
\end{aligned}$$

and the solution is

$$\begin{aligned}
u(x,t) &= A + x(B-A) + \sum_{n=1}^{\infty} w_n(t)\sin n\pi x \\
&= -f(x) + \sum_{n=1}^{\infty} w_n(t)\sin n\pi x \\
&= \sum_{n=1}^{\infty} [w_n(t) - f_n]\sin n\pi x \\
&= \sum_{n=1}^{\infty} \left[\left(f_n - \frac{F_n}{n^2\pi^2}\right)e^{-n^2\pi^2 t} + \frac{F_n}{n^2\pi^2} - f_n\right]\sin n\pi x \\
&= \sum_{n=1}^{\infty} \left(f_n - \frac{F_n}{n^2\pi^2}\right)\left(e^{-n^2\pi^2 t} - 1\right)\sin n\pi x
\end{aligned}$$

for $0 < x < 1,\ t > 0$.

Does this agree with the result of Example 3.3? To see that it does, recall that

$$-v(x) = \sum_{n=1}^{\infty} b_n \sin n\pi x,$$

where the b_n's are given by (3.2). Thus, we have

$$f_n - \frac{F_n}{n^2\pi^2} = \frac{2}{n\pi}[A - (-1)^n B] - \frac{2\gamma}{n^3\pi^3}[1 - (-1)^n] = b_n,$$

and therefore,

$$\begin{aligned}
u(x,t) &= \sum_{n=1}^{\infty} b_n\left(e^{-n^2\pi^2 t} - 1\right)\sin n\pi x \\
&= -\sum_{n=1}^{\infty} b_n \sin n\pi x + \sum_{n=1}^{\infty} b_n e^{-n^2\pi^2 t}\sin n\pi x \\
&= v(x) + \sum_{n=1}^{\infty} b_n e^{-n^2\pi^2 t}\sin n\pi x.
\end{aligned}$$

∎

As another example using the method of eigenfunction expansions, we present the following:

Example 3.5. Solve the nonhomogeneous heat equation with time-dependent source

$$\frac{\partial u}{\partial t} = k\frac{\partial^2 u}{\partial x^2} + Q(x,t), \quad 0 < x < \ell, \quad t > 0, \tag{3.7}$$

$$u(x,0) = f(x)$$

and nonhomogeneous boundary conditions

$$u(0,t) = A(t) \qquad \text{and} \qquad \frac{\partial u}{\partial x}(\ell,t) = B(t).$$

Solution. First we reduce the problem to one with homogeneous boundary conditions by writing

$$u(x,t) = w(x,t) + v(x,t)$$

and assuming that v satisfies just the boundary conditions (and nothing else), then use the method of eigenfunction expansions to solve for $w(x,t)$. If $u(x,t)$ is a solution to the problem (3.7), we write

$$u(x,t) = v(x,t) + w(x,t),$$

where $v(x,t)$ satisfies only the boundary conditions

$$\begin{aligned} v(0,t) &= A(t), \\ \frac{\partial v}{\partial x}(\ell,t) &= B(t) \end{aligned} \tag{3.8}$$

for $t > 0$. We take the simplest possible such function, namely,

$$v(x,t) = B(t)\,x + A(t);$$

then

$$u(x,t) = w(x,t) + B(t)\,x + A(t)$$

and

$$\frac{\partial u}{\partial t} = \frac{\partial w}{\partial t} + \frac{dB(t)}{dt}\,x + \frac{dA(t)}{dt}$$

and

$$\frac{\partial^2 u}{\partial x^2} = \frac{\partial^2 w}{\partial x^2}.$$

Therefore,

$$\frac{\partial w}{\partial t} = k\frac{\partial^2 w}{\partial x^2} - \frac{dB(t)}{dt}\,x - \frac{dA(t)}{dt} + Q(x,t).$$

Also,

$$A(t) = u(0,t) = w(0,t) + A(t), \qquad \text{so that} \qquad w(0,t) = 0,$$

while

$$B(t) = \frac{\partial u}{\partial x}(\ell, t) = \frac{\partial w}{\partial x}(\ell, t) + B(t), \qquad \text{so that} \qquad \frac{\partial w}{\partial x}(\ell, t) = 0.$$

Therefore, $w(x, t)$ satisfies the nonhomogeneous problem with homogeneous boundary conditions given by

$$\frac{\partial w}{\partial t} = k\frac{\partial^2 w}{\partial x^2} - \frac{dB(t)}{dt}\,x - \frac{dA(t)}{dt} + Q(x, t), \quad 0 < x < \ell, \quad t > 0, \tag{3.9}$$

$$w(0, t) = 0,$$

$$\frac{\partial w}{\partial x}(\ell, t) = 0,$$

$$w(x, 0) = f(x) - B(0)\,x - A(0).$$

The boundary value–initial value problem for $w(x, t)$ now consists of a nonhomogeneous partial differential equation, but with homogeneous boundary conditions. Again, as is usual with nonhomogeneous equations, we first find the solution to the homogeneous problem

$$\frac{\partial w}{\partial t} = k\frac{\partial^2 w}{\partial x^2}, \quad 0 < x < \ell, \quad t > 0,$$

$$w(0, t) = 0,$$

$$\frac{\partial w}{\partial x}(\ell, t), = 0,$$

using separation of variables. Assuming a solution of the form $w(x, t) = \phi(x) \cdot T(t)$, we get two ordinary differential equations:

$$\begin{aligned} &\phi''(x) + \lambda\phi(x) = 0, \quad 0 < x < \ell, \qquad T'(t) + \lambda k\, T(t) = 0, \quad t > 0, \\ &\phi(0) = 0, \\ &\phi'(\ell) = 0. \end{aligned}$$

The eigenvalues are

$$\lambda_n = \left(\frac{(2n-1)\pi}{2\ell}\right)^2$$

with corresponding eigenfunctions

$$\phi_n(x) = \sin\frac{(2n-1)\pi x}{2\ell}$$

for $n \geqslant 1$.

Again, we are not solving the T equation and finding the general solution to the homogeneous problem; instead, we use the method of eigenfunction expansions to write the solution $w(x,t)$ to (3.9), the nonhomogeneous problem, in terms of the eigenfunctions of the homogeneous problem:

$$w(x,t) = \sum_{n=1}^{\infty} a_n(t) \sin \frac{(2n-1)\pi x}{2\ell}, \tag{3.10}$$

where, like the method of variation of parameters, the coefficients $a_n(t)$ depend on the time t. Next, we force this to be a solution to equation (3.9) by requiring that each $a_n(t)$ satisfy a first-order ordinary differential equation together with an initial condition. We look at the initial conditions first; when $t = 0$ we want

$$w(x,0) = f(x) - B(0)\,x - A(0) = \sum_{n=1}^{\infty} a_n(0) \sin \frac{(2n-1)\pi x}{2\ell},$$

and from the orthogonality of the eigenfunctions on the interval $[0, \ell]$, we find the coefficients

$$a_n(0) = \frac{2}{\ell} \int_0^{\ell} [f(x) - B(0)\,x - A(0)] \sin \frac{(2n-1)\pi x}{2\ell}\, dx$$

for $n \geqslant 1$. Now from (3.10) we have

$$\frac{\partial w}{\partial t} = \sum_{n=1}^{\infty} \frac{da_n(t)}{dt} \sin \frac{(2n-1)\pi}{2\ell} x$$

and

$$\frac{\partial^2 w}{\partial x^2} = -\sum_{n=1}^{\infty} a_n(t) \left(\frac{(2n-1)\pi}{2\ell} \right)^2 \sin \frac{(2n-1)\pi x}{2\ell},$$

and substituting these expressions into equation (3.9), after some simplification we obtain

$$\sum_{n=1}^{\infty} \left(\frac{da_n}{dt} + k\lambda_n\, a_n \right) \sin \frac{(2n-1)\pi x}{2\ell} = -\frac{dB(t)}{dt}\, x - \frac{dA(t)}{dt} + Q(x,t).$$

The left-hand side of this equation is just the generalized Fourier series of the function

$$F(x,t) = -\frac{dB(t)}{dt}\, x - \frac{dA(t)}{dt} + Q(x,t),$$

so that

$$\frac{da_n}{dt} + k\lambda_n\, a_n = \frac{2}{\ell} \int_0^{\ell} F(x,t) \sin \frac{(2n-1)\pi x}{2\ell}\, dx = F_n(t), \tag{3.11}$$

and $a_n(t)$ satisfies the initial value problem

$$\frac{da_n(t)}{dt} + k\lambda_n\, a_n(t) = F_n(t), \quad t > 0,$$

$$a_n(0) = \frac{2}{\ell}\int_0^\ell [f(x) - B(0)\,x - A(0)] \sin\frac{(2n-1)\pi x}{2\ell}\, dx.$$

Multiplying by the integrating factor $e^{\lambda_n kt}$, we can solve this first-order linear equation to get

$$a_n(t) = a_n(0)e^{-\lambda_n kt} + e^{-\lambda_n kt}\int_0^t F_n(s)e^{\lambda_n ks}\, ds, \quad t > 0 \tag{3.12}$$

for $n \geqslant 1$.

The solution to the original equation is

$$u(x,t) = B(t)\,x + A(t) + \sum_{n=1}^{\infty} a_n(t) \sin\sqrt{\lambda_n}\, x$$

for $0 < x < \ell,\ t > 0$, where

$$\lambda_n = \left(\frac{(2n-1)\pi}{2\ell}\right)^2$$

and $a_n(t)$ is given by (3.12) for $n \geqslant 1$. ∎

We now have a relatively systematic way to handle nonhomogeneities that may arise in our PDEs and/or boundary conditions. The crucial part is being able to solve the corresponding homogeneous problem. In the examples that we have considered thus far, the homogeneous PDEs have been relatively easy to solve. Suppose now that we consider some slightly more complicated homogeneous PDEs.

Recall the general heat equation,

$$\frac{\partial u}{\partial t} = \frac{1}{\rho c}\nabla\cdot(K\nabla u) + \frac{h}{\rho c},$$

where ρ is the density, c the heat capacity, K the thermal conductivity, and h the rate of internal heat generation. Let us consider the one-dimensional case with $h = h_1 u$:

$$\rho(x)c(x)\frac{\partial u}{\partial t} = \frac{\partial}{\partial x}\left[K(x)\frac{\partial u}{\partial x}\right] + h_1(x)u$$

$$+ \text{ auxiliary conditions}$$

If we try to solve this equation using separation of variables, by setting

$$u(x,t) = X(x)\cdot T(t),$$

we obtain

$$\rho c X T' = \frac{d}{dx}(KX')T + h_1 XT,$$

which separates to

$$\frac{T'}{T} = \frac{(KX')'}{\rho c X} + \frac{h_1}{\rho c} = -\lambda.$$

The equation for X is

$$(KX')' + (h_1 + \lambda \rho c)X = 0,$$

a second-order ODE with nonconstant coefficients, and is, of course, much more difficult to solve than

$$X'' + \lambda X = 0,$$

the equation we had before.

In general, we need to consider ODEs of the type

$$(p(x)\phi')' + [q(x) + \lambda\sigma(x)]\phi = 0, \quad x \in (a, b),$$

$$+ \text{ boundary conditions.}$$

Notice that for the case

$$p(x) \equiv 1, \qquad q(x) \equiv 0, \qquad \text{and} \qquad \sigma(x) \equiv 1,$$

this reduces to

$$\phi'' + \lambda\phi = 0,$$

which is the simple ODE that we had before.

Now the question is: For what values of λ will nontrivial solutions exist? These **eigenvalue problems**, called **Sturm-Liouville problems**, are a subject of study in the next chapter.

3.3 SUMMARY

This chapter is at the heart of any PDE text and shows the power of the method of separation of variables. Most authors avoid giving a stepwise scheme for the separation method, since there always exists an example that must be treated differently. Nevertheless, we attempt such a stepwise summary here, but keep in mind that it may not be applicable to all the problems encountered.

Separation, a practical guide:

1. Identify the problem and see that the PDE and the boundary conditions are linear and homogeneous.

2. Assume a product representation where all variables are separated: for example, $u(x, t) = X(x) \cdot T(t)$.

3. Isolate separate problems for $X(x)$ and $T(t)$ and then introduce a separation constant λ.

4. One of the problems found will be complete, that is, will have the correct number of side conditions. Solve this problem first to find the **eigenvalues**, λ_n, and the **eigenfunctions**, $X_n(x)$.

5. Solve the second of these problems, using only the eigenvalues λ_n for the values of the separation constant, and write $u_n(x, t) = X_n(x) \cdot T_n(t)$.

6. Use the **superposition principle** to write $u(x, t) = \sum a_n u_n(x, t)$.

7. Determine the coefficients a_n from the remaining side conditions, usually initial conditions.

The method of separation can be extended to treat **nonhomogeneous linear equations**. There are two strategies for this:

Strategy 1: Try to split the original nonhomogeneous problem into two problems by writing $u(x, t) = v(x, t) + w(x, t)$, where the problem for one of the unknown functions becomes extremely simple, for example, an ODE; and the problem for the other unknown function is homogeneous. Solve the simple problem directly, and then solve the homogeneous problem using steps 1 to 7 above. Finally, add these solutions to find $u(x, t)$.

Strategy 2: Use the **method of eigenfunction expansions**. This method can be used only if the boundary conditions are homogeneous. If the boundary conditions are not homogeneous, first use strategy 1 to get a problem with homogeneous boundary conditions. For the nonhomogeneous problem with homogeneous boundary conditions, solve the homogeneous problem to find the eigenfunctions. Expand all terms in the nonhomogeneous problem in terms of these eigenfunctions, and use orthogonality to derive equations for the coefficients. These are typically ODEs, and ODE methods are needed to solve them. Thus, a solid knowledge of these methods is necessary to master this chapter.

3.3.1 Problems and Notes

Exercise 3.1. Problems from Part II:

Exercise	13.2	13.3	13.4	13.5	13.6	13.7
Notes						
Exercise	13.8	13.9	13.13	13.15	14.1	14.2
Notes						
Exercise	14.3					
Notes						

Midterm exam questions:

Exercise	18.9
Notes	

Final exam questions:

Exercise	19.5	19.7	19.13
Notes			

CHAPTER 4

STURM LIOUVILLE THEORY

4.1 FORMULATION

The homogeneous second-order linear ODE

$$(p(x)\phi')' + [q(x) + \lambda\sigma(x)]\phi = 0, \quad a < x < b, \tag{4.1}$$

where $p'(x)$, $q(x)$, and $\sigma(x)$ are all continuous on the interval $a \leqslant x \leqslant b$, and $p(x) > 0$ and $\sigma(x) > 0$ for $a \leqslant x \leqslant b$, is said to be in **self-adjoint form** or **Sturm-Liouville form**. We show below that this form is fairly general.

Theorem 4.1. Any homogeneous second-order linear ODE

$$a_0(x)\phi'' + a_1(x)\phi' + [a_2(x) + \lambda]\phi = 0, \quad a < x < b,$$

where the coefficient functions $a_0(x)$, $a_1(x)$, $a_2(x)$ are continuous and $a_0(x) > 0$ on the interval $a \leqslant x \leqslant b$, can be put into self-adjoint form.

Proof. Multiply the equation by the integrating factor $\mu(x) = p(x)/a_0(x)$ to get

$$p(x)\phi'' + \frac{p(x)a_1(x)}{a_0(x)}\phi' + \frac{p(x)a_2(x)}{a_0(x)}\phi + \frac{\lambda p(x)}{a_0(x)}\phi = 0,$$

Partial Differential Equations: Theory and Completely Solved Problems, First Edition. By T. Hillen, I. E. Leonard, H. van Roessel

and choose $p(x)$ so that

$$p'(x) = p(x)\,\frac{a_1(x)}{a_0(x)}.$$

We can solve this first-order linear ODE for $p(x)$ to get

$$p(x) = \exp\left\{\int \frac{a_1(x)}{a_0(x)}\,dx\right\},$$

and if we let

$$q(x) = \frac{p(x)a_2(x)}{a_0(x)} \qquad \text{and} \qquad \sigma(x) = \frac{p(x)}{a_0(x)},$$

the equation is in the self-adjoint form (4.1).

□

Definition 4.2. A **regular Sturm-Liouville problem** denotes the problem of finding an eigenfunction-eigenvalue pair (ϕ, λ) which solves the problem

$$\begin{aligned} &(p(x)\phi')' + [q(x) + \lambda\sigma(x)]\phi = 0, \quad a < x < b, \\ &\alpha_1\phi(a) + \beta_1\phi'(a) = 0, \\ &\alpha_2\phi(b) + \beta_2\phi'(b) = 0, \end{aligned} \tag{4.2}$$

where

(i) $p(x)$, $p'(x)$, $q(x)$, and $\sigma(x)$ are real valued and continuous for $a \leqslant x \leqslant b$;

(ii) $p(x) > 0$ and $\sigma(x) > 0$ for $a \leqslant x \leqslant b$; and

(iii) α_1, α_2, β_1, β_2 are real valued, $\alpha_1^2 + \beta_1^2 \neq 0$ and $\alpha_2^2 + \beta_2^2 \neq 0$.

Note: The function $\phi(x) \equiv 0$ is always a solution, the **trivial solution**. We want to determine the values of the parameter λ, if any, for which **nontrivial solutions** exist. The parameter λ is called an **eigenvalue**, the nontrivial solution ϕ is called a corresponding **eigenfunction**, and the set of all eigenvalues for the problem is called the **spectrum**. Such problems are generally known as **eigenvalue problems**, and the pair (ϕ, λ) that satisfies the ODE and the boundary conditions is called an **eigenpair**.

Definition 4.3. A Sturm-Liouville problem is said to be **singular** if at least one of the conditions (i), (ii), or (iii) in Definition 4.2 fails, or if the interval is infinite. In the case where the interval is infinite, or one or both of the functions $p(x)$ and $\sigma(x)$ approach 0 or ∞ at an endpoint of the interval, one or more of the boundary conditions are usually replaced by **boundedness conditions** on ϕ.

Example 4.1. Consider the boundary value problem

$$\begin{aligned} &\phi'' + \left(\frac{1+\lambda x}{x}\right)\phi = 0, \\ &\phi(1) = 0, \\ &\phi(2) = 0, \end{aligned}$$

on the interval $[1,2]$. We will put the equation in Sturm-Liouville form and determine whether the problem is regular or singular.

An equation is in Sturm-Liouville form if it has the form

$$(p(x)\,\phi')' + q(x)\,\phi + \lambda\sigma(x)\,\phi = 0$$

for $a < x < b$. We can rewrite the boundary value problem above in the form

$$\begin{aligned} &(1\cdot\phi')' + \frac{1}{x}\,\phi + \lambda\,\phi = 0, \quad 1 < x < 2, \\ &\phi(1) = 0, \\ &\phi(2) = 0. \end{aligned}$$

Here $p(x) = 1$, $p'(x) = 0$, $q(x) = 1/x$, and $\sigma(x) = 1$ are all continuous on the interval $[1,2]$, with $p(x) > 0$ and $\sigma(x) > 0$ for all $x \in [1,2]$. The boundary conditions are of the form

$$\begin{aligned} \alpha_1\phi(1) + \beta_1\phi'(1) = 0, \\ \alpha_2\phi(2) + \beta_2\phi'(2) = 0, \end{aligned}$$

where $\alpha_1 = \alpha_2 = 1$ and $\beta_1 = \beta_2 = 0$, so that

$$\alpha_1^2 + \beta_1^2 = 1 \neq 0 \qquad \text{and} \qquad \alpha_2^2 + \beta_2^2 = 1 \neq 0.$$

Thus, they are Sturm-Liouville-type boundary conditions, and therefore, this is a regular Sturm-Liouville problem on the interval $[1,2]$. ∎

Example 4.2. Consider the following boundary value problem, which we have solved several times before:

$$\begin{aligned} &\phi'' + \lambda\phi = 0, \quad 0 < x < \ell, \\ &\phi(0) = 0, \\ &\phi(\ell) = 0. \end{aligned}$$

This is a regular Sturm-Liouville problem with

$$p(x) \equiv 1, \qquad \sigma(x) \equiv 1, \qquad \text{and} \qquad q(x) \equiv 0$$

and

$$\alpha_1 = \alpha_2 = 1 \qquad \text{and} \qquad \beta_1 = \beta_2 = 0.$$

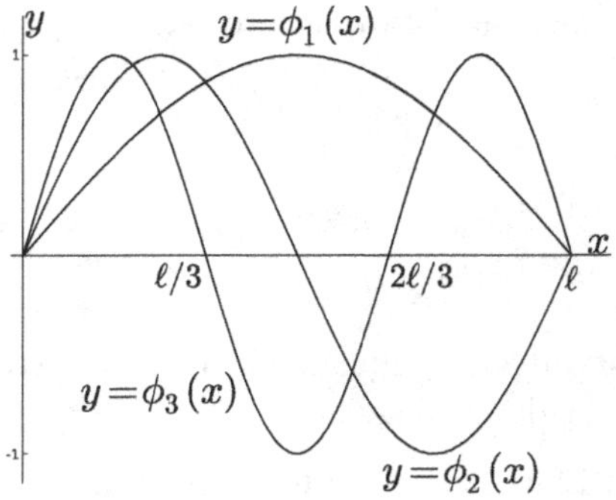

Figure 4.1 First three eigenfunctions, $\phi_1(x)$, $\phi_2(x)$, $\phi_3(x)$, of Example 4.2.

Nontrivial solutions exist only for the eigenvalues

$$\lambda_n = \frac{n^2\pi^2}{\ell^2},$$

and the corresponding eigenfunctions are

$$\phi_n(x) = \sin\frac{n\pi x}{\ell}$$

for $n = 1, 2, 3, \ldots$. See Figure 4.1 for the first three eigenfunctions. Clearly, $\lambda_n \geqslant 0$ for all $n \geqslant 1$, and

$$\lim_{n\to\infty} \lambda_n = \lim_{n\to\infty} \frac{n^2\pi^2}{\ell^2} = +\infty.$$

∎

The most common singular Sturm-Liouville problem encountered is that where $p(x) > 0$ for $a < x < b$, but either $p(a) = 0$ or $p(b) = 0$.

Example 4.3. (*Legendre's Equation*)
Consider the boundary value problem for Legendre's equation,

$$\begin{aligned}
&((1-x^2)\phi')' + \lambda\phi = 0, \quad -1 < x < 1,\\
&\alpha_1\phi(-1) + \beta_1\phi'(-1) = 0,\\
&\alpha_2\phi(1) + \beta_2\phi'(1) = 0.
\end{aligned}$$

Here $p(x) = 1 - x^2$, $p'(x) = -2x$, $q(x) = 0$, and $\sigma(x) = 1$ are continuous on the closed interval $[-1, 1]$; however, $p(1) = p(-1) = 0$. Together with the condition that a solution ϕ be *bounded* on the interval $(-1, 1)$, this is an example of a singular Sturm-Liouville problem. We show in Chapter 7 that the eigenvalues are $\lambda_n = n(n+1)$, where n is a nonnegative integer, and the corresponding eigenfunctions are real polynomials, $P_n(x)$, called the **Legendre polynomials**. ∎

Example 4.4. (*Bessel's Equation*)
For fixed n, Bessel's equation on the interval $a < r < b$,

$$
\begin{aligned}
(r\phi')' + \left(\lambda r - \frac{n^2}{r}\right)\phi &= 0, \\
\phi(a) &= 0, \\
\phi(b) &= 0,
\end{aligned}
$$

is a regular Sturm-Liouville problem with $p(r) = r$, $q(r) = -n^2/r$, and $\sigma(r) = r$, provided that $0 < a < b$. When $a = 0$, $p(r)$ is zero at $r = 0$, and we have a singular Sturm-Liouville problem, which we discuss in Chapter 6. ∎

4.2 PROPERTIES OF STURM-LIOUVILLE PROBLEMS

The eigenvalues and eigenfunctions of a regular Sturm-Liouville problem exhibit many of the properties of Example 4.2. We give some of these properties here, and we state the first property without proof.

Theorem 4.4. The spectrum of a regular Sturm-Liouville problem is a countably infinite set with no limit points, that is, an infinite discrete set.

Theorem 4.5. If λ_m and λ_n are distinct eigenvalues of a regular Sturm-Liouville problem, that is, $\lambda_m \neq \lambda_n$, the corresponding eigenfunctions ϕ_m and ϕ_n are orthogonal relative to the inner product

$$\langle f, g \rangle = \int_a^b f(x)g(x)\sigma(x)\,dx.$$

Proof. Since ϕ_m and ϕ_n are solutions to the Sturm-Liouville problem, they satisfy the boundary conditions

$$
\begin{aligned}
\alpha_1\phi_m(a) + \beta_1\phi_m'(a) &= 0, \\
\alpha_2\phi_m(b) + \beta_2\phi_m'(b) &= 0
\end{aligned}
$$

and

$$
\begin{aligned}
\alpha_1\phi_n(a) + \beta_1\phi_n'(a) &= 0, \\
\alpha_2\phi_n(b) + \beta_2\phi_n'(b) &= 0.
\end{aligned}
$$

The boundary condition at the left-hand boundary, $x = a$, can be written in matrix form as

$$\begin{pmatrix} \phi_m(a) & \phi_m'(a) \\ \phi_n(a) & \phi_n'(a) \end{pmatrix} \begin{pmatrix} \alpha_1 \\ \beta_1 \end{pmatrix} = \begin{pmatrix} 0 \\ 0 \end{pmatrix},$$

and since $\alpha_1^2 + \beta_1^2 \neq 0$, the determinant of the coefficient matrix must be zero; that is, the coefficient matrix must be singular. Thus,

$$\begin{vmatrix} \phi_m(a) & \phi_m'(a) \\ \phi_n(a) & \phi_n'(a) \end{vmatrix} = 0,$$

and therefore,

$$\phi_m(a)\phi_n'(a) - \phi_m'(a)\phi_n(a) = 0. \tag{4.3}$$

A similar argument at the right-hand boundary, $x = b$, yields

$$\phi_m(b)\phi_n'(b) - \phi_m'(b)\phi_n(b) = 0. \tag{4.4}$$

Since ϕ_m and ϕ_n also satisfy the ODE, we have

$$(p(x)\phi_m')' + [q(x) + \lambda_m\sigma(x)]\phi_m = 0$$

and

$$(p(x)\phi_n')' + [q(x) + \lambda_n\sigma(x)]\phi_n = 0.$$

Multiplying the first equation by ϕ_n and the second equation by ϕ_m and subtracting, we have

$$\phi_n(p\phi_m')' - \phi_m(p\phi_n')' + (\lambda_m - \lambda_n)\sigma\,\phi_m\phi_n = 0.$$

Integrating over the interval $[a, b]$,

$$\int_a^b [\phi_n(p\phi_m')' - \phi_m(p\phi_n')']\,dx = (\lambda_n - \lambda_m)\int_a^b \phi_m\phi_n\,\sigma\,dx,$$

and integrating by parts yields

$$\begin{aligned}
(\lambda_n - \lambda_m)\,\langle\phi_m, \phi_n\rangle &= \int_a^b [\phi_n(p\phi_m')' - \phi_m(p\phi_n')']\,dx \\
&= p\phi_n\phi_m'\Big|_a^b - \int_a^b p\phi_m'\phi_n'\,dx - p\phi_m\phi_n'\Big|_a^b + \int_a^b p\phi_m'\phi_n'\,dx \\
&= p(\phi_n\phi_m' - \phi_m\phi_n')\Big|_a^b - \int_a^b p(\phi_m'\phi_n' - \phi_m'\phi_n')\,dx \\
&= p(b)[\phi_n(b)\phi_m'(b) - \phi_m(b)\phi_n'(b)] \\
&\qquad - p(a)[\phi_n(a)\phi_m'(a) - \phi_m(a)\phi_n'(a)] \\
&= 0,
\end{aligned}$$

where we used (4.3) and (4.4). Since the eigenvalues are distinct, $\lambda_n \neq \lambda_m$, and we must have $\langle\phi_m, \phi_n\rangle = 0$. Therefore, the eigenfunctions corresponding to distinct eigenvalues are orthogonal with respect to the weight function $\sigma(x)$.

□

Using the Theorem 4.5, we can show the following:

Theorem 4.6. If λ is an eigenvalue of a regular Sturm-Liouville problem:

(a) λ is real, and

(b) if ϕ and ψ are eigenfunctions corresponding to λ,

$$\psi(x) = k\phi(x), \quad a \leqslant x \leqslant b,$$

where k is a nonzero constant, and each eigenfunction can be made real-valued by multiplying it by an appropriate nonzero constant.

Proof.

(a) Suppose that λ is an eigenvalue of the regular Sturm-Liouville problem

$$\begin{aligned} &(p(x)\phi')' + [q(x) + \lambda\sigma(x)]\phi = 0, \quad a < x < b, \qquad (4.5)\\ &\alpha_1\phi(a) + \beta_1\phi'(a) = 0,\\ &\alpha_2\phi(b) + \beta_2\phi'(b) = 0, \end{aligned}$$

with corresponding eigenfunction $\phi(x)$. Taking the complex conjugate of equation (4.5), since the coefficient functions are all real-valued, we have

$$\begin{aligned} &\left(p(x)\overline{\phi}\,'\right)' + [q(x) + \overline{\lambda}\sigma(x)]\overline{\phi} = 0, \quad a < x < b, \qquad (4.6)\\ &\alpha_1\overline{\phi}(a) + \beta_1\overline{\phi}\,'(a) = 0,\\ &\alpha_2\overline{\phi}(b) + \beta_2\overline{\phi}\,'(b) = 0, \end{aligned}$$

and therefore $\overline{\lambda}$ is an eigenvalue with corresponding eigenfunction $\overline{\phi}$. If λ is complex-valued, $\overline{\lambda} \neq \lambda$, and from Theorem 4.5, the eigenfunctions ϕ and $\overline{\phi}$ are orthogonal on the interval $a \leqslant x \leqslant b$ with respect to the weight function $\sigma(x)$. Multiplying Equation (4.5) by $\overline{\phi}$ and Equation (4.6) by ϕ, subtracting, integrating over the interval $[a, b]$, and integrating by parts yields

$$\begin{aligned} (\overline{\lambda} - \lambda)\int_a^b \overline{\phi}\,\phi\,\sigma\,dx &= \int_a^b (p\,\phi')'\,\overline{\phi}\,dx - \int_a^b \left(p\,\overline{\phi}\,'\right)'\,\phi\,dx\\ &= p\,\overline{\phi}\,\phi'\Big|_a^b - \int_a^b p\,\phi'\,\overline{\phi}\,'\,dx - p\,\phi\,\overline{\phi}\,'\Big|_a^b + \int_a^b p\,\phi'\,\overline{\phi}\,'\,dx\\ &= p\,\overline{\phi}\,\phi'\Big|_a^b - p\,\phi\,\overline{\phi}\,'\Big|_a^b, \end{aligned}$$

that is,

$$\begin{aligned} (\overline{\lambda} - \lambda)\int_a^b |\phi|^2\,\sigma\,dx &= p(b)\,\overline{\phi}(b)\,\phi'(b) - p(a)\,\overline{\phi}(a)\,\phi'(a)\\ &\qquad - p(b)\,\phi(b)\,\overline{\phi}\,'(b) + p(a)\,\phi(a)\,\overline{\phi}\,'(a),\\ &= p(b)\left[\overline{\phi}(b)\,\phi'(b) - \phi(b)\,\overline{\phi}\,'(b)\right]\\ &\qquad - p(a)\left[\overline{\phi}(a)\,\phi'(a) - \phi(a)\,\overline{\phi}\,'(a)\right], \end{aligned}$$

and since $\overline{\phi}(x)$ and $\phi(x)$ satisfy the same boundary conditions,

$$(\overline{\lambda} - \lambda) \int_a^b |\phi(x)|^2 \, \sigma(x) \, dx = 0.$$

However, the eigenfunctions are nontrivial, so that

$$\int_a^b |\phi(x)|^2 \, \sigma(x) \, dx \neq 0,$$

and we must have $\overline{\lambda} - \lambda = 0$; thus, the eigenvalue λ is real.

(b) Let $\phi_1(x)$ and $\phi_2(x)$ be eigenfunctions corresponding to the same eigenvalue λ, and define

$$\phi(x) = \phi_2'(a)\phi_1(x) - \phi_1'(a)\phi_2(x);$$

then from the superposition principle, $\phi(x)$ satisfies equation (4.5) also. Since

$$\phi'(x) = \phi_2'(a)\phi_1'(x) - \phi_1'(a)\phi_2'(x)$$

for $a \leqslant x \leqslant b$, clearly, $\phi'(a) = 0$. Also, since $\phi_1(x)$ and $\phi_2(x)$ satisfy the same boundary condition at $x = a$,

$$\begin{aligned} \alpha_1\phi_1(a) + \beta_1\phi_1'(a) &= 0, \\ \alpha_1\phi_2(a) + \beta_1\phi_2'(a) &= 0, \end{aligned}$$

and since α_1 and β_1 are not both zero, the determinant of the coefficient matrix must be zero; that is,

$$\phi_2'(a)\phi_1(a) - \phi_1'(a)\phi_2(a) = 0,$$

and thus, $\phi(a) = 0$ also. Therefore, $\phi(x)$ satisfies the initial value problem

$$\begin{aligned} &(p(x)u')' + [q(x) + \lambda\sigma(x)]u = 0, \quad a < x < b, \\ &u(a) = 0, \\ &u'(a) = 0. \end{aligned} \tag{4.7}$$

Since every point x in the interval $[a, b]$ is an ordinary point of the initial value problem (4.7), it has a unique solution, so that $\phi(x) \equiv 0$ on $[a, b]$; that is,

$$\phi_2'(a)\phi_1(x) - \phi_1'(a)\phi_2(x) = 0$$

for $a \leqslant x \leqslant b$.

Now, if $\phi_1'(a) \neq 0$ and $\phi_2'(a) \neq 0$,

$$\phi_2(x) = \frac{\phi_2'(a)}{\phi_1'(a)} \, \phi_1(x) = k \, \phi_1(x)$$

for $a \leqslant x \leqslant b$. On the other hand, if $\phi_1'(a) = 0$, then since

$$\phi_2'(a)\phi_1(x) - \phi_1'(a)\phi_2(x) = 0$$

for $a \leqslant x \leqslant b$, and since the eigenfunction $\phi_1(x)$ is not identically zero, we must have $\phi_2'(a) = 0$ also, and from the uniqueness of the solutions to equation (4.7), we must have both $\phi_1(a) \neq 0$ and $\phi_2(a) \neq 0$. Now we apply the same reasoning to the function

$$\psi(x) = \phi_2(a)\phi_1(x) - \phi_1(a)\phi_2(x), \quad a \leqslant x \leqslant b,$$

to show that

$$\phi_2(x) = \frac{\phi_2(a)}{\phi_1(a)}\,\phi_1(x) = k_1\,\phi_1(x)$$

for $a \leqslant x \leqslant b$.

Finally, let (ϕ, λ) be an eigenpair of the regular Sturm-Liouville problem (4.5). Since λ is real, if we write

$$\phi(x) = u(x) + iv(x),$$

where u and v are real-valued functions, and separate real and imaginary parts in (4.5), u and v are both solutions to the same regular Sturm-Liouville problem (4.5). From the above, there is a constant γ such that

$$v(x) = \gamma u(x)$$

for all $a \leqslant x \leqslant b$. Note that γ is real, since u and v are real valued; therefore,

$$\phi(x) = u(x) + iv(x) = (1 + i\gamma)u(x),$$

so that ϕ is a nonzero constant times a real-valued eigenfunction. Thus,

$$u(x) = \frac{1}{1 + i\gamma}\,\phi(x)$$

for all $a \leqslant x \leqslant b$. □

The proof above is essentially that given in Brown and Churchill [11].

Example 4.5. (*Cauchy-Euler Equation*)
The boundary value problem

$$\begin{aligned}
&(x\phi')' + \frac{\lambda}{x}\phi = 0, \quad 1 < x < \ell, \\
&\phi(1) = 0, \\
&\phi(\ell) = 0,
\end{aligned}$$

is a regular Sturm-Liouville problem with

$$p(x) = x, \qquad \sigma(x) = 1/x, \qquad \text{and} \qquad q(x) = 0$$

and with

$$\alpha_1 = \alpha_2 = 1 \qquad \text{and} \qquad \beta_1 = \beta_2 = 0.$$

Making the transformation

$$x = e^s, \qquad u(s) = \phi(x) = \phi(e^s),$$

from the chain rule we have

$$u'(s) = \phi'(e^s)\, e^s = x\phi'(x)$$

and

$$u''(s) = \left(e^s\phi'(e^s)\right)' = x\left(x\phi'(x)\right)';$$

hence,

$$u'' + \lambda u = x\left(x\phi'\right)' + \lambda\phi = 0.$$

Therefore, $u = u(s)$ satisfies the boundary value problem

$$\begin{aligned} &u'' + \lambda u = 0, \quad 0 < s < \log \ell, \\ &u(0) = 0, \\ &u(\log \ell) = 0. \end{aligned}$$

As we have seen earlier, the equation for u has nontrivial solutions only for $\lambda > 0$, say, $\lambda = \mu^2$, where $\mu \neq 0$, and the general solution is

$$u(s) = A\cos\mu s + A\sin\mu s$$

for $0 < s < \log \ell$. Therefore,

$$\phi(x) = A\cos(\mu\log x) + B\sin(\mu\log x)$$

for $1 < x < \ell$.
Applying the boundary conditions yields

$$\lambda_n = \frac{n^2\pi^2}{(\log \ell)^2}$$

and

$$\phi_n(x) = \sin\left(n\pi\frac{\log x}{\log \ell}\right)$$

for $n = 1, 2, \ldots$. For nonnegative integers $m \neq n$, we have

$$\begin{aligned}\langle \phi_m, \phi_n \rangle &= \int_1^\ell \sin\left(\frac{m\pi \log x}{\log \ell}\right) \sin\left(\frac{n\pi \log x}{\log \ell}\right) \cdot \frac{1}{x}\, dx \\ &= \frac{\log \ell}{\pi} \int_0^\pi \sin mt \sin nt \, dt,\end{aligned}$$

after making the substitution $t = \pi \log x / \log \ell$, and we have

$$\begin{aligned}\langle \phi_m, \phi_n \rangle &= \frac{\log \ell}{2\pi} \int_0^\pi [\cos(n-m)t - \cos(n+m)t]\, dt \\ &= \frac{\log \ell}{2\pi} \left[\frac{\sin(n-m)t}{n-m} - \frac{\sin(n+m)t}{n+m}\right]_0^\pi = 0.\end{aligned}$$

Thus, as predicted by Theorem 4.5, the eigenfunctions corresponding to distinct eigenvalues are orthogonal with respect to the inner product

$$\langle f, g \rangle = \int_1^\ell f(x)\, g(x)\, \frac{dx}{x}\,.$$

∎

Theorem 4.7. Given the regular Sturm-Liouville problem,

$$\begin{aligned}&(p(x)\phi')' + [q(x) + \lambda\sigma(x)]\phi = 0, \quad a < x < b, \\ &\alpha_1\phi(a) + \beta_1\phi'(a) = 0, \\ &\alpha_2\phi(b) + \beta_2\phi'(b) = 0,\end{aligned} \tag{4.8}$$

with eigenvalues λ_n and corresponding eigenfunctions ϕ_n.

(a) The regular Sturm-Liouville problem has an infinite spectrum

$$S = \{\lambda_1, \lambda_2, \ldots, \lambda_n, \ldots\}$$

and $\lim\limits_{n\to\infty} \lambda_n = +\infty$.

(b) If $\alpha_1\beta_1 \leqslant 0$ and $\alpha_2\beta_2 \geqslant 0$, the spectrum is bounded below and the eigenvalues may be ordered as

$$\lambda_1 < \lambda_2 < \cdots < \lambda_n < \cdots.$$

Moreover, if $q(x) \leqslant 0$ for $a \leqslant x \leqslant b$, then $\lambda_n \geqslant 0$ for all $n \geqslant 1$.

(c) If the eigenvalues are ordered as $\lambda_1 < \lambda_2 < \cdots < \lambda_n < \cdots$, the eigenfunction corresponding to λ_n has exactly $(n-1)$ zeros in the interval $a < x < b$.

Proof. We give the proof of the second statement in part (b) only; the proofs of parts (a) and (c) are beyond the scope of this book.

The inner product we use is one with weight function $\sigma(x)$:

$$\langle f, g \rangle = \int_a^b f(x)\, g(x)\, \sigma(x)\, dx.$$

If λ_n is an eigenvalue with corresponding eigenfunction ϕ_n,

$$(p(x)\phi_n')' + [q(x) + \lambda_n \sigma(x)]\phi_n = 0.$$

Multiplying this equation by ϕ_n and integrating over the interval $[a, b]$,

$$\int_a^b \phi_n(x)(p(x)\phi_n')'\, dx + \int_a^b \phi_n(x)^2 [q(x) + \lambda_n \sigma(x)]\, dx = 0;$$

that is,

$$\lambda_n \int_a^b \phi_n(x)^2 \sigma(x)\, dx + \int_a^b \phi_n(x)^2 q(x)\, dx + \int_a^b \phi_n(x)(p(x)\phi_n')'\, dx = 0.$$

Integrating the last term by parts, we have

$$\int_a^b \phi_n(x)(p(x)\phi_n')'\, dx = p(x)\phi_n'(x)\phi_n(x)\Big|_a^b - \int_a^b p(x)\phi_n'(x)^2\, dx,$$

so that

$$\lambda_n = \frac{-p(x)\phi_n'(x)\phi_n(x)\Big|_a^b + \displaystyle\int_a^b \left[p(x)\phi_n'(x)^2 - q(x)\phi_n^2(x)\right]\, dx}{\displaystyle\int_a^b \phi_n^2(x)\sigma(x)\, dx} \tag{4.9}$$

(this important formula will be introduced later as the **Rayleigh quotient**). Since $p(a) > 0$, from the boundary condition at the end $x = a$,

$$p(a)\phi_n'(a)\phi_n(a) = \begin{cases} -p(a)\dfrac{\beta_1}{\alpha_1}\, \phi_n'(a)^2, & \text{if } \alpha_1 \neq 0, \\[2ex] -p(a)\dfrac{\alpha_1}{\beta_1}\, \phi_n(a)^2, & \text{if } \beta_1 \neq 0, \end{cases}$$

and since $\alpha_1 \beta_1 \leqslant 0$,

$$p(a)\phi_n'(a)\phi_n(a) \geqslant 0.$$

Since $p(b) > 0$, from the boundary condition at the end $x = b$,

$$p(b)\phi_n'(b)\phi_n(b) = \begin{cases} -p(b)\dfrac{\beta_2}{\alpha_2}\, \phi_n'(b)^2, & \text{if } \alpha_2 \neq 0, \\[2ex] -p(b)\dfrac{\alpha_2}{\beta_2}\, \phi_n(b)^2, & \text{if } \beta_2 \neq 0, \end{cases}$$

and since $\alpha_2\beta_2 \geqslant 0$,

$$p(b)\phi_n'(b)\phi_n(b) \leqslant 0.$$

Therefore,

$$-p(x)\phi_n'(x)\phi_n(x)\Big|_a^b = p(a)\phi_n'(a)\phi_n(a) - p(b)\phi_n'(b)\phi_n(b) \geqslant 0,$$

and from equation (4.9),

$$\lambda_n \geqslant \frac{-\displaystyle\int_a^b \phi_n(x)^2 q(x)\,dx}{\displaystyle\int_a^b \phi_n(x)^2\sigma(x)\,dx} \tag{4.10}$$

for all $n \geqslant 1$. Finally, if $q(x) \leqslant 0$ for $a \leqslant x \leqslant b$, from (4.10) we have $\lambda_n \geqslant 0$ for $n \geqslant 1$.

□

4.3 EIGENFUNCTION EXPANSIONS

Again, consider the regular Sturm-Liouville problem

$$\begin{aligned} &(p(x)\phi')' + [q(x) + \lambda\sigma(x)]\phi = 0, \quad a < x < b, \\ &\alpha_1\phi(a) + \beta_1\phi'(a) = 0, \\ &\alpha_2\phi(b) + \beta_2\phi'(b) = 0, \end{aligned} \tag{4.11}$$

where we denote the eigenvalues by λ_n, the corresponding eigenfunctions by ϕ_n, and where we define the inner product as

$$\langle f, g\rangle = \int_a^b f(x)g(x)\sigma(x)\,dx.$$

We know from Section 4.2 that eigenfunctions corresponding to distinct eigenvalues are orthogonal, so that $\langle \phi_m, \phi_n\rangle = 0$ for $m \neq n$.

Can we represent a function f defined on $[a, b]$ by a series of eigenfunctions? That is, does there exist a series expansion of the form

$$f(x) = \sum_{n=1}^{\infty} c_n\phi_n(x)$$

valid for $a \leqslant x \leqslant b$? Assuming, for the moment, that this is the case, the coefficients of the series may be calculated as follows:

$$\langle f, \phi_m\rangle = \left\langle \sum_{n=1}^{\infty} c_n\phi_n, \phi_m \right\rangle = c_m \langle \phi_m, \phi_m\rangle$$

and therefore,

$$c_m = \frac{\langle f, \phi_m \rangle}{\|\phi_m\|^2}$$

for $m \geqslant 1$.

Definition 4.8. Let f be piecewise continuous on the interval $[a, b]$. The eigenfunction expansion

$$f(x) \sim \sum_{n=1}^{\infty} c_n \phi_n(x)$$

with coefficients

$$c_n = \frac{\langle f, \phi_n \rangle}{\|\phi_n\|^2},$$

where the inner product has weight function $\sigma(x)$, is called a **generalized Fourier series** of f.

If the function f is piecewise continuous on the interval $[a, b]$, the **generalized Fourier coefficients** of f with respect to the eigenfunctions $\{\phi_n\}_{n \geqslant 1}$ exist, and we have a convergence theorem analogous to Dirichlet's theorem:

Theorem 4.9. (*Dirichlet's Theorem*)
If f is piecewise smooth on $[a, b]$, the generalized Fourier series,

$$f(x) \sim \sum_{n=1}^{\infty} c_n \phi_n(x),$$

where

$$c_n = \frac{\langle f, \phi_n \rangle}{\|\phi_n\|^2} = \frac{1}{\|\phi_n\|^2} \int_a^b f(x) \phi_n(x) \sigma(x)\, dx$$

for $n \geqslant 1$, converges pointwise to $[f(x^+) + f(x^-)]/2$ for each $x \in (a, b)$.

Note: This theorem says that if f is continuous at x, that is,

$$f(x^+) = f(x^-) = f(x),$$

the eigenfunction expansion converges to $f(x)$, and if f has a jump discontinuity at x, that is,

$$f(x^+) \neq f(x^-),$$

the eigenfunction expansion converges to a point midway between the limiting values. A proof may be found in Brown and Churchill [11].

Example 4.6. Consider the regular Sturm-Liouville problem

$$\begin{aligned}
&\phi'' + \lambda \phi = 0, \quad 0 < x < 1,\\
&\phi(0) = 0,\\
&2\phi(1) - \phi'(1) = 0.
\end{aligned}$$

Here

$$p(x) = 1, \qquad q(x) = 0, \qquad \text{and} \qquad \sigma(x) = 1$$

and

$$\alpha_1 = 1, \qquad \beta_1 = 0, \qquad \alpha_2 = 2, \qquad \text{and} \qquad \beta_2 = -1.$$

Note that, in this case,

$$\alpha_1\beta_1 = 0, \qquad \text{but} \qquad \alpha_2\beta_2 = -2 < 0.$$

Hence, Theorem 4.7 does **not** guarantee nonnegative eigenvalues. However, the theorem does imply that there are at most a finite number of negative eigenvalues.

- Case (i): $\lambda < 0$. Let $\lambda = -\mu^2$, where $\mu \neq 0$, then the differential equation becomes

  $$\phi'' - \mu^2\phi = 0,$$

  with general solution

  $$\phi(x) = a\cosh\mu x + b\sinh\mu x.$$

  At the left-hand boundary, we have $\phi(0) = 0$, and thus $a = 0$, and the solution is now

  $$\phi(x) = b\sinh\mu x.$$

  At the right-hand boundary, we have $2\phi(1) - \phi'(1) = 0$, so that

  $$2b\sinh\mu - b\mu\cosh\mu = 0,$$

  and hence

  $$\tanh\mu = \frac{\mu}{2}.$$

  As can be seen from Figure 4.2, this transcendental equation has solutions only at $\mu = 0$ and $\mu = \pm\mu_0$, where μ_0 is the positive root of the equation $\tanh\mu = \mu/2$; therefore, only μ_0 yields a nontrivial solution:

  $$\phi_0(x) = \sinh\mu_0 x$$

  [the root $-\mu_0$ yields the solution $\sinh(-\mu_0 x) = -\sinh\mu_0 x$].

- Case(ii): $\lambda = 0$. In this case the differential equation becomes

  $$\phi''(x) = 0,$$

  with general solution

  $$\phi(x) = ax + b.$$

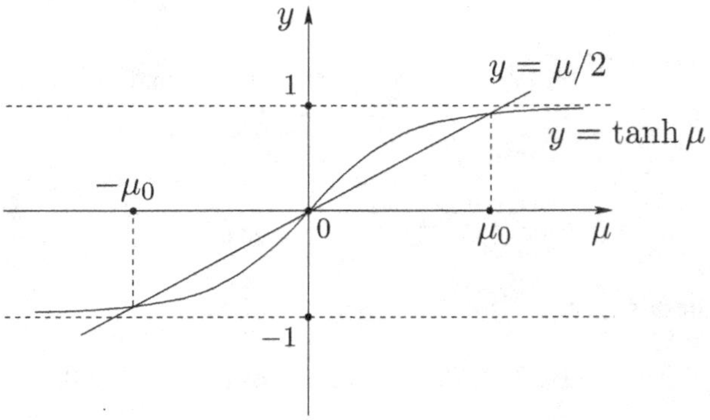

Figure 4.2 Intersection of $y = \tanh \mu$ and $y = \mu/2$.

At the left-hand boundary, we have $\phi(0) = 0$, and thus $b = 0$, and the solution is now

$$\phi(x) = ax.$$

At the right-hand boundary, we have $2\phi(1) - \phi'(1) = 0$, so that

$$2a - a = 0,$$

and hence $a = 0$. Thus, the solution is $\phi(x) = 0$ for all $0 < x < 1$, and there are no nontrivial solutions in this case.

- Case (iii): $\lambda > 0$. Let $\lambda = \mu^2$, where $\mu \neq 0$; then the differential equation becomes

$$\phi'' + \mu^2 \phi = 0,$$

with general solution

$$\phi(x) = a \cos \mu x + b \sin \mu x.$$

At the left-hand boundary, we have $\phi(0) = 0$, and thus $a = 0$, and the solution is now

$$\phi(x) = b \sin \mu x.$$

At the right-hand boundary, we have $2\phi(1) - \phi'(1) = 0$, so that

$$2b \sin \mu - b\mu \cos \mu = 0.$$

If $b = 0$, we get only the trivial solution; hence, to obtain nontrivial solutions, we must have

$$\tan \mu = \frac{\mu}{2}.$$

As can be seen in Figure 4.3, this transcendental equation has infinitely many solutions

$$\mu_1 < \mu_2 < \cdots < \mu_n < \cdots$$

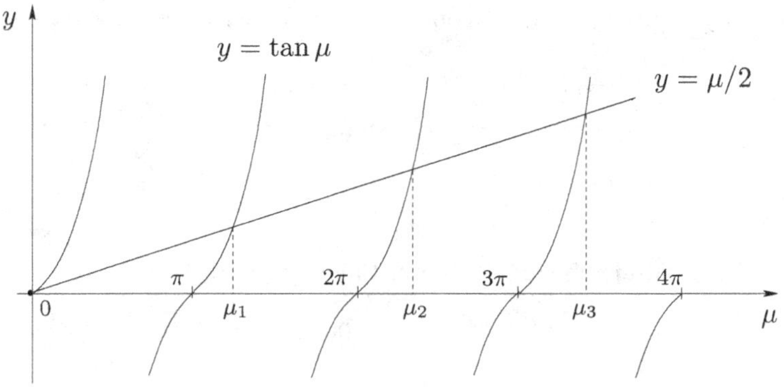

Figure 4.3 Intersection of $y = \tan \mu$ and $y = \mu/2$.

with corresponding nontrivial solutions

$$\phi_n(x) = \sin \mu_n x$$

for $n \geqslant 1$.

In summary, we have an infinite set of eigenvalues:

$$-\mu_0^2 < \mu_1^2 < \mu_2^2 < \cdots < \mu_n^2 < \cdots,$$

where

$$\tanh \mu_0 = \frac{\mu_0}{2} \qquad \text{and} \qquad \tan \mu_n = \frac{\mu_n}{2},$$

with corresponding eigenfunctions

$$\phi_0(x) = \sinh \mu_0 x \qquad \text{and} \qquad \phi_n(x) = \sin \mu_n x$$

for $n \geqslant 1$. The norms of the eigenfunctions are given by

$$\|\phi_0\|^2 = \langle \phi_0, \phi_0 \rangle = \int_0^1 \sinh^2 \mu_0 x \, dx = \frac{1}{2}\left(1 - \frac{\sinh 2\mu_0}{2\mu_0}\right),$$

$$\|\phi_n\|^2 = \langle \phi_n, \phi_n \rangle = \int_0^1 \sin^2 \mu_n x \, dx = \frac{1}{2}\left(1 - \frac{\sin 2\mu_n}{2\mu_n}\right)$$

for $n \geqslant 1$.

Now, given a piecewise continuous function f on $[0, 1]$, the eigenfunction expansion of f with respect to the eigenfunctions $\{\phi_n(x)\}_{n\geqslant 0}$ is given by

$$f(x) \sim \sum_{n=0}^{\infty} c_n \phi_n(x), \quad 0 < x < 1,$$

where $c_n = \langle f, \phi_n \rangle / \|\phi_n\|^2$ for $n \geqslant 1$.

Moreover, if f is piecewise smooth on $[0, 1]$, then from Theorem 4.9, we have

$$\frac{f(x^+) + f(x^-)}{2} = \sum_{n=0}^{\infty} c_n \phi_n(x)$$

for $0 < x < 1$.

∎

Example 4.7. Consider the regular Sturm-Liouville problem

$$\phi'' + \lambda^2 \phi = 0, \quad 0 < x < \pi,$$

$$\phi'(0) = 0,$$

$$\phi(\pi) = 0.$$

(a) Find the eigenvalues λ_n^2 and the corresponding eigenfunctions ϕ_n for this problem.

(b) Show directly, by integration, that eigenfunctions corresponding to distinct eigenvalues are orthogonal.

(c) Given the function $f(x) = \pi^2 - x^2/2, \quad 0 < x < \pi$, find the eigenfunction expansion of f.

(d) Show that

$$\frac{\pi^3}{32} = 1 - \frac{1}{3^3} + \frac{1}{5^3} - \frac{1}{7^3} + \frac{1}{9^3} - + \cdots .$$

Solution. Since the eigenvalues are real, we are assuming that they are given by $\lambda^2 \geqslant 0$.

(a) Case(i): $\lambda = 0$. The general solution to the equation $\phi'' + \lambda^2\phi = 0$ in this case is

$$\phi(x) = c_1\, x + c_2,$$

and differentiating gives us $\phi'(x) = c_1$. The condition $\phi'(0) = 0$ implies that $c_1 = 0$. Since the condition $\phi(\pi) = 0$ implies that $c_2 = 0$, there are no nontrivial solutions in this case.

Case(ii): $\lambda \neq 0$. The general solution to the equation $\phi'' + \lambda^2\phi = 0$ in this case is

$$\phi(x) = c_1 \cos \lambda x + c_2 \sin \lambda x,$$

and differentiating gives us

$$\phi'(x) = -c_1 \lambda \sin \lambda x + c_2 \lambda \cos \lambda x.$$

The condition $\phi'(0) = 0$ implies that $c_2\, \lambda = 0$; thus, $c_2 = 0$. The solution is then

$$\phi(x) = c_1 \cos \lambda x,$$

and the condition $\phi(\pi) = 0$ implies that $\cos \lambda\pi = 0$. Therefore, the eigenvalues are

$$\lambda_n^2 = \left(\frac{2n-1}{2}\right)^2,$$

and the corresponding eigenfunctions are

$$\phi_n(x) = \cos \frac{(2n-1)x}{2}$$

for $n \geqslant 1$.

(b) Let $\lambda_n = (2n-1)/2$ for $n = 1, 2, 3, \ldots$. For $m \neq n$ we have

$$\begin{aligned}
\int_0^\pi \phi_m(x)\phi_n(x)\,dx &= \int_0^\pi \cos \lambda_m x \,\cos \lambda_n x\,dx \\
&= \frac{1}{2}\int_0^\pi \{\cos(\lambda_m + \lambda_n)x + \cos(\lambda_m - \lambda_n)x\}\,dx \\
&= \frac{1}{2(\lambda_m + \lambda_n)} \sin(\lambda_m + \lambda_n)x \Big|_0^\pi \\
&\quad + \frac{1}{2(\lambda_m - \lambda_n)} \sin(\lambda_m - \lambda_n)x \Big|_0^\pi \\
&= \frac{1}{2(\lambda_m + \lambda_n)} \sin(\lambda_m + \lambda_n)\pi \\
&\quad + \frac{1}{2(\lambda_m - \lambda_n)} \sin(\lambda_m - \lambda_n)\pi \\
&= 0,
\end{aligned}$$

since $(\lambda_m + \lambda_n)\pi = (m+n-1)\pi$ and $(\lambda_m - \lambda_n)\pi = (m-n)\pi$. For $m = n$, we have

$$\begin{aligned}
\int_0^\pi \phi_n(x)\phi_n(x)\,dx &= \int_0^\pi \cos^2 \lambda_n x\,dx \\
&= \int_0^\pi \left(\frac{1}{2} + \frac{1}{2}\cos 2\lambda_n x\right) dx \\
&= \frac{1}{2}\int_0^\pi dx + \frac{1}{2}\int_0^\pi \cos 2\lambda_n x\,dx \\
&= \frac{\pi}{2} + \frac{1}{4\lambda_n} \sin 2\lambda_n x \Big|_0^\pi \\
&= \frac{\pi}{2}.
\end{aligned}$$

(c) If we write

$$f(x) = \frac{\pi^2 - x^2}{2} \sim \sum_{n=1}^{\infty} c_n \phi_n(x),$$

the coefficients c_n in the eigenfunction expansion can be found using the orthogonality of the eigenfunctions on $[0, \pi]$:

$$\begin{aligned} c_n &= \frac{2}{\pi} \int_0^{\pi} \left(\frac{\pi^2 - x^2}{2} \right) \cos \lambda_n x \, dx \\ &= \frac{2}{\pi} \frac{\sin \lambda_n \pi}{\lambda_n^3} = \frac{16}{\pi(2n-1)^3} \sin \frac{(2n-1)\pi}{2} \\ &= \frac{16(-1)^{n+1}}{\pi(2n-1)^3}, \end{aligned}$$

where the integral is evaluated by repeated integration by parts. Therefore, the eigenfunction expansion of f is given by

$$\frac{\pi^2 - x^2}{2} \sim \frac{16}{\pi} \sum_{n=1}^{\infty} \frac{(-1)^{n+1}}{(2n-1)^3} \cos \frac{(2n-1)x}{2}.$$

(d) In this particular problem, the eigenfunction expansion is actually the Fourier cosine series for f. Since the function f is piecewise smooth on the interval $[0, \pi]$ and since the even extension of f to $[-\pi, \pi]$ is continuous at $x = 0$, then, by Dirichlet's theorem, the series converges to $f(0) = \pi^2/2$ when $x = 0$; therefore,

$$\frac{\pi^3}{32} = 1 - \frac{1}{3^3} + \frac{1}{5^3} - \frac{1}{7^3} + \frac{1}{9^3} - + \cdots.$$

∎

4.4 RAYLEIGH QUOTIENT

Often, it is necessary to estimate the first (**leading**) eigenvalue of a Sturm-Liouville problem. If the leading eigenvalue is positive, the system modeled by this problem is stable, in the sense that small perturbations are damped, and the system converges to the equilibrium state or steady state. If the leading eigenvalue is negative, the system is unstable, and small perturbations are amplified. This instability can have adverse consequences for the system at hand, for example, collapsing bridges or buildings.

The Rayleigh quotient is a simple and elegant method for estimating the leading eigenvalue.

Theorem 4.10. If (ϕ_n, λ_n) is an eigenpair for the regular Sturm-Liouville problem

$$
\begin{aligned}
&(p(x)\phi')' + [q(x) + \lambda\sigma(x)]\phi = 0, \quad a < x < b,\\
&\alpha_1\phi(a) + \beta_1\phi'(a) = 0,\\
&\alpha_2\phi(b) + \beta_2\phi'(b) = 0;
\end{aligned}
$$

then λ_n can be calculated from the **Rayleigh quotient**:

$$
\lambda_n = \frac{-p(x)\phi_n(x)\phi_n'(x)\Big|_a^b + \displaystyle\int_a^b \left(p(x)\phi_n'(x)^2 - q(x)\phi_n(x)^2\right)\,dx}{\displaystyle\int_a^b \phi_n(x)^2\sigma(x)\,dx}. \tag{4.12}
$$

Proof. Multiplying the differential equation in the Sturm-Liouville problem by ϕ_n gives

$$
\phi_n\frac{d}{dx}(p\phi_n') + q\phi_n^2 + \lambda_n\sigma\phi_n^2 = 0.
$$

Integrating over the interval $[a, b]$ and then integrating by parts yields

$$
\begin{aligned}
0 &= \int_a^b \phi_n\frac{d}{dx}(p\phi_n')\,dx + \int_a^b q\phi_n^2\,dx + \int_a^b \lambda_n\sigma\phi_n^2\,dx\\
&= \phi_n(p\phi_n')\Big|_a^b - \int_a^b \phi_n'p\phi_n'\,dx + \int_a^b q\phi_n^2\,dx + \int_a^b \lambda_n\sigma\phi_n^2\,dx,
\end{aligned}
$$

and equation (4.12) follows after rearrangement. □

Corollary 4.11. Let (ϕ_n, λ_n) be an eigenpair for the regular Sturm-Liouville problem

$$
\begin{aligned}
&(p(x)\phi')' + [q(x) + \lambda\sigma(x)]\phi = 0, \quad a < x < b,\\
&\alpha_1\phi(a) + \beta_1\phi'(a) = 0,\\
&\alpha_2\phi(b) + \beta_2\phi'(b) = 0.
\end{aligned}
$$

If

$$
-p(x)\phi_n(x)\phi_n'(x)\Big|_a^b = -\left[p(b)\phi_n(b)\phi_n'(b) - p(a)\phi_n(a)\phi_n'(a)\right] \geq 0
$$

and $q(x) \leq 0$ for $a \leqslant x \leqslant b$, then $\lambda_n > 0$.

We can write down the Rayleigh quotient for **any** function for which it makes sense. For example, for any piecewise smooth function $u = u(x)$ defined on the interval

$[a,b]$ (it does not have to be an eigenfunction), we define

$$\mathcal{R}(u) = \frac{-p(x)u(x)u'(x)\Big|_a^b + \int_a^b \left(p(x)u'(x)^2 - q(x)u(x)^2\right)\,dx}{\int_a^b u(x)^2\sigma(x)\,dx}.$$

It can be shown that the leading eigenvalue is the smallest of all the Rayleigh quotients for functions that satisfy the correct boundary conditions.

Theorem 4.12. Given the regular Sturm-Liouville problem

$$\begin{aligned}&(p(x)\phi')' + [q(x) + \lambda\sigma(x)]\phi = 0, \quad a < x < b,\\ &\alpha_1\phi(a) + \beta_1\phi'(a) = 0,\\ &\alpha_2\phi(b) + \beta_2\phi'(b) = 0,\end{aligned}$$

with spectrum

$$\lambda_1 < \lambda_2 < \cdots < \lambda_n < \cdots,$$

let $\mathcal{D}(L)$ denote the set of all continuous functions u that satisfy the boundary conditions

$$\alpha_1 u(a) + \beta_1 u'(a) = 0 \qquad \text{and} \qquad \alpha_2 u(b) + \beta_2 u'(b) = 0.$$

The leading eigenvalue is

$$\lambda_1 = \min_{u\in\mathcal{D}(L)} \mathcal{R}(u).$$

A proof of this result may be found in Haberman [25].

By restricting the minimum to continuous functions u that satisfy the boundary conditions and are orthogonal to the leading eigenfunction ϕ_1, we get the next eigenvalue, λ_2, and so on.

Example 4.8. Find good upper and lower estimates for the leading eigenvalue of the regular Sturm-Liouville problem,

$$\begin{aligned}&2\phi''(x) + \lambda\phi(x) = 0, \quad 0 < x < \ell,\\ &\phi'(0) = 0,\\ &\phi'(\ell) = 0.\end{aligned}$$

Here we have

$$p(x) = 2, \qquad q(x) = 0, \qquad \text{and} \qquad \sigma(x) = 1,$$

and the boundary term in $\mathcal{R}$ is

$$-p(x)\phi(x)\phi'(x)\Big|_0^\ell = -2\phi(\ell)\phi'(\ell) + 2\phi(0)\phi'(0) = 0.$$

Hence, the Rayleigh quotient becomes

$$\mathcal{R}(\phi) = \frac{\displaystyle\int_0^\ell 2\phi'^2(x)\,dx}{\displaystyle\int_0^\ell \phi^2(x)\,dx} \geqslant 0,$$

and a lower estimate is

$$\lambda_1 \geqslant 0.$$

To find an upper estimate, we choose a test function that satisfies the boundary conditions (but not necessarily the equation); for example, $u(x) = 5$. For this function we find

$$\mathcal{R}(u) = \frac{\displaystyle\int_0^\ell 2\cdot 0\,dx}{\displaystyle\int_0^\ell 25\,dx} = 0;$$

hence, an upper estimate is

$$\lambda_1 \leqslant 0.$$

Combining these lower and upper estimates, we find that

$$\lambda_1 = 0.$$

In this particular case, we can find the exact solution to this Sturm-Liouville problem: If we let $\lambda/2 = \mu$, the Sturm-Liouville problem above becomes

$$\begin{aligned}
&\phi'' + \mu\phi = 0, \quad 0 < x < \ell,\\
&\phi'(0) = 0,\\
&\phi'(\ell) = 0,
\end{aligned}$$

which is one of our standard problems. The eigenvalues are λ_n^2, where

$$\lambda_n = 2\mu_n = 2\left(\frac{n\pi}{\ell}\right)^2$$

for $n \geqslant 0$; hence, the leading eigenvalue is $\lambda_0 = 0$. Notice that here the leading eigenvalue is called λ_0, just for convenience, since the index n starts at 0.

∎

Example 4.9. Find good upper and lower bounds for the leading eigenvalue of the regular Sturm-Liouville problem

$$\begin{aligned}
&\phi'' - x\phi + \lambda\phi = 0, \quad 0 < x < 1,\\
&\phi'(0) = 0,\\
&2\phi(1) + \phi'(1) = 0.
\end{aligned}$$

Here we have

$$p(x) = 1, \qquad q(x) = -x, \qquad \text{and} \qquad \sigma(x) = 1,$$

and the boundary term in $\mathcal{R}$ is

$$-p(x)\phi(x)\phi'(x)\Big|_0^1 = -\phi(1)\phi'(1) + \phi(0)\phi'(0) = 2\phi(1)^2 \geqslant 0.$$

Since $q(x) = -x \leq 0$ for $x \in [0,1]$, we have $\mathcal{R}(\phi) \geqslant 0$. Hence,

$$\lambda_1 \geqslant 0.$$

For an upper estimate, we choose a test function that satisfies the boundary conditions. Here it is not so easy to guess the form of the test function, so we try something simple, a quadatic function:

$$u(x) = Ax^2 + Bx + C.$$

We leave it to the reader to check that a test function which is linear and satisfies the boundary conditions would be identically zero.

Substituting the function u above into the boundary conditions, we get

$$B = 0 \qquad \text{and} \qquad 2A + 2A + 2C = 0;$$

hence, $B = 0$ and $C = -2A$. If we choose $A = 1$, then $u(x) = x^2 - 2$ is a test function that satisfies the boundary conditions. Now, we compute the Rayleigh quotient

$$\mathcal{R}(u) = \frac{-u(x)u'(x)\Big|_0^1 + \int_0^1 \left(u'(x)^2 + xu(x)^2\right)\,dx}{\int_0^1 u(x)^2\,dx}.$$

The first term is

$$-u(x)u'(x)\Big|_0^1 = 2,$$

and the integrals are

$$\int_0^1 \left(u'(x)^2 + xu(x)^2\right)\,dx = \int_0^1 4\left(x^2 + x(x^4 - 4x^2 + 4)\right)\,dx = \frac{5}{2}$$

and

$$\int_0^1 u(x)^2\,dx = \int_0^1 \left(x^4 - 4x^2 + 4\right)\,dx = \frac{43}{15}.$$

Therefore,

$$\mathcal{R}(u) = \frac{135}{86} \approx 1.57,$$

and thus

$$0 \leqslant \lambda_1 \leqslant 1.57.$$

∎

Example 4.10. Find the generalized Fourier series solution to the homogeneous Neumann problem for the wave equation. Use the Rayleigh quotient to show that $\lambda_1 > 0$.

$$\alpha(x)\frac{\partial^2 u}{\partial t^2} = \frac{\partial}{\partial x}\left(\tau(x)\frac{\partial u}{\partial x}\right) - \beta(x)u, \quad 0 < x < \ell, \quad t > 0,$$

$$\frac{\partial u}{\partial x}(0,t) = 0,$$

$$\frac{\partial u}{\partial x}(\ell,t) = 0,$$

$$u(x,0) = f(x),$$

$$\frac{\partial u}{\partial t}(x,0) = g(x),$$

where $\alpha(x) > 0$, $\tau(x) > 0$, and $\beta(x) > 0$ for $0 \leqslant x \leqslant \ell$.
Solution. Separation of variables with $u(x,t) = X(x) \cdot T(t)$ leads to the time problem

$$T''(t) + \lambda T(t) = 0, \quad t > 0$$

and the spatial problem

$$\begin{aligned}
&(\tau(x)X'(x))' - \beta(x)X(x) + \lambda\alpha(x)X(x) = 0, \quad 0 < x < \ell,\\
&X'(0) = 0,\\
&X'(\ell) = 0,
\end{aligned}$$

which is a regular Sturm-Liouville problem with

$$p(x) = \tau(x), \qquad q(x) = -\beta(x) < 0, \qquad \text{and} \qquad \sigma(x) = \alpha(x).$$

From the general theory, it follows that we have a complete set of eigenvalues,

$$\lambda_1 < \lambda_2 < \cdots < \lambda_n < \cdots,$$

and corresponding eigenfunctions,

$$\phi_1, \phi_2, \ldots, \phi_n, \ldots,$$

where $\phi_n = X_n$ is the solution to the spatial problem when $\lambda = \lambda_n$.

The boundary term in the Rayleigh quotient $\mathcal{R}(\phi)$ is

$$-p(x)\phi(x)\phi'(x)\Big|_0^\ell = \tau(\ell)\phi(\ell)\phi'(\ell) - \tau(0)\phi(0)\phi'(0) = 0.$$

Since $q(x) < 0$, we have $\lambda_1 > 0$, which is the information we need to solve the time problem

$$T'' + \lambda_n T = 0.$$

For $\lambda_n = \mu_n^2 > 0$, the solution is

$$T_n(t) = a_n \cos \mu_n t + b_n \sin \mu_n t.$$

Using the superposition principle, we find the solution in the form of a generalized Fourier series:

$$u(x,t) = \sum_{n=1}^{\infty} \left(a_n \cos \mu_n t + b_n \sin \mu_n t\right) \phi_n(x).$$

Finally, we use the initial conditions to determine the constants a_n and b_n. At $t = 0$, the initial displacement is

$$f(x) = u(x,0) = \sum_{n=1}^{\infty} a_n \phi_n(x),$$

which is the generalized Fourier series of $f(x)$ with coefficients

$$a_n = \frac{\langle f(x), \phi_n(x)\rangle}{\|\phi_n\|^2}$$

for $n \geqslant 1$. Similarly, the initial velocity is

$$g(x) = \frac{\partial u}{\partial t}(x,0) = \sum_{n=1}^{\infty} \mu_n b_n \phi_n(x),$$

which is the generalized Fourier series of $g(x)$ with coefficients

$$\mu_n b_n = \frac{\langle g(x), \phi_n(x)\rangle}{\|\phi_n\|^2};$$

that is,

$$b_n = \frac{\langle g(x), \phi_n(x)\rangle}{\mu_n \|\phi_n\|^2}$$

for $n \geqslant 1$.

∎

4.5 SUMMARY

Sturm-Liouville theory is the central focus of this book. The spectral properties of Sturm-Liouville problems are quite amazing. And even more, they are typical of a wide class of problems that involve compact operators. For example, the spectral theorem for positive definite matrices gives similar results for matrices, and the spectral theorem for compact operators can be applied to integral operators. Furthermore, the spectral decomposition provides a natural motivation of Fourier series expansions, as they can now be seen as special cases of generalized eigenfunction expansions. A good understanding of Sturm-Liouville theory not only helps you solve PDEs, it also serves as a preparation for more complicated mathematical theories.

We summarize the basic properties of Sturm-Liouville properties in the following theorem.

Theorem 4.13. (*Spectral Theorem for Sturm-Liouville Problems*)
Consider a regular Sturm-Liouville problem as given in Definition 4.2.

(a) The spectrum of a regular Sturm-Liouville problem with self-adjoint operator L is a countably infinite set

$$\sigma(L) = \{\lambda_1, \lambda_2, \ldots, \lambda_n, \ldots\},$$

where the eigenvalues are real and can be arranged in increasing order as

$$\lambda_1 < \lambda_2 < \cdots < \lambda_n < \cdots$$

with $\lim_{n\to\infty} \lambda_n = \infty$.

(b) The eigenfunctions $\phi_n(x)$ can be chosen to be real, and they are unique up to multiplication by a constant. The eigenfunction corresponding to the eigenvalue λ_n has exactly $(n-1)$ zeros in the interval $a < x < b$.

(c) If λ_m and λ_n are distinct eigenvalues, that is, $\lambda_m \neq \lambda_n$, the corresponding eigenfunctions ϕ_m and ϕ_n are orthogonal with respect to the inner product

$$\langle f, g\rangle = \int_a^b f(x)g(x)\sigma(x)\,dx.$$

(d) The orthogonal system of eigenfunctions is complete in the sense that every piecewise smooth function f on the interval (a, b) has a *generalized Fourier series expansion* of the form

$$f(x) = \sum_{n=1}^{\infty} c_n\phi_n(x), \qquad c_n = \frac{\langle f, \phi_n\rangle}{\|\phi_n\|^2},$$

and we have a convergence theorem similar to Dirichlet's theorem; the series above converges to

$$\frac{1}{2}\left[f(x^+) + f(x^-)\right]$$

for each $x \in (a, b)$.

(e) Each eigenvalue λ can be represented by the Rayleigh quotient

$$\lambda = \frac{-p(x)\phi(x)\phi'(x)\Big|_a^b + \int_a^b \left[p(x)\phi'(x)^2 - q(x)\phi(x)^2\right] dx}{\int_a^b \phi(x)^2 \sigma(x)\, dx},$$

where $\phi(x)$ is the unique eigenfunction corresponding to λ.

(f) Let $\mathcal{D}(L)$ denote the subset of piecewise continuous functions on $[a, b]$ that satisfy the Sturm-Liouville boundary conditions

$$\alpha_1\phi(a) + \beta_1\phi'(a) = 0,$$
$$\alpha_2\phi(b) + \beta_2\phi'(b) = 0;$$

then the smallest eigenvalue λ_1 is given by the minimum of the Rayleigh quotient over all functions $\phi \in \mathcal{D}(L)$:

$$\lambda_1 = \inf_{\phi \in \mathcal{D}(L)} \frac{-p(x)\phi(x)\phi'(x)\Big|_a^b + \int_a^b \left[p(x)\phi'(x)^2 - q(x)\phi(x)^2\right] dx}{\int_a^b \phi(x)^2 \sigma(x)\, dx}.$$

In this chapter, and particularly in the solved problems, we see that many Sturm-Liouville problems have specific names, and the eigenfunctions carry those names as well. For example, the eigenfunctions of **Bessel's equation** are called **Bessel functions**, the eigenfunctions of **Hermite's equation** are called **Hermite polynomials**. Additionally, there are **Laguerre polynomials**, **Legendre polynomials**, **Tschebycheff polynomials**, **spherical harmonics**, and many more. All these relate to specific problems that arise in physical or engineering applications, and we will study Bessel functions and Legendre polynomials in detail later. At this point it is important to understand that they can be used to expand a given piecewise smooth function defined on a finite interval as a linear combination (infinite series) of eigenfunctions, that is, as a **generalized Fourier series**.

Example 4.11. Here, we summarize the four most frequently occurring Sturm-Liouville problems. The underlying computations are given in Exercise 12.1.

Model Type	S L Problem	Spectrum	Eigenfunctions
Homogeneous	$\phi''(x) + \lambda\phi(x) = 0$	$\lambda_n = \left(\frac{n\pi}{\ell}\right)^2$	$\phi_n = \sin\frac{n\pi x}{\ell}$
Dirichlet B.C.	$\phi(0) = \phi(\ell) = 0$	$n = 1, 2, \ldots$	$n = 1, 2, \ldots$
Homogeneous	$\phi''(x) + \lambda\phi(x) = 0$	$\lambda_n = \left(\frac{n\pi}{\ell}\right)^2$	$\phi_n = \cos\frac{n\pi x}{\ell}$
Neumann B.C.	$\phi'(0) = \phi'(\ell) = 0$	$n = 0, 1, \ldots$	$n = 0, 1, \ldots$
Mixed B.C.	$\phi''(x) + \lambda\phi(x) = 0$	$\lambda_n = \left(\frac{(2n-1)\pi}{2\ell}\right)^2$	$\phi_n = \sin\frac{(2n-1)\pi x}{2\ell}$
Type I	$\phi(0) = 0,\ \phi'(\ell) = 0$	$n = 1, 2, \ldots$	$n = 1, 2, \ldots$
Mixed B.C.	$\phi''(x) + \lambda\phi(x) = 0$	$\lambda_n = \left(\frac{(2n-1)\pi}{2\ell}\right)^2$	$\phi_n = \cos\frac{(2n-1)\pi x}{2\ell}$
Type II	$\phi'(0) = 0,\ \phi(\ell) = 0$	$n = 1, 2, \ldots$	$n = 1, 2, \ldots$

∎

4.5.1 Problems and Notes

Problems from Part II:

Exercise	12.1	12.2	12.3	12.4	12.5	12.6
Notes						
Exercise	12.7	12.8	12.9	12.10	12.11	12.12
Notes						
Exercise	12.13	12.14	12.15	12.16	12.17	12.18
Notes						
Exercise	12.19					
Notes						

Midterm exam questions:

Exercise	18.1	18.4	18.7	18.10
Notes				

Final exam questions:

Exercise	19.2	19.4	19.9	19.11	19.16
Notes					

You should now be able to do **Midterm Exam 1**.

CHAPTER 5

HEAT, WAVE, AND LAPLACE EQUATIONS

5.1 ONE-DIMENSIONAL HEAT EQUATION

In this section we consider the problem of heat flow in a conducting rod of length ℓ, with internal heat generation; where the left-hand boundary, $x = 0$, is insulated and there is a loss/gain of heat energy by convection across the right-hand boundary, $x = \ell$. This problem is modeled by the following boundary value–initial value problem for the one-dimensional heat equation:

$$
\begin{aligned}
&\frac{\partial u}{\partial t} = k\left(\frac{\partial^2 u}{\partial x^2} + \gamma u\right), \qquad 0 < x < \ell, \quad t > 0,\\
&\frac{\partial u}{\partial x}(0,t) = 0,\\
&\kappa u(\ell,t) + \frac{\partial u}{\partial x}(\ell,t) = 0,\\
&u(x,0) = f(x).
\end{aligned}
$$

Here we assume that the material properties governing heat generation γ, thermal diffusivity k, and thermal conductivity κ are all independent of x and t and are all

Partial Differential Equations: Theory and Completely Solved Problems, First Edition. By T. Hillen, I. E. Leonard, H. van Roessel

positive constants. We will find the solution to this problem, and then analyze the long-term behavior of heat flow in the rod.

Since this is a linear homogeneous PDE, with linear and homogeneous boundary conditions, as before, we assume a separated solution of the form

$$u(x,t) = X(x) \cdot T(t).$$

Substituting this into the PDE and separating variables, we obtain

$$\frac{T'}{kT} - \gamma = \frac{X''}{X} = -\lambda,$$

where λ is the separation constant. This gives the following problems for X and T:

$$\begin{aligned} &X'' + \lambda X = 0, \quad 0 < x < \ell, \qquad T' + (\lambda - \gamma)kT = 0, \quad t > 0,\\ &X'(0) = 0,\\ &\kappa X(\ell) + X'(\ell) = 0. \end{aligned}$$

The problem for X is a regular Sturm-Liouville problem with

$$p(x) = 1, \qquad q(x) = 0, \qquad \sigma(x) = 1$$

for $0 \leqslant x \leqslant \ell$, and

$$\alpha_1 = 0, \qquad \beta_1 = 1, \qquad \alpha_2 = \kappa, \qquad \beta_2 = 1.$$

Since $\alpha_1\beta_1 = 0$ and $\alpha_2\beta_2 \geqslant 0$, according to Theorem 4.7, part (b), all of the eigenvalues are nonnegative; thus, $\lambda = \mu^2$.

- If $\mu = 0$, the equation

  $$X'' = 0$$

  has general solution

  $$X(x) = ax + b,$$

  and since $X'(0) = a = 0$, the solution becomes

  $$X(x) = b.$$

  Since $\kappa X(\ell) + X'(\ell) = 0$, then $\kappa b = 0$, so that $b = 0$, and there are no nontrivial solutions. Thus, $\lambda = 0$ is not an eigenvalue.

- If $\mu \neq 0$, the equation becomes

  $$X'' + \mu^2 X = 0$$

  and has general solution

  $$X(x) = a\cos\mu x + b\sin\mu x$$

with

$$X'(x) = -a\mu \sin \mu x + b\mu \cos \mu x.$$

Applying the boundary condition at the left-hand boundary, $x = 0$, we have $X'(0) = 0$, and hence $b = 0$. The solution is now

$$X(x) = a \cos \mu x$$

with

$$X'(x) = -a\mu \sin \mu x.$$

Applying the boundary condition at the right-hand boundary, $x = \ell$, we have

$$\kappa X(\ell) + X'(\ell) = 0,$$

which implies that

$$a\kappa \cos \mu\ell - a\mu \sin \mu\ell = 0;$$

that is,

$$\tan \mu\ell = \frac{\kappa}{\mu}.$$

Since $\tan \mu\ell$ is a periodic function of μ, with singularities at odd multiples of $\pi/2\ell$, and since κ/μ is a decreasing function of μ, we have an infinite sequence of solutions, and the eigenvalues are

$$\lambda_n = \mu_n^2,$$

where $\tan \mu_n\ell = \kappa/\mu_n$ (Fig. 5.1). The corresponding eigenfunctions are

$$X_n(x) = \cos \mu_n x,$$

and the corresponding solutions to the time equation are

$$T_n(t) = e^{-(\mu_n^2 - \gamma)kt}$$

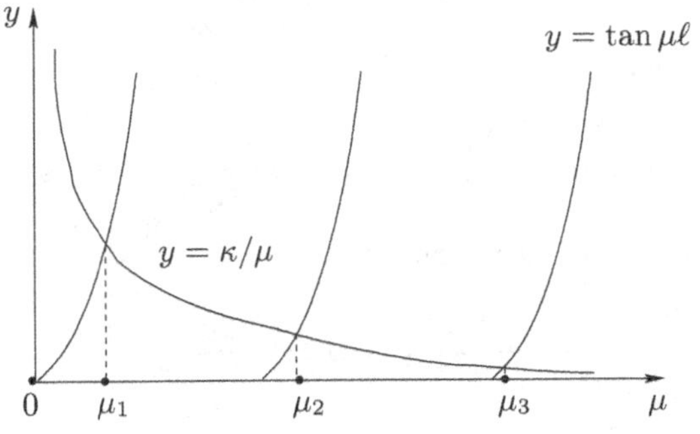

Figure 5.1 Intersection of $y = \tanh \mu\ell$ and $y = \kappa/\mu$.

for $n \geqslant 1$. The norm of the eigenfunctions, relative to the appropriate inner product, is

$$\|X_n\|^2 = \langle X_n, X_n \rangle = \int_0^\ell X_n^2(x)\, dx = \int_0^\ell \cos^2 \mu_n x\, dx = \frac{1}{2}\left(\ell + \frac{\sin 2\mu_n \ell}{2\mu_n}\right).$$

Applying the principle of superposition, we obtain

$$u(x,t) = \sum_{n=1}^{\infty} c_n X_n(x) \cdot T_n(t) = \sum_{n=1}^{\infty} c_n e^{-(\mu_n^2 - \gamma)kt} \cos \mu_n x, \tag{5.1}$$

and from the initial condition $u(x,0) = f(x)$, we get

$$f(x) = \sum_{n=1}^{\infty} c_n \cos \mu_n x.$$

Thus, the problem is solved, provided that we can find the expansion of f above in terms of the eigenfunctions. If $f \in PWC(0,\ell)$, the expansion exists with coefficients given by

$$c_n = \frac{\langle f, X_n \rangle}{\|X_n\|^2} = \frac{4\mu_n}{2\mu_n \ell + \sin 2\mu_n \ell} \int_0^\ell f(x) \cos \mu_n x\, dx. \tag{5.2}$$

We can interpret the long term behavior of these results as follows. The eigenvalues are the positive solutions to the transcendental equation

$$\tan \mu\ell = \frac{\kappa}{\mu},$$

and hence the eigenvalue μ_n is the nth intersection point of the graphs of $y = \tan \mu\ell$ and $y = \kappa/\mu$.

Therefore,

$$\frac{(n-1)\pi}{\ell} \leqslant \mu_n \leqslant \frac{(2n-1)\pi}{2\ell}$$

for $n \geqslant 1$, and we have

$$0 \leqslant \mu_1 \leqslant \frac{\pi}{2\ell}, \quad \frac{\pi}{\ell} \leqslant \mu_2 \leqslant \frac{3\pi}{2\ell}, \quad \ldots, \quad \frac{(n-1)\pi}{\ell} \leqslant \mu_n \leqslant \frac{(2n-1)\pi}{2\ell}, \quad \ldots.$$

Clearly, μ_n depends on κ, so we can write

$$\frac{(n-1)\pi}{\ell} \leqslant \mu_n(\kappa) \leqslant \frac{(2n-1)\pi}{2\ell}$$

for $n \geqslant 1$.

We consider two cases.

Case (i): Perfect insulation. If $\kappa \to 0$, we have

$$\lim_{\kappa \to 0} \mu_n(\kappa) = \frac{(n-1)\pi}{\ell},$$

and, in particular,

$$\lim_{\kappa \to 0} \mu_1(\kappa) = 0.$$

Other quantities simplify as follows:

$$\|X_n\|^2 = \frac{1}{2}\left(\ell + \frac{\sin 2\mu_n \ell}{2\mu_n}\right) = \frac{1}{2}\left(\ell + \frac{\sin 2(n-1)\pi}{2(n-1)\pi/\ell}\right) = \begin{cases} \dfrac{\ell}{2}, & n \neq 1, \\ \ell, & n = 1, \end{cases}$$

$$c_n = \frac{2}{\ell}\int_0^\ell f(x)\cos\left((n-1)\pi x/\ell\right)\,dx,$$

$$u(x,t) = e^{\gamma k t}\left\{c_1 + \sum_{n=2}^{\infty} c_n e^{-\mu_n^2 k t}\cos \mu_n x\right\}.$$

Taking the limit as $t \to \infty$, and assuming that $c_1 > 0$ and $\gamma > 0$, we get

$$\lim_{t \to \infty} u(x,t) = +\infty.$$

Does this result make sense? Yes it does. Remember that $\gamma > 0$ represents internal heat generation, and as $\kappa \to 0$, the boundary conditions become

$$\frac{\partial u}{\partial x}(0,t) = 0,$$
$$\frac{\partial u}{\partial x}(\ell,t) = 0,$$

which are homogeneous Neumann conditions. Therefore, with perfect insulation, no heat can escape; thus, the temperature must continue to rise indefinitely.

Case (ii): Imperfect insulation. If $\kappa > 0$, in this case, the solution is

$$u(x,t) = c_1 e^{(\gamma - \mu_1^2)kt}\cos \mu_1 x + \sum_{n=2}^{\infty} c_n e^{(\gamma - \mu_n^2)kt}\cos \mu_n x,$$

and letting $t \to \infty$, we get

$$\lim_{t \to \infty} u(x,t) = \begin{cases} +\infty, & \text{for } \gamma > \mu_1^2, \\ c_1 \cos \mu_1 x, & \text{for } \gamma = \mu_1^2, \\ 0, & \text{for } \gamma < \mu_1^2. \end{cases}$$

With imperfect insulation there is heat loss. What this result is saying is that if the internal heat generation is sufficiently large, that is, if $\gamma > \mu_1^2$, heat is generated at a rate faster than it can escape through the boundary. Therefore, the temperature rises indefinitely. On the other hand, if internal heat generation is very low, heat escapes faster than it is generated internally, and the temperature eventually goes to zero. But if the rate of internal heat generation is just right, that is, if $\gamma = \mu_1^2$, there is a balance between internal heat generation and heat loss through the boundary, and an equilibrium temperature distribution is reached. The critical value of internal heat generation is precisely the value of the smallest eigenvalue λ_1.

5.2 TWO-DIMENSIONAL HEAT EQUATION

In this section we consider the flow of heat energy in a two-dimensional rectangular plate occupying the region $0 < x < a,\ 0 < y < b$. This problem is modeled by the following boundary value–initial value problem for the two-dimensional heat equation, where we assume that three sides of the plate are insulated and the fourth side is kept at zero temperature:

$$\begin{aligned}
&\frac{\partial u}{\partial t} = k\left(\frac{\partial^2 u}{\partial x^2} + \frac{\partial^2 u}{\partial y^2}\right), \quad 0 < x < a, \quad 0 < y < b, \quad t > 0,\\
&u(0, y, t) = 0,\\
&\frac{\partial u}{\partial x}(a, y, t) = 0,\\
&\frac{\partial u}{\partial y}(x, 0, t) = 0,\\
&\frac{\partial u}{\partial y}(x, b, t) = 0,\\
&u(x, y, 0) = f(x, y).
\end{aligned}$$

As in Section 5.1, we will find the solution to this problem and then analyze the long-term behavior of heat flow in the rectangular plate; that is, we examine the solution as $t \to \infty$. Since the equation and the boundary conditions are linear and homogeneous, we can use separation of variables. Assuming a solution of the form

$$u(x, y, t) = X(x) \cdot Y(y) \cdot T(t)$$

and substituting this into the partial differential equation, we have

$$T'(t) \cdot X(x) \cdot Y(y) = k\left[X''(x) \cdot Y(y) \cdot T(t) + Y''(y) \cdot X(x) \cdot T(t)\right],$$

and separating variables yields

$$\frac{T'(t)}{kT(t)} = \frac{X''(x)}{X(x)} + \frac{Y''(y)}{Y(y)} = -\lambda,$$

where λ is the separation constant. This gives

$$\frac{X''(x)}{X(x)} = -\lambda - \frac{Y''(y)}{Y(y)} = -\tau,$$

where τ is another separation constant.

We can satisfy the boundary conditions by requiring that

$$X(0) = X'(a) = 0 \qquad \text{and} \qquad Y'(0) = Y'(b) = 0,$$

and therefore X and Y satisfy the boundary value problems

$$\begin{aligned} &X''(x) + \tau X(x) = 0, \quad 0 < x < a, && Y''(y) + \alpha Y(y) = 0, \quad 0 < y < b,\\ &X(0) = 0, && Y'(0) = 0,\\ &X'(a) = 0, && Y'(b) = 0, \end{aligned}$$

where $\alpha = \lambda - \tau$. The function T satisfies the differential equation

$$T' + \lambda k T = 0, \quad t > 0.$$

The eigenvalues and corresponding eigenfunctions for the X-problem are (see Exercise 12.1)

$$\tau_n = \left(\frac{(2n-1)\pi}{2a}\right)^2 \qquad \text{and} \qquad X_n(x) = \sin\frac{(2n-1)\pi x}{2a}$$

for $n = 1, 2, 3, \ldots$, while the eigenvalues and corresponding eigenfunctions for the Y-problem are

$$\alpha_m = \left(\frac{m\pi}{b}\right)^2 \qquad \text{and} \qquad Y_m(y) = \cos\frac{m\pi y}{b}$$

for $m = 0, 1, 2, 3, \ldots$. The corresponding solutions to the time equation

$$T'(t) + \lambda k T(t) = 0$$

are

$$T_{nm} = e^{-\lambda_{nm} kt},$$

where

$$\lambda_{nm} = \tau_n + \alpha_m = \left(\frac{(2n-1)\pi}{2a}\right)^2 + \left(\frac{m\pi}{b}\right)^2$$

for $n = 1, 2, 3, \ldots$ and $m = 0, 1, 2, 3, \ldots$.

The products

$$u_{nm}(x, y, t) = X_n(x) \cdot Y_m(y) \cdot T_{nm}(t) = \sin \frac{(2n-1)\pi x}{2a} \cos \frac{m\pi y}{b} e^{-\lambda_{nm} kt}$$

for $n \geqslant 1$ and $m \geqslant 0$ satisfy the partial differential equation and the boundary conditions, and by the superposition principle, the function

$$u(x, y, t) = \sum_{n=1}^{\infty} \sum_{m=0}^{\infty} C_{nm} \sin \frac{(2n-1)\pi x}{2a} \cos \frac{m\pi y}{b} e^{-\lambda_{nm} kt}$$

also satisfies the partial differential equation and all the boundary conditions.

To satisfy the initial condition, we could use the fact that the eigenfunctions

$$\left\{ \sin \frac{(2n-1)\pi x}{2a} \cos \frac{m\pi y}{b} \right\}_{n \geqslant 1, m \geqslant 0}$$

form an orthogonal set on the rectangle $[0, a] \times [0, b]$ in $\mathbb{R}^2$. However, we use another method, which is similar to the methods used for one-dimensional Fourier series expansions. Setting $t = 0$ in the expression above for $u(x, y, t)$, we want

$$f(x, y) = u(x, y, 0) = \sum_{n=1}^{\infty} \sum_{m=0}^{\infty} C_{nm} \sin \frac{(2n-1)\pi x}{2a} \cos \frac{m\pi y}{b}$$

for $0 \leqslant x \leqslant a,\ 0 \leqslant y \leqslant b$, and writing this as

$$f(x, y) = \sum_{n=1}^{\infty} \underbrace{\left(\sum_{m=0}^{\infty} C_{nm} \cos \frac{m\pi y}{b} \right)}_{B_n(y)} \sin \frac{(2n-1)\pi x}{2a},$$

we have

$$f(x, y) = \sum_{n=1}^{\infty} B_n(y) \sin \frac{(2n-1)\pi x}{2a}.$$

This is a Fourier sine series expansion of $f(x, y)$ on the interval $[0, a]$ holding y fixed, and therefore

$$B_n(y) = \frac{2}{a} \int_0^a f(x, y) \sin \frac{(2n-1)\pi x}{2a}\, dx,$$

for $n \geqslant 1$. However, for $n \geqslant 1$,

$$B_n(y) = \sum_{m=0}^{\infty} C_{nm} \cos \frac{m\pi y}{b}$$

is a Fourier cosine series expansion of $B_n(y)$ on the interval $[0, b]$, so that

$$C_{n0} = \frac{1}{b} \int_0^b B_n(y)\, dy$$

$$= \frac{2}{ab} \int_0^b \int_0^a f(x, y) \sin \frac{(2n-1)\pi x}{2a}\, dx\, dy$$

for $n \geqslant 1$, and

$$\begin{aligned} C_{nm} &= \frac{2}{b}\int_0^b B_n(y)\cos\frac{m\pi y}{b}\,dy \\ &= \frac{4}{ab}\int_0^b\int_0^a f(x,y)\sin\frac{(2n-1)\pi x}{2a}\cos\frac{m\pi y}{b}\,dx\,dy \end{aligned}$$

for $n, m \geqslant 1$.

Finally, note that in the solution

$$u(x,y,t) = \sum_{n=1}^{\infty}\sum_{m=0}^{\infty} C_{nm}\sin\frac{(2n-1)\pi x}{2a}\cos\frac{m\pi y}{b}e^{-\lambda_{nm}kt},$$

all terms in the sum for which either $n \geqslant 1$ or $m \geqslant 1$ contain a factor of

$$e^{-\lambda_{nm}kt},$$

where $\lambda_{nm} > 0$, and as $t \to \infty$, all these terms tend to 0, and therefore,

$$\lim_{t\to\infty} u(x,y,t) = 0.$$

5.3 ONE-DIMENSIONAL WAVE EQUATION

The general wave equation derived earlier is

$$\frac{\partial^2 u}{\partial t^2} = c^2\nabla^2 u + F,$$

and the one-dimensional homogeneous version is

$$\frac{\partial^2 u}{\partial t^2} = c^2\frac{\partial^2 u}{\partial x^2},$$

where c is a constant. The one-dimensional wave equation is one of the rare PDEs for which the "general" solution can actually be found. How this is done is not immediately obvious, so again we apply separation of variables to find a solution.

Example 5.1. Consider the following one-dimensional wave equation:

$$\begin{aligned} &\frac{\partial^2 u}{\partial t^2} = c^2\frac{\partial^2 u}{\partial x^2}, \quad 0 < x < \ell, \quad t > 0, \\ &u(0,t) = 0, \\ &u(\ell,t) = 0, \\ &u(x,0) = f(x), \\ &\frac{\partial u}{\partial t}(x,0) = g(x). \end{aligned}$$

This problem governs the vertical displacement of a string, with its endpoints fixed at $x = 0$ and $x = \ell$. At time $t = 0$, we give the string an initial displacement $f(x)$ and an initial velocity $g(x)$, and then the solution to the problem governs the subsequent motion of the string.

The boundary conditions are linear and homogeneous, so we try separation of variables. Let

$$u(x,t) = X(x) \cdot T(t),$$

and substitute this into the equation to obtain

$$X \cdot T'' = c^2 X'' \cdot T;$$

that is,

$$\frac{T''}{c^2 T} = \frac{X''}{X} = -\lambda,$$

where λ is the separation constant. The boundary conditions imply that

$$X(0) = 0 \qquad \text{and} \qquad X(\ell) = 0,$$

and we have the following problems for X and T:

$$\begin{aligned} & X'' + \lambda X = 0, \quad 0 < x < \ell, \qquad T'' + \lambda c^2 T = 0, \quad t > 0, \\ & X(0) = 0, \\ & X(\ell) = 0. \end{aligned}$$

The problem for X is a regular Sturm-Liouville problem, and Theorem 4.7 implies that all of the eigenvalues are nonnegative. Letting $\lambda = \mu^2$, the eigenvalues and eigenfunctions are

$$\lambda_n = \mu_n^2 = \frac{n^2\pi^2}{\ell^2},$$

$$X_n(x) = \sin \mu_n x$$

for $n \geqslant 1$.
The corresponding equation for T becomes

$$T'' + c^2 \mu_n^2 T = 0$$

with general solution

$$T_n(t) = \alpha_n \cos \mu_n ct + \beta_n \sin \mu_n ct.$$

Since both the PDE and the boundary conditions are linear and homogeneous, we may apply the principle of superposition to obtain

$$u(x,t) = \sum_{n=1}^{\infty} X_n(x) \cdot T_n(t) = \sum_{n=1}^{\infty} \sin \mu_n x (\alpha_n \cos \mu_n ct + \beta_n \sin \mu_n ct)$$

and

$$\frac{\partial u}{\partial t}(x,t) = \sum_{n=1}^{\infty} c\mu_n \sin \mu_n x(-\alpha_n \sin \mu_n ct + \beta_n \cos \mu_n ct).$$

We determine the coefficients α_n and β_n from the initial conditions. The initial displacement is $u(x,0) = f(x)$, and setting $t = 0$ in the first equation, we obtain

$$f(x) = \sum_{n=1}^{\infty} \alpha_n \sin \mu_n x,$$

the Fourier sine series for $f(x)$.

The initial velocity is $\partial u(x,0)/\partial t = g(x)$, and setting $t = 0$ in the second equation, we obtain

$$g(x) = \sum_{n=1}^{\infty} c\beta_n \mu_n \sin \mu_n x,$$

the Fourier sine series for $g(x)$. Let $\bar{f}_{\text{odd}}$ and $\bar{g}_{\text{odd}}$ be the odd 2ℓ-periodic extensions of f and g, respectively; then

$$\bar{f}_{\text{odd}}(x) = \sum_{n=1}^{\infty} \alpha_n \sin \mu_n x \qquad \text{and} \qquad \bar{g}_{\text{odd}}(x) = \sum_{n=1}^{\infty} c\beta_n \mu_n \sin \mu_n x,$$

where

$$\alpha_n = \frac{2}{\ell}\int_0^{\ell} f(x) \sin \mu_n x \, dx \qquad \text{and} \qquad \beta_n = \frac{2}{c\mu_n \ell}\int_0^{\ell} g(x) \sin \mu_n x \, dx.$$

For convenience we define

$$G(x) = \int \bar{g}_{\text{odd}}(x)\, dx = -\sum_{n=1}^{\infty} c\beta_n \cos \mu_n x$$

and note that G is even and 2ℓ-periodic. Looking more closely at the solution $u(x,t)$, we see that

$$\begin{aligned}
u(x,t) &= \sum_{n=1}^{\infty} \sin \mu_n x \big[\alpha_n \cos \mu_n ct + \beta_n \sin \mu_n ct\big] \\
&= \frac{1}{2}\sum_{n=1}^{\infty} \Big\{\alpha_n \big[\sin \mu_n(x-ct) + \sin \mu_n(x+ct)\big] \\
&\qquad\qquad + \beta_n \big[\cos \mu_n(x-ct) - \cos \mu_n(x+ct)\big]\Big\} \\
&= \frac{1}{2}\Big\{\bar{f}_{\text{odd}}(x-ct) + \bar{f}_{\text{odd}}(x+ct) - \frac{1}{c}\big[G(x-ct) - G(x+ct)\big]\Big\},
\end{aligned}$$

and thus,

$$u(x,t) = \frac{1}{2}\left[\bar{f}_{\text{odd}}(x-ct) + \bar{f}_{\text{odd}}(x+ct)\right] + \frac{1}{2c}\int_{x-ct}^{x+ct} \bar{g}_{\text{odd}}(\xi)\, d\xi.$$

We have the solution in closed form! In other words, the series has been summed, and the solution above is called **d'Alembert's solution to the initial value problem for the one-dimensional wave equation**, found by d'Alembert in 1746.

Now we give a physical interpretation of the solution; for simplicity, we consider the case of zero initial velocity, that is, $g = 0$. The solution in this case is

$$u(x,t) = \frac{1}{2}\left[\bar{f}_{\text{odd}}(x-ct) + \bar{f}_{\text{odd}}(x+ct)\right]. \tag{5.3}$$

Define $\xi = x - ct$, called a *phase shift*; if ξ is constant, $d\xi/dt = 0$, that is, $dx/dt = c$. Therefore, the function $\bar{f}_{\text{odd}}(\xi) = \bar{f}_{\text{odd}}(x-ct)$ represents a wave traveling to the right with speed or phase velocity c. Similarly, if we define another phase shift $\eta = x + ct$, $\eta =$ constant implies that $dx/dt = -c$. Therefore, the function $\bar{f}_{\text{odd}}(x+ct)$ represents a wave traveling to the left with speed or phase velocity c. Suppose that f is defined as the "hat" function, shown in Figure 5.2.

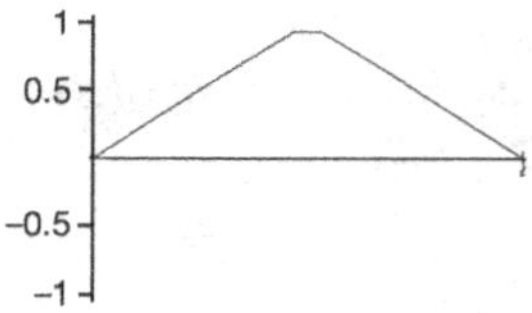

Figure 5.2 The hat function.

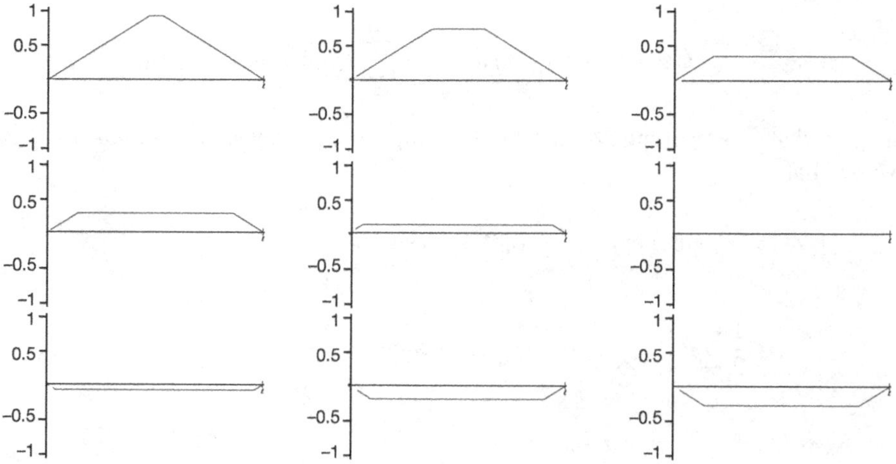

Figure 5.3 Snapshots of the string at various times.

Figure 5.3 shows how a string, with $y = f(x)$ as its initial displacement, evolves with time. Specifically, the figure exhibits snapshots of the string's displacement at various times.

∎

5.3.1 d'Alembert's Solution

Since the solution to the wave equation in Example 5.1 represents two waves, one traveling to the left and one traveling to the right, we consider going back to the beginning and making a change of coordinates:

$$\xi = x + ct, \quad \eta = x - ct$$

and

$$u(x,t) = u\left(\frac{\xi+\eta}{2}, \frac{\xi-\eta}{2c}\right) = w(\xi,\eta).$$

From the chain rule, we obtain

$$\frac{\partial u}{\partial x} = \frac{\partial w}{\partial \xi} + \frac{\partial w}{\partial \eta}, \qquad \frac{\partial^2 u}{\partial x^2} = \frac{\partial^2 w}{\partial \xi^2} + 2\frac{\partial^2 w}{\partial \xi\,\partial \eta} + \frac{\partial^2 w}{\partial \eta^2},$$

$$\frac{\partial u}{\partial t} = c\left(\frac{\partial w}{\partial \xi} - \frac{\partial w}{\partial \eta}\right), \qquad \frac{\partial^2 u}{\partial t^2} = c^2\left(\frac{\partial^2 w}{\partial \xi^2} - 2\frac{\partial^2 w}{\partial \xi\,\partial \eta} + \frac{\partial^2 w}{\partial \eta^2}\right).$$

Substituting this into the wave equation, we have

$$\begin{aligned} 0 &= \frac{\partial^2 u}{\partial t^2} - c^2\frac{\partial^2 u}{\partial x^2} \\ &= c^2\left(\frac{\partial^2 w}{\partial \xi^2} - 2\frac{\partial^2 w}{\partial \xi\,\partial \eta} + \frac{\partial^2 w}{\partial \eta^2}\right) - c^2\left(\frac{\partial^2 w}{\partial \xi^2} + 2\frac{\partial^2 w}{\partial \xi\,\partial \eta} + \frac{\partial^2 w}{\partial \eta^2}\right) \\ &= -4c^2\frac{\partial^2 w}{\partial \xi\,\partial \eta}. \end{aligned}$$

Therefore,

$$\frac{\partial^2 w}{\partial \xi\,\partial \eta} = 0,$$

and this is easily solved, since

$$\frac{\partial}{\partial \eta}\left(\frac{\partial w}{\partial \xi}\right) = 0$$

implies that $\partial w/\partial \xi$ equals a function that depends only on ξ. For convenience we denote this function by $\phi'(\xi)$,

$$\frac{\partial w}{\partial \xi} = \phi'(\xi).$$

This, in turn, implies that

$$w(\xi, \eta) = \phi(\xi) + \psi(\eta),$$

where ϕ and ψ are arbitrary twice continuously differentiable functions of a single real variable. Thus,

$$u(x, t) = \phi(x + ct) + \psi(x - ct), \tag{5.4}$$

which is **d'Alembert's general solution to the one-dimensional wave equation** found by d'Alembert in 1747.

We now apply the initial conditions to determine ϕ and ψ. Differentiating d'Alembert's solution with respect to t, we obtain

$$\frac{\partial u}{\partial t}(x, t) = c[\phi'(x + ct) - \psi'(x - ct)],$$

and at $t = 0$, we have

$$u(x, 0) = f(x) \qquad \text{and} \qquad \frac{\partial u}{\partial t}(x, 0) = g(x),$$

which implies that

$$\phi(x) + \psi(x) = f(x) \qquad \text{and} \qquad c[\phi'(x) - \psi'(x)] = g(x);$$

that is,

$$\phi(x) + \psi(x) = f(x) \qquad \text{and} \qquad \phi'(x) - \psi'(x) = \frac{1}{c}g(x).$$

Therefore,

$$\phi(x) = \frac{1}{2}\left[f(x) + \frac{1}{c}G(x)\right] \qquad \text{and} \qquad \psi(x) = \frac{1}{2}\left[f(x) - \frac{1}{c}G(x)\right],$$

where $G'(x) = g(x)$, for $0 < x < \ell$.

The functions ϕ and ψ are defined only for $0 < x < \ell$; however, $x - ct$ and $x + ct$ go beyond the interval $(0, \ell)$, so we let $\widetilde{f}$ and $\widetilde{G}$ be extensions of f and G to the entire real line. We have

$$\phi(x) = \frac{1}{2}\left[\widetilde{f}(x) + \frac{1}{c}\widetilde{G}(x)\right] \qquad \text{and} \qquad \psi(x) = \frac{1}{2}\left[\widetilde{f}(x) - \frac{1}{c}\widetilde{G}(x)\right]$$

for all $x \in \mathbb{R}$. Now we apply the boundary conditions. At the left-hand boundary, $x = 0$, we have

$$u(0, t) = 0,$$

and this implies that

$$\phi(ct) + \psi(-ct) = 0;$$

that is,

$$\widetilde{f}(ct) + \widetilde{f}(-ct) + \frac{1}{c}[\widetilde{G}(ct) - \widetilde{G}(-ct)] = 0.$$

Since f and g are linearly independent functions in $PWC(0, \ell)$, we must have

$$\widetilde{f}(ct) + \widetilde{f}(-ct) = 0 \qquad \text{and} \qquad \widetilde{G}(ct) - \widetilde{G}(-ct) = 0.$$

This means that

$$\widetilde{f}(ct) = -\widetilde{f}(-ct)$$

for all t, and therefore $\widetilde{f}$ is an odd function. Similarly,

$$\widetilde{G}(ct) = \widetilde{G}(-ct)$$

for all t, and therefore $\widetilde{G}$ is an even function.

At the right-hand boundary, $x = \ell$, we have $u(\ell, t) = 0$, which implies that

$$\phi(\ell + ct) + \psi(\ell - ct) = 0;$$

that is,

$$\widetilde{f}(\ell + ct) + \widetilde{f}(\ell - ct) + \frac{1}{c}[\widetilde{G}(\ell + ct) - \widetilde{G}(\ell - ct)] = 0.$$

Again, since f and g are linearly independent functions, we must have

$$\widetilde{f}(\ell + ct) = -\widetilde{f}(\ell - ct) \qquad \text{and} \qquad \widetilde{G}(\ell + ct) = \widetilde{G}(\ell - ct).$$

Also, since $\widetilde{f}$ is an odd function,

$$\widetilde{f}(\ell + ct) = -\widetilde{f}(\ell - ct) = \widetilde{f}(-\ell + ct) = \widetilde{f}(\ell + ct - 2\ell),$$

and since $\widetilde{G}$ is an even function,

$$\widetilde{G}(\ell + ct) = \widetilde{G}(\ell - ct) = \widetilde{G}(-\ell + ct) = \widetilde{G}(\ell + ct - 2\ell).$$

Therefore, $\widetilde{f}$ and $\widetilde{G}$ are 2ℓ-periodic; hence,

$$\widetilde{f} = \bar{f}_{\text{odd}} \qquad \text{and} \qquad \widetilde{G} = \int \bar{g}_{\text{odd}}(x)\, dx = G.$$

Thus,

$$\begin{aligned} u(x,t) &= \phi(x + ct) + \psi(x - ct) \\ &= \frac{1}{2}[\bar{f}_{\text{odd}}(x - ct) + \bar{f}_{\text{odd}}(x + ct)] + \frac{1}{2c}[G(x + ct) - G(x - ct)], \end{aligned}$$

which is exactly the solution obtained previously by the method of separation of variables, that is, d'Alembert's solution to the initial value problem for the one-dimensional wave equation.

In the next example we consider the one-dimensional wave equation on an infinite domain with time-dependent boundary conditions.

Example 5.2. Consider the following boundary value–initial value problem:

$$\begin{aligned}
&\frac{\partial^2 u}{\partial t^2} = c^2 \frac{\partial^2 u}{\partial x^2}, \quad x > 0, \quad t > 0, \\
&u(0,t) = h(t), \\
&u(x,0) = 0, \\
&\frac{\partial u}{\partial t}(x,0) = 0,
\end{aligned}$$

where c is a constant. This problem governs the vertical displacement of a "semi-infinite" string with the motion at its (one and only) boundary, $x = 0$, prescribed. The general solution, that is, d'Alembert's solution, to the equation is

$$u(x,t) = \phi(x+ct) + \psi(x-ct),$$

where ϕ and ψ are arbitrary twice continuously differentiable functions of one real variable. Differentiating with respect to t, we have

$$\frac{\partial u}{\partial t}(x,t) = c[\phi'(x+ct) - \psi'(x-ct)].$$

Applying the initial conditions at $t = 0$ gives

$$\phi(x) + \psi(x) = 0 \qquad \text{and} \qquad \phi'(x) - \psi'(x) = 0$$

for $x > 0$. We have

$$\psi(x) = -\phi(x),$$

which implies that $\phi'(x) = -\psi'(x)$ and $\phi'(x) = \psi'(x)$ for $x > 0$; hence,

$$\phi'(x) = 0;$$

that is,

$$\phi(x) = A \qquad \text{and} \qquad \psi(x) = -A,$$

for $x > 0$, where A is a constant.

At the boundary $x = 0$, we have

$$\phi(ct) + \psi(-ct) = h(t),$$

which implies that

$$A + \psi(-ct) = h(t);$$

that is,

$$\psi(-ct) = h(t) - A,$$

for $t > 0$. Thus, we have

$$\psi(\xi) = \begin{cases} -A, & \text{for } \xi > 0, \\ h(-\xi/c) - A, & \text{for } \xi < 0, \end{cases}$$

and the solution becomes

$$u(x,t) = \phi(x+ct) + \psi(x-ct) = A + \psi(x-ct) = \widetilde{\psi}(x-ct),$$

where

$$\widetilde{\psi}(\xi) = A + \psi(\xi) = \begin{cases} 0, & \text{for } \xi > 0, \\ h(-\xi/c), & \text{for } \xi < 0. \end{cases}$$

Therefore,

$$u(x,t) = \widetilde{\psi}(x-ct) = \begin{cases} 0, & \text{for } x > ct, \\ h(t - x/c), & \text{for } x < ct, \end{cases}$$

and this solution represents a traveling wave moving to the right with speed c.

A physical interpretation is given with the initial displacement

$$h(t) = \begin{cases} t, & \text{for } 0 \leqslant t \leqslant 1, \\ 2-t, & \text{for } 1 \leqslant t \leqslant 2, \\ 0, & \text{for } t < 0 \text{ or } t > 2. \end{cases}$$

The displacement as a function of time at various positions is given by

$$u(0,t) = \begin{cases} 0, & \text{for } t < 0, \\ h(t), & \text{for } t > 0, \end{cases}$$

$$u(c,t) = \begin{cases} 0, & \text{for } t < 1, \\ h(t-1), & \text{for } t > 1, \end{cases}$$

$$u(2c,t) = \begin{cases} 0, & \text{for } t < 2, \\ h(t-2), & \text{for } t > 2, \end{cases}$$

$$u(3c,t) = \begin{cases} 0, & \text{for } t < 3, \\ h(t-3), & \text{for } t > 3, \end{cases}$$

$$u(4c,t) = \begin{cases} 0, & \text{for } t < 4, \\ h(t-4), & \text{for } t > 4. \end{cases}$$

In Figure 5.4 we show snapshots of the string at various time intervals:

$$u(x,0)=0, \qquad u(x,1)=\begin{cases} 0, & \text{for } x>c, \\ h(1-x/c), & \text{for } x<c, \end{cases}$$

$$u(x,2)=\begin{cases} 0, & \text{for } x>2c, \\ h(2-x/c), & \text{for } x<2c, \end{cases} \qquad u(x,3)=\begin{cases} 0, & \text{for } x>3c, \\ h(3-x/c), & \text{for } x<3c, \end{cases}$$

$$u(x,4)=\begin{cases} 0, & \text{for } x>4c, \\ h(4-x/c), & \text{for } x<4c. \end{cases}$$

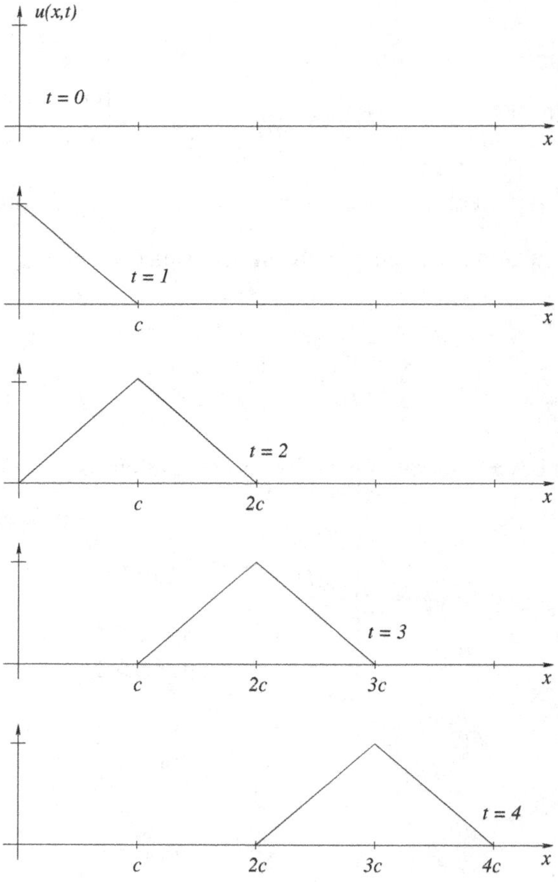

Figure 5.4 Snapshots of the wave front at various times.

∎

5.4 LAPLACE'S EQUATION

Laplace's equation, also called the **potential equation**, may be written as

$$\text{div}(\text{grad } u) = \nabla \cdot \nabla u = \nabla^2 u = 0.$$

Written out explicitly, we have

$$\frac{\partial^2 u}{\partial x^2} + \frac{\partial^2 u}{\partial y^2} = 0,$$

which is the two-dimensional Laplace equation. Also,

$$\frac{\partial^2 u}{\partial x^2} + \frac{\partial^2 u}{\partial y^2} + \frac{\partial^2 u}{\partial z^2} = 0,$$

$$\frac{\partial^2 u}{\partial x_1^2} + \cdots + \frac{\partial^2 u}{\partial x_n^2} = 0,$$

are the three- and n-dimensional Laplace equations, respectively.

Solutions of Laplace's equation are called **harmonic functions**, and solutions of either the heat equation or the wave equation that are time-independent are harmonic functions. For Laplace's equation on a region Ω, a **complete well-posed problem** consists of the potential equation together with appropriate boundary conditions:

$$\nabla^2 u = 0 \quad \text{in } \Omega,$$

$$\alpha u + \beta \frac{\partial u}{\partial n} = f \quad \text{on } \partial\Omega,$$

where

$$\frac{\partial u}{\partial n} = \nabla u \cdot \mathbf{n},$$

with $\mathbf{n}$ being the outward unit normal to Ω.

5.4.1 Potential in a Rectangle

Consider Laplace's equation defined on the rectangle $[0, a] \times [0, b]$:

$$\begin{aligned}
&\frac{\partial^2 u}{\partial x^2} + \frac{\partial^2 u}{\partial y^2} = 0, \quad 0 < x < a, \quad 0 < y < b,\\
&u(x, 0) = f_1(x),\\
&u(x, b) = f_2(x),\\
&u(0, y) = 0,\\
&u(a, y) = 0.
\end{aligned}$$

Notice that according to the rule of thumb for side conditions, we need all four boundary conditions. Two are homogeneous boundary conditions on one pair of opposite sides. Since the PDE and the boundary conditions are linear and homogeneous, we can try separation of variables. We let

$$u(x, y) = X(x) \cdot Y(y),$$

and substitute this expression into the partial differential equation to obtain

$$\frac{X''}{X} = -\frac{Y''}{Y} = -\lambda,$$

where λ is the separation constant. Using the homogeneous boundary conditions, we get

$$u(0, y) = 0 \qquad \text{and} \qquad u(a, y) = 0,$$

which implies that

$$X(0) \cdot Y(y) = 0 \qquad \text{and} \qquad X(a) \cdot Y(y) = 0.$$

To obtain a nontrivial solution, we need

$$X(0) = 0 \qquad \text{and} \qquad X(a) = 0,$$

and we obtain the following ODEs for X and Y:

$$\begin{aligned} &X'' + \lambda X = 0, \quad 0 < x < a, \qquad && Y'' - \lambda Y = 0, \quad 0 < y < b, \\ &X(0) = 0, && \\ &X(a) = 0. && \end{aligned}$$

Clearly, the qualitative nature of the solutions of the X-equation will be different than that of the Y-equation. The problem for X is a regular Sturm-Liouville problem, and from Theorem 4.7, the eigenvalues for the X-problem are all nonnegative. Therefore, we set $\lambda = \mu^2$ and get the problem

$$\begin{aligned} &X'' + \mu^2 X = 0, \quad 0 < x < a, \\ &X(0) = 0, \\ &X(a) = 0, \end{aligned}$$

with eigenvalues

$$\lambda_n = \mu_n^2 = \frac{n^2\pi^2}{a^2}$$

and eigenfunctions

$$X_n(x) = \sin\frac{n\pi x}{a}$$

for $n \geqslant 1$. For these values of λ, the corresponding equation for Y becomes

$$Y'' - \mu_n^2 Y = 0$$

with general solution

$$Y_n(y) = \alpha_n \cosh \mu_n y + \beta_n \sinh \mu_n y$$

for $n \geqslant 1$. For each $n \geqslant 1$, the product

$$u_n(x, y) = X_n(x) \cdot Y_n(y)$$

is a solution to Laplace's equation and also satisfies the homogeneous boundary conditions.

Since Laplace's equation, and the boundary conditions, are linear and homogeneous, we can apply the principle of superposition to get

$$u(x, y) = \sum_{n=1}^{\infty} X_n(x) \cdot Y_n(y) = \sum_{n=1}^{\infty} \left(\alpha_n \cosh \mu_n y + \beta_n \sinh \mu_n y\right) \sin \mu_n x.$$

None of the products that make up the individual terms in the sum can be made to satisfy the nonhomogeneous boundary conditions

$$u(x, 0) = f_1(x),$$
$$u(x, b) = f_2(x)$$

for all $0 < x < a$; however, we can apply these remaining boundary conditions to the series. At $y = 0$, we have

$$u(x, 0) = f_1(x),$$

and using the orthogonality of the eigenfunctions,

$$f_1(x) = \sum_{n=1}^{\infty} \alpha_n \sin \mu_n x.$$

Therefore, the α_n's must be the Fourier sine coefficients of the function $f_1(x)$; hence,

$$\alpha_n = \frac{2}{a} \int_0^a f_1(x) \sin \frac{n\pi x}{a}\, dx.$$

At $y = b$, we have

$$u(x, b) = f_2(x),$$

and using the orthogonality of the eigenfunctions again yields

$$f_2(x) = \sum_{n=1}^{\infty} \left(\alpha_n \cosh \mu_n b + \beta_n \sinh \mu_n b\right) \sin \mu_n x,$$

from which we conclude that

$$\alpha_n \cosh \mu_n b + \beta_n \sinh \mu_n b = \frac{2}{a} \int_0^a f_2(x) \sin \frac{n\pi x}{a}\, dx = \gamma_n,$$

so that

$$\beta_n = \frac{\gamma_n - \alpha_n \cosh \mu_n b}{\sinh \mu_n b}.$$

Hence,

$$Y_n(y) = \alpha_n \cosh \mu_n y + \beta_n \sinh \mu_n y = \alpha_n \cosh \mu_n y + \frac{\gamma_n - \alpha_n \cosh \mu_n b}{\sinh \mu_n b} \sinh \mu_n y$$

$$= \frac{\alpha_n}{\sinh \mu_n b}\big(\sinh \mu_n b \cosh \mu_n y - \cosh \mu_n b \sinh \mu_n y\big) + \frac{\gamma_n}{\sinh \mu_n b} \sinh \mu_n y$$

$$= \frac{1}{\sinh \mu_n b}[\alpha_n \sinh \mu_n (b-y) + \gamma_n \sinh \mu_n y].$$

The final solution is

$$u(x,y) = \sum_{n=1}^{\infty} \frac{1}{\sinh \mu_n b}[\alpha_n \sinh \mu_n (b-y) + \gamma_n \sinh \mu_n y] \sin \mu_n x, \tag{5.5}$$

where

$$\alpha_n = \frac{2}{a}\int_0^a f_1(x) \sin \frac{n\pi x}{a}\, dx,$$

$$\gamma_n = \frac{2}{a}\int_0^a f_2(x) \sin \frac{n\pi x}{a}\, dx,$$

$$\mu_n = \frac{n\pi}{a}$$

for $n \geqslant 1$.

Consider the following more general problem:

$$\frac{\partial^2 u}{\partial x^2} + \frac{\partial^2 u}{\partial y^2} = 0, \quad 0 < x < a, \quad 0 < y < b,$$
$$u(x,0) = f_1(x), \quad u(0,y) = g_1(y),$$
$$u(x,b) = f_2(x), \quad u(a,y) = g_2(y),$$

where none of the boundary conditions is homogeneous. To solve this problem we split it into two problems, each of which is in a form similar to the one in the preceding example. Consider the following problems for v and w:

$$\frac{\partial^2 v}{\partial x^2} + \frac{\partial^2 v}{\partial y^2} = 0, \qquad \frac{\partial^2 w}{\partial x^2} + \frac{\partial^2 w}{\partial y^2} = 0,$$

$$v(x,0) = f_1(x), \quad v(0,y) = 0, \qquad w(x,0) = 0, \quad w(0,y) = g_1(y),$$

$$v(x,b) = f_2(x), \quad v(a,y) = 0, \qquad w(x,b) = 0, \quad w(a,y) = g_2(y);$$

then $u(x,y) = v(x,y) + w(x,y)$ is the solution to the original problem.

5.5 MAXIMUM PRINCIPLE

Consider a closed and bounded region Ω in $\mathbb{R}^2$ with boundary $\partial\Omega$.

Theorem 5.1. Let $u(x, y)$ be any twice continuously differentiable solution of Poisson's equation,

$$\nabla^2 u = F(x, y),$$

in Ω; then

(i) if $F(x, y) > 0$ for all $(x, y) \in \Omega$, the maximum of u in Ω occurs on the boundary $\partial\Omega$;

(ii) if $F(x, y) < 0$ for all $(x, y) \in \Omega$, the minimum of u in Ω occurs on the boundary $\partial\Omega$.

Proof. (by contradiction). We prove only part (i) with $F(x, y) > 0$ for all $(x, y) \in \Omega$. The proof of part (ii) is similar.

Since u is a continuous solution in the closed and bounded set Ω, it has a maximum at some point $(x_0, y_0) \in \Omega$. Suppose that $(x_0, y_0) \in \text{int}(\Omega)$, the interior of Ω, that is, $(x_0, y_0) \notin \partial\Omega$; then

$$\frac{\partial u}{\partial x}(x_0, y_0) = 0 \qquad \text{and} \qquad \frac{\partial u}{\partial y}(x_0, y_0) = 0.$$

Since u has a maximum at (x_0, y_0), its graph is concave down at (x_0, y_0), and we have

$$\frac{\partial^2 u}{\partial x^2}(x_0, y_0) \leqslant 0 \qquad \text{and} \qquad \frac{\partial^2 u}{\partial y^2}(x_0, y_0) \leqslant 0.$$

Adding these inequalities yields

$$\frac{\partial^2 u}{\partial x^2}(x_0, y_0) + \frac{\partial^2 u}{\partial y^2}(x_0, y_0) \leqslant 0,$$

which contradicts the assumption that $u_{xx} + u_{yy} = F > 0$ at all points of Ω. Therefore, the point $(x_0, y_0) \in \Omega$ at which u attains its maximum is not in the interior of Ω; hence, this point must lie on the boundary $\partial\Omega$.

□

Theorem 5.2. If u is a twice continuously differentiable solution of $\nabla^2 u = 0$ in the closed and bounded region Ω, the maximum and minimum of u occur on the boundary $\partial\Omega$.

Proof. Since $\partial\Omega$ is a closed and bounded subset of $\mathbb{R}^2$ and u is continuous on $\partial\Omega$, u attains a maximum on $\partial\Omega$. Let M be the maximum of u on $\partial\Omega$, so that

$$|u(x, y)| \leqslant M$$

for all $(x, y) \in \partial\Omega$. We will show that this inequality holds not only for $(x, y) \in \partial\Omega$, but for all $(x, y) \in \Omega$. To this end, consider a square of length 2ℓ, where ℓ is large enough so that the square contains all of Ω; then

$$|x| \leqslant \ell$$

for all $(x, y) \in \Omega$. Now let $\epsilon > 0$ be arbitrary, and define

$$v(x, y) = u(x, y) + \epsilon x^2.$$

For $(x, y) \in \partial\Omega$, we have

$$v(x, y) = u(x, y) + \epsilon x^2 \leqslant M + \epsilon\ell^2.$$

Differentating, we have

$$\nabla^2 v = \nabla^2 u + \epsilon\nabla^2(x^2) = 2\epsilon,$$

and since $\nabla^2 v = 2\epsilon > 0$, Theorem 5.1 implies that v attains its maximum on $\partial\Omega$. Therefore,

$$v(x, y) \leqslant M + \epsilon\ell^2$$

for all $(x, y) \in \Omega$. From the definition of v, we have

$$u(x, y) = v(x, y) - \epsilon x^2 \leqslant v(x, y) \leqslant M + \epsilon\ell^2$$

for all $(x, y) \in \Omega$. However, $\epsilon > 0$ was arbitrary, and letting $\epsilon \to 0$, we obtain

$$u(x, y) \leqslant M$$

for all $(x, y) \in \Omega$. Thus, M, the global maximum for u, occurs on the boundary $\partial\Omega$. The proof for the case of the minimum value is similar and is omitted.

□

5.6 TWO-DIMENSIONAL WAVE EQUATION

Recall the general wave equation derived earlier:

$$\frac{\partial^2 u}{\partial t^2} = \nabla^2 u + f.$$

We now consider the two-dimensional wave equation on a rectangle.

Example 5.3. Consider the following two-dimensional, homogeneous wave equation defined on a rectangle;

$$\frac{\partial^2 u}{\partial t^2} = c^2\left(\frac{\partial^2 u}{\partial x^2} + \frac{\partial^2 u}{\partial y^2}\right), \quad 0 < x < a, \quad 0 < y < b, \quad t > 0,$$

$$u(0, y, t) = 0, \quad u(x, 0, t) = 0,$$

$$u(a, y, t) = 0, \quad u(x, b, t) = 0,$$

$$u(x, y, 0) = f(x, y),$$

$$\frac{\partial u}{\partial t}(x, y, 0) = 0.$$

This problem governs the vertical displacement u of a membrane stretched over a rectangle and fastened at the edges (think of a rectangular drum).

Since the PDE and the boundary conditions are linear and homogeneous, we try separation of variables. Notice that according to our rule of thumb, we need six side conditions. First we look for a solution where the spatial variables x and y can be separated from the temporal variable t. We assume a solution of the form

$$u(x, y, t) = \phi(x, y) \cdot T(t)$$

and substitute this expression into the PDE to obtain

$$\phi \cdot T'' = c^2\left(\frac{\partial^2 \phi}{\partial x^2} + \frac{\partial^2 \phi}{\partial y^2}\right) \cdot T,$$

so that

$$\frac{1}{c^2}\frac{T''}{T} = \frac{1}{\phi}\left(\frac{\partial^2 \phi}{\partial x^2} + \frac{\partial^2 \phi}{\partial y^2}\right) = -\lambda,$$

where λ is a separation constant. Substituting the expression $u(x, y, t) = \phi(x, y) \cdot T(t)$ into the homogeneous boundary conditions, we find the boundary conditions satisfied by ϕ in order to obtain a nontrivial solution, and ϕ and T satisfy the following boundary value problems:

$$\frac{\partial^2 \phi}{\partial x^2} + \frac{\partial^2 \phi}{\partial y^2} = -\lambda\phi, \quad 0 < x < a, \quad 0 < y < b, \qquad T'' + \lambda c^2 T = 0, \quad t > 0,$$

$$\phi(0, y) = 0, \quad \phi(x, 0) = 0, \qquad T'(0) = 0,$$

$$\phi(a, y) = 0, \quad \phi(x, b) = 0.$$

The problem for ϕ consists of a linear and homogeneous PDE with linear and homogeneous boundary conditions, so we try separation of variables again to separate the spatial variables. We assume a solution of the form

$$\phi(x, y) = X(x) \cdot Y(y),$$

and substituting this expression into the PDE, the equation for ϕ becomes

$$X'' \cdot Y + X \cdot Y'' = -\lambda X \cdot Y,$$

so that

$$\frac{X''}{X} = -\frac{Y''}{Y} - \lambda = -\gamma,$$

where γ is another separation constant. Substituting the expression $\phi(x, y) = X(x) \cdot Y(y)$ into the boundary conditions, we find the boundary conditions for X and Y in order to obtain nontrivial solutions, and X and Y satisfy the following boundary value problems:

$$\begin{aligned} &X'' + \gamma X = 0, \quad 0 < x < a, \qquad &&Y'' + (\lambda - \gamma)Y = 0, \quad 0 < y < b, \\ &X(0) = X(a) = 0, &&Y(0) = Y(b) = 0. \end{aligned}$$

The problem for X is a regular Sturm-Liouville problem, and from Theorem 4.7, the eigenvalues are all nonnegative. Setting $\gamma = \mu^2$, the eigenvalues are

$$\gamma_n = \mu_n^2 = \frac{n^2\pi^2}{a^2},$$

and the corresponding eigenfunctions are

$$X_n(x) = \sin \mu_n x$$

for $n = 1, 2, \ldots$. The problem for Y is also a regular Sturm-Liouville problem, and from Theorem 4.7 the eigenvalues are all nonnegative. Setting $\lambda - \gamma = \omega^2$, the eigenvalues are

$$\omega_m^2 = \frac{m^2\pi^2}{b^2},$$

and the corresponding eigenfunctions are

$$Y_m(y) = \sin \omega_m y$$

for $m = 1, 2, \ldots$.

The product solutions for the boundary value problem satisfied by ϕ can be written as

$$\phi_{mn}(x, y) = X_n(x) \cdot Y_m(y) = \sin \mu_n x \sin \omega_m y,$$

and since

$$\lambda = \gamma + \omega^2 = \mu^2 + \omega^2,$$

we have nontrivial solutions only for

$$\lambda_{mn} = \mu_n^2 + \omega_m^2 = \pi^2 \left(\frac{n^2}{a^2} + \frac{m^2}{b^2} \right).$$

The equation for T becomes

$$T'' + c^2\alpha_{mn}^2 T = 0,$$

where $\lambda_{mn} = \alpha_{mn}^2$, with solution

$$T_{mn}(t) = A_{mn}\cos\alpha_{mn}ct + B_{mn}\sin\alpha_{mn}ct.$$

From the problem for T, we can satisfy the initial condition $T'(0) = 0$ by requiring that $T'_{mn}(0) = 0$, and this implies that $B_{mn} = 0$ for all $m \geqslant 1$ and $n \geqslant 1$. Thus, the corresponding nontrivial solutions to the time equation are given by

$$T_{mn}(t) = A_{mn}\cos\alpha_{mn}ct$$

for $m \geqslant 1$ and $n \geqslant 1$. Applying the superposition principle, we can write the solution as

$$u(x,y,t) = \sum_{m=1}^{\infty}\sum_{n=1}^{\infty}\phi_{mn}(x,y)\cdot T_{mn}(t) = \sum_{m=1}^{\infty}\sum_{n=1}^{\infty} A_{mn}\sin\mu_n x\sin\omega_m y\cos\alpha_{mn}ct,$$

and this satisfies the boundary conditions and the initial condition $u_t(x,y,0) = 0$.

All that remains is to satisfy the last initial condition: $u(x,y,0) = f(x,y)$, and setting $t = 0$, we have

$$f(x,y) = \sum_{m=1}^{\infty}\sum_{n=1}^{\infty} A_{mn}\sin\mu_n x\sin\omega_m y = \sum_{m=1}^{\infty} F_m(x)\sin\omega_m y, \tag{5.6}$$

where we have defined

$$F_m(x) = \sum_{n=1}^{\infty} A_{mn}\sin\mu_n x. \tag{5.7}$$

From (5.6) it is clear that the F_m's are just the Fourier sine coefficients of f, and from (5.7) it is clear that the A_{mn}'s are just the Fourier sine coefficients of the F_m's. Therefore, we have

$$F_m(x) = \frac{2}{b}\int_0^b f(x,y)\sin\omega_m y\,dy,$$

$$A_{mn} = \frac{2}{a}\int_0^a F_m(x)\sin\mu_n x\,dx,$$

and the final solution to the problem is given by

$$u(x,y,t) = \sum_{m=1}^{\infty}\sum_{n=1}^{\infty} A_{mn}\sin\mu_n x\sin\omega_m y\cos\alpha_{mn}ct,$$

where

$$A_{mn} = \frac{4}{ab}\int_0^a \int_0^b f(x,y)\sin\mu_n x \sin\omega_m y\, dy\, dx.$$

■

For the one-dimensional problems, we found the coefficients in the series solution by means of inner products. Can the same be done in this case? Can the formula for the A_{mn}'s be expressed in terms of inner products? If we examine the formula for A_{mn}'s more closely, we see that

$$A_{mn} = \frac{4}{ab}\int_0^a \int_0^b f(x,y)\phi_{mn}(x,y)\, dy\, dx,$$

which resembles $K \cdot \langle f, \phi_{mn}\rangle$ for some constant K and "appropriate" inner product $\langle f, \phi_{mn}\rangle$. In the next section we shall see how this can be done in a systematic manner.

5.7 EIGENFUNCTIONS IN TWO DIMENSIONS

Consider the homogeneous version of either the two-dimensional heat equation or the two-dimensional wave equation:

$$\frac{\partial u}{\partial t} = k\nabla^2 u \qquad \text{or} \qquad \frac{\partial^2 u}{\partial t^2} = c^2\nabla^2 u,$$

together with appropriate boundary conditions and initial conditions. If we apply separation of variables and assume that

$$u(x,y,t) = \phi(x,y)\cdot T(t),$$

we obtain

$$\frac{T'}{kT} = \frac{\nabla^2\phi}{\phi} = -\lambda \qquad \text{or} \qquad \frac{T''}{c^2 T} = \frac{\nabla^2\phi}{\phi} = -\lambda.$$

In either case, we end up with a two-dimensional eigenvalue problem for ϕ, that is, the **Helmholtz equation**,

$$\nabla^2\phi = -\lambda\phi,$$

together with the appropriate boundary conditions.

Typically, problems of this type will have a spectrum that is a countably infinite set with no limit points, that is, an infinite discrete set. In other words, nontrivial solutions exist for an infinite but discrete set of values of λ. The main result of this section will be to show that with the appropriate inner product, eigenfunctions corresponding to distinct eigenvalues are orthogonal. To accomplish these results, we will need **Green's theorem**, along with what are known as **Green's identities**.

Theorem 5.3. (*Green's Theorem*)
If Ω is a domain in the xy-plane with boundary $\partial\Omega$, and P and Q are continuous with continuous partial derivatives in Ω, then

$$\oint_{\partial\Omega} P(x,y)\,dx + Q(x,y)\,dy = \iint\limits_{\Omega} \left(\frac{\partial Q}{\partial x} - \frac{\partial P}{\partial y}\right) dA.$$

The pertinent results are

(a) Green's theorem,

(b) Green's first identity,

(c) Green's second identity,

(d) eigenfunctions are orthogonal.

We will not prove (a), since it is usually proven in a second-year calculus course; we will show that (a) implies (b) implies (c) implies (d).

Theorem 5.4. (*Green's First Identity*)
If u and v are twice continuously differentiable in the region $\Omega \subset \mathbb{R}^2$, then

$$\iint\limits_{\Omega} (u\nabla^2 v + \nabla u \cdot \nabla v)\,dA = \oint_{\partial\Omega} u\frac{\partial v}{\partial n}\,ds.$$

In the equation above, $\partial v/\partial n$ refers to the directional derivative of v in the direction of $\mathbf{n}$, the outward-pointing unit normal to $\partial\Omega$, and s refers to the arc length along $\partial\Omega$.

Proof. Parameterize the boundary $\partial\Omega$ in terms of arc length as follows:

$$\partial\Omega = \{(x,y) \in \mathbb{R}^2 \,:\, x = x(s),\ y = y(s),\ 0 \leqslant s \leqslant L\},$$

where L is the length of the boundary $\partial\Omega$. Let $\mathbf{r}$ be the position vector to a point $(x,y) \in \partial\Omega$, $\mathbf{T}$ be the unit tangent vector at (x,y), and $\mathbf{n}$ be the outward-pointing unit normal vector to $\partial\Omega$ at (x,y); then

$$\mathbf{r} = (x,y), \quad \mathbf{T} = \frac{d\mathbf{r}}{ds} = \left(\frac{dx}{ds}, \frac{dy}{ds}\right), \quad \mathbf{n} = \left(\frac{dy}{ds}, -\frac{dx}{ds}\right).$$

Letting

$$P = u\frac{\partial v}{\partial y}, \qquad Q = -u\frac{\partial v}{\partial x},$$

then

$$\frac{\partial P}{\partial y} = u\frac{\partial^2 v}{\partial y^2} + \frac{\partial u}{\partial y}\frac{\partial v}{\partial y}, \qquad \frac{\partial Q}{\partial x} = -u\frac{\partial^2 v}{\partial x^2} - \frac{\partial u}{\partial x}\frac{\partial v}{\partial x}$$

and

$$\iint_{\Omega} (u\nabla^2 v + \nabla u \cdot \nabla v)\, dA = \iint_{\Omega} \left[u\left(\frac{\partial^2 v}{\partial x^2} + \frac{\partial^2 v}{\partial y^2}\right) + \left(\frac{\partial u}{\partial x}, \frac{\partial u}{\partial y}\right) \cdot \left(\frac{\partial v}{\partial x}, \frac{\partial v}{\partial y}\right)\right] dA$$

$$= \iint_{\Omega} \left[\left(u\frac{\partial^2 v}{\partial x^2} + \frac{\partial u}{\partial x}\frac{\partial v}{\partial x}\right) + \left(u\frac{\partial^2 v}{\partial y^2} + \frac{\partial u}{\partial y}\frac{\partial v}{\partial y}\right)\right] dA,$$

so that

$$\begin{aligned}
\iint_{\Omega} (u\nabla^2 v + \nabla u \cdot \nabla v)\, dA &= \iint_{\Omega} \left(-\frac{\partial Q}{\partial x} + \frac{\partial P}{\partial y}\right) dA \\
&= -\oint_{\partial\Omega} P(x,y)\, dx + Q(x,y)\, dy \\
&= -\oint_{\partial\Omega} \left(u\frac{\partial v}{\partial y}\, dx - u\frac{\partial v}{\partial x}\, dy\right) \\
&= \oint_{\partial\Omega} u\left(-\frac{\partial v}{\partial y}\frac{dx}{ds} + \frac{\partial v}{\partial x}\frac{dy}{ds}\right) ds \\
&= \oint_{\partial\Omega} u\left(\frac{\partial v}{\partial x}, \frac{\partial v}{\partial y}\right) \cdot \left(\frac{dy}{ds}, -\frac{dx}{ds}\right) ds \\
&= \oint_{\partial\Omega} u\,(\nabla v \cdot \mathbf{n})\, ds \\
&= \oint_{\partial\Omega} u\frac{\partial v}{\partial n}\, ds.
\end{aligned}$$

□

The next result is much easier to prove.

Theorem 5.5. (*Green's Second Identity*)
If u and v are twice continuously differentialble in the region $\Omega \subset \mathbb{R}^2$, then

$$\iint_{\Omega} (u\nabla^2 v - v\nabla^2 u)\, dA = \oint_{\partial\Omega} \left(u\frac{\partial v}{\partial n} - v\frac{\partial u}{\partial n}\right) ds.$$

Proof. From Green's first identity, we have

$$\iint_{\Omega} (u\nabla^2 v + \nabla u \cdot \nabla v)\, dA = \oint_{\partial\Omega} u\frac{\partial v}{\partial n}\, ds,$$

and rewriting this with u and v interchanged yields

$$\iint_{\Omega} (v\nabla^2 u + \nabla v \cdot \nabla u)\, dA = \oint_{\partial\Omega} v\frac{\partial u}{\partial n}\, ds.$$

Subtraction gives the result.

□

We now come to one of the two main results of this section.

Theorem 5.6. (*Orthogonality of Eigenfunctions*)
Given the eigenvalue problem for the Helmholtz equation

$$\nabla^2\phi = -\lambda\phi$$

in $\Omega \subseteq \mathbb{R}^2$, with Dirichlet or Neumann boundary conditions, that is,

$$\phi = 0 \qquad \text{or} \qquad \frac{\partial\phi}{\partial n} = 0$$

on $\partial\Omega$, eigenfunctions corresponding to distinct eigenvalues are orthogonal relative to the inner product:

$$\langle f, g\rangle = \iint_{\Omega} f(x,y)g(x,y)\, dA.$$

Proof. Let λ and μ be distinct eigenvalues to one of the problems above, that is, $\lambda \neq \mu$, with corresponding eigenfunctions ϕ and ψ, respectively; then

$$\nabla^2\phi = -\lambda\phi \qquad \text{and} \qquad \nabla^2\psi = -\mu\psi,$$

and therefore,

$$\begin{aligned}\iint_{\Omega} (\psi\nabla^2\phi - \phi\nabla^2\psi)\, dA &= \iint_{\Omega} [\psi(-\lambda\phi) - \phi(-\mu\psi)]\, dA \\ &= (\lambda - \mu)\iint_{\Omega} \phi\psi\, dA \\ &= (\lambda - \mu)\langle \phi, \psi\rangle .\end{aligned}$$

Using Green's second identity, we get

$$\begin{aligned}(\lambda - \mu)\langle \phi, \psi\rangle &= \iint_{\Omega} (\phi\nabla^2\psi - \psi\nabla^2\phi)\, dA \\ &= \oint_{\partial\Omega} \left(\phi\frac{\partial\psi}{\partial n} - \psi\frac{\partial\phi}{\partial n}\right) ds \\ &= 0.\end{aligned}$$

Since $\lambda \neq \mu$, it follows that $\langle \phi, \psi \rangle = 0$.

□

And our final result in this section:

Theorem 5.7. (*Nonnegativity of Eigenvalues*)
The eigenvalues of either of the problems

$$\begin{aligned} \nabla^2 \phi &= -\lambda \phi \quad \text{in} \quad \Omega \\ \phi &= 0 \quad \text{on} \quad \partial\Omega \end{aligned}$$

or

$$\begin{aligned} \nabla^2 \phi &= -\lambda \phi \quad \text{in} \quad \Omega \\ \frac{\partial \phi}{\partial n} &= 0 \quad \text{on} \quad \partial\Omega \end{aligned}$$

are nonnegative.

Proof. Let λ be an eigenvalue of either of the problems above, with corresponding eigenfunction ϕ; then from Green's first identity with $u = v = \phi$, we have

$$\begin{aligned} \lambda \|\phi\|^2 = \lambda \langle \phi, \phi \rangle = \langle \lambda\phi, \phi \rangle = \left\langle -\nabla^2 \phi, \phi \right\rangle &= -\iint_\Omega \phi \nabla^2 \phi \, dA \\ &= -\oint_{\partial\Omega} \phi \frac{\partial \phi}{\partial n} \, ds + \iint_\Omega \nabla\phi \cdot \nabla\phi \, dA. \end{aligned}$$

Since either $\phi = 0$ or $\partial\phi/\partial n = 0$ on $\partial\Omega$, then

$$\lambda \|\phi\|^2 = 0 + \iint_\Omega |\nabla \phi|^2 \, dA \geqslant 0,$$

and since $\|\phi\| > 0$, it follows that $\lambda \geqslant 0$.

□

5.8 SUMMARY

In this chapter we considered more examples for the method of separation of variables in domains with cartesian coordinates. This is done for each of the three classical second-order linear equations. One example covers **Robin boundary conditions** and **production terms** for the heat equation. For the one-dimensional wave equation, we derive **d'Alembert's solution**, which is an important concept for the wave equation. The Fourier series representation of the solution to the one-dimensional wave equation allows us to hear the frequencies of vibration. However, d'Alembert's formula gives an explicit solution to the wave equation, and shows us the wave forms and how the waves move and interact. We give at least two more derivations of this important formula later in the book.

An interesting extension is the study of the method of separation of variables in two-dimensional spatial domains. For Laplace's equation in two dimensions we assume a solution of the form $u(x, y) = X(x) \cdot Y(y)$, and proceed as usual. For the higher-dimensional wave equation and heat equation, we assume a solution of the form $u(x, y, t) = X(x) \cdot Y(y) \cdot T(t)$, and here we have two choices. We may prefer to find a separated solution of the form $u(x, y, t) = \Phi(x, y) \cdot T(t)$ and then separate variables again by assuming a solution for $\Phi(x, y)$ of the form $\Phi(x, y) = X(x) \cdot Y(y)$. Alternatively, we may prefer to find a separated solution of the form $u(x, y, t) = X(x) \cdot Y(y) \cdot T(t)$ directly. The choice is purely a matter of taste, since in either case, we obtain two separation constants. Applying the boundary conditions, we find nontrivial solutions to the resulting ODEs; that is, we find the eigenvalues λ_{nm} and eigenfunctions $\Phi(x, y) = X_n(x) \cdot Y_m(y)$, where now the eigenvalues and eigenfunctions depend on two indices, n and m. The solution is then a generalized Fourier series, where we sum over all combinations of n and m. Finally, we find the coefficients a_{nm} from the initial conditions, and again we have two ways to do this. One way is to consider the $X_n(x)$ as basis functions, write down the corresponding coefficients, and then use $Y_m(y)$ as basis functions. Alternatively, we can consider $\Phi_{nm}(x, y)$ as two-dimensional basis functions and compute the coefficients a_{nm} directly. All this sounds rather complicated, but careful bookkeeping of indices and coefficients will help in mastering these problems.

5.8.1 Problems and Notes

Problems from Part II:

Exercise	13.10	13.11	13.12	13.16	13.17	14.4
Notes						
Exercise	14.5	14.8	14.9	14.12	14.13	14.19
Notes						
Exercise	15.5	15.6	15.7	15.8	15.9	
Notes						

Midterm exam questions:

Exercise	18.6	18.8
Notes		

Final exam questions:

Exercise	19.17
Notes	

You should now be able to do **Midterm Exams 2 and 3**.

CHAPTER 6

POLAR COORDINATES

6.1 INTERIOR DIRICHLET PROBLEM FOR A DISK

There are certain geometries, such as those shown in Figure 6.1, for which polar coordinates are more useful than Cartesian coordinates.

Polar coordinates are defined as follows: Given a point P with Cartesian coodinates (x, y), where $(x, y) \neq (0, 0)$, the **polar coordinates** of P are (r, θ), where

$$x = r\cos\theta,$$
$$y = r\sin\theta$$

(Fig. 6.2). Here r is the distance from the pole O, that is, the origin, to the point P, and θ is the angle between the directed segment $\overline{OP}$ and the positive x-axis.

The Jacobian determinant for the transformation is

$$\frac{\partial(x, y)}{\partial(r, \theta)} = \begin{vmatrix} \cos\theta & -r\sin\theta \\ \sin\theta & r\cos\theta \end{vmatrix} = r,$$

Partial Differential Equations: Theory and Completely Solved Problems, First Edition. By T. Hillen, I. E. Leonard, H. van Roessel

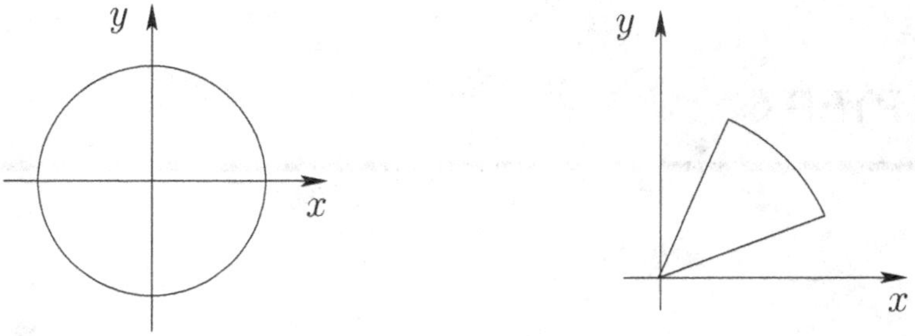

Figure 6.1 Geometries more suited to polar coordinates.

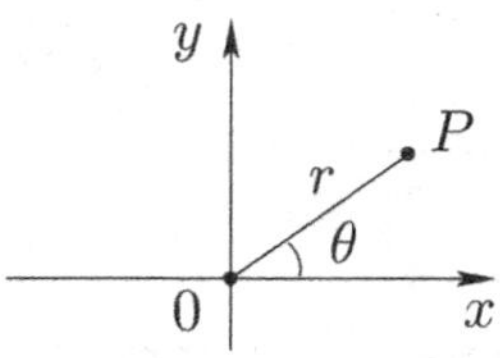

Figure 6.2 Polar coordinate transformation.

which indicates that the transformation is singular at $r = 0$, that is, at the origin. The Laplacian in polar coordinates is

$$\nabla^2 u = \frac{1}{r}\frac{\partial}{\partial r}\left(r\frac{\partial u}{\partial r}\right) + \frac{1}{r^2}\frac{\partial^2 u}{\partial \theta^2}.$$

Solutions of Laplace's equation $\nabla^2 u = 0$ are called **harmonic functions**. Specifically, a function u is **harmonic at a point** P in $\mathbb{R}^2$ if there exists an open disk Ω centered at P such that u has continuous second derivatives and satisfies Laplace's equation at each point in Ω. A function u is **harmonic in a region** R if it is harmonic at each point of R. Similar definitions are made for harmonic functions defined on $\mathbb{R}^n$. Solutions of Laplace's equation are also called **potential functions**, because of the physical significance of the solution.

In the example below we consider Laplace's equation defined in a circular region. The interior Dirichlet problem for Laplace's equation on the disk $D(a)$ models, among other things, the steady-state temperature distribution of a circular plate with top and bottom perfectly insulated and boundaries held at the temperatures given.

Example 6.1. (*Potential in a Disk*)
We want to solve the Dirichlet problem for Laplace's equation,

$$\begin{aligned} &\nabla^2 u = 0 \quad \text{on } D(a), \\ &u(a, \theta) = f(\theta) \quad \text{on } \partial D(a), \end{aligned}$$

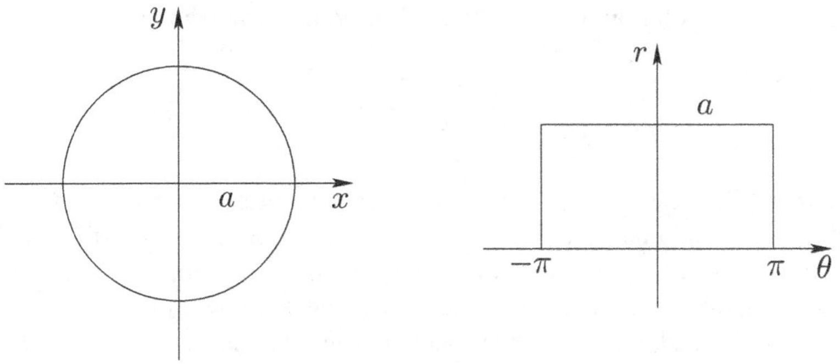

Figure 6.3 Representations of Ω in the xy and $r\theta$ planes.

where $D(a) = \{ (x, y) \mid x^2 + y^2 \leqslant a^2 \}$. Here the appropriate coordinate system consists of plane polar coordinates r and θ, where $x = r\cos\theta$ and $y = r\sin\theta$. The disk above can then be described as $D(a) = \{ (r, \theta) \mid 0 \leqslant r \leqslant a,\ -\pi \leqslant \theta \leqslant \pi \}$. When we consider the domain in the $r\theta$-plane (see (Fig. 6.3), it appears as a rectangle, and three of the boundaries of this rectangle are not real boundaries, or physical boundaries. Nevertheless, we treat the region as a rectangle in the $r\theta$-plane and apply our rule of thumb for side conditions. Hence, we need to impose boundary conditions on these "nonphysical boundaries."

A formal statement of the problem is as follows:

(i) The function $u(r, \theta)$ must satisfy Laplace's equation in polar coordinates r, θ; that is,

$$\nabla^2 u = \frac{\partial^2 u}{\partial r^2} + \frac{1}{r}\frac{\partial u}{\partial r} + \frac{1}{r^2}\frac{\partial^2 u}{\partial \theta^2} = \frac{1}{r}\frac{\partial}{\partial r}\left(r\frac{\partial u}{\partial r}\right) + \frac{1}{r^2}\frac{\partial^2 u}{\partial \theta^2} = 0 \qquad (6.1)$$

for $(r, \theta) \in D(a)$.

(ii) To ensure that the solution is single-valued, $u(r, \theta)$ must satisfy **periodicity conditions** at $\theta = \pm\pi$; that is,

$$u(r, -\pi) = u(r, \pi),$$

$$\frac{\partial u}{\partial \theta}(r, -\pi) = \frac{\partial u}{\partial \theta}(r, \pi)$$

for $0 \leqslant r \leqslant a$.

(iii) To ensure that the solution is continuous, $u(r, \theta)$ must satisfy **boundedness conditions** at $r = 0$; that is,

$$\lim_{r \to 0+} u(r, \theta) = u(0, \theta) \quad \text{(finite)}$$

for $-\pi \leqslant \theta \leqslant \pi$.

(iv) Finally, the solution must satisfy the **boundary condition** at $r = a$; that is,

$$u(a, \theta) = f(\theta)$$

for $-\pi \leqslant \theta \leqslant \pi$.

Assumption (iii) is made to preclude the occurrence of singular solutions that may arise due to the singular nature of the transformation to polar coordinates. While these boundary and boundedness conditions are quite different from those we have experienced before, and are not homogeneous, they do allow us to use separation of variables. To this end, we look for a separable solution, that is, a solution of the form

$$u(r, \theta) = R(r) \cdot \Theta(\theta),$$

and substituting this into Laplace's equation (6.1), we obtain

$$\frac{1}{r}\frac{d}{dr}\left(r\frac{dR}{dr}\right) \cdot \Theta + R \cdot \frac{1}{r^2}\frac{d^2\Theta}{d\theta^2} = 0;$$

that is,

$$r^2 R'' \cdot \Theta + r\,R' \cdot \Theta + R \cdot \Theta'' = 0.$$

Separating variables, we have

$$\frac{\Theta''}{\Theta} = -\frac{r^2 R'' + r\,R'}{R} = -\lambda,$$

where λ is the separation constant, and we have two ordinary differential equations:

- The **angular problem**:

$$\Theta'' + \lambda\Theta = 0.$$

- The **radial problem**:

$$r^2 R'' + r\,R' - \lambda R = 0.$$

Note that the periodicity conditions (ii) imply that

$$R(r) \cdot \Theta(-\pi) = R(r) \cdot \Theta(\pi) \qquad \text{and} \qquad R(r) \cdot \Theta'(-\pi) = R(r) \cdot \Theta'(\pi)$$

for all $0 \leqslant r \leqslant a$, and to obtain a nontrivial solution, we must have

$$\Theta(-\pi) = \Theta(\pi),$$
$$\Theta'(-\pi) = \Theta'(\pi).$$

Therefore, Θ satisfies the boundary value problem

$$\Theta'' + \lambda\Theta = 0,$$
$$\Theta(-\pi) = \Theta(\pi),$$
$$\Theta'(-\pi) = \Theta'(\pi).$$

Here the differential equation is of self-adjoint or Sturm-Liouville form; however, the boundary conditions are not, since each of the boundary conditions involves **both** of the boundary points.

We solve the angular problem first, and as usual we consider three cases:

(a) If $\lambda = 0$, the equation $\Theta'' = 0$ has the general solution

$$\Theta(\theta) = A\theta + B$$

with $\Theta'(\theta) = A$. The first periodicity condition gives

$$-A\pi + B = A\pi + B,$$

so that $A = 0$. The second periodicity condition is then satisfied automatically, so there is one nontrivial solution in this case. The eigenvalue is $\lambda_0 = 0$, with corresponding eigenfunction $\Theta_0 = 1$.

(b) If $\lambda < 0$, say $\lambda = -\mu^2$, where $\mu \neq 0$, the differential equation becomes

$$\Theta'' - \mu^2\Theta = 0$$

and has the general solution

$$\Theta(\theta) = A\cosh\mu\theta + B\sinh\mu\theta$$

with $\Theta'(\theta) = \mu A\sinh\mu\theta + \mu B\cosh\mu\theta$. The first periodicity condition gives

$$A\cosh\mu\pi - B\sinh\mu\pi = A\cosh\mu\pi + B\sinh\mu\pi,$$

since $\cosh\mu\theta$ is an even function and $\sinh\mu\theta$ is an odd function. Therefore, $2B\sinh\mu\pi = 0$, and since $\sinh\mu\pi \neq 0$, $B = 0$, and the solution is $\Theta(\theta) = A\cosh\mu\theta$. The second periodicity condition gives

$$-\mu A\sinh\mu\pi = \mu A\sinh\mu\pi,$$

so that $2\mu A\sinh\mu\pi = 0$. Since $\mu \neq 0$, $\sinh\mu\pi \neq 0$, and we must have $A = 0$. Therefore, there are no nontrivial solutions in this case.

(c) If $\lambda > 0$, say $\lambda = \mu^2$, where $\mu \neq 0$, the differential equation becomes

$$\Theta'' + \mu^2\Theta = 0$$

and has the general solution

$$\Theta(\theta) = A\cos\mu\theta + B\sin\mu\theta$$

with $\Theta'(\theta) = -A\mu\sin\mu\theta + B\mu\cos\mu\theta$. Applying the first periodicity condition, we have

$$\Theta(-\pi) = A\cos\mu\pi - B\sin\mu\pi = A\cos\mu\pi + B\sin\mu\pi = \Theta(\pi),$$

so that $2B \sin \mu\pi = 0$. Applying the second periodicity condition, we have

$$\Theta'(-\pi) = A\mu \sin \mu\pi + B\mu \cos \mu\pi = -A\mu \sin \mu\pi + B\mu \cos \mu\pi = \Theta'(\pi),$$

so that $2A \sin \mu\pi = 0$. Therefore, the following equations must hold simultaneously:

$$A \sin \mu\pi = 0$$
$$B \sin \mu\pi = 0.$$

To get a nontrivial solution, we must have either $A \neq 0$ or $B \neq 0$, and if both equations hold, we must have $\sin \mu\pi = 0$. Therefore, μ must be an integer, so that the eigenvalues are

$$\lambda_n = \mu_n^2 = n^2$$

for $n \geqslant 1$, and the eigenfunctions corresponding to these eigenvalues are $\sin n\theta$ and $\cos n\theta$ for $n \geqslant 1$.

Note that for periodic boundary conditions, we do not have a unique eigenfunction for each λ_n; instead, we have two linearly independent eigenfunctions for each positive eigenvalue $\lambda_n = n^2$.

The eigenvalues and corresponding eigenfunctions for the Θ problem are given by

$$\begin{aligned} \lambda_0 &= 0, & \Theta_0(\theta) &= a_0, \\ \lambda_n &= n^2, & \Theta_n(\theta) &= a_n \cos n\theta + b_n \sin n\theta \end{aligned}$$

for $n \geqslant 1$.

The corresponding problem for R,

$$r^2 R'' + r R' - \lambda R = 0,$$

is a **Cauchy-Euler equation**, and we assume a solution of the form $R(r) = r^s$, so that

$$R'(r) = s r^{s-1} \qquad \text{and} \qquad R''(r) = s(s-1) r^{s-2}.$$

Substituting this into the equation, we have

$$s(s-1)r^s + sr^s - \lambda r^s = 0,$$

and assuming that $r \neq 0$, we get the characteristic equation

$$s(s-1) + s - \lambda = 0;$$

that is, $s^2 = \lambda$, and therefore $s = \pm\sqrt{\lambda}$.
Now we have to consider two cases:

(a) If $n = 0$, then $\lambda_n = 0$ and $s = 0$ is a double root of the characteristic equation, and one solution to the Cauchy-Euler equation is $R(r) = c_1$, that is, a constant solution. To find a second linearly independent solution, we consider the original differential equation for $\lambda = 0$,

$$r \frac{d}{dr}\left(r \frac{dR}{dr}\right) = 0;$$

integrating gives us

$$r \frac{dR}{dr} = c_2,$$

so that

$$\frac{dR}{dr} = \frac{c_2}{r},$$

and a second independent solution is

$$R(r) = c_2 \log r.$$

The general solution to the radial problem for $\lambda_0 = 0$ is then

$$R_0(r) = c_1 + c_2 \log r$$

for $0 < r \leqslant a$.

(b) If $n > 0$, then $\lambda_n = n^2$ and $s = \pm n$, and the general solution to the radial problem for $\lambda_n = n^2$ is

$$R_n(r) = c_3 r^n + c_4 r^{-n}$$

for $0 < r \leqslant a$. From the boundedness condition (iii), we need

$$|u(r, \theta)| < \infty$$

as $r \to 0^+$, so we must have $c_2 = 0$ and $c_4 = 0$, and thus

$$R_0(r) = c_1 \qquad \text{and} \qquad R_n(r) = r^n$$

for $n \geqslant 1$.

Using the superposition principle, we write the final solution as

$$u(r, \theta) = \sum_{n=0}^{\infty} r^n \left(a_n \cos n\theta + b_n \sin n\theta\right) \tag{6.2}$$

and determine the constants from the boundary condition (iv), so that

$$f(\theta) = u(a, \theta) = \sum_{n=0}^{\infty} a^n \left(a_n \cos n\theta + b_n \sin n\theta\right),$$

where

$$a_0 = \frac{1}{2\pi}\int_{-\pi}^{\pi} f(\varphi)\,d\varphi,$$

$$a_n = \frac{1}{\pi a^n}\int_{-\pi}^{\pi} f(\varphi)\cos n\varphi\,d\varphi,$$

$$b_n = \frac{1}{\pi a^n}\int_{-\pi}^{\pi} f(\varphi)\sin n\varphi\,d\varphi$$

for $n \geqslant 1$.

A couple of observations:

- At $r = 0$, we have

$$u(0,\theta) = a_0 = \frac{1}{2\pi}\int_{-\pi}^{\pi} f(\varphi)\,d\varphi = \frac{1}{2\pi}\int_{-\pi}^{\pi} u(a,\varphi)\,d\varphi,$$

  and the value of u at the center of the disk is just the average of u over the boundary.

- If $f(\theta) \equiv f_0$ is a constant, $a_0 = f_0$ and $a_n = b_n = 0$ for all $n \geqslant 1$, so that

$$u(r,\theta) \equiv f_0;$$

  that is, if u is constant on the boundary of the disk, u is constant throughout the disk. This should not be too surprising, since the maximum principle implies that the maximum and minimum of u both occur on the boundary, so if $u = f_0$ on the boundary, then

$$\begin{aligned} f_0 = \min\{u(r,\theta)\,:\,(r,\theta)\in D(a)\} &\leqslant f(r,\theta) \\ &\leqslant \max\{u(r,\theta)\,:\,(r,\theta)\in D(a)\} = f_0 \end{aligned}$$

  for all $(r,\theta) \in D(a)$.

∎

6.1.1 Poisson Integral Formula

We have the solution u of the interior Dirichlet problem for a disk in the form of an infinite series. As it happens, we can express the solution (6.2) as an integral over the boundary of the disk. Substituting the values of a_n and b_n into (6.2), we have

$$u(r,\theta) = \frac{1}{2\pi}\int_{-\pi}^{\pi} f(\varphi)\,d\varphi + \sum_{n=1}^{\infty}\frac{1}{\pi a^n}\int_{-\pi}^{\pi} f(\varphi)r^n \cos n\theta \cos n\varphi\,d\varphi$$

$$+ \sum_{n=1}^{\infty}\frac{1}{\pi a^n}\int_{-\pi}^{\pi} f(\varphi)r^n \sin n\theta \sin n\varphi\,d\varphi$$

$$= \frac{1}{2\pi}\int_{-\pi}^{\pi} f(\varphi)\left(1 + 2\sum_{n=1}^{\infty}\frac{r^n}{a^n}(\cos n\theta\cos n\varphi + \sin n\theta \sin n\varphi)\right)d\varphi$$

$$= \frac{1}{2\pi}\int_{-\pi}^{\pi} f(\varphi)\left(1 + 2\sum_{n=1}^{\infty}\frac{r^n}{a^n}\cos n(\theta-\varphi)\right)d\varphi$$

and

$$u(r,\theta) = \frac{1}{2\pi}\int_{-\pi}^{\pi} f(\varphi)\left(1 + 2\sum_{n=1}^{\infty}\frac{r^n}{a^n}\cos n(\theta-\varphi)\right)d\varphi \tag{6.3}$$

for $0 \leqslant r \leqslant a,\ -\pi \leqslant \theta \leqslant \theta$. We can evaluate the series inside the integral by noting that

$$2\cos n\theta\cos\theta = \cos(n+1)\theta + \cos(n-1)\theta,$$

and if $|b| < 1$, then

$$2\sum_{n=1}^{\infty} b^n \cos n\theta \cos\theta = \frac{1}{b}\sum_{n=1}^{\infty} b^{n+1}\cos(n+1)\theta + b\sum_{n=1}^{\infty} b^{n-1}\cos(n-1)\theta \tag{6.4}$$

$$= \frac{1}{b}\left(\sum_{n=1}^{\infty} b^n\cos n\theta - b\cos\theta\right) + b\left(\sum_{n=1}^{\infty} b^n \cos n\theta + 1\right) \tag{6.5}$$

$$= \left(\frac{1}{b} + b\right)\sum_{n=1}^{\infty} b^n \cos n\theta + b - \cos\theta, \tag{6.6}$$

so that

$$(1 - 2b\cos\theta + b^2)\sum_{n=1}^{\infty} b^n\cos n\theta = b\cos\theta - b^2$$

and

$$\sum_{n=1}^{\infty} b^n \cos n\theta = \frac{b\cos\theta - b^2}{1 - 2b\cos\theta + b^2}.$$

Replacing b by r/a and θ by $\theta - \varphi$ in this sum, (6.3) becomes

$$u(r,\theta) = \frac{1}{2\pi}\int_{-\pi}^{\pi} f(\varphi)\left(1 + 2\frac{ar\cos(\theta-\varphi) - r^2}{a^2 - 2ar\cos(\theta-\varphi) + r^2}\right)d\varphi;$$

that is,

$$u(r,\theta) = \frac{a^2 - r^2}{2\pi} \int_{-\pi}^{\pi} \frac{f(\varphi)}{a^2 - 2ar\cos(\theta - \varphi) + r^2}\, d\varphi \tag{6.7}$$

for $0 \leqslant r \leqslant a,\ -\pi \leqslant \theta \leqslant \pi$. This is called **Poisson's integral formula for the disk** $D(a)$, and gives the unique solution to the interior Dirichlet problem for Laplace's equation on the disk.

To justify the convergence of the series and the interchange of summation and integration, we need f to be piecewise smooth on the interval $[-\pi, \pi]$. However, the integral in (6.7) exists and defines a differentiable function if all we assume is continuity of f. In fact, the Poisson integral formula for u is harmonic in the open disk and satisfies the boundedness and boundary conditions when f is only assumed to be continuous on $[-\pi, \pi]$.

6.2 VIBRATING CIRCULAR MEMBRANE

We use the following boundary value–initial value problem satisfied by a vibrating circular membrane in the plane to introduce Bessel's equation and its solutions.

$$\frac{\partial^2 u}{\partial t^2} = c^2 \left(\frac{\partial^2 u}{\partial r^2} + \frac{1}{r}\frac{\partial u}{\partial r} + \frac{1}{r^2}\frac{\partial^2 u}{\partial \theta^2} \right),$$

$$u(r, -\pi, t) = u(r, \pi, t),$$

$$\frac{\partial u}{\partial \theta}(r, -\pi, t) = \frac{\partial u}{\partial \theta}(r, \pi, t),$$

$$u(a, \theta, t) = 0,$$

$$|u(r,\theta,t)| < \infty \text{ as } r \to 0^+,$$

$$u(r, \theta, 0) = f(r, \theta),$$

$$\frac{\partial u}{\partial t}(r, \theta, 0) = g(r, \theta),$$

for $0 < r < a$, $-\pi < \theta < \pi$, and $t > 0$. We look for a separated solution of the form

$$u(r,\theta,t) = R(r) \cdot \phi(\theta) \cdot G(t),$$

and substituting this into the wave equation, we have

$$R \cdot \phi \cdot G'' = c^2 \left(R'' + \frac{1}{r} R' \right) \cdot \phi \cdot G + \frac{c^2}{r^2} R \cdot \phi'' \cdot G;$$

dividing by $c^2 R \cdot \phi \cdot G$, we have

$$\frac{G''}{c^2 G} = \frac{rR'' + R'}{rR} + \frac{1}{r^2}\frac{\phi''}{\phi} = -\lambda,$$

where λ is a constant, since the left-hand side depends only on t and the right-hand side depends only on r and θ. Thus, we have the two differential equations

$$G'' + \lambda c^2 G = 0 \qquad \text{and} \qquad r^2 R'' \cdot \phi + rR' \cdot \phi + \lambda r^2 R \cdot \phi = -R \cdot \phi'',$$

and separating variables again in the second equation, we have

$$\frac{r^2 R'' + rR' + \lambda r^2 R}{R} = -\frac{\phi''}{\phi} = \mu,$$

where μ is another separation constant, so that

$$r^2 R'' + rR' + (\lambda r^2 - \mu)R = 0 \qquad \text{and} \qquad \phi'' + \mu\phi = 0.$$

We can meet the periodicity conditions and the boundary and boundedness conditions by requiring that

$$\phi(-\pi) = \phi(\pi) \qquad \text{and} \qquad \phi'(-\pi) = \phi'(\pi)$$

and

$$R(a) = 0 \qquad \text{and} \qquad |R(r)| < \infty \quad \text{as} \quad r \to 0^+.$$

Thus, we have the following ordinary differential equations:

- **Angular problem**:

$$\begin{aligned} &\phi'' + \mu\phi = 0, \quad -\pi < \theta < \pi, \\ &\phi(-\pi) = \phi(\pi), \\ &\phi'(-\pi) = \phi'(\pi), \end{aligned}$$

  with eigenvalues and corresponding eigenfunctions

$$\mu_n = n^2 \qquad \text{and} \qquad \phi_n(\theta) = a_n \cos n\theta + b_n \sin n\theta$$

  for $n \geqslant 0$.

- **Radial problem**:

$$\begin{gathered} r^2 R'' + rR' + (\lambda r^2 - n^2)R = 0, \\ R(a) = 0, \\ |R(r)| < \infty \quad \text{as} \quad r \to 0^+. \end{gathered}$$

- **Temporal problem**:

$$G'' + \lambda c^2 G = 0.$$

We can put the radial equation into Sturm-Liouville form:

$$\frac{d}{dr}\left(r\frac{dR}{dr}\right)+\left(\lambda r-\frac{n^2}{r}\right)R=0$$

with

$$p(r)=r,\qquad q(r)=-\frac{n^2}{r},\qquad \sigma(r)=r,$$

and this eigenvalue problem is a singular Sturm-Liouville problem since

$$p(0)=\sigma(0)=0 \quad\text{and}\quad q(r)\to-\infty \quad\text{as}\quad r\to 0^+,$$

and as in Example 6.1, the boundary conditions are not of Sturm-Liouville form. However, we can still find the eigenvalues and eigenfunctions.

If (R,λ) is an eigenpair of the radial equation, the Rayleigh quotient is

$$\lambda=\frac{-rR(r)R'(r)\Big|_0^a+\displaystyle\int_0^a\left[rR'(r)^2+\frac{n^2}{r}R(r)^2\right]dr}{\displaystyle\int_0^a R(r)^2 r\,dr},$$

and from the boundary condition and the boundedness condition, we have

$$-rR(r)R'(r)\Big|_0^a=-aR(a)R'(a)+\lim_{r\to0^+}rR(r)R'(r)=0,$$

and therefore all the eigenvalues are nonnegative. For $\lambda=0$, the radial problem is

$$\begin{gathered}r^2R''+rR'-n^2R=0,\quad 0<r<a,\\ R(a)=0,\\ |R(r)|<\infty \quad\text{as}\quad r\to0^+,\end{gathered}$$

and the differential equation is a Cauchy-Euler equation or an equidimensional equation with solution

$$R_0(r)=A_0\log r+B_0 \quad\text{for}\quad n=0$$

and

$$R_n(r)=A_nr^n+\frac{B_n}{r^n} \quad\text{for}\quad n\geqslant 1.$$

For $\lambda>0$, we can transform the radial equation into an equation that is independent of λ by letting $x=\sqrt{\lambda}\,r$; then

$$\frac{dR}{dr}=\sqrt{\lambda}\frac{dR}{dx}\qquad\text{and}\qquad\frac{d^2R}{dr^2}=\lambda\frac{d^2R}{dx^2},$$

and the equation becomes

$$r^2\frac{d^2R}{dr^2}+r\frac{dR}{dr}+(\lambda r^2-n^2)R=x^2\frac{d^2R}{dx^2}+x\frac{dR}{dx}+(x^2-n^2)R=0,$$

that is,

$$x^2\frac{d^2R}{dx^2}+x\frac{dR}{dx}+(x^2-n^2)R=0,$$

which is **Bessel's equation** of order n (nonparametric).

6.3 BESSEL'S EQUATION

In this section we indicate a method, called the **method of Frobenius**, for solving the nonparametric form of Bessel's equation,

$$x^2 \frac{d^2R}{dx^2} + x\frac{dR}{dx} + (x^2 - \lambda^2)R = 0, \tag{6.8}$$

where λ is a real number, not necessarily an integer. However, we first recall a few results from ODEs.

6.3.1 Series Solutions of ODEs

First we note that the general second-order linear homogeneous ODE can be written in the form

$$\frac{d^2u}{dx^2} + P(x)\frac{du}{dx} + Q(x)u = 0, \tag{6.9}$$

where P and Q are real-valued functions. A point x_0 is called an **ordinary point** of this equation if P and Q are **analytic** at x_0, that is, P and Q can be expanded in a Taylor series valid in an open interval centered at x_0; otherwise, the point x_0 is called a **singular point**. If x_0 is an ordinary point of equation (6.9), two linearly independent solutions of the form

$$u(x) = \sum_{n=0}^{\infty} a_n(x - x_0)^n$$

can be found. If, however, x_0 is a singular point of the equation, this is not possible.

In this case, we can get an idea of what happens by taking a closer look at the Cauchy-Euler equation,

$$x^2 \frac{d^2u}{dx^2} + px\frac{du}{dx} + qu = 0, \tag{6.10}$$

where p and q are constant. If we look for a solution of the form $u = x^s$, substituting this into equation (6.10), we find

$$[s(s-1) + ps + q]x^s = 0,$$

and assuming that $x \neq 0$, we get the quadratic equation, or **characteristic equation**,

$$s^2 + (p-1)s + q = 0. \tag{6.11}$$

Thus, $u = x^s$ is a solution to equation (6.10) if and only if s is a root of the quadratic equation (6.11).

Example 6.2. For the ODE

$$3x^2\frac{d^2u}{dx^2} + 11x\frac{du}{dx} - 3u = 0,$$

the characteristic equation (6.11) becomes

$$s^2 + \tfrac{8}{3}s - 1 = 0,$$

which leads to two linearly independent solutions

$$u_1(x) = x^{\frac{1}{3}} \qquad \text{and} \qquad u_2(x) = \frac{1}{x^3}.$$

In general, the solution to the Cauchy-Euler equation (6.10) is

$$u(x) = \begin{cases} c_1 x^{s_1} + c_2 x^{s_2}, & \text{if } s_1 \neq s_2, \\ c_1 x^s + c_2 x^s \log x, & \text{if } s_1 = s_2 = s, \\ x^\alpha [c_1 \cos(\beta \log x) + c_2 \sin(\beta \log x)], & \text{if } s_1, s_2 = \alpha \pm i\beta. \end{cases} \tag{6.12}$$

∎

So how does this help? When written in standard form,

$$\frac{d^2u}{dx^2} + \frac{p}{x}\frac{du}{dx} + \frac{q}{x^2}u = 0, \tag{6.13}$$

it is clear that $x = 0$ is a singular point of the Cauchy-Euler equation. The solutions will also usually be singular at $x = 0$, as was the case in Example 6.2. This prompts us to make the following definition.

Definition 6.1. Suppose that x_0 is a singular point of the equation

$$\frac{d^2u}{dx^2} + P(x)\frac{du}{dx} + Q(x)u = 0.$$

If $(x - x_0)P(x)$ and $(x - x_0)^2 Q(x)$ are analytic at x_0, x_0 is called a **regular singular point** of the differential equation; otherwise, it is called an **irregular singular point.**

If x_0 is a regular singular point of equation (6.9), P and Q can be written as

$$P(x) = \frac{A(x)}{x - x_0}, \qquad Q(x) = \frac{B(x)}{(x - x_0)^2},$$

where A and B are analytic at x_0. equation (6.9) now becomes

$$\frac{d^2u}{dx^2} + \frac{A(x)}{x - x_0}\frac{du}{dx} + \frac{B(x)}{(x - x_0)^2}u = 0,$$

or alternatively,

$$(x - x_0)^2\frac{d^2u}{dx^2} + A(x)(x - x_0)\frac{du}{dx} + B(x)u = 0. \tag{6.14}$$

Now, equation (6.14) resembles the Cauchy-Euler equation, so we look for a solution of the form

$$u(x) = (x - x_0)^s U(x),$$

where U is analytic at x_0. Since U is assumed to be analytic at x_0, it has a Taylor series expansion valid in an open interval containing x_0, so the form of solution we seek is

$$u(x) = (x - x_0)^s \sum_{n=0}^{\infty} a_n (x - x_0)^n = \sum_{n=0}^{\infty} a_n (x - x_0)^{n+s}, \qquad a_0 \neq 0. \quad (6.15)$$

The series in (6.15) is called a **Frobenius series** and the exponent s is called the **indicial exponent**.

In fact, if x_0 is a regular singular point of (6.9), a solution in the form of a Frobenius series always exists, and a second linearly independent solution will be of the form

$$u(x) = (x - x_0)^r V(x) \quad \text{or} \quad u(x) = (x - x_0)^s U(x) \log(x - x_0) + (x - x_0)^r V(x),$$

where V is analytic at x_0.

Before studying the solutions to Bessel's equation, we note that many differential equations are really Bessel's equation in disguised form.

Example 6.3. Consider the equation

$$x^2 \frac{d^2 u}{dx^2} + (2c+1)x \frac{du}{dx} + \left[a^2 b^2 x^{2b} + (c^2 - \mu^2 b^2)\right] u = 0, \qquad (6.16)$$

where a, b, c, μ are constants (μ is not an integer).

(a) Show that the change of variables defined by

$$s = ax^b \qquad \text{and} \qquad w(s) = x^c u(x)$$

transforms equation (6.16) into Bessel's equation for $w(s)$.

(b) Write the general solution of equation (6.16) in terms of Bessel functions.

(c) Find the solution to **Airy's equation** $u'' + xu = 0$.

Solution.

(a) Let $s = ax^b$ and $w(s) = x^c u(x)$; then

$$x = \left(\frac{s}{a}\right)^{\frac{1}{b}} \qquad \text{and} \qquad u(x) = \frac{w(s)}{\left(\frac{s}{a}\right)^{\frac{c}{b}}}.$$

Now

$$\frac{ds}{dx} = abx^{b-1} = ab\left(\frac{s}{a}\right)^{\frac{b-1}{b}} = a^{\frac{1}{b}} b s^{1-\frac{1}{b}}.$$

Define

$$A = a^{\frac{c}{b}}, \quad \lambda = \frac{1}{b}, \quad B = a^{\frac{1}{b}}b;$$

then

$$\frac{ds}{dx} = Bs^{1-\lambda}, \quad u(x) = As^{-c\lambda}w(s),$$

and

$$\frac{du}{dx} = \frac{d}{ds}\big(As^{-c\lambda}w(s)\big)\,\frac{ds}{dx} = AB\big[s^{1-(1+c)\lambda}w' - c\lambda s^{-(1+c)\lambda}w\big],$$

so that

$$\begin{aligned}\frac{d^2u}{dx^2} &= \frac{d}{ds}\big\{AB\big[s^{1-(1+c)\lambda}w' - c\lambda s^{-(1+c)\lambda}w\big]\big\}\,\frac{ds}{dx}\\ &= AB^2\big\{s^{2-(2+c)\lambda}w'' + [1-(1+2c)\lambda]s^{1-(2+c)\lambda}w'\\ &\qquad\qquad + c(1+c)\lambda^2 s^{-(2+c)\lambda}w\big\}.\end{aligned}$$

Therefore,

$$x^2\,\frac{d^2u}{dx^2} = a^{c\lambda}b^2 s^{-c\lambda}\big[s^2w'' + [1-(1+2c)\lambda]sw' + c(1+c)\lambda^2 w\big]$$

and

$$(2c+1)x\frac{du}{dx} = (1+2c)a^{c\lambda}bs^{-c\lambda}(sw' - c\lambda$$

From (6.16) we have

$$\begin{aligned}&a^{c\lambda}b^2s^{-c\lambda}\{s^2w'' + [1-(1+2c)\lambda]sw' + c(1+c)\lambda^2w\}\\ &+a^{c\lambda}bs^{-c\lambda}(1+2c)(sw'-c\lambda w)+[a^2b^2a^{-2}s^2+c^2-\mu^2b^2]a^{c\lambda}s^{-c\lambda}w = 0;\end{aligned}$$

that is,

$$\begin{aligned}a^{c\lambda}b^2s^{-c\lambda}\Bigg\{&s^2w'' + \left[1-(1+2c)\lambda + \frac{1}{b}(1+2c)\right]sw'\\ &+\left[c(c+1)\lambda^2 - \frac{1}{b}(1+2c)c\lambda + s^2 + \frac{c^2}{b^2} - \mu^2\right]w\Bigg\} = 0.\end{aligned}$$

Therefore,

$$s^2w'' + sw' + (s^2-\mu^2)w = 0, \tag{6.17}$$

which is Bessel's equation of order μ.

(b) As we will see in the next section, since μ is not an integer, the general solution to (6.17) is

$$w(s) = c_1 J_\mu(s) + c_2 J_{-\mu}(s),$$

where J_μ is the Bessel function of the first kind of order μ. Therefore, the solution to the original equation (6.16) is

$$u(x) = x^{-c} w(s) = x^{-c} w(ax^b) = x^{-c}[c_1 J_\mu(ax^b) + c_2 J_{-\mu}(ax^b)].$$

(c) Now we can use the result in (a) and (b) to obtain the general solution of Airy's equation,

$$u'' + xu = 0.$$

The first step is to solve $u'' + xu = 0$ by choosing a, b, c, μ so that (6.16) looks like $u'' + xu = 0$. Since

$$\frac{d^2u}{dx^2} + (2c+1)\frac{1}{x}\frac{du}{dx} + \left[a^2b^2x^{2(b-1)} + (c^2 - \mu^2b^2)x^{-2}\right]u = 0,$$

we need

$$\begin{aligned} 2c + 1 &= 0, \\ 2(b-1) &= 1, \\ a^2b^2 &= 1, \\ c^2 - \mu^2 b^2 &= 0; \end{aligned}$$

that is,

$$c = -\frac{1}{2}, \qquad b = \frac{3}{2}, \qquad a = \frac{2}{3}, \qquad \mu = \frac{1}{3}.$$

The solution to Airy's equation $u'' + xu = 0$ is

$$u(x) = x^{1/2}\left[c_1 J_{\frac{1}{3}}\left(\frac{2}{3}x^{\frac{3}{2}}\right) + c_2 J_{-\frac{1}{3}}\left(\frac{2}{3}x^{\frac{3}{2}}\right)\right].$$

■

6.4 BESSEL FUNCTIONS

Now we find the solutions to Bessel's equation,

$$x^2\frac{d^2u}{dx^2} + x\frac{du}{dx} + (x^2 - \lambda^2)u = 0, \tag{6.18}$$

where λ is a constant (not necessarily an integer). Putting the equation into standard form, we have

$$\frac{d^2u}{dx^2} + \frac{1}{x}\frac{du}{dx} + \frac{x^2 - \lambda^2}{x^2}u = 0$$

with

$$P(x) = \frac{1}{x} \qquad \text{and} \qquad Q(x) = \frac{x^2 - \lambda^2}{x^2}.$$

Clearly, $x_0 = 0$ is a singular point of (6.18); it is, in fact, a regular singular point since

$$xP(x) = 1 \qquad \text{and} \qquad x^2 Q(x) = x^2 - \lambda^2$$

are both analytic at $x_0 = 0$. Thus, we look for a solution to (6.18) in the form of a Frobenius series,

$$u(x) = a_0 x^m + a_1 x^{m+1} + a_2 x^{m+2} + \cdots + a_k x^{m+k} + \cdots,$$

where m is the indicial exponent; that is, we look for a series solution

$$u(x) = x^m \sum_{k=0}^{\infty} a_k x^k = \sum_{k=0}^{\infty} a_k x^{m+k},$$

where $a_0 \neq 0$. Differentiating the series, we have

$$\frac{du}{dx} = \sum_{k=0}^{\infty} (m+k) a_k x^{m+k-1}$$

and

$$\frac{d^2u}{dx^2} = \sum_{k=0}^{\infty} (m+k)(m+k-1) a_k x^{m+k-2}.$$

Substituting this into the differential equation, we have

$$\sum_{k=0}^{\infty} \left[(m+k)(m+k-1) + (m+k) - \lambda^2\right] a_k x^{m+k} + \sum_{k=0}^{\infty} a_k x^{m+k+2} = 0,$$

and reindexing the last sum, we have

$$\sum_{k=0}^{\infty} \left[(m+k)(m+k-1) + (m+k) - \lambda^2\right] a_k x^{m+k} + \sum_{k=2}^{\infty} a_{k-2} x^{m+k} = 0.$$

Therefore,

$$\sum_{k=0}^{\infty} \left[(m+k)^2 - \lambda^2\right] a_k x^{m+k} + \sum_{k=2}^{\infty} a_{k-2} x^{m+k} = 0;$$

that is,

$$(m^2 - \lambda^2) a_0 x^m + \left[(m+1)^2 - \lambda^2\right] a_1 x^{m+1} + \sum_{k=2}^{\infty} \left\{\left[(m+k)^2 - \lambda^2\right] a_k + a_{k-2}\right\} x^{m+k} = 0.$$

This must be an identity in x; hence, all coefficients must vanish, so that

$$(m^2 - \lambda^2)a_0 = 0$$
$$\left[(m+1)^2 - \lambda^2\right] a_1 = 0$$
$$\left[(m+k)^2 - \lambda^2\right] a_k + a_{k-2} = 0$$

for $k \geqslant 2$. Since $a_0 \neq 0$, we get the **indicial equation** from the first equation,

$$m^2 - \lambda^2 = 0,$$

so that $m = \pm\lambda$, and for the moment we choose $m = \lambda$. Also, with this choice of m, $(m+1)^2 - \lambda^2 \neq 0$, and from the second equation we must have $a_1 = 0$ (unless $\lambda = -1/2$).

Finally, from the third equation we get the **recurrence relation**

$$a_k = -\frac{1}{(k+\lambda)^2 - \lambda^2} a_{k-2} = -\frac{1}{k(2\lambda + k)} a_{k-2}$$

for $k \geqslant 2$. Using the fact that $a_1 = 0$, from the recurrence relation, we find that

$$a_3 = 0, \quad a_5 = 0, \quad a_7 = 0, \quad a_9 = 0, \quad \ldots;$$

that is, $a_{2k+1} = 0$ for $k \geqslant 0$. Also, from the recurrence relation,

$$a_2 = -\frac{1}{1 \cdot (\lambda+1)} \cdot \frac{1}{2^2} a_0,$$

$$a_4 = -\frac{1}{2 \cdot (\lambda+2)} \cdot \frac{1}{2^2} a_2,$$

$$a_6 = -\frac{1}{3 \cdot (\lambda+3)} \cdot \frac{1}{2^2} a_4,$$

$$\vdots$$

$$a_{2k} = -\frac{1}{k \cdot (\lambda+k)} \cdot \frac{1}{2^2} a_{2k-2}.$$

If we multiply the terms on the left and multiply the terms on the right, and cancel the common factor $a_2 \cdot a_4 \cdot a_6 \cdots a_{2k-2}$, we get

$$a_{2k} = \frac{(-1)^k}{k!(\lambda+1)(\lambda+2)\cdots(\lambda+k)} \cdot \frac{1}{2^{2k}} a_0$$

for $k \geqslant 1$. Therefore, the solution is (recall that $m = \lambda$)

$$u(x) = a_0 x^\lambda + \sum_{k=1}^{\infty} a_{2k} x^{\lambda+2k},$$

and using the ratio test we see that this series converges absolutely and uniformly for all real numbers x.

Note that the constant a_0 can have any nonzero value, and if we choose

$$a_0 = \frac{1}{2^\lambda \Gamma(1+\lambda)},$$

where $\Gamma(x)$ is the **gamma function**, (see Appendix A)

$$\Gamma(x) = \int_0^\infty t^{x-1}\, e^{-t}\, dt,$$

we get

$$a_{2k} = \frac{(-1)^k}{2^{\lambda+2k}\, k!\, \Gamma(1+\lambda)(1+\lambda)(2+\lambda)(3+\lambda)(4+\lambda)\cdots(k+\lambda)}$$

$$= \frac{(-1)^k}{2^{\lambda+2k}\, k!\, \Gamma(k+\lambda+1)}.$$

The solution becomes

$$u(x) = \frac{x^\lambda}{2^\lambda \Gamma(1+\lambda)} + \sum_{k=1}^\infty \frac{(-1)^k x^{\lambda+2k}}{2^{\lambda+2k}\, k!\, \Gamma(k+\lambda+1)},$$

and we denote this function by $J_\lambda(x)$, so that

$$J_\lambda(x) = \frac{1}{\Gamma(1+\lambda)} \left(\frac{x}{2}\right)^\lambda + \sum_{k=1}^\infty \frac{(-1)^k}{k!\, \Gamma(k+\lambda+1)} \left(\frac{x}{2}\right)^{\lambda+2k};$$

that is,

$$J_\lambda(x) = \sum_{k=0}^\infty \frac{(-1)^k}{k!\, \Gamma(k+\lambda+1)} \left(\frac{x}{2}\right)^{\lambda+2k}, \tag{6.19}$$

and the function $J_\lambda(x)$, which is one solution to Bessel's equation (6.18), is called the **Bessel function of the first kind of order** λ.

If λ is not an integer, a second linearly independent solution (recall that the indicial exponents are $m = \pm\lambda$) is given by $J_{-\lambda}$. In this case the general solution is

$$u(x) = AJ_\lambda(x) + BJ_{-\lambda}(x),$$

where A and B are arbitrary constants. If λ is a nonnegative integer, say $\lambda = n$, the Bessel function of order n is given by

$$J_n(x) = \sum_{k=0}^\infty \frac{(-1)^k}{k!(n+k)!} \left(\frac{x}{2}\right)^{n+2k}$$

for $n \geqslant 0$. Note that in this case, the roots of the indicial equation $\lambda_1 = n$ and $\lambda_2 = -n$ differ by an integer, and J_n and J_{-n} are linearly dependent; in fact,

$$J_{-n}(x) = (-1)^n J_n(x)$$

for $n \geqslant 0$. If we replace λ by the integer $-n$ in equation (6.19), and use the fact that since the gamma function is infinite at the negative integers (see Appendix A), we may consider

$$\frac{1}{\Gamma(-n+k+1)} = 0$$

for $0 \leqslant k \leqslant n-1$, and thus

$$\begin{aligned} J_{-n}(x) &= \sum_{k=n}^{\infty} \frac{(-1)^k}{k!\,\Gamma(-n+k+1)} \left(\frac{x}{2}\right)^{-n+2k} \\ &= \sum_{k=0}^{\infty} \frac{(-1)^{k+n}}{(k+n)!\,\Gamma(k+1)} \left(\frac{x}{2}\right)^{n+2k} \\ &= (-1)^n \sum_{k=0}^{\infty} \frac{(-1)^k}{k!\,(n+k)!} \left(\frac{x}{2}\right)^{n+2k}, \end{aligned}$$

where we reindexed the sum in the first equation. Thus, $J_{-n}(x) = (-1)^n J_n(x)$. From the above it follows that

$J_n(x)$ is an **even** function if $n = 0, 2, 4, \ldots$

$J_n(x)$ is an **odd** function if $n = 1, 3, 5, \ldots$

and $J_0(0) = 1$, while $J_n(0) = 0$ for $n \geqslant 1$. The Bessel function of order 0 is given by

$$J_0(x) = \sum_{k=0}^{\infty} \frac{(-1)^k}{(k!)^2} \left(\frac{x}{2}\right)^{2k},$$

and since $2^{2k}(k!)^2 = 2^2 \cdot 4^2 \cdot 6^2 \cdots (2k)^2$, we can rewrite this as

$$J_0(x) = 1 + \sum_{k=1}^{\infty} \frac{(-1)^k\, x^{2k}}{2^2 4^2 6^2 \cdots (2k)^2} = 1 - \frac{x^2}{2^2} + \frac{x^4}{2^2 4^2} - \frac{x^6}{2^2 4^2 6^2} + \cdots,$$

which looks similar to the Maclaurin series for $\cos x$.

There are also similarities between the derivatives of $J_0(x)$ and $\cos x$; for example,

$$\begin{aligned} J_0{}'(x) &= \sum_{k=0}^{\infty} \frac{(-1)^k}{(k!)^2} \frac{(2k)}{2} \left(\frac{x}{2}\right)^{2k-1} = \sum_{k=1}^{\infty} \frac{(-1)^k}{k!(k-1)!} \left(\frac{x}{2}\right)^{2k-1} \\ &= \sum_{\ell=0}^{\infty} \frac{(-1)^{\ell+1}}{\ell!(\ell+1)!} \left(\frac{x}{2}\right)^{2\ell+1}; \end{aligned}$$

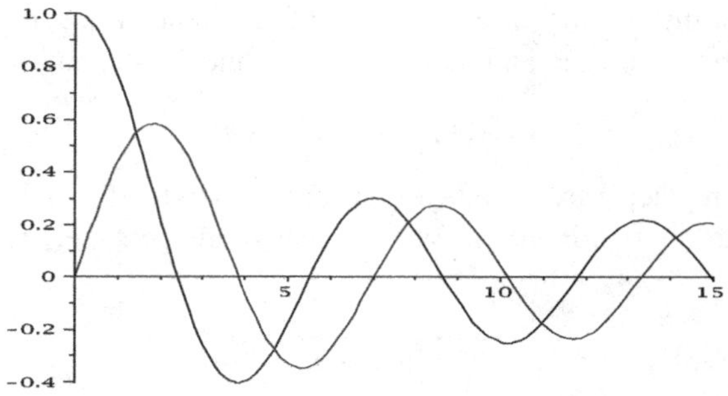

Figure 6.4 $J_0(x)$ and $J_1(x)$

that is,

$$J_0{}'(x) = -\sum_{k=0}^{\infty} \frac{(-1)^k}{k!(k+1)!} \left(\frac{x}{2}\right)^{2k+1} = -J_1(x)$$

for all x.

Graphs of $J_0(x)$ and $J_1(x)$ are shown in Figure 6.4.

When $\lambda = n$ is an integer, a second linearly independent solution can be obtained as follows: If we let

$$Y_\nu(x) = \frac{J_\nu(x)\cos(\nu\pi) - J_{-\nu}(x)}{\sin(\nu\pi)},$$

a second linearly independent solution to Bessel's equation is given by

$$Y_n(x) = \lim_{\nu \to n} Y_\nu(x),$$

and using L'Hospital's rule, we have

$$Y_n(x) = \lim_{\nu \to n} \left(\frac{\dfrac{\partial J_\nu(x)}{\partial \nu}\cos\nu\pi - \pi J_\nu(x)\sin\nu\pi - \dfrac{\partial J_{-\nu}(x)}{\partial \nu}}{\pi\cos\nu\pi} \right)$$

$$= \frac{1}{\pi}\left[\frac{\partial J_\nu(x)}{\partial \nu} - (-1)^n \frac{\partial J_{-\nu}(x)}{\partial \nu}\right]_{\nu=n}.$$

It can be shown that we can write this second linearly independent solution as

$$Y_n(x) = \frac{2}{\pi}\left[J_n(x)\left(\gamma + \log\frac{x}{2}\right) - \frac{1}{2}\sum_{k=0}^{n-1} \frac{(n-k-1)!}{k!}\left(\frac{x}{2}\right)^{2k-n}\right.$$

$$\left. - \frac{1}{2}\sum_{k=1}^{\infty} \frac{(-1)^k\left[\phi(k) + \phi(k+n)\right]\left(\frac{x}{2}\right)^{n+2k}}{k!(n+k)!}\right],$$

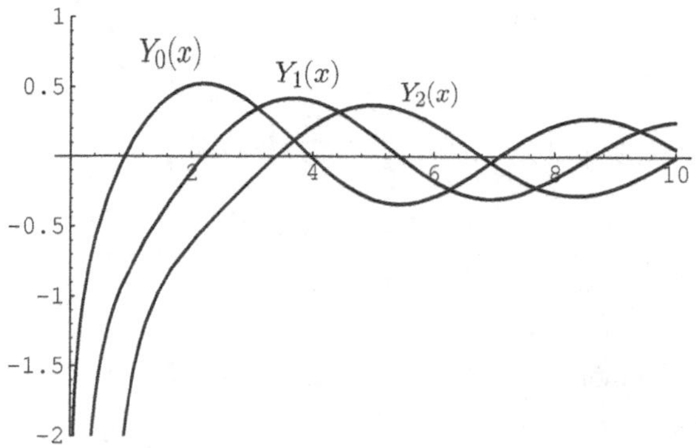

Figure 6.5 Bessel functions of the second kind of integer order.

where

$$\phi(k) = 1 + \frac{1}{2} + \frac{1}{3} + \cdots + \frac{1}{k}$$

and γ is the **Euler-Mascheroni** constant,

$$\gamma = \lim_{k\to\infty}\left(1 + \frac{1}{2} + \frac{1}{3} + \cdots + \frac{1}{k} - \log k\right) = 0.5772157\ldots.$$

The function $Y_n(x)$ is called a **Bessel function of the second kind of order** n, or a **Neumann function**, or a **Weber function**. The functions $Y_0(x), Y_1(x)$, and $Y_2(x)$ are plotted in Figure 6.5.

It can be shown that J_n and Y_n are linearly independent, and the general solution of Bessel's equation of integer order $\lambda = n$ is

$$u(x) = A\,J_n(x) + B\,Y_n(x),$$

where A and B are arbitrary constants.

6.4.1 Properties of Bessel Functions

The asymptotic behavior of J_λ and Y_λ is described below. Here, the ratio of two quantities connected by the symbol $\sim$ approaches 1 as x tends to the relevant limit (see Hildebrand [28], p. 148).

- For small values of x, that is, as $x \to 0$,

$$\begin{aligned}
J_0(x) &\sim 1,\\
J_\lambda(x) &\sim \frac{1}{2^\lambda\,\Gamma(1+\lambda)}\,x^\lambda,\\
Y_0(x) &\sim \frac{2}{\pi}\,\log x,\\
Y_\lambda(x) &\sim -\frac{2^\lambda\,\Gamma(\lambda)}{\pi}\,\frac{1}{x^\lambda} \text{ for } \lambda \neq 0.
\end{aligned}$$

- For large values of x, that is, as $x \to \infty$,

$$J_\lambda(x) \sim \sqrt{\frac{2}{\pi x}} \cos\left(x - \frac{\pi}{4} - \frac{\lambda\pi}{2}\right),$$

$$Y_\lambda(x) \sim \sqrt{\frac{2}{\pi x}} \sin\left(x - \frac{\pi}{4} - \frac{\lambda\pi}{2}\right).$$

Note that only the functions $J_\lambda(x)$ are finite at $x = 0$ (when $\lambda \geqslant 0$), while the functions $Y_\lambda(x)$ are singular at the origin.

We also note that the Bessel functions are oscillatory and the spacing between the zeros approaches π. Let x_n and x_{n+1} be two consecutive zeros of $J_\lambda(x)$. If n is large, the quantities

$$x_n - \frac{\pi}{4} - \frac{\lambda\pi}{2} \qquad \text{and} \qquad x_{n+1} - \frac{\pi}{4} - \frac{\lambda\pi}{2}$$

are close to two consecutive zeros of the cosine function, say

$$(2k+1)\frac{\pi}{2} \qquad \text{and} \qquad (2k+3)\frac{\pi}{2};$$

therefore,

$$x_{n+1} - x_n \sim [(2k+3) - (2k+1)]\,\frac{\pi}{2} = \pi$$

as $n \to \infty$ (see Hildebrand [28], p. 148). Recall that for a nonnegative integer n,

$$J_n(x) = \sum_{k=0}^{\infty} \frac{(-1)^k}{k!(n+k)!} \left(\frac{x}{2}\right)^{n+2k} \tag{6.20}$$

for $-\infty < x < \infty$.

The following relations hold among Bessel functions and their derivatives, and are true for $J_n(x)$ as well as $Y_n(x)$, whether or not n is an integer.

Theorem 6.2. We have

(i) $\dfrac{d}{dx}\left[x^n J_n(x)\right] = x^n J_{n-1}(x) \quad$ for $\quad n \geqslant 1$, and

(ii) $\dfrac{d}{dx}\left[x^{-n} J_n(x)\right] = -x^{-n} J_{n+1}(x) \quad$ for $\quad n \geqslant 0$.

Proof. We prove part (i) and leave part (ii) as an exercise. Differentiating (6.20), we have

$$\frac{d}{dx}\left[x^n J_n(x)\right] = \frac{d}{dx}\left[\sum_{k=0}^{\infty} \frac{(-1)^k}{k!(n+k)!}\left(\frac{x}{2}\right)^{2n+2k} \cdot 2^n\right]$$

$$= \sum_{k=0}^{\infty} \frac{(-1)^k 2(n+k)}{k!(n+k)!}\left(\frac{x}{2}\right)^{2n+2k-1} \cdot 2^{n-1}$$

$$= x^n \sum_{k=0}^{\infty} \frac{(-1)^k}{k!(n-1+k)!}\left(\frac{x}{2}\right)^{2k+n-1}$$

$$= x^n J_{n-1}(x).$$

□

We also have the following recurrence relations for $J_n(x)$ and $J_n{}'(x)$:

Theorem 6.3. We have

(i) $J_n{}'(x) + \frac{n}{x} J_n(x) = J_{n-1}(x)$;

(ii) $J_n{}'(x) - \frac{n}{x} J_n(x) = -J_{n+1}(x)$;

(iii) $2J_n{}'(x) = J_{n-1}(x) - J_{n+1}(x)$;

(iv) $\frac{2n}{x} J_n(x) = J_{n-1}(x) + J_{n+1}(x)$.

Proof. From Theorem 6.2 we have

$$\frac{d}{dx}\left[x^n J_n(x)\right] = x^n J_n{}'(x) + n x^{n-1} J_n(x) = x^n J_{n-1}(x)$$

and

$$\frac{d}{dx}\left[x^{-n} J_n(x)\right] = x^{-n} J_n{}'(x) - n x^{-n-1} J_n(x) = x^{-n} J_{n+1}(x).$$

Dividing the first equation by x^n and multiplying the second equation by x^n, we get (i) and (ii). Adding and subtracting (i) and (ii), we get (iii) and (iv).

□

Note that from the above it is clear that every Bessel function $J_n(x)$ with n an integer can be expressed in terms of $J_0(x)$ and $J_1(x)$; for example, taking $n = 1$ in (iv), we have

$$J_2(x) = \frac{2}{x} J_1(x) - J_0(x),$$

and taking $n = 2$ in (iv), we have

$$J_3(x) = -\frac{4}{x} J_0(x) - \left(1 - \frac{8}{x^2}\right) J_1(x).$$

Also, we can write the differentiation formulas

$$\frac{d}{dx}\left[x^n J_n(x)\right] = x^n J_{n-1}(x) \quad \text{for} \quad n \geqslant 1$$

and

$$\frac{d}{dx}\left[x^{-n} J_n(x)\right] = -x^{-n} J_{n+1}(x) \quad \text{for} \quad n \geqslant 0$$

as integral formulas,

$$\int x^n J_{n-1}(x)\, dx = x^n J_n(x) + C$$

and

$$\int x^{-n} J_{n+1}(x)\, dx = -x^{-n} J_n(x) + C,$$

where C is constant. For example, when $n = 1$, the first equation yields

$$\int x\, J_0(x)\, dx = x\, J_1(x) + C,$$

and when $n = 0$, the second equation yields

$$\int J_1(x)\, dx = -J_0(x) + C.$$

6.4.2 Integral Representation of Bessel Functions

We will show that the following integral representation holds for Bessel functions:

Theorem 6.4.

$$J_n(x) = \frac{1}{\pi} \int_0^{\pi} \cos(n\theta - x\sin\theta)\, d\theta$$

for all integers n.

Proof. If n is a positive integer, since

$$J_{-n}(x) = (-1)^n J_n(x)$$

for all x, we can restrict our proof to nonnegative integers. We have

$$\frac{1}{\pi} \int_0^{\pi} \underbrace{\cos(n\theta - x\sin\theta)}_{\text{even}}\, d\theta = \frac{1}{2\pi} \int_{-\pi}^{\pi} \cos(n\theta - x\sin\theta)\, d\theta,$$

and replacing θ by $-\theta$, we have

$$\begin{aligned}\frac{1}{\pi}\int_0^\pi \cos(n\theta - x\sin\theta)\,d\theta &= \frac{1}{2\pi}\int_{-\pi}^{\pi} \cos(x\sin\theta - n\theta)\,d\theta \\ &= \frac{1}{2\pi}\int_{-\pi}^{\pi} \Big[\cos(x\sin\theta - n\theta) \\ &\qquad\qquad + i\underbrace{\sin(x\sin\theta - n\theta)}_{\text{odd}}\Big]\,d\theta \\ &= \frac{1}{2\pi}\int_{-\pi}^{\pi} e^{i(x\sin\theta - n\theta)}\,d\theta.\end{aligned}$$

From the Maclaurin series expansion for the exponential function, we have

$$e^{ix\sin\theta} = \sum_{k=0}^{\infty}\frac{(ix\sin\theta)^k}{k!},$$

and for each fixed x, this series converges absolutely and uniformly for $-\pi \leqslant \theta \leqslant \pi$, and we can integrate term by term to get

$$\int_{-\pi}^{\pi} e^{ix\sin\theta}e^{-in\theta}\,d\theta = \sum_{k=0}^{\infty}\frac{(ix)^k}{k!}\int_{-\pi}^{\pi}\sin^k\theta\, e^{-in\theta}\,d\theta. \tag{6.21}$$

We use Euler's formula to evaluate the integral inside the summation; we have

$$\sin^k\theta = \left(\frac{e^{i\theta} - e^{-i\theta}}{2i}\right)^k$$

and from the binomial theorem,

$$\sin^k\theta = \frac{1}{(2i)^k}\sum_{m=0}^{k}(-1)^{k-m}\binom{k}{m}e^{im\theta}e^{-i(k-m)\theta};$$

that is,

$$\sin^k\theta = \frac{1}{(2i)^k}\sum_{m=0}^{k}(-1)^{k-m}\binom{k}{m}e^{i(2m-k)\theta}. \tag{6.22}$$

Note that this is just the complex form of the Fourier series for $\sin^k\theta$ on the interval $[-\pi,\pi]$. Now multiply (6.22) by $e^{-in\theta}$ and integrate over the interval $[-\pi,\pi]$ to get

$$\int_{-\pi}^{\pi}\sin^k\theta\, e^{-in\theta}\,d\theta = \frac{1}{(2i)^k}\sum_{m=0}^{k}(-1)^{k-m}\binom{k}{m}\int_{-\pi}^{\pi}e^{i(2m-k-n)\theta}\,d\theta, \tag{6.23}$$

and by orthogonality, all the integrals are zero except when $2m - k - n = 0$, and in this case the integral has the value 2π. Therefore,

$$\frac{1}{2\pi}\int_{-\pi}^{\pi} \sin^k\theta\, e^{-in\theta}\, d\theta = \begin{cases} \dfrac{(-1)^{k-m}}{(2i)^k}\dbinom{k}{m}, & \text{for } \ 0 \leqslant k+n = 2m \leqslant 2k \\[2ex] 0, & \text{for } \ k+n \quad \text{odd.} \end{cases} \tag{6.24}$$

For a fixed n, the nonzero coefficients are obtained when

$$k = n,\ n+2,\ n+4,\ n+6,\ \ldots,$$

since then $k = n + 2j$, and $k + n = 2(n + j)$ is even and $k + n = k + k - 2j \leqslant 2k$. Therefore, from (6.21), we have

$$\begin{aligned} \frac{1}{2\pi}\int_{-\pi}^{\pi} e^{ix\sin\theta} e^{-in\theta}\, d\theta &= \sum_{k=0}^{\infty} \frac{(ix)^k}{k!}\int_{-\pi}^{\pi} \sin^k\theta\, e^{-in\theta}\, d\theta \\ &= \sum_{j=0}^{\infty} \frac{(ix)^{n+2j}}{(n+2j)!}\int_{-\pi}^{\pi} \sin^{n+2j}\theta\, e^{-in\theta}\, d\theta, \end{aligned}$$

where in (6.24), we put $k = n + 2j$ and $m = n + j$, so that

$$\begin{aligned} \frac{1}{\pi}\int_{0}^{\pi} \cos(n\theta - x\sin\theta)\, d\theta &= \frac{1}{2\pi}\int_{-\pi}^{\pi} e^{i(x\sin\theta - n\theta)}\, d\theta \\ &= \sum_{j=0}^{\infty} \frac{(ix)^{n+2j}}{(n+2j)!}\, \frac{(-1)^{n+2j-(n+j)}}{(2i)^{n+2j}}\binom{n+2j}{n+j} \\ &= \sum_{j=0}^{\infty} (-1)^j \frac{x^{n+2j}}{2^{n+2j}\, j!\,(n+j)!} \\ &= \sum_{j=0}^{\infty} \frac{(-1)^j}{j!(n+j)!}\left(\frac{x}{2}\right)^{n+2j} \\ &= J_n(x). \end{aligned}$$

□

From this we find the generating function for the Bessel functions.

Corollary 6.5.

$$e^{ix\sin\theta} = \sum_{n=-\infty}^{\infty} J_n(x)\, e^{in\theta}$$

for all $x,\ \theta \in \mathbb{R}$.

Proof. If we define

$$f(\theta) = e^{ix\sin\theta} = \cos(x\sin\theta) + i\sin(x\sin\theta)$$

for $-\pi \leqslant x \leqslant \pi$, the complex Fourier coefficients of $f(\theta)$ are

$$c_n = \frac{1}{2\pi}\int_{-\pi}^{\pi} f(\theta)\, e^{-in\theta}\, d\theta;$$

that is,

$$c_n = \frac{1}{2\pi}\int_{-\pi}^{\pi} e^{i(x\sin\theta - n\theta)}\, d\theta = J_n(x).$$

From Dirichlet's theorem, since the function $f(\theta)$ is a continuously differentiable 2π-periodic function, the Fourier series converges to $f(\theta)$ for all θ, so that

$$e^{ix\sin\theta} = \sum_{n=-\infty}^{\infty} c_n\, e^{in\theta} = \sum_{n=-\infty}^{\infty} J_n(x)\, e^{in\theta}$$

for all $x,\ \theta \in \mathbb{R}$.

□

Corollary 6.6. For $n \geqslant 0$ we have

$$|J_n(x)| \leqslant 1$$

for all real numbers x.

Proof. We have

$$\begin{aligned} |J_n(x)| &= \left|\frac{1}{2\pi}\int_{-\pi}^{\pi} e^{i(x\sin\theta - n\theta)}\, d\theta\right| \\ &\leqslant \frac{1}{2\pi}\int_{-\pi}^{\pi} \left|e^{i(x\sin\theta - n\theta)}\right| d\theta \\ &= \frac{1}{2\pi}\int_{-\pi}^{\pi} 1\, dx = 1, \end{aligned}$$

since the function $e^{i(x\sin\theta - n\theta)}$ has absolute value 1.

□

Corollary 6.7. For $n \geqslant 0$ we have

$$\sin x = 2\sum_{n=0}^{\infty}(-1)^n J_{2n+1}(x),$$

$$\cos x = J_0(x) + 2\sum_{n=1}^{\infty}(-1)^n J_{2n}(x)$$

for all real numbers x.

Proof. From Corollary 6.5, since $J_{-n}(x) = (-1)^n J_n(x)$, we have

$$
\begin{aligned}
e^{ix\sin\theta} &= \sum_{n=-\infty}^{\infty} J_n(x)\, e^{in\theta} \\
&= \sum_{n=-\infty}^{-1} J_n(x)e^{in\theta} + J_0(x) + \sum_{n=1}^{\infty} J_n(x)e^{in\theta} \\
&= \sum_{m=1}^{\infty} J_{-m}(x)e^{-im\theta} + J_0(x) + \sum_{n=1}^{\infty} J_n(x)e^{in\theta} \\
&= J_0(x) + \sum_{n=1}^{\infty} J_n(x)\left[e^{in\theta} + (-1)^n e^{-in\theta}\right] \\
&= J_0(x) + \sum_{n=1}^{\infty} [1 + (-1)^n]\, J_n(x)\cos n\theta \\
&\qquad + i\sum_{n=1}^{\infty} [1 - (-1)^n]\, J_n(x)\sin n\theta \\
&= J_0(x) + 2\sum_{n=1}^{\infty} J_{2n}(x)\cos 2n\theta \\
&\qquad + 2i\sum_{n=0}^{\infty} J_{2n+1}(x)\sin(2n+1)\theta.
\end{aligned}
$$

Taking real and imaginary parts, we have

$$\cos(x\sin\theta) = J_0(x) + 2\sum_{n=1}^{\infty} J_{2n}(x)\cos 2n\theta$$

and

$$\sin(x\sin\theta) = 2\sum_{n=0}^{\infty} J_{2n+1}(x)\sin(2n+1)\theta$$

for all real numbers x. Putting $\theta = \pi/2$, we get

$$\cos x = J_0(x) + 2\sum_{n=1}^{\infty} J_{2n}(x)\cos n\pi = J_0(x) + 2\sum_{n=1}^{\infty}(-1)^n J_{2n}(x)$$

and

$$\sin x = 2\sum_{n=0}^{\infty} J_{2n+1}(x)\sin\frac{(2n+1)\pi}{2} = 2\sum_{n=0}^{\infty}(-1)^n J_{2n+1}(x)$$

for all real numbers x.

□

Corollary 6.8. For $n = 0, 1, 2, \ldots$ we have

$$J_{2n}(x) = \frac{1}{\pi}\int_0^{\pi} \cos(x\sin\theta)\cos 2n\theta\, d\theta$$

and

$$J_{2n+1}(x) = \frac{1}{\pi}\int_0^{\pi} \sin(x\sin\theta)\sin(2n+1)\theta\, d\theta.$$

In particular,

$$J_0(x) = \frac{1}{\pi}\int_0^{\pi} \cos(x\sin\theta)\, d\theta \qquad \text{and} \qquad J_1(x) = \frac{1}{\pi}\int_0^{\pi} \sin(x\sin\theta)\sin\theta\, d\theta.$$

Proof. From the proof of Corollary 6.7, we have

$$e^{ix\sin\theta} = J_0(x) + \sum_{n=1}^{\infty}\left[1+(-1)^n\right]J_n(x)\cos n\theta + i\sum_{n=1}^{\infty}\left[1-(-1)^n\right]J_n(x)\sin n\theta,$$

and equating real and imaginary parts, we have

$$\cos(x\sin\theta) = J_0(x) + \sum_{n=1}^{\infty}\left[1+(-1)^n\right]J_n(x)\cos n\theta;$$

that is,

$$\cos(x\sin\theta) = J_0(x) + 2\sum_{n=1}^{\infty} J_{2n}(x)\cos 2n\theta \tag{6.25}$$

and

$$\sin(x\sin\theta) = \sum_{n=1}^{\infty}\left[1-(-1)^n\right]J_n(x)\sin n\theta = 2\sum_{n=1}^{\infty} J_{2n-1}(x)\sin(2n-1)\theta. \tag{6.26}$$

Since $\cos(x\sin\theta)$ is continuously differentiable on the interval $0 \leqslant x \leqslant \pi$, (6.25) is the Fourier cosine series expansion of this function, and

$$a_0 = \frac{1}{\pi}\int_0^{\pi} \cos(x\sin\theta)\, d\theta = J_0(x)$$

$$a_{2n} = \frac{2}{\pi}\int_0^{\pi} \cos(x\sin\theta)\cos 2n\theta\, d\theta = 2J_{2n}(x)$$

$$a_{2n-1} = 0$$

for $n \geqslant 1$.

Thus,

$$J_{2n}(x) = \frac{1}{\pi} \int_0^{\pi} \cos(x \sin \theta) \cos 2n\theta \, d\theta$$

for all $n \geqslant 0$. Also, since the function $\sin(x \sin \theta)$ is continuously differentiable on the interval $0 \leqslant x \leqslant \pi$ (in fact, on the whole real line), (6.26) is the Fourier sine series expansion of the function, and

$$b_{2n-1} = \frac{2}{\pi} \int_0^{\pi} \sin(x \sin \theta) \sin(2n-1)\theta \, d\theta = 2J_{2n-1}(x)$$

$$b_{2n} = 0$$

for $n \geqslant 1$; thus,

$$J_{2n+1}(x) = \frac{1}{\pi} \int_0^{\pi} \sin(x \sin \theta) \sin(2n+1)\theta \, d\theta$$

for all $n \geqslant 0$.

□

6.5 FOURIER-BESSEL SERIES

Given a fixed nonnegative integer m, the function $J_m(x)$ has an infinite number of positive zeros: z_{mn}, $n = 1, 2, 3, \ldots$, so that

$$J_m(z_{mn}) = 0$$

for $n = 1, 2, 3, \ldots$. Suppose that we want to expand a given function $f(x)$ in terms of a fixed Bessel function; that is,

$$f(x) = \sum_{n=1}^{\infty} a_n J_m(z_{mn}x) = a_1 J_m(z_{m1}x) + a_2 J_m(z_{m2}x) + \cdots + a_n J_m(z_{mn}x) + \cdots,$$

where $f(x)$ is defined for $0 \leqslant x \leqslant 1$ and the z_{mn}'s are the positive zeros of $J_m(x)$. To determine the coefficients a_n, we need an **orthogonality relation** just as we did for Fourier series. In this case, however, we are expanding the function $f(x)$ in terms of a fixed Bessel function $J_m(x)$, and the series is summed over the positive zeros of $J_m(x)$. Recall that the Bessel function $J_m(x)$ was a solution to a singular Sturm-Liouville problem; we show that the eigenfunctions $J_m(z_{mn}x)$ for $n = 1, 2, 3, \ldots$ are orthogonal on the interval $[0, 1]$ with respect to the weight function $\sigma(x) = x$.

Theorem 6.9. For a fixed integer $m \geqslant 0$,

$$\int_0^1 x J_m(z_{mn}x) J_m(z_{mk}x) \, dx = 0$$

for $n \neq k$, and

$$\int_0^1 x J_m(z_{mn}x)^2\,dx = \frac{1}{2}\,J_{m+1}(z_{mn})^2,$$

where z_{mn} is the nth positive zero of $J_m(x)$. Thus, the functions $J_m(z_{mn}x)$ are orthogonal on the interval $0 \leqslant x \leqslant 1$ with respect to the weight function $\sigma(x) = x$.

Proof. Note that $y = J_m(x)$ is a solution to the differential equation

$$y'' + \frac{1}{x}y' + \left(1 - \frac{m^2}{x^2}\right)y = 0,$$

and if a and b are distinct positive constants, the functions

$$u(x) = J_m(ax) \qquad \text{and} \qquad v(x) = J_m(bx)$$

satisfy the differential equations

$$u'' + \frac{1}{x}u' + \left(a^2 - \frac{m^2}{x^2}\right)u = 0$$

and

$$v'' + \frac{1}{x}v' + \left(b^2 - \frac{m^2}{x^2}\right)v = 0,$$

respectively. Multiplying the first of these equations by v and the second by u, and subtracting, we have

$$\frac{d}{dx}\left[u'v - v'u\right] + \frac{1}{x}\left(u'v - v'u\right) + (a^2 - b^2)uv = 0,$$

and multiplying by x, we have

$$\frac{d}{dx}\left[x\left(u'v - v'u\right)\right] = (b^2 - a^2)xuv.$$

Integrating from 0 to 1 gives us

$$(b^2 - a^2)\int_0^1 xu(x)v(x)\,dx = \left[x(u'v - v'u)\right]\Big|_0^1,$$

and since

$$u(1) = J_m(a) \qquad \text{and} \qquad v(1) = J_m(b),$$

if a and b are distinct positive zeros of $J_m(x)$, say z_{mn} and z_{mk}, then

$$\left(z_{mn}^2 - z_{mk}^2\right)\int_0^1 xJ_m(z_{mn}x)J_m(z_{mk}x)\,dx = 0,$$

and since $z_{mn} \neq z_{mk}$,

$$\int_0^1 J_m(z_{mn}x) J_m(z_{mk}x)\, x\, dx = 0;$$

that is, the functions $J_m(z_{mn}x)$ are orthogonal on the interval $0 \leqslant x \leqslant 1$ with respect to the weight function $\sigma(x) = x$.

To find the normalization constant, we note again that for $a > 0$, the function $J_m(ax)$ satisfies the differential equation

$$u'' + \frac{1}{x}u' + \left(a^2 - \frac{m^2}{x^2}\right)u = 0,$$

and multiplying this equation by $2x^2u'$, we get

$$2x^2u'u'' + 2xu'u' + 2a^2x^2uu' - 2m^2uu' = 0;$$

that is,

$$\frac{d}{dx}\left[x^2u'^2\right] + \frac{d}{dx}\left[a^2x^2u^2\right] - 2a^2xu^2 - \frac{d}{dx}\left[m^2u^2\right] = 0.$$

Integrating this equation from 0 to 1, we have

$$2a^2\int_0^1 xu^2\, dx = \left[x^2u'^2 + (a^2x^2 - m^2)u^2\right]\Big|_0^1,$$

and since $u(0) = J_m(0)$ for $m \geqslant 1$, then at $x = 0$,

$$x^2u'^2 + (a^2x^2 - m^2)u^2 = 0$$

for all $m \geqslant 0$. Also, at $x = 1$,

$$u'(1) = \frac{d}{dx}\left[J_m(ax)\right]\Big|_{x=1} = aJ_m{}'(a),$$

so that

$$\int_0^1 J_m(ax)^2 x\, dx = \frac{1}{2}J_m{}'(a)^2 + \frac{1}{2}\left(1 - \frac{m^2}{a^2}\right)J_m(a)^2$$

for $m \geqslant 0$. Now put $a = z_{mn}$, the n th positive zero of $J_m(x)$; then from Theorem 6.3, part (ii), we have

$$\int_0^1 J_m(z_{mn}x)^2 x\, dx = \frac{1}{2}J_m{}'(z_{mn})^2 = \frac{1}{2}J_{m+1}(z_{mn})^2$$

for $m \geqslant 0$.

□

Now suppose that we have a function f defined on the interval $0 \leqslant x \leqslant 1$, and that it has a **Fourier-Bessel series expansion** given by

$$f(x) = \sum_{k=1}^{\infty} a_k J_m(z_{mk} x).$$

We can find the coefficients in this expansion by multiplying the equation by $x J_m(z_{mn} x)$ and integrating, to get

$$\int_0^1 f(x) J_m(z_{mn} x)\, x\, dx = \frac{a_n}{2} J_{m+1}(z_{mn})^2.$$

Therefore, if

$$f(x) = \sum_{n=1}^{\infty} a_n J_m(z_{mn} x),$$

then

$$a_n = \frac{2}{J_{m+1}(z_{mn})^2} \int_0^1 f(x) J_m(z_{mn} x)\, x\, dx$$

for $n \geqslant 1$.

We have a convergence theorem for Fourier-Bessel series similar to Dirichlet's theorem (see [11] or [28]).

Theorem 6.10. (*Fourier-Bessel Expansion Theorem*)

If f and f' are piecewise continuous on the interval $0 \leqslant x \leqslant 1$, then for $0 < x_0 < 1$, the series

$$\sum_{n=1}^{\infty} a_n J_m(z_{mn} x_0)$$

converges to $\left[f(x_0^+) + f(x_0^-)\right]/2$. At $x_0 = 1$, the series converges to 0, since every $J_m(z_{mn}) = 0$. At $x_0 = 0$, the series converges to 0 if $m \geqslant 1$, and to $f(0^+)$ if $m = 0$.

6.6 SOLUTION TO THE VIBRATING MEMBRANE PROBLEM

We now return to the vibrating circular membrane problem we started with:

$$\frac{\partial^2 u}{\partial t^2} = c^2 \left(\frac{\partial^2 u}{\partial r^2} + \frac{1}{r}\frac{\partial u}{\partial r} + \frac{1}{r^2}\frac{\partial^2 u}{\partial \theta^2} \right),$$

$$u(r, -\pi, t) = u(r, \pi, t),$$

$$\frac{\partial u}{\partial \theta}(r, -\pi, t) = \frac{\partial u}{\partial \theta}(r, \pi, t),$$

$$u(a, \theta, t) = 0,$$

$$|u(r, \theta, t)| < \infty \text{ as } r \to 0^+,$$

$$u(r, \theta, 0) = f(r, \theta),$$

$$\frac{\partial u}{\partial t}(r, \theta, 0) = g(r, \theta)$$

for $0 < r < a$, $-\pi < \theta < \pi$, and $t > 0$. Separating variables, we assumed that

$$u(r, \theta, t) = R(r) \cdot \phi(\theta) \cdot G(t)$$

and obtained three ordinary differential equations:

- **Temporal equation**:

$$G'' + \lambda c^2 G = 0, \quad t > 0,$$

- **Angular equation**:

$$\begin{aligned} &\phi'' + \mu\phi = 0, \quad -\pi < \theta < \pi \\ &\phi(-\pi) = \phi(\pi), \\ &\phi'(-\pi) = \phi'(\pi), \end{aligned}$$

- **Radial equation**:

$$\begin{aligned} &r^2 R'' + rR' + (\lambda r^2 - \mu)R = 0, \quad 0 < r < a \\ &R(a) = 0, \\ &|R(r)| < \infty \quad \text{as} \quad r \to 0^+, \end{aligned}$$

where λ and μ are separation constants. For the angular equation, the eigenvalues and eigenfunctions are

$$\mu_m = m^2 \qquad \text{and} \qquad \phi_m(\theta) = a_m \cos m\theta + b_m \sin m\theta$$

for $m \geqslant 0$.

We saw earlier, using the Rayleigh quotient, that all the eigenvalues of the radial problem are nonnegative. For $\lambda = 0$, the radial equation is

$$\begin{gathered} r^2 R'' + rR' - m^2 R = 0, \quad 0 < r < a, \\ R(a) = 0, \\ |R(r)| < \infty \quad \text{as} \quad r \to 0^+, \end{gathered}$$

and the differential equation is a Cauchy-Euler equation or an equidimensional equation with solution

$$R_0(r) = A_0 \log r + B_0, \quad \text{for} \quad m = 0,$$

and

$$R_m(r) = A_m r^m + \frac{B_m}{r^m}, \quad \text{for} \quad m \geqslant 1.$$

Since the solutions must be bounded at $r = 0$, we have $A_0 = 0$ and $B_m = 0$ for all $m \geqslant 1$, so that

$$R_0(r) = B_0 \qquad \text{and} \qquad R_m(r) = A_m r^m,$$

and applying the boundary condition $R(a) = 0$, we see that $B_0 = 0$ and $A_m = 0$ for all $m \geqslant 1$. Thus, $\lambda = 0$ is not an eigenvalue.

For $\lambda > 0$, the radial equation is

$$\begin{gathered} r^2 R'' + rR' + (\lambda r^2 - m^2) R = 0, \\ R(a) = 0, \\ |R(r)| < \infty \quad \text{as} \quad r \to 0^+. \end{gathered}$$

Using the transformation $x = \sqrt{\lambda}\, r$, this can be transformed into the nonparametric form of Bessel's equation (6.8),

$$x^2 \frac{d^2 R}{dx^2} + x \frac{dR}{dx} + (x^2 - m^2) R = 0,$$

with general solution

$$R(r) = A J_m(\sqrt{\lambda}\, r) + B Y_m(\sqrt{\lambda}\, r),$$

and since Y_m is not bounded as $r \to 0^+$, we must have $B = 0$, and the solution is

$$R(r) = A J_m(\sqrt{\lambda}\, r).$$

The boundary condition $R(a) = 0$ gives us

$$A J_m(\sqrt{\lambda}\, a) = 0,$$

and in order to get a nontrivial solution we must have

$$J_m(\sqrt{\lambda}\, a) = 0.$$

The eigenvalues are those values of λ for which $J_m(\sqrt{\lambda}\,a) = 0$, that is, for which $\sqrt{\lambda}\,a = z_{mn}$, where z_{mn} is the nth positive zero of J_m; hence,

$$\lambda_{mn} = \left(\frac{z_{mn}}{a}\right)^2$$

for $n \geqslant 1$ and $m \geqslant 0$. The corresponding eigenfunctions are

$$R_{mn}(r) = J_m\left(\frac{z_{mn}}{a}\, r\right)$$

for $n \geqslant 1$ and $m \geqslant 0$.

Note that for a fixed $m \geqslant 0$, these eigenfunctions form an orthogonal basis for the linear space *PWS* $[0, a]$ of piecewise smooth functions on $[0, a]$ with weight function $\sigma(r) = r$, so that

$$\int_0^a J_m\left(\frac{z_{mn}}{a}\, r\right) J_m\left(\frac{z_{mk}}{a}\, r\right)\, r\, dr = 0$$

for $n \neq k$. For $w \in PWS[0, a]$, the generalized Fourier series, now called a Fourier-Bessel series, is given by

$$w(r) = \sum_{n=1}^{\infty} a_n J_m\left(\frac{z_{mn}}{a}\, r\right),$$

where

$$a_n = \frac{\displaystyle\int_0^a f(r) J_m\left(\frac{z_{mn}}{a}\, r\right) r\, dr}{\displaystyle\int_0^a J_m\left(\frac{z_{mn}}{a}\, r\right)^2 r\, dr}$$

for $n \geqslant 1$.

We only need to solve the time equation for those values of λ for which we have a nontrivial solution to the radial equation, that is, $\lambda = \lambda_{mn}$, and the time equation is

$$G'' + c^2 \lambda_{mn} G = 0,$$

with general solution

$$G_{mn}(t) = A\cos(\sqrt{\lambda_{mn}}\, ct) + B\sin(\sqrt{\lambda_{mn}}\, ct)$$

for $n \geqslant 1$ and $m \geqslant 0$.

For each $m \geqslant 0,\ n \geqslant 1$, the product solution

$$u_{mn}(r, \theta, t) = R_{mn}(r) \cdot \phi_m(\theta) \cdot G_{mn}(t)$$

satisfies the PDE, the boundary condition, and the boundedness condition.

To satisfy the initial conditions we use the superposition principle to write

$$
\begin{aligned}
u(r,\theta,t) = \sum_{m=0}^{\infty}\sum_{n=1}^{\infty}\Big\{J_m\big(\sqrt{\lambda_{mn}}\,r\big)\Big[a_{mn}\cos m\theta\cos\big(\sqrt{\lambda_{mn}}\,ct\big) \\
+ b_{mn}\cos m\theta\sin\big(\sqrt{\lambda_{mn}}\,ct\big) + c_{mn}\sin m\theta\cos\big(\sqrt{\lambda_{mn}}\,ct\big) \\
+ d_{mn}\sin m\theta\sin\big(\sqrt{\lambda_{mn}}\,ct\big)\Big]\Big\}
\end{aligned}
$$

for $0 < r < a,\ -\pi < \theta < \pi,\ t > 0$. Now we apply the initial conditions to determine the constants a_{mn}, b_{mn}, c_{mn}, and d_{mn}. For simplicity, we assume that

$$
\frac{\partial u}{\partial t}(r,\theta,0) = g(r,\theta) = 0,
$$

and this implies, after differentiating term by term and setting $t = 0$, that

$$
b_{mn} = 0 \qquad \text{and} \qquad d_{mn} = 0
$$

for $m \geqslant 0$ and $n \geqslant 1$, so that the solution becomes

$$
u(r,\theta,t) = \sum_{m=0}^{\infty}\sum_{n=1}^{\infty} J_m\big(\sqrt{\lambda_{mn}}\,r\big)\cos\big(\sqrt{\lambda_{mn}}\,ct\big)\big(a_{mn}\cos m\theta + c_{mn}\sin m\theta\big).
$$

Setting $t = 0$, we have

$$
f(r,\theta) = u(r,\theta,0) = \sum_{m=0}^{\infty}\sum_{n=1}^{\infty} J_m\big(\sqrt{\lambda_{mn}}\,r\big)\big(a_{mn}\cos m\theta + c_{mn}\sin m\theta\big),
$$

which is a Fourier series for $f(r,\theta)$ on the interval $[-\pi,\pi]$ holding r fixed. Therefore,

$$
\sum_{n=1}^{\infty} a_{0n}J_0\big(\sqrt{\lambda_{0n}}\,r\big) = \frac{1}{2\pi}\int_{-\pi}^{\pi} f(r,\theta)\,d\theta, \quad \text{for } m = 0,
$$

$$
\sum_{n=1}^{\infty} a_{mn}J_m\big(\sqrt{\lambda_{mn}}\,r\big) = \frac{1}{\pi}\int_{-\pi}^{\pi} f(r,\theta)\cos m\theta\,d\theta, \quad \text{for } m \geqslant 1,
$$

$$
\sum_{n=1}^{\infty} c_{mn}J_m\big(\sqrt{\lambda_{mn}}\,r\big) = \frac{1}{\pi}\int_{-\pi}^{\pi} f(r,\theta)\sin m\theta\,d\theta, \quad \text{for } m \geqslant 1.
$$

Note that these Fourier series coefficients are actually Fourier-Bessel series, so that

$$a_{0n} = \frac{\dfrac{1}{2\pi}\displaystyle\int_{-\pi}^{\pi}\int_0^a f(r,\theta)J_0\big(\sqrt{\lambda_{0n}}\,r\big)\,r\,dr\,d\theta}{\displaystyle\int_0^a J_0\big(\sqrt{\lambda_{0n}}\,r\big)^2 r\,dr}, \qquad \text{for } m = 0,\ n \geqslant 1,$$

$$a_{mn} = \frac{\dfrac{1}{\pi}\displaystyle\int_{-\pi}^{\pi}\int_0^a f(r,\theta)J_m\big(\sqrt{\lambda_{mn}}\,r\big)\cos m\theta\,r\,dr\,d\theta}{\displaystyle\int_0^a J_m\big(\sqrt{\lambda_{mn}}\,r\big)^2 r\,dr}, \qquad \text{for } m \geqslant 1,\ n \geqslant 1,$$

$$c_{mn} = \frac{\dfrac{1}{\pi}\displaystyle\int_{-\pi}^{\pi}\int_0^a f(r,\theta)J_m\big(\sqrt{\lambda_{mn}}\,r\big)\sin m\theta\,r\,dr\,d\theta}{\displaystyle\int_0^a J_m\big(\sqrt{\lambda_{mn}}\,r\big)^2 r\,dr}, \qquad \text{for } m \geqslant 1,\ n \geqslant 1,$$

where the normalization constants are given by

$$\int_0^a J_m\big(\sqrt{\lambda_{mn}}\,r\big)^2 r\,dr = \int_0^a J_m\left(\frac{z_{mn}}{a}\,r\right)^2 r\,dr = \frac{a^2}{2}\,J_{m+1}(z_{mn})$$

by Theorem 6.9.

6.7 SUMMARY

Polar coordinates or **cylindrical coordinates** are used in domains that have radial symmetry in two dimensions or cylindrical symmetry in three dimensions. If we study a three-dimensional problem, the z-coordinate is unchanged in the transformation to cylindrical coordinates. Hence here we focus on the transformation of the two dimensional situation.

The Cartesian coordinates (x, y) are transformed to a radial coordinate $r > 0$ and an angular coordinate $\theta \in (-\pi, \pi)$ according to $x = r\cos\theta, y = r\sin\theta$. The Laplacian transforms into

$$\nabla^2 u = \Delta u = \frac{1}{r}\frac{\partial}{\partial r}\left(r\frac{\partial u}{\partial r}\right) + \frac{1}{r^2}\frac{\partial^2 u}{\partial \theta^2}.$$

If we consider an open disk of radius a, for example, we can write

$$D(a) = \{\,(x,y)\,|\,x^2+y^2 < a^2\,\} = \{\,(r,\theta)\,|\,0 < r < a, -\pi < \theta \leqslant \pi\,\} \cup \{\,(0,0)\,\};$$

hence, the boundary of the disk $D(a)$, which is a circle in cartesian coordinates, appears to be a rectangle in polar coordinates. According to the rule of thumb for side conditions, this allows us to find the appropriate number of boundary conditions.

We use periodicity conditions for u and for $\partial u/\partial\theta$ at $\theta = -\pi$ and $\theta = \pi$ in order that the solution will be single-valued and continuous as we move across this artificial boundary. Since the transformation into polar coordinates is singular at $r = 0$, we impose a boundedness condition in order that the solution be finite on the artificial boundary $r = 0$.

If the problem is linear and homogeneous, we can use the method of separation of variables. For Laplace's equation, we look for separated solutions of the form $u(r,\theta) = R(r)\cdot\Theta(\theta)$. For the heat equation and wave equation, we look for separated solutions of the form $u(r,\theta,t) = R(r)\cdot\Theta(\theta)\cdot T(t)$. Separating variables, this results in a **radial problem**, an **angular problem**, and in the latter cases, a **time problem**. The radial problem is either a **Cauchy-Euler equation**,

$$r^2R'' + rR' - \lambda R = 0,$$

or the nonparametric form of **Bessel's equation**,

$$x^2R'' + xR' + (x^2 - n^2)R = 0,$$

where $x = \sqrt{\lambda}\, r$. The solutions to the Cauchy-Euler equation are r^n and r^{-n}, where $\lambda = n^2$, while the solutions to Bessel's equation are **Bessel functions of the first kind** $J_m(x)$ and **Bessel functions of the second kind** $Y_m(x)$, $m = 0, 1, 2, \ldots$. In this part we rely heavily on solution methods from ODEs, and we suggest that you review the methods of integrating factors, variation of parameters, and series solutions of ODEs.

We derive an extensive menu of **Bessel functions properties**, which include series representations, recurrence relations, integral representations, and asymptotics for large and small values of x. Of particular interest are the **positive zeros of Bessel functions**, that is, points z_{mn} such that $J_m(z_{mn}) = 0$ for $n = 1, 2, \ldots$. For a fixed Bessel function $J_m(x)$, these zeros can be used to define an orthogonal basis in $PWC(0,1)$. Given $J_m(x)$ and its zeros z_{mn}, the set of functions

$$\{\, J_m(z_{mn}x) \,\}_{n=1,2,\ldots}$$

forms an orthogonal set with weight function $\sigma(x) = x$. Functions in $PWC(0,1)$ can be expanded in terms of this basis, and a solution to the PDE will then be represented in the form of a **Fourier-Bessel series**. The coefficients can be determined from the initial conditions, and there is also a convergence theorem similar to Dirichlet's theorem for Fourier series.

6.7.1 Problems and Notes

Problems from Part II:

Exercise	13.14	13.18	14.14	14.15	14.16	14.17
Notes						
Exercise	14.18	15.3	15.4	15.10	15.11	
Notes						

Midterm exam questions:

Exercise	18.11
Notes	

You should now be able to do **Midterm Exam 4**.

CHAPTER 7

SPHERICAL COORDINATES

7.1 SPHERICAL COORDINATES

As with polar coordinates in the plane, there are certain geometries for which spherical coordinates are more useful than Cartesian coordinates (Fig. 7.1). Spherical coordinates are defined as follows: Given a point P with Cartesian coordinates (x, y, z) where $(x, y) \neq (0, 0)$, the **spherical coordinates** of P are (r, θ, ϕ), where

$$\begin{aligned} x &= r\cos\phi\sin\theta, \\ y &= r\sin\phi\sin\theta, \\ z &= r\cos\theta, \end{aligned}$$

and where $r \geqslant 0$, $0 \leqslant \theta \leqslant \pi$, and $-\pi \leqslant \phi \leqslant \pi$. Here r is the distance from the origin O to the point P, θ is the angle the radius OP makes with the positive z-axis, and ϕ is the angle between the projection of OP onto the xy-plane and the positive x-axis (Fig. 7.2). The Jacobian determinant for the transformation is

Partial Differential Equations: Theory and Completely Solved Problems, First Edition. By T. Hillen, I. E. Leonard, H. van Roessel

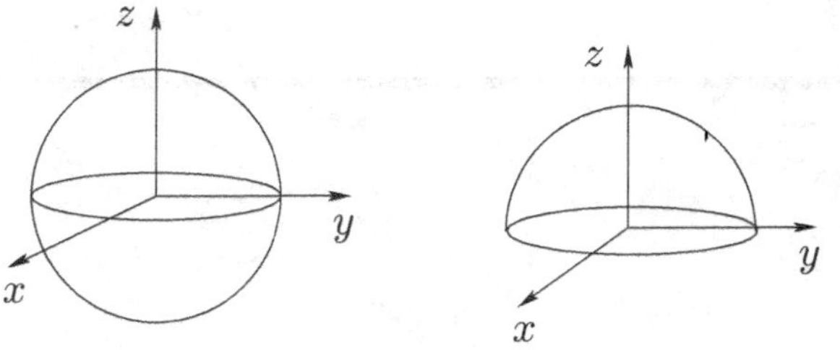

Figure 7.1 Geometries more suited to spherical coordinates.

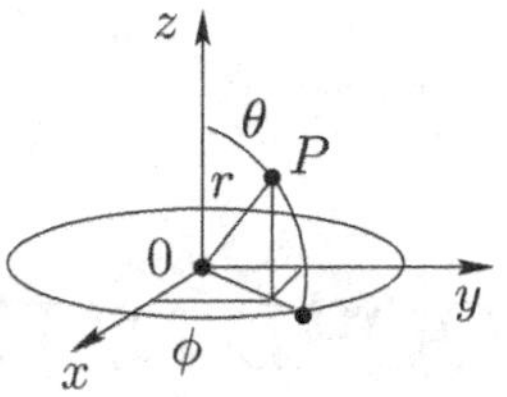

Figure 7.2 Spherical coordinate transformation.

$$\frac{\partial(x,y,z)}{\partial(r,\theta,\phi)} = \begin{vmatrix} \sin\theta\cos\phi & r\cos\theta\cos\phi & -r\sin\theta\sin\phi \\ \sin\theta\sin\phi & r\cos\theta\sin\phi & r\sin\theta\cos\phi \\ \cos\theta & -r\sin\theta & 0 \end{vmatrix} = r^2\sin\theta,$$

which indicates that the transformation is singular at $r = 0$ and also for $\theta = 0$ and $\theta = \pi$, that is, along the entire z-axis. The Laplacian in spherical coordinates is

$$\begin{aligned}\nabla^2 u &= \frac{\partial^2 u}{\partial x^2} + \frac{\partial^2 u}{\partial y^2} + \frac{\partial^2 u}{\partial z^2} \\ &= \frac{\partial^2 u}{\partial r^2} + \frac{2}{r}\frac{\partial u}{\partial r} + \frac{1}{r^2}\frac{\partial^2 u}{\partial \theta^2} + \frac{\cos\theta}{r^2\sin\theta}\frac{\partial u}{\partial \theta} + \frac{1}{r^2\sin^2\theta}\frac{\partial^2 u}{\partial \phi^2};\end{aligned}$$

that is,

$$\nabla^2 u = \frac{1}{r^2}\frac{\partial}{\partial r}\left(r^2\frac{\partial u}{\partial r}\right) + \frac{1}{r^2\sin\theta}\frac{\partial}{\partial\theta}\left(\sin\theta\frac{\partial u}{\partial\theta}\right) + \frac{1}{r^2\sin^2\theta}\frac{\partial^2 u}{\partial\phi^2}. \tag{7.1}$$

An equivalent expression for the Laplacian in spherical coordinates is

$$\nabla^2 u = \frac{\partial^2 u}{\partial r^2} + \frac{2}{r}\frac{\partial u}{\partial r} + \frac{1}{r^2}\left(\frac{\partial^2 u}{\partial\theta^2} + \cot\theta\frac{\partial u}{\partial\theta} + \csc^2\theta\frac{\partial^2 u}{\partial\phi^2}\right).$$

7.1.1 Derivation of the Laplacian

To derive the form of the Laplacian, we follow the method used in Pinsky [39] or Strauss [48]. We consider the transformation

$$(x, y, z) \longrightarrow (r, \theta, \phi)$$

as the composition of transformations

$$(x, y, z) \longrightarrow (s, \phi, z) \longrightarrow (r, \theta, \phi),$$

where, as in Figure 7.3,

$$\begin{aligned} s &= \sqrt{x^2+y^2}, & r &= \sqrt{s^2+z^2}, \\ x &= s\cos\phi, & z &= r\cos\theta, \\ y &= s\sin\phi, & s &= r\sin\theta. \end{aligned}$$

That is, we change to a polar coordinate system (s, ϕ) in the xy-plane and then to a polar coordinate system (r, θ) in the zs-plane. From the first transformation, we have

$$\frac{\partial^2 u}{\partial x^2} + \frac{\partial^2 u}{\partial y^2} = \frac{\partial^2 u}{\partial s^2} + \frac{1}{s}\frac{\partial u}{\partial s} + \frac{1}{s^2}\frac{\partial^2 u}{\partial \phi^2},$$

while from the second transformation, we have

$$\frac{\partial^2 u}{\partial z^2} + \frac{\partial^2 u}{\partial s^2} = \frac{\partial^2 u}{\partial r^2} + \frac{1}{r}\frac{\partial u}{\partial r} + \frac{1}{r^2}\frac{\partial^2 u}{\partial \theta^2}.$$

Adding these two equations, we get

$$\begin{aligned} \nabla^2 u &= \frac{\partial^2 u}{\partial x^2} + \frac{\partial^2 u}{\partial y^2} + \frac{\partial^2 u}{\partial z^2} \\ &= \frac{\partial^2 u}{\partial r^2} + \frac{1}{r}\frac{\partial u}{\partial r} + \frac{1}{r^2}\frac{\partial^2 u}{\partial \theta^2} + \frac{1}{s}\frac{\partial u}{\partial s} + \frac{1}{s^2}\frac{\partial^2 u}{\partial \phi^2}. \end{aligned}$$

Replacing s^2 in the last term by $r^2 \sin^2\theta$ and differentiating with respect to s, we get

$$\begin{aligned} \frac{\partial u}{\partial s} &= \frac{\partial u}{\partial r}\frac{\partial r}{\partial s} + \frac{\partial u}{\partial \theta}\frac{\partial \theta}{\partial s} + \frac{\partial u}{\partial \phi}\frac{\partial \phi}{\partial s} \\ &= \frac{\partial u}{\partial r}\cdot\frac{s}{r} + \frac{\partial u}{\partial \theta}\cdot\frac{\cos\theta}{r} + \frac{\partial u}{\partial \phi}\cdot 0. \end{aligned}$$

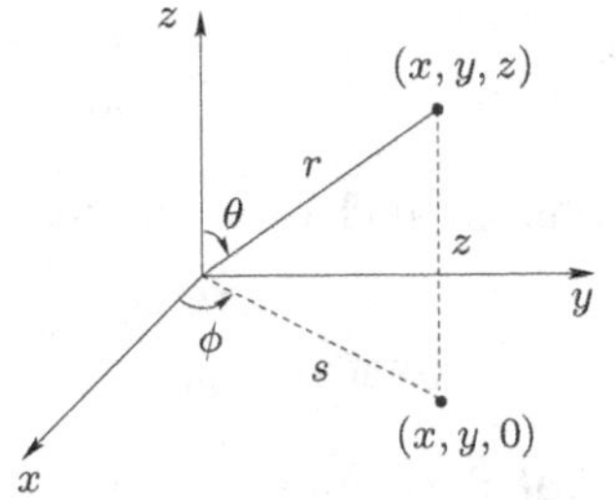

Figure 7.3 Composition of polar coordinate transformations.

Since in the transformation from cylindrical coordinates (s, ϕ, z) to spherical coordinates (r, θ, ϕ), we have $r = \sqrt{s^2 + z^2}$,

$$\frac{\partial r}{\partial s} = \frac{s}{\sqrt{s^2 + z^2}} = \frac{s}{r}.$$

Also, $\partial\phi/\partial s = 0$, while $\tan\theta = s/z$, so that

$$\sec^2\theta \frac{\partial\theta}{\partial s} = \frac{1}{z},$$

and thus

$$\frac{\partial\theta}{\partial s} = \frac{\cos^2\theta}{z} = \frac{\cos\theta}{r}.$$

Therefore,

$$\nabla^2 u = \frac{\partial^2 u}{\partial r^2} + \frac{2}{r}\frac{\partial u}{\partial r} + \frac{1}{r^2}\left(\frac{\partial^2 u}{\partial\theta^2} + \cot\theta \frac{\partial u}{\partial\theta} + \frac{1}{\sin^2\theta}\frac{\partial^2 u}{\partial\phi^2}\right),$$

or, equivalently,

$$\nabla^2 u = \frac{1}{r^2}\frac{\partial}{\partial r}\left(r^2 \frac{\partial u}{\partial r}\right) + \frac{1}{r^2 \sin\theta}\frac{\partial}{\partial\theta}\left(\sin\theta \frac{\partial u}{\partial\theta}\right) + \frac{1}{r^2\sin^2\theta}\frac{\partial^2 u}{\partial\phi^2}.$$

7.2 LEGENDRE'S EQUATION

Now we consider the homogeneous version of the three-dimensional heat equation and the three-dimensional wave equation:

$$\frac{\partial u}{\partial t} = k\nabla^2 u \qquad \text{and} \qquad \frac{\partial^2 u}{\partial t^2} = c^2\nabla^2 u,$$

together with the appropriate boundary conditions and initial conditions. If we apply separation of variables by letting

$$u(r, \theta, \phi, t) = \Psi(r, \theta, \phi) \cdot T(t),$$

we get

$$\frac{T'}{kT} = \frac{\nabla^2\Psi}{\Psi} = -\eta \qquad \text{and} \qquad \frac{T''}{c^2 T} = \frac{\nabla^2\Psi}{\Psi} = -\eta,$$

where η is a separation constant. In both cases, we end up with a three-dimensional eigenvalue problem for Ψ,

$$\nabla^2\Psi = -\eta\Psi, \tag{7.2}$$

called the **homogeneous Helmholtz equation**. As was the case for two-dimensional eigenvalue problems, this three-dimensional eigenvalue problem, with either Dirichlet, Neumann, or Robin boundary conditions, will have only nonnegative eigenvalues.

Theorem 7.1. Given the Helmholtz equation $\nabla^2\phi = -\eta\phi$ in the domain $\Omega \in \mathbb{R}^3$, with smooth boundary $\partial\Omega$, if (ϕ, η) is an eigenpair of the boundary value problem with

(a) Dirichlet conditions: $\phi = 0$ on $\partial\Omega$,

(b) Neumann conditions: $\partial\phi/\partial n = 0$ on $\partial\Omega$,

(c) Robin conditions: $\partial\phi/\partial n + \alpha\,\phi = 0$ on $\partial\Omega$, where $\alpha > 0$,

then $\eta \geqslant 0$.

Proof. Suppose that (ϕ, η) is an eigenpair of one of the problems above. Multiplying the differential equation by ϕ and integrating over Ω, we have

$$\iiint_\Omega \phi\nabla^2\phi\, dV = -\eta \iiint_\Omega \phi^2\, dV.$$

Since

$$\iiint_\Omega \nabla\cdot(\phi\nabla\phi)\; dV = \iiint_\Omega \phi\nabla^2\phi\, dV + \iiint_\Omega (\nabla\phi\cdot\nabla\phi)\; dV,$$

then

$$\eta\iiint_\Omega \phi^2\, dV = -\iiint_\Omega \nabla\cdot(\phi\nabla\phi)\; dV + \iiint_\Omega (\nabla\phi\cdot\nabla\phi)\; dV.$$

From the divergence theorem,

$$\iiint_\Omega \nabla\cdot(\phi\nabla\phi)\; dV = \iint_{\partial\Omega} \phi\,\frac{\partial\phi}{\partial n}\, dS,$$

where $\partial\phi/\partial n$ is the outward normal derivative on $\partial\Omega$. Therefore, we have a formula similar to the Rayleigh quotient for η,

$$\eta = \frac{-\displaystyle\iint_{\partial\Omega} \phi\,\frac{\partial\phi}{\partial n}\, dS + \iiint_\Omega (\nabla\phi\cdot\nabla\phi)\; dV}{\displaystyle\iiint_\Omega \phi^2\, dV}.$$

For both the homogeneous Dirichlet and homogeneous Neumann boundary conditions, the boundary term

$$\iint_{\partial\Omega} \phi\,\frac{\partial\phi}{\partial n}\, dS = 0,$$

and the eigenvalue η is nonnegative. For the homogeneous Robin boundary condition, we have

$$\iint_{\partial\Omega} \phi \frac{\partial \phi}{\partial n}\, dS = -\alpha \iint_{\partial\Omega} \phi^2\, dS \leqslant 0,$$

and again, the eigenvalue η is nonnegative.

□

Thus, in equation (7.2), we set $\eta = \mu^2$ and consider the problem

$$\nabla^2 \Psi = -\mu^2 \Psi. \tag{7.3}$$

We look for separated solutions of the form

$$\Psi(r, \theta, \phi) = R(r) \cdot Y(\theta, \phi),$$

and substituting this into equation (7.3), we get

$$\frac{1}{r^2}\left\{\frac{\partial}{\partial r}\left(r^2 \frac{\partial R}{\partial r}\right) Y + \frac{R}{\sin\theta}\frac{\partial}{\partial \theta}\left(\sin\theta \frac{\partial Y}{\partial \theta}\right) + \frac{R}{\sin^2\theta}\frac{\partial^2 Y}{\partial \phi^2}\right\} = -\mu^2 R \cdot Y,$$

and separating variables yields

$$\frac{1}{R}\frac{d}{dr}\left(r^2 \frac{dR}{dr}\right) + \mu^2 r^2 = -\frac{1}{Y \sin\theta}\left[\frac{\partial}{\partial \theta}\left(\sin\theta \frac{\partial Y}{\partial \theta}\right) + \frac{1}{\sin\theta}\frac{\partial^2 Y}{\partial \phi^2}\right] = \lambda,$$

where λ is another separation constant. Thus, we obtain an ordinary differential equation for R,

$$\frac{d}{dr}\left(r^2 \frac{dR}{dr}\right) + (\mu^2 r^2 - \lambda) R = 0, \quad 0 < r < a, \tag{7.4}$$

called the **spherical Bessel equation**, whose solutions are called **spherical Bessel functions**.

We also obtain a partial differential equation for Y,

$$\frac{1}{\sin\theta}\frac{\partial}{\partial \theta}\left(\sin\theta \frac{\partial Y}{\partial \theta}\right) + \frac{1}{\sin^2\theta}\frac{\partial^2 Y}{\partial \phi^2} + \lambda Y = 0, \tag{7.5}$$

called the **surface Helmholtz equation**, whose solutions are called the **surface harmonics** or **spherical harmonics**. Applying separation of variables to the second of these equations, we set

$$Y(\theta, \phi) = S(\theta) \cdot Q(\phi),$$

and substitute this into the equation for Y to get

$$\frac{1}{\sin\theta}\frac{\partial}{\partial \theta}\left(\sin\theta \frac{\partial S}{\partial \theta}\right) \cdot Q + \frac{1}{\sin^2\theta} S \cdot \frac{\partial^2 Q}{\partial \phi^2} + \lambda S \cdot Q = 0;$$

that is,

$$\frac{1}{S}\sin\theta\frac{d}{d\theta}\left(\sin\theta\frac{dS}{d\theta}\right)+\lambda\sin^2\theta=-\frac{1}{Q}\frac{d^2Q}{d\phi^2}=\nu,$$

where ν is yet another separation constant. We get the following ordinary differential equations for S and Q:

$$\sin\theta\frac{d}{d\theta}\left(\sin\theta\frac{dS}{d\theta}\right)+(\lambda\sin^2\theta-\nu)S=0,\quad 0<\theta<\pi \tag{7.6}$$

and

$$\frac{d^2Q}{d\phi^2}+\nu Q=0,\quad -\pi<\phi<\pi. \tag{7.7}$$

The equation for Q, equation (7.7), is one of our standard examples and has been solved many times.

In equation (7.6), if we make the transformation $x=\cos\theta$ and let $v(x)=S(\theta)$, then $0<\theta<\pi$ implies that $-1<x<1$, and from the chain rule

$$\sin\theta\frac{dS}{d\theta}=\sin\theta\frac{dv}{dx}\cdot\frac{dx}{d\theta}=-\sin^2\theta\frac{dv}{dx}=-(1-x^2)\frac{dv}{dx}$$

and

$$\sin\theta\frac{d}{d\theta}\left(\sin\theta\frac{dS}{d\theta}\right)=-(1-x^2)\frac{d}{dx}\left(-(1-x^2)\frac{dv}{dx}\right).$$

Therefore, the transformation of the independent variable θ in equation (7.6) allows us to write the equation as a second-order ODE,

$$(1-x^2)^2\frac{d^2v}{dx^2}-2x(1-x^2)\frac{dv}{dx}+\left(\lambda(1-x^2)-\nu\right)v=0,$$

that is,

$$(1-x^2)\frac{d^2v}{dx^2}-2x\frac{dv}{dx}+\left(\lambda-\frac{\nu}{1-x^2}\right)v=0,\quad -1<x<1, \tag{7.8}$$

called the **associated Legendre equation**, which we can solve using power series. For the special case $\nu=0$, this reduces to

$$(1-x^2)\frac{d^2v}{dx^2}-2x\frac{dv}{dx}+\lambda v=0,\quad -1<x<1, \tag{7.9}$$

which is **Legendre's equation.**

7.3 LEGENDRE FUNCTIONS

Before we can go any further in trying to solve the heat equation or wave equation in spherical coordinates, we need to get some understanding of the behavior of the solutions to Legendre's equation and the spherical Bessel equation.

7.3.1 Legendre Polynomials

In Legendre's equation,

$$(1-x^2)\frac{d^2v}{dx^2} - 2x\frac{dv}{dx} + \lambda v = 0, \quad -1 < x < 1, \tag{7.10}$$

it is clear that $x = 0$ is an ordinary point and $x = \pm 1$ are both regular singular points. In applications, we need solutions to be bounded at $x = \pm 1$, which corresponds to $\theta = 0$ and $\theta = \pi$, that is, the positive z-axis and negative z-axis, respectively. Also, if we put the equation in self-adjoint form,

$$\frac{d}{dx}\left[(1-x^2)\frac{dv}{dx}\right] + \lambda v = 0,$$

then

$$p(x) = 1 - x^2, \quad q(x) = 0, \quad \text{and} \quad \sigma(x) = 1.$$

Since $p(1) = p(-1) = 0$, any Sturm-Liouville problem involving Legendre's equation will be a singular Sturm-Liouville problem, and to get an orthogonal set of eigenfunctions we need to impose boundedness conditions such as

$$|v(x)| \quad \text{and} \quad |v'(x)| \quad \text{bounded}$$

as $x \to -1^+$ and $x \to 1^-$.

All solutions will be analytic at $x = 0$, since this is an ordinary point of Legendre's equation, hence will have a Taylor series expansion about $x = 0$. Thus, if $v(x)$ is a solution,

$$v(x) = \sum_{k=0}^{\infty} a_k x^k,$$

valid for all x in some open interval centered at 0. Differentiating and substituting the series into equation (7.10), we have

$$\sum_{k=2}^{\infty} k(k-1)a_k x^{k-2} - \sum_{k=0}^{\infty} k(k-1)a_k x^k - 2\sum_{k=0}^{\infty} k a_k x^k + \lambda \sum_{k=0}^{\infty} a_k x^k = 0,$$

and reindexing the first sum, we have

$$\sum_{k=0}^{\infty}(k+2)(k+1)a_{k+2}x^k - \sum_{k=0}^{\infty} k(k-1)a_k x^k - 2\sum_{k=0}^{\infty} k a_k x^k + \lambda \sum_{k=0}^{\infty} a_k x^k = 0,$$

that is,

$$\sum_{k=0}^{\infty}\{(k+2)(k+1)a_{k+2} + [\lambda - k(k+1)]a_k\}x^k = 0,$$

and therefore the coefficients satisfy the following recurrence relation:

$$a_{k+2} = -\frac{\lambda - k(k+1)}{(k+2)(k+1)}a_k$$

for $k = 0, 1, 2, \ldots$. Writing out the first few coefficients explicitly, we have

$$a_2 = -\frac{\lambda}{2 \cdot 1} a_0,$$

$$a_3 = -\frac{\lambda - 1 \cdot 2}{3 \cdot 2} a_1,$$

$$a_4 = -\frac{\lambda - 2 \cdot 3}{4 \cdot 3} a_2 = \frac{\lambda(\lambda - 2 \cdot 3)}{4 \cdot 3 \cdot 2} a_0,$$

$$a_5 = -\frac{\lambda - 3 \cdot 4}{5 \cdot 4} a_3 = \frac{(\lambda - 1 \cdot 2)(\lambda - 3 \cdot 4)}{5!} a_1,$$

$$a_6 = -\frac{\lambda - 4 \cdot 5}{6 \cdot 5} a_4 = -\frac{\lambda(\lambda - 2 \cdot 3)(\lambda - 4 \cdot 5)}{6!} a_0,$$

$$a_7 = -\frac{(\lambda - 1 \cdot 2)(\lambda - 3 \cdot 4)(\lambda - 5 \cdot 6)}{7!} a_1$$

$$\vdots$$

Therefore,

$$\begin{aligned} v(x) = a_0 &\left[1 - \frac{\lambda}{2!}x^2 + \frac{\lambda(\lambda - 2 \cdot 3)}{4!}x^4 - \frac{\lambda(\lambda - 2 \cdot 3)(\lambda - 4 \cdot 5)}{6!}x^6 + \cdots\right] \\ &+ a_1 \left[x - \frac{\lambda - 1 \cdot 2}{3!}x^3 + \frac{(\lambda - 1 \cdot 2)(\lambda - 3 \cdot 4)}{5!}x^5 \right. \\ &\left. \qquad - \frac{(\lambda - 1 \cdot 2)(\lambda - 3 \cdot 4)(\lambda - 5 \cdot 6)}{7!}x^7 + \cdots\right], \end{aligned}$$

where a_0 and a_1 are arbitrary. Note that the even and odd parts are linearly independent, and thus this is the general solution to Legendre's equation. The quantities in square brackets,

$$v_1(x) = 1 - \frac{\lambda}{2!}x^2 + \frac{\lambda(\lambda - 2 \cdot 3)}{4!}x^4 - \frac{\lambda(\lambda - 2 \cdot 3)(\lambda - 4 \cdot 5)}{6!}x^6 + \cdots$$

and

$$\begin{aligned} v_2(x) = x &- \frac{\lambda - 1 \cdot 2}{3!}x^3 + \frac{(\lambda - 1 \cdot 2)(\lambda - 3 \cdot 4)}{5!}x^5 \\ &- \frac{(\lambda - 1 \cdot 2)(\lambda - 3 \cdot 4)(\lambda - 5 \cdot 6)}{7!}x^7 + \cdots \end{aligned}$$

are multiples of what are called **Legendre functions**.

It can be shown that for general values of λ, both series in the general solution are unbounded at $x = \pm 1$, and in fact, we get bounded solutions to Legendre's equation if and only if $\lambda = n(n+1)$, where $n = 0, 1, 2, \ldots$. See, for example, Brown and Churchill [11].

In this case, the recurrence relation becomes

$$a_{k+2} = -\frac{n(n+1) - k(k+1)}{(k+2)(k+1)}\, a_k,$$

for $0 \leqslant k \leqslant n-1$, and $a_{n+2} = 0$. Hence, $a_{n+2} = 0$, $a_{n+4} = 0$, $a_{n+6} = 0$, and in general, $a_{n+2k} = 0$ for $k = 1, 2, 3, \ldots$. Thus, one of the series terminates, giving a polynomial, while the other series does not terminate.

To see what is happening, the recurrence relation can by rewitten as

$$a_{k+2} = \frac{(k-n)(k+1+n)}{(k+2)(k+1)}\, a_k$$

for $k \geqslant 0$, and

$$a_2 = -\frac{n(n+1)}{2!}\, a_0,$$

$$a_3 = -\frac{(n-1)(n+2)}{3!}\, a_1,$$

$$a_4 = -\frac{(n-2)(n+3)}{4 \cdot 3}\, a_2 = \frac{(n-2)n(n+1)(n+3)}{4!}\, a_0,$$

$$a_5 = -\frac{(n-3)(n+4)}{5 \cdot 4}\, a_3 = \frac{(n-3)(n-1)(n+2)(n+4)}{5!}\, a_1,$$

$$\vdots$$

from which the pattern is clear. For example, the factor $(n-2)$ appears in all the coefficients a_n with even $n \geqslant 4$.

Now we can write the general solution as

$$v(x) = a_0 v_1(x) + a_1 v_2(x),$$

where

$$v_1(x) = 1 - \frac{n(n+1)}{2!}\, x^2 + \frac{(n-2)n(n+1)(n+3)}{4!}\, x^4 + \cdots$$

and

$$v_2(x) = x - \frac{(n-1)(n+2)}{3!}\, x^3 + \frac{(n-3)(n-1)(n+2)(n+4)}{5!}\, x^5 - \cdots$$

and if n is even, $v_1(x)$ reduces to a polynomial of degree n, while if n is odd, $v_2(x)$ reduces to a polynomial of degree n. These polynomials are called **Legendre polynomials**, denoted by $P_n(x)$, and are scaled so that $P_n(1) = 1$ for $n \geqslant 0$; that is,

$$P_n(x) = \begin{cases} \dfrac{v_1(x)}{v_1(1)}, & \text{if n is even,} \\[2ex] \dfrac{v_2(x)}{v_2(1)}, & \text{if n is odd.} \end{cases}$$

Theorem 7.2. The singular Sturm-Liouville problem given by

$$\begin{aligned} &(1-x^2)\frac{d^2v}{dx^2} - 2x\frac{dv}{dx} + \lambda v = 0, \quad -1 < x < 1, \\ &|v(x)| \text{ and } |v'(x)| \text{ bounded as } x \to -1^+ \text{ and } x \to 1^- \end{aligned}$$

has eigenvalues

$$\lambda_n = n(n+1)$$

with corresponding eigenfunctions

$$\phi_n(x) = P_n(x)$$

for $n \geqslant 0$.

Note: For Legendre's equation we have $p(x) = (1-x^2)$, $p(x)$ is zero at both endpoints of the interval, and both boundary conditions are replaced by boundedness conditions. Thus, the eigenvalues and eigenfunctions are completely determined by these boundedness conditions.

The first few Legendre polynomials are

$$\begin{aligned} P_0(x) &= 1, \\ P_1(x) &= x, \\ P_2(x) &= \frac{1}{2}(3x^2 - 1), \\ P_3(x) &= \frac{1}{2}(5x^3 - 3x), \\ P_4(x) &= \frac{1}{8}(35x^4 - 30x^2 + 3), \\ P_5(x) &= \frac{1}{8}(63x^5 - 70x^3 + 15x), \end{aligned}$$

and their graphs are shown in Figure 7.4. In general it can be shown (see, e.g., Wylie [52]) that

$$P_n(x) = \sum_{k=0}^{\lfloor \frac{n}{2} \rfloor} \frac{(-1)^k (2n-2k)!\, x^{n-2k}}{2^n\, k!\, (n-k)!\, (n-2k)!} \tag{7.11}$$

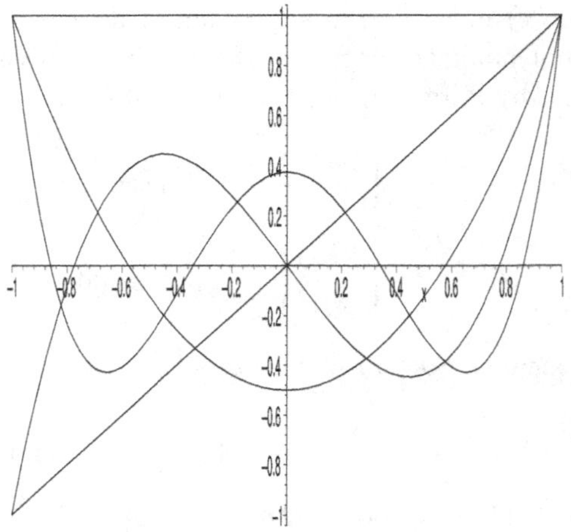

Figure 7.4 Graphs of Legendre polynomials

for $n = 0, 1, 2, \ldots$, where $\lfloor n/2 \rfloor$ is the greatest integer less than or equal to $n/2$. Note that $P_n(-x) = (-1)^n P_n(x)$, so that $P_n(-1) = (-1)^n P_n(1) = (-1)^n$ for $n \geqslant 0$.

We should note here that each $P_n(x)$ is continuous on $[-1, 1]$, and hence is bounded there, and that these are the only bounded solutions to Legendre's equation. Later in this chapter we show that, in fact, $|P_n(x)| \leqslant 1$ for all $x \in [-1, 1]$.
We can show that the eigenfunctions are orthogonal on the interval $[-1, 1]$ directly from Legendre's equation.

Theorem 7.3. (*Orthogonality of Legendre Polynomials*)
If m and n are nonnegative integers with $m \neq n$,

$$\int_{-1}^{1} P_m(x)\, P_n(x)\, dx = 0.$$

Proof. Since P_m and P_n are both eigenfunctions,

$$\left[(1 - x^2)P_m'\right]' + m(m+1)P_m = 0 \tag{7.12}$$

and

$$\left[(1 - x^2)P_n'\right]' + n(n+1)P_n = 0. \tag{7.13}$$

Multiplying (7.12) by P_n and (7.13) by P_m, integrating, and subtracting, we have

$$\int_{-1}^{1} \left\{P_n\left[(1 - x^2)P_m'\right]' - P_m\left[(1 - x^2)P_n'\right]'\right\} dx$$

$$+ \left[m(m+1) - n(n+1)\right] \int_{-1}^{1} P_m\, P_n\, dx = 0.$$

Integrating the two integrals on the left by parts, we have

$$\int_{-1}^{1} P_n\big[(1-x^2)P_m'\big]'\,dx = (1-x^2)P_nP_m'\Big|_{-1}^{1} - \int_{-1}^{1}(1-x^2)P_m'P_n'\,dx,$$

that is,

$$\int_{-1}^{1} P_n\big[(1-x^2)P_m'\big]'\,dx = -\int_{-1}^{1}(1-x^2)P_m'P_n'\,dx,$$

and similarly,

$$\int_{-1}^{1} P_m\big[(1-x^2)P_n'\big]'\,dx = -\int_{-1}^{1}(1-x^2)P_n'P_m'\,dx.$$

Therefore,

$$\big[m(m+1)-n(n+1)\big]\int_{-1}^{1} P_m\,P_n\,dx = 0,$$

and since $m \neq n$, $m(m+1)-n(n+1) \neq 0$, and

$$\int_{-1}^{1} P_m\,P_n\,dx = 0.$$

□

Another important identity involving Legendre polynomials is given in the following theorem.

Theorem 7.4. (*Rodrigues' Formula*)
For a nonnegative integer n, we have

$$P_n(x) = \frac{1}{2^n\,n!}\,\frac{d^n}{dx^n}\left[(x^2-1)^n\right].$$

Proof. We let $u = (x^2-1)^n$; then $du/dx = 2nx(x^2-1)^{n-1}$, and multiplying by x^2-1, we have

$$(x^2-1)\frac{du}{dx} = 2nx(x^2-1)^n,$$

so that

$$(1-x^2)\frac{du}{dx} + 2nx\,u = 0.$$

If we differentiate this equation k times, we get

$$\begin{gathered}
(1-x^2)u'' + 2(n-1)x\,u' + 2n\,u = 0\\
(1-x^2)u''' + 2(n-2)x\,u'' + 2(2n-1)u' = 0\\
(1-x^2)u^{(4)} + 2(n-3)x\,u''' + 3(2n-2)u'' = 0\\
\vdots\\
(1-x^2)u^{(k+2)} + 2(n-k-1)x\,u^{(k+1)} + (k+1)(2n-k)\,u^{(k)} = 0.
\end{gathered}$$

Now take $k = n$ and let $v = u^{(n)}$; then the last equation becomes

$$(1 - x^2)v'' - 2x\,v' + n(n+1)v = 0,$$

which is Legendre's equation. However,

$$v = u^{(n)} = \frac{d^n}{dx^n}\left[(x^2 - 1)^n\right]$$

is a polynomial of degree n, bounded on the interval $[-1, 1]$, and is an eigenfunction of the singular Sturm-Liouville problem involving Legendre's equation, with eigenvalue $n(n+1)$. Thus, $v(x)$ is a multiple of $P_n(x)$; that is,

$$\frac{d^n}{dx^n}\left[(x^2 - 1)^n\right] = c\,P_n(x).$$

We can determine the constant c as follows: The coefficient of x^n on the left-hand side arises from differentiating the term of highest degree, x^{2n}, exactly n times, and is

$$(2n)(2n-1)\cdots[2n-(n-1)] = \frac{(2n)!}{n!}.$$

On the other hand, from Equation (7.11), the coefficient of the highest power of x in $P_n(x)$ is

$$\frac{(2n)!}{2^n\,(n!)^2}.$$

Therefore,

$$c\,\frac{(2n)!}{2^n\,(n!)^2} = \frac{(2n)!}{n!},$$

so that

$$c = 2^n\,n!$$

and

$$P_n(x) = \frac{1}{2^n\,n!}\,\frac{d^n}{dx^n}\left[(x^2 - 1)^n\right]$$

for $n \geqslant 0$. □

The next theorem gives a few of the recurrence relations satisfied by the Legendre polynomials.

Theorem 7.5. (*Recurrence Relations*)

(i) $P'_{n+1}(x) - P'_{n-1}(x) = (2n+1)P_n(x)$;

(ii) $(n+1)P_{n+1}(x) + nP_{n-1}(x) = (2n+1)x\,P_n(x)$

for $n = 1, 2, 3, \ldots$.

Proof.

(i) From Rodrigues' formula we have

$$P'_{n+1} - P'_{n-1} = \frac{1}{2^{n+1}(n+1)!}\frac{d^{n+2}}{dx^{n+2}}\left[(x^2-1)^{n+1}\right] - \frac{1}{2^{n-1}(n-1)!}\frac{d^n}{dx^n}\left[(x^2-1)^{n-1}\right];$$

that is,

$$\begin{aligned}
P'_{n+1} - P'_{n-1} &= \frac{1}{2^{n+1}(n+1)!}\frac{d^n}{dx^n}\left[\frac{d^2}{dx^2}(x^2-1)^{n+1}\right] - \frac{1}{2^{n-1}(n-1)!}\frac{d^n}{dx^n}\left[(x^2-1)^{n-1}\right] \\
&= \frac{1}{2^{n+1}(n+1)!}\frac{d^n}{dx^n}\left[2(n+1)(x^2-1)^n + 4n(n+1)x^2(x^2-1)^{n-1}\right] - \frac{1}{2^{n-1}(n-1)!}\frac{d^n}{dx^n}\left[(x^2-1)^{n-1}\right] \\
&= \frac{1}{2^{n+1}(n+1)!}\frac{d^n}{dx^n}\left[2(n+1)(x^2-1)^n\right] + \frac{1}{2^{n-1}(n-1)!}\frac{d^n}{dx^n}\left[(x^2-1)^{n-1}x^2 - (x^2-1)^{n-1}\right] \\
&= \frac{1}{2^{n+1}(n+1)!}\frac{d^n}{dx^n}\left[2(n+1)(x^2-1)^n\right] + \frac{1}{2^{n-1}(n-1)!}\frac{d^n}{dx^n}\left[(x^2-1)^n\right] \\
&= \left(\frac{1}{2^n n!} + \frac{1}{2^{n-1}(n-1)!}\right)\frac{d^n}{dx^n}\left[(x^2-1)^n\right] \\
&= \frac{(2n+1)}{2^n n!}\frac{d^n}{dx^n}\left[(x^2-1)^n\right],
\end{aligned}$$

and thus,

$$P'_{n+1} - P'_{n-1} = (2n+1)P_n$$

for $n \geqslant 1$.

(ii) For $n \geqslant 1$, P_{n+1} and P_{n-1} are eigenfunctions of Legendre's equation with eigenvalues $n(n+1)$ and $n(n-1)$, respectively, so that

$$\left[(1-x^2)P'_{n+1}\right]' + (n+1)(n+2)P_{n+1} = 0$$

and

$$\left[(1-x^2)P_{n-1}'\right]' + n(n-1)P_{n-1} = 0.$$

Differentiating gives us

$$\left[(1-x^2)P_{n+1}'\right]'' + (n+1)(n+2)P_{n+1}' = 0$$

and

$$\left[(1-x^2)P_{n-1}'\right]'' + n(n-1)P_{n-1}' = 0.$$

Subtracting, we get

$$\left[(1-x^2)\left(P_{n+1}' - P_{n-1}'\right)\right]'' + (n+1)(n+2)P_{n+1}' - n(n-1)P_{n-1}' = 0,$$

and from (i),

$$\left[(2n+1)(1-x^2)P_n\right]'' + (n+1)(n+2)P_{n+1}' - n(n-1)P_{n-1}' = 0;$$

that is,

$$\left[(1-x^2)(2n+1)P_n'\right]' - \left[2(2n+1)xP_n\right]' \\ + (n+1)(n+2)P_{n+1}' - n(n-1)P_{n-1}' = 0,$$

and from Legendre's equation

$$-(2n+1)n(n+1)P_n - \left[2(2n+1)xP_n\right]' \\ + (n+1)(n+2)P_{n+1}' - n(n-1)P_{n-1}' = 0.$$

Using (i) again, we have

$$-n(n+1)\left(P_{n+1}' - P_{n-1}'\right) - \left[2(2n+1)xP_n\right]' \\ + (n+1)(n+2)P_{n+1}' - n(n-1)P_{n-1}' = 0;$$

that is,

$$\left(2(2n+1)xP_n - 2(n+1)P_{n+1} - 2nP_{n-1}\right)' = 0.$$

Therefore,

$$(2n+1)xP_n - (n+1)P_{n+1} - nP_{n-1} = c,$$

where c is a constant. Finally, since $P_k(1) = 1$ for all $k \geqslant 1$, then $c = 0$, and

$$(2n+1)xP_n(x) = (n+1)P_{n+1}(x) + nP_{n-1}(x)$$

for all x.

□

We can use the previous results to evaluate the integral

$$\int_{-1}^{1} P_n(x)^2\,dx.$$

Theorem 7.6. (*Norm of Legendre Polynomials*)
For $n \geqslant 0$, we have

$$a_n = \int_{-1}^{1} P_n(x)^2\,dx = \frac{2}{2n+1}.$$

Proof. From the recurrence relation (ii), we have

$$P_n(x) = \frac{2n-1}{n}\,x\,P_{n-1}(x) - \frac{n-1}{n}\,P_{n-2}(x)$$

for $n \geqslant 2$, and hence

$$a_n = \int_{-1}^{1} P_n(x)\left[\frac{2n-1}{n}\,x\,P_{n-1}(x) - \frac{n-1}{n}\,P_{n-2}(x)\right] dx.$$

Using the orthogonality relation, we have

$$a_n = \frac{2n-1}{n}\int_{-1}^{1} x\,P_n(x)\,P_{n-1}(x)\,dx,$$

while from (ii) we have

$$xP_n(x) = \frac{1}{2n+1}\left[(n+1)P_{n+1}(x) + n\,P_{n-1}(x)\right]$$

for $n \geqslant 1$, so that

$$\frac{n}{2n-1}\,a_n = \frac{n}{2n+1}\int_{-1}^{1} P_{n-1}^2\,dx = \frac{n}{2n+1}a_{n-1}.$$

Therefore,

$$(2n+1)a_n = (2n-1)a_{n-1}$$

for all $n \geqslant 2$. Since $P_0(x) = 1$ and $P_1(x) = x$,

$$a_0 = \int_{-1}^{1} dx = 2 \qquad \text{and} \qquad a_1 = \int_{-1}^{1} x^2\,dx = \frac{2}{3}.$$

An easy induction argument shows that the quantity $(2n+1)a_n$ is constant for all $n \geqslant 0$. Thus,

$$(2n+1)a_n = a_0 = 2;$$

that is,

$$a_n = \int_{-1}^{1} P_n(x)^2\,dx = \frac{2}{2n+1}$$

for all $n \geqslant 0$. □

A final identity involving the Legendre polynomials is the generating function for the sequence $\{P_n(x)\}_{n\geqslant 0}$.

Theorem 7.7. (*Generating Function for Legendre Polynomials*)

$$\frac{1}{\sqrt{1-xt+t^2}} = P_0(x) + P_1(x)\,t + P_2(x)\,t^2 + \cdots + P_n(x)\,t^n + \cdots$$

for $-1 \leqslant x \leqslant 1$ and $-1 < t < 1$.

Proof. We expand the denominator on the left-hand side using the binomial theorem to get

$$\begin{aligned}\big[1 - t(2x-t)\big]^{-1/2} &= 1 + \frac{1}{2}t(2x-t) + \frac{1\cdot 3}{2^2\,2!}t^2(2x-t)^2 + \cdots \\ &\quad + \frac{1\cdot 3\cdots(2n-3)}{2^{n-1}\,(n-1)!}t^{n-1}(2x-t)^{n-1} \qquad (7.14)\\ &\quad + \frac{1\cdot 3\cdots(2n-1)}{2^n\,n!}t^n(2x-t)^n + \cdots\end{aligned}$$

The t^n can only occur in the terms out to and including the one containing $t^n(2x-t)^n$, and expanding the various powers of $2x-t$ in these terms, the coefficient of t^n is

$$\frac{1\cdot 3\cdots(2n-1)}{2^n\,n!}(2x)^n - \frac{1\cdot 3\cdots(2n-3)}{2^{n-1}\,(n-1)!}(2x)^{n-2} + \frac{1\cdot 3\cdot(2n-5)}{2^{n-2}\,(n-2)!}\frac{(n-2)(n-3)}{2!}(2x)^{n-4} - \cdots,$$

that is,

$$\frac{(2n)!}{2^n\,n!\,n!}x^n - \frac{(2n-2)!}{2^{n-1}\,1!(n-1)!\,(n-2)!}x^{n-2} + \frac{(2n-4)!}{2^{n-2}\,2!\,(n-2)!\,(n-4)!}x^{n-4} - \cdots,$$

which is exactly the expanded form of $P_n(x)$. Therefore,

$$(1-2xt+t^2)^{-1/2} = \sum_{n=0}^{\infty} P_n(x)\,t^n$$

for $|x| \leqslant 1$, and $|t| < 1$. □

Example 7.1. (*Potential Theory*)
Let Q be a fixed point on the z-axis a distance $h > 1$ from the origin, and let P be any point in the unit sphere $0 < r < 1$. If we apply the law of cosines to the triangle OPQ, in spherical coordinates the distance from P to Q is given by

$$\|P - Q\| = \sqrt{r^2 + h^2 - 2rh\cos\theta}$$

as in Figure 7.5. It is easy to show that the potential function

$$u(r, \theta) = \frac{1}{\sqrt{r^2 + h^2 - 2rh\cos\theta}}$$

satisfies Laplace's equation in the region $0 < r < 1$; that is,

$$\frac{\partial^2 u}{\partial r^2} + \frac{2}{r}\frac{\partial u}{\partial r} + \frac{1}{r^2 \sin\theta}\frac{\partial}{\partial\theta}\left(\sin\theta\frac{\partial u}{\partial\theta}\right) + \frac{1}{r^2\sin^2\theta}\frac{\partial^2 u}{\partial\phi^2} = 0$$

for $0 < r < 1, 0 < \theta < \pi$. Since u is independent of ϕ, we write $u(r,\theta) = R(r)\cdot S(\theta)$ as before, and separate variables to get

$$\frac{1}{R}\left(r^2\frac{d^2R}{dr^2} + 2r\frac{dR}{dr}\right) = -\frac{1}{S\sin\theta}\frac{d}{d\theta}\left(\sin\theta\frac{dS}{d\theta}\right) = \lambda,$$

where λ is the separation constant, and this leads to two ordinary differential equations:

$$r^2\frac{d^2R}{dr^2} + 2r\frac{dR}{dr} - \lambda R = 0, \quad 0 < r < 1, \tag{7.15}$$

$$\sin\theta\frac{d}{d\theta}\left(\sin\theta\frac{dS}{d\theta}\right) + \lambda S\sin^2\theta = 0, \quad 0 < \theta < \pi. \tag{7.16}$$

Again, making the substitution $x = \cos\theta$ and $v(x) = S(\theta)$, equation (7.16) becomes

$$(1 - x^2)\frac{d^2v}{dx^2} - 2x\frac{dv}{dx} + \lambda v = 0, \quad -1 < x < 1, \tag{7.17}$$

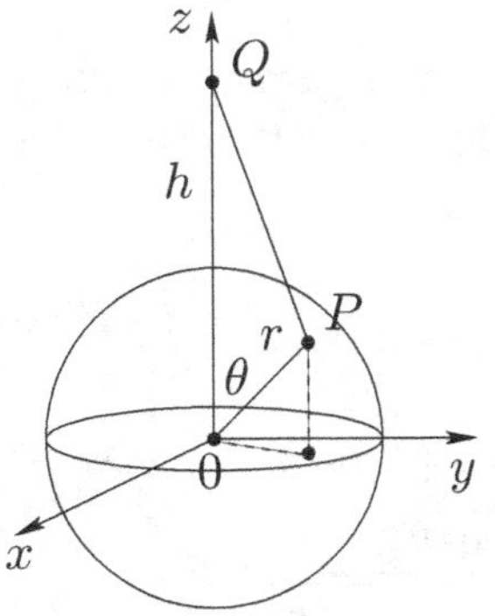

Figure 7.5 Law of cosines.

which is Legendre's equation, and this has bounded solutions if and only if

$$\lambda = n(n+1)$$

for some nonnegative integer n, namely,

$$S_n(\theta) = v_n(x) = P_n(x) = P_n(\cos\theta)$$

for $n \geqslant 0$.

The corresponding radial equation

$$r^2\frac{d^2R}{dr^2} + 2r\frac{dR}{dr} - n(n+1)R = 0, \quad 0 < r < 1 \tag{7.18}$$

is a Cauchy-Euler equation with general solution

$$R_0(r) = A_0 + B_0 \log r, \quad 0 < r < 1$$

for $n = 0$, and

$$R_n(r) = A_n r^n + B_n \frac{1}{r^{n+1}}, \quad 0 < r < 1$$

for $n \geqslant 1$. Thus, the only bounded solutions to Laplace's equation are the **spherical harmonics**

$$u_n(r,\theta) = r^n P_n(\cos\theta);$$

for $n \geqslant 0$. Using the superposition principle, we write

$$u(r,\theta) = \sum_{k=0}^{\infty} a_k r^k P_k(\cos\theta),$$

that is,

$$\frac{1}{\sqrt{r^2 + h^2 - 2rh\cos\theta}} = \sum_{k=0}^{\infty} a_k r^k P_k(\cos\theta) \tag{7.19}$$

for $0 \leqslant r \leqslant 1, 0 \leqslant \theta \leqslant \pi$.

We can determine the coefficients a_k from the orthogonality of the Legendre polynomials, multiplying (7.19) by $P_n(\cos\theta)\sin\theta$, setting $r = 1$, and integrating, we have

$$\int_0^\pi \frac{P_n(\cos\theta)\sin\theta}{\sqrt{1+h^2-2h\cos\theta}}\,d\theta = \sum_{k=0}^{\infty} a_k \int_0^\pi P_n(\cos\theta)P_k(\cos\theta)\sin\theta\,d\theta,$$

and hence from Theorem 7.6 we have

$$a_n = \frac{2n+1}{2}\int_0^\pi \frac{P_n(\cos\theta)\sin\theta}{\sqrt{1+h^2-2h\cos\theta}}\,d\theta = \frac{2n+1}{2}\int_{-1}^1 \frac{P_n(x)}{\sqrt{1-2xh+h^2}}\,dx$$

for $n \geqslant 0$.

However, in this case it is much easier to use the generating function for the sequence of Legendre polynomials than to evaluate the integral. If we let $t = r/h$, then $0 \leqslant t < 1$, and from (7.19) we have

$$\frac{1}{\sqrt{1-2xt+t^2}} = \sum_{n=0}^{\infty} a_n P_n(x) h^{n+1} t^n,$$

and therefore,

$$a_n h^{n+1} = 1,$$

and the solution is

$$u(r,\theta) = \frac{1}{\sqrt{r^2+h^2-2rh\cos\theta}} = \frac{1}{h}\sum_{n=0}^{\infty}\left(\frac{r}{h}\right)^n P_n(\cos\theta) \tag{7.20}$$

for $0 \leqslant r \leqslant 1$, $0 \leqslant \theta \leqslant \pi$. ∎

We mentioned earlier that the Legendre polynomials $P_n(x)$ for $n \geqslant 0$, being continuous functions on the closed and bounded interval $[-1,1]$, were the only bounded solutions to Legendre's equation. Now we show that, in fact, they are all bounded in absolute value by 1 on the interval $[-1,1]$. We need the following additional recurrence relations, which follow from Theorem 7.5.

Lemma 7.8. (*More Recurrence Relations*)

(i) $P'_{n+1} - xP'_n - (n+1)P_n = 0$

(ii) $xP'_n - P'_{n-1} = nP_n$

(iii) $(1-x^2)P'_n + nxP_n - nP_{n-1} = 0$

(iv) $P_n = xP_{n-1} + \dfrac{x^2-1}{n} P'_{n-1}$

for $n = 1, 2, 3, \ldots$.

Proof.

(i) From Theorem 7.5, part (ii), we have

$$(n+1)P_{n+1} + nP_{n-1} = (2n+1)xP_n,$$

and differentiating yields

$$(n+1)P'_{n+1} + nP'_{n-1} = (2n+1)P_n + (2n+1)xP'_n,$$

while from Theorem 7.5, part (i), we have

$$nP'_{n+1} - nP'_{n-1} = n(2n+1)P_n.$$

Adding these two equations, we have

$$(2n+1)P'_{n+1} = (2n+1)(n+1)P_n + (2n+1)xP'_n;$$

that is,

$$P'_{n+1} = (n+1)P_n + xP'_n$$

for $n \geqslant 1$, which is (i).

(ii) From (i) we have

$$P'_{n+1} = xP'_n + (n+1)P_n,$$

and from Theorem 7.5, part (i), we have

$$P'_{n+1} - P'_{n-1} = (2n+1)P_n,$$

so that

$$P'_{n-1} + (2n+1)P_n = xP'_n + (n+1)P_n;$$

that is,

$$xP'_n = P'_{n-1} + nP_n,$$

which is (ii).

(iii) From (ii) we have

$$x^2P'_n - xP'_{n-1} = nxP_n,$$

and thus,

$$(1-x^2)P'_n = P'_n - xP'_{n-1} - nxP_n.$$

However,

$$xP'_{n-1} - P'_{n-2} = (n-1)P_{n-1},$$

so that

$$(1-x^2)P'_n = P'_n - P'_{n-2} - (n-1)P_{n-1} - nxP_n,$$

and since $P'_n - P'_{n-2} = (2n-1)P_{n-1}$, then

$$(1-x^2)P'_n = (2n-1)P_{n-1} - (n-1)P_{n-1} - nxP_n;$$

that is,

$$(1-x^2)P'_n = nP_{n-1} - nxP_n$$

for $n \geqslant 2$, but it is also true for $n = 1$, hence (iii) holds.

(iv) From (iii) we have

$$(1-x^2)P'_{n-1} = (n-1)P_{n-2} - (n-1)xP_{n-1},$$

and from Theorem 7.5, part (ii), we have

$$nP_n + (n-1)P_{n-2} = (2n-1)xP_{n-1},$$

so that

$$(1-x^2)P'_{n-1} = (2n-1)xP_{n-1} - nP_n - (n-1)xP_{n-1} = nxP_{n-1} - nP_n;$$

that is,

$$P_n = xP_{n-1} + \frac{x^2-1}{n} P'_{n-1}$$

for all $n \geqslant 1$, which is (iv).

□

Theorem 7.9. (*Boundedness of Legendre Polynomials*)

For $n \geqslant 0$, $|P_n(x)| \leqslant 1$ for all $|x| \leqslant 1$.

Proof. From Lemma 7.8, part (iv), we have

$$(P_n)^2 = x^2 (P_{n-1})^2 + \frac{(x^2-1)^2}{n^2} (P'_{n-1})^2 + \frac{2x(x^2-1)}{n} P_{n-1}P'_{n-1}, \tag{7.21}$$

and from Lemma 7.8, part (i), we have

$$P'_n = xP'_{n-1} + nP_{n-1},$$

so that

$$(P'_n)^2 = x^2 (P'_{n-1})^2 + n^2 (P_{n-1})^2 + 2xnP_{n-1}P'_{n-1}. \tag{7.22}$$

From (7.21) and (7.22), we have

$$\begin{aligned}\frac{1-x^2}{n^2} (P'_n)^2 + (P_n)^2 &= \frac{1-x^2}{n^2} x^2 (P'_{n-1})^2 + (1-x^2)(P_{n-1})^2 \\ &\quad + \frac{2x(1-x^2)}{n} P_{n-1}P'_{n-1} + x^2 (P_{n-1})^2 \\ &\quad + \frac{(x^2-1)^2}{n^2} (P'_{n-1})^2 + \frac{2x(x^2-1)}{n} P_{n-1}P'_{n-1} \\ &= \frac{(1-x^2)}{n^2} (P'_{n-1})^2 + (P_{n-1})^2.\end{aligned}$$

Therefore,

$$\frac{1-x^2}{n^2} (P'_n)^2 + (P_n)^2 = \frac{1-x^2}{n^2} (P'_{n-1})^2 + (P_{n-1})^2, \tag{7.23}$$

and since $P_0(x) \equiv 1$, an easy induction argument shows that

$$0 \leqslant \frac{1-x^2}{n^2} (P'_n)^2 + (P_n)^2 \leqslant 1 \tag{7.24}$$

for all $n \geqslant 1$. Thus,

$$|P_n(x)|^2 \leqslant 1$$

whenever $|x| \leqslant 1$ and $n \geqslant 1$.

□

Example 7.2. Show that there exists a positive constant M such that

$$\int_{-1}^{1} |P_n(x)|\, dx < \frac{M}{\sqrt{n}}$$

for all $n \geqslant 1$.

Solution. From the Cauchy-Schwarz inequality we have

$$\int_{-1}^{1} |P_n(x)|\, dx \leqslant \left(\int_{-1}^{1} P_n(x)^2\, dx\right)^{\frac{1}{2}} \left(\int_{-1}^{1} 1^2\, dx\right)^{\frac{1}{2}},$$

but

$$\int_{-1}^{1} P_n(x)^2\, dx = \frac{2}{2n+1},$$

so that

$$\int_{-1}^{1} |P_n(x)|\, dx \leqslant \sqrt{\frac{2}{2n+1}}\,\sqrt{2} = \frac{2}{\sqrt{2n+1}} < \frac{2}{\sqrt{2n}}$$

and

$$\int_{-1}^{1} |P_n(x)|\, dx < \frac{\sqrt{2}}{\sqrt{n}}$$

for $n \geqslant 1$. ∎

We summarize in the following theorem the properties of the Legendre polynomials that we have proven. Note that these properties followed directly from the results on Sturm-Liouville theory or from Legendre's equation.

Theorem 7.10. The Legendre polynomials satisfy the following:

1. For $n \geqslant 0$, $P_n(1) = 1$, $P_n(-1) = (-1)^n$, and $|P_n(x)| \leqslant 1$ for $-1 \leqslant x \leqslant 1$.

2. For $n \geqslant 0$ and $m \geqslant 0$, $\langle P_n, P_m \rangle = 0$ if $n \neq m$, where

$$\langle f, g \rangle = \int_{-1}^{1} f(x)g(x)\, dx.$$

3. $\|P_n\|^2 = \dfrac{2}{2n+1}$ for $n \geqslant 0$.

4. Rodrigues' formula: $P_n(x) = \dfrac{1}{2^n\, n!}\dfrac{d^n}{dx^n}\left[(x^2-1)^n\right]$ for $n \geqslant 0$.

5. Generating function: $\dfrac{1}{\sqrt{1-2xt+t^2}} = \sum\limits_{n=0}^{\infty} P_n(x)t^n$ for $|x| \leqslant 1$ and $|t| < 1$.

6. $(n+1)P_{n+1}(x) = (2n+1)xP_n(x) - nP_{n-1}(x)$ for $n \geqslant 1$.

7. $(2n+1)P_n(x) = P'_{n+1}(x) - P'_{n-1}(x)$ for $n \geqslant 1$.

7.3.2 Fourier-Legendre Series

Suppose that we have a function $f(x)$ defined on the interval $[-1,1]$, and that it has a **Fourier-Legendre expansion** given by

$$f(x)=\sum_{n=0}^{\infty}a_nP_n(x)=a_0P_0(x)+a_1P_1(x)+a_2P_2(x)+\cdots+a_nP_n(x)+\cdots.$$

To determine the coefficients a_n, we use the orthogonality relation for $P_n(x)$ in Theorem 7.10. The Legendre polynomials $P_n(x)$ are solutions to a singular Sturm-Liouville problem, and we know that the eigenfunctions $P_n(x)$ are orthogonal on the interval $[-1,1]$ with respect to the weight function $\sigma(x)=1$. Multiplying the equation above by $P_n(x)$ and integrating, we have

$$\int_{-1}^{1}f(x)P_n(x)\,dx=a_n\int_{-1}^{1}P_n(x)^2\,dx=\frac{2}{2n+1}\,a_n$$

for $n\geqslant 0$, and thus

$$a_n=\frac{2n+1}{2}\int_{-1}^{1}f(x)P_n(x)\,dx$$

for $n\geqslant 0$.

We have a convergence theorem for Fourier-Legendre series similar to Dirichlet's theorem.

Theorem 7.11. (*Fourier-Legendre Expansion Theorem*)

If f and f' are piecewise continuous on the interval $[-1,1]$, and

$$a_n=\frac{2n+1}{2}\int_{-1}^{1}f(x)P_n(x)\,dx,$$

then for $-1<x_0<1$, the series

$$\sum_{n=0}^{\infty}a_nP_n(x_0)$$

converges to

$$\frac{1}{2}\left[f(x_0^+)+f(x_0^-)\right].$$

For $x_0=-1$, the series converges to $f(-1^+)$, while for $x_0=1$, the series converges to $f(1^-)$.

Note that if f is only defined on the interval $[0,1]$, then by considering the even and odd extensions of f to $[-1,1]$, we obtain similar results.

Example 7.3. Find the Fourier-Legendre expansion of the function

$$f(x) = x^3$$

for $-1 \leqslant x \leqslant 1$.

Solution. Since $P_1(x) = x$ and $P_3(x) = \frac{5}{2}x^3 - \frac{3}{2}x$, then

$$x^3 = \frac{2}{5}P_3(x) + \frac{3}{5}P_1(x). \tag{7.25}$$

If the Fourier-Legendre series expansion of x^3 is

$$x^3 = \sum_{n=0}^{\infty} a_n P_n(x)$$

for $-1 \leqslant x \leqslant 1$, then

$$a_n = \frac{2n+1}{2}\int_{-1}^{1} x^3 P_n(x)\,dx = \frac{2n+1}{2}\int_{-1}^{1}\left[\frac{2}{5}P_3(x) + \frac{3}{5}P_1(x)\right] P_n(x)\,dx,$$

and from the orthogonality of the Legendre polynomials, $a_n = 0$ unless $n = 1$ or $n = 3$. For $n = 1$ we have

$$a_1 = \frac{3}{2}\int_{-1}^{1}\frac{3}{5}P_1(x)^2\,dx = \frac{9}{10}\int_{-1}^{1}P_1(x)^2\,dx = \frac{9}{10}\frac{2}{3} = \frac{3}{5},$$

while for $n = 3$ we have

$$a_3 = \frac{7}{2}\int_{-1}^{1}\frac{2}{5}P_3(x)^2\,dx = \frac{7}{5}\int_{-1}^{1}P_3(x)^2\,dx = \frac{7}{5}\frac{2}{7} = \frac{2}{5}.$$

This result was not totally unexpected, since (7.25) is the Fourier-Legendre expansion of the function x^3 on the interval $[-1, 1]$. ∎

Example 7.4. Let $u(x,t)$ represent the temperature in a nonhomogeneous insulated bar lying along the x-axis, where $-1 \leqslant x \leqslant 1$. Suppose that the thermal conductivity K is proportional to $1 - x^2$, so that

$$K = a(1 - x^2),$$

where $a > 0$ is a constant.

If the linear density ρ and the specific heat c of the bar are both constant, the heat equation becomes

$$\frac{\partial u}{\partial t} = \frac{1}{\rho c}\frac{\partial}{\partial x}\left(K\frac{\partial u}{\partial x}\right) = b\frac{\partial}{\partial x}\left((1-x^2)\frac{\partial u}{\partial x}\right),$$

where $b = a/\rho c > 0$ is a constant. Note that since the conductivity is zero at both ends of the bar, from Fourier's law there is no heat flow through the ends $x = -1$ and $x = 1$.

If the initial temperature distribution is

$$u(x,0) = f(x), \quad -1 \leqslant x \leqslant 1,$$

the boundary value–initial value problem for the temperature in the bar is

$$\frac{\partial u}{\partial t} = b\frac{\partial}{\partial x}\left((1-x^2)\frac{\partial u}{\partial x}\right), \quad -1 \leqslant x \leqslant 1, \quad t \geqslant 0,$$

$$u(x,0) = f(x).$$

The PDE and the boundary conditions are both linear and homogeneous, so we may apply separation of variables.

Assuming a solution of the form

$$u(x,t) = X(x)\cdot T(t),$$

and substituting this expression into the PDE, we have

$$X \cdot T' = b\frac{d}{dx}\left((1-x^2)\frac{dX}{dx}\right)\cdot Tx,$$

and

$$\frac{1}{X}\frac{d}{dx}\left((1-x^2)\frac{dX}{dx}\right) = \frac{T'}{bT} = -\lambda.$$

Therefore, we have the ODEs for X and T,

$$\frac{d}{dx}\left((1-x^2)\frac{dX}{dx}\right) + \lambda X = 0, \quad -1 \leqslant x \leqslant 1,$$

$$T' + \lambda b T = 0, \quad t \geqslant 0.$$

The first equation is Legendre's equation, and we have bounded solutions if and only if $\lambda = n(n+1)$ for $n \geqslant 0$, with corresponding eigenfunctions $P_n(x)$ for $n \geqslant 0$. For these values of λ, the solution to the second equation is

$$T_n(t) = e^{-n(n+1)bt}.$$

Applying the superposition principle, we write

$$u(x,t) = \sum_{n=0}^{\infty} a_n P_n(x)\, e^{-n(n+1)bt},$$

where, as seen earlier, the coefficients in the Fourier-Legendre expansion are given by

$$a_n = \frac{2n+1}{2} \int_{-1}^{1} f(x) P_n(x)\, dx$$

for $n \geqslant 0$. ∎

7.3.3 Legendre Functions of the Second Kind

As noted, for the case $\lambda = n(n+1)$, one of the series in the general solution to Legendre's equation terminates but the other one does not, and the Legendre polynomial $P_n(x)$ is only one solution to Legendre's equation: It corresponds to $v_1(x)$ if n is even and $v_2(x)$ if n is odd. The functions represented by the other series, when suitably scaled, are denoted by Q_n and define a second linearly independent solution called **Legendre functions of the second kind**.

We define

$$Q_n(x) = \begin{cases} v_1(1)v_2(x), & \text{if n is even} \\ -v_2(1)v_1(x), & \text{if n is odd,} \end{cases} \tag{7.26}$$

and we can write the general solution of Legendre's equation as

$$v(x) = A\,P_n(x) + B\,Q_n(x).$$

Note that if we set $n = 0$ in (7.26), then $v_1(1) = 1$, and hence

$$Q_0(x) = v_2(x) = x + \frac{x^3}{3} + \frac{x^5}{5} + \frac{x^7}{7} + \cdots$$

for $|x| < 1$. For $|t| < 1$, we have

$$\sum_{n=0}^{\infty} t^{2n} = \frac{1}{1-t^2} = \frac{1}{2}\frac{1}{1+t} + \frac{1}{2}\frac{1}{1-t},$$

and integrating from 0 to x, where $|x| < 1$, we have

$$\sum_{n=0}^{\infty} \frac{x^{2n+1}}{2n+1} = \frac{1}{2}\log(1+x) - \frac{1}{2}\log(1-x) = \frac{1}{2}\log\frac{1+x}{1-x}.$$

Thus,

$$Q_0(x) = x + \frac{x^3}{3} + \frac{x^5}{5} + \frac{x^7}{7} + \cdots = \frac{1}{2}\log\frac{1+x}{1-x}$$

for $-1 < x < 1$.

Similarly, the first few Legendre functions of the second kind are

$$Q_0(x) = \frac{1}{2}\log\frac{1+x}{1-x},$$

$$Q_1(x) = \frac{x}{2}\log\frac{1+x}{1-x} - 1,$$

$$Q_2(x) = \frac{1}{4}(3x^2-1)\log\frac{1+x}{1-x} - \frac{3}{2}x.$$

In fact, the general formula is

$$Q_n(x) = \frac{1}{2}P_n(x)\log\frac{1+x}{1-x} - W_{n-1}(x),$$

where

$$W_{n-1}(x) = \sum_{k=1}^{n}\frac{1}{k}P_{k-1}(x)P_{n-k}(x).$$

Notice that Legendre functions of the second kind are singular at ± 1 since

$$\lim_{x\to\pm 1}|Q_n(x)| = \infty.$$

In fact, the only bounded solutions are the Legendre polynomials. We made the following comment earlier, and now we state it as a theorem.

Theorem 7.12. Legendre's equation,

$$(1-x^2)\frac{d^2v}{dx^2} - 2x\frac{dv}{dx} + \lambda v = 0,$$

has bounded solutions on the interval $[-1, 1]$ if and only if

$$\lambda = n(n+1)$$

for $n = 0, 1, 2, \ldots$.

7.3.4 Associated Legendre Functions

We can derive the solutions to the associated Legendre equation

$$(1-x^2)\frac{d^2w}{dx^2} - 2x\frac{dw}{dx} + \left(\lambda - \frac{\nu}{1-x^2}\right)w = 0 \tag{7.27}$$

from the solutions to Legendre's equation,

$$(1-x^2)\frac{d^2v}{dx^2} - 2x\frac{dv}{dx} + \lambda v = 0, \tag{7.28}$$

by the following rather convoluted argument. Using Leibniz's rule for derivatives of a product,

$$\frac{d^m}{dx^m}(fg) = \sum_{j=0}^{m} \binom{m}{j} f^{(j)} g^{(m-j)},$$

we differentiate (7.28) m times, to get

$$(1-x^2)v^{(m+2)} - (2m+2)xv^{(m+1)} + [\lambda - m(m+1)]v^{(m)} = 0. \tag{7.29}$$

Now let $w(x) = (1-x^2)^{\frac{m}{2}} v^{(m)}$; then

$$w' = (1-x^2)^{\frac{m}{2}} v^{(m+1)} - mx(1-x^2)^{\frac{m}{2}-1} v^{(m)},$$

$$(1-x^2)w' = (1-x^2)^{\frac{m}{2}+1} v^{(m+1)} - mx(1-x^2)^{\frac{m}{2}} v^{(m)},$$

and

$$\begin{aligned}
\left[(1-x^2)w'\right]' &= (1-x^2)^{\frac{m}{2}+1} v^{(m+2)} - (2m+2)x(1-x^2)^{\frac{m}{2}} v^{(m+1)} \\
&\quad + \left[m^2x^2(1-x^2)^{\frac{m}{2}-1} - m(1-x^2)^{\frac{m}{2}}\right] v^{(m)} \\
&= (1-x^2)^{\frac{m}{2}} \bigg\{(1-x^2)v^{(m+2)} - (2m+2)xv^{(m+1)} \\
&\quad + \left(\frac{m^2x^2}{1-x^2} - m\right) v^{(m)}\bigg\} \\
&= (1-x^2)^{\frac{m}{2}} \left\{ -\lambda + m(m+1) + \frac{m^2x^2}{1-x^2} - m \right\} v^{(m)} \\
&= \left(-\lambda + \frac{m^2}{1-x^2} \right) w.
\end{aligned}$$

Therefore, w satisfies the equation

$$\left[(1-x^2)w'\right]' + \left(\lambda - \frac{m^2}{1-x^2}\right) w = 0;$$

that is,

$$(1-x^2)w'' - 2xw' + \left(\lambda - \frac{m^2}{1-x^2}\right) w = 0, \tag{7.30}$$

and equation (7.30) is the associated Legendre equation with $\nu = m^2$.

Thus, we have shown that if $v(x)$ is a solution to Legendre's equation (7.10),

$$w(x) = (1-x^2)^{\frac{m}{2}} v^{(m)}(x)$$

is a solution to the associated Legendre equation (7.30).

As with Legendre's equation, we have the following result for the associated Legendre's equation.

Theorem 7.13. The associated Legendre equation has bounded solutions on the interval $[-1, 1]$ if and only if

$$\lambda = n(n+1)$$

for $n = 0, 1, 2, \ldots$.

The **associated Legendre functions of degree n and order m of the first kind** are defined as follows:

$$P_n^m(x) = (1-x^2)^{\frac{m}{2}} \frac{d^m}{dx^m}\left[P_n(x)\right],$$

for $0 \leqslant m \leqslant n$. Note that $P_n^0(x) = P_n(x)$ for all n, and $P_n^m(x) \equiv 0$ if $m > n$. The first few associated Legendre functions are

$$P_1^1(x) = (1-x^2)^{\frac{1}{2}},$$

$$P_2^1(x) = 3x(1-x^2)^{\frac{1}{2}}, \qquad P_2^2(x) = 3(1-x^2),$$

$$P_3^1(x) = \frac{3}{2}(1-x^2)^{\frac{1}{2}}(5x^2-1), \quad P_3^2(x) = 15x(1-x^2), \quad P_3^3(x) = 15(1-x^2)^{\frac{3}{2}}.$$

We may extend the definition of associated Legendre functions to negative values of m by using Rodrigues' formula in the definition above. Thus, we redefine the associated Legendre functions of the first kind as

$$P_n^m(x) = \frac{1}{2^n\, n!}(1-x^2)^{\frac{m}{2}} \frac{d^{m+n}}{dx^{m+n}}(x^2-1)^n$$

for $-n \leqslant m \leqslant n$.

The **associated Legendre functions of degree n and order m of the second kind** are defined as follows:

$$Q_n^m(x) = (1-x^2)^{\frac{m}{2}} \frac{d^m}{dx^m}(Q_n(x))$$

for $0 \leqslant m \leqslant n$.

Lemma 7.14. The associated Legendre functions of the first kind satisfy the following properties:

1. $\|P_n^m\|^2 = \displaystyle\int_{-1}^{1} P_n^m(x)^2\,dx = \frac{2}{2n+1}\cdot\frac{(n+m)!}{(n-m)!}.$

2. $\langle P_k^m, P_l^m\rangle = \displaystyle\int_{-1}^{1} P_k^m(x)P_l^m(x)\,dx = 0;\quad \text{if } k \neq l.$

7.4 SPHERICAL BESSEL FUNCTIONS

Consider the radial equation (7.4) obtained earlier, with $\lambda = n(n+1)$:

$$r^2 R'' + 2rR' + (\mu^2 r^2 - n(n+1))R = 0,$$

and let $x = \mu r$ and $u(x) = R(r)$; then we get

$$x^2 u'' + 2xu' + (x^2 - n(n+1))u = 0, \tag{7.31}$$

which is the **spherical Bessel equation.** This resembles Bessel's equation, but differs from it because of the 2 in the second term; however, we can transform it into Bessel's equation as follows: Let $u(x) = x^\gamma w(x)$, and substitute this into equation (7.31) to get

$$x^2 w'' + 2(\gamma+1)xw' + [x^2 - n(n+1) + \gamma(\gamma+1)]w = 0.$$

Now set $2(\gamma + 1) = 1$, so that $\gamma = -1/2$; then the equation becomes

$$x^2 w'' + xw' + [x^2 - \left(n + \tfrac{1}{2}\right)^2]w = 0,$$

which is Bessel's equation of order $n + \frac{1}{2}$. One solution is

$$w(x) = J_{n+\frac{1}{2}}(x),$$

and therefore

$$u(x) = \frac{w(x)}{\sqrt{x}} = \frac{1}{\sqrt{x}} J_{n+\frac{1}{2}}(x).$$

The **spherical Bessel functions of order n of the first kind** are defined as

$$j_n(x) = \sqrt{\frac{\pi}{2x}}\, J_{n+\frac{1}{2}}(x),$$

and the **spherical Bessel functions of order n of the second kind** are defined as

$$y_n(x) = \sqrt{\frac{\pi}{2x}}\, Y_{n+\frac{1}{2}}(x).$$

The first few spherical Bessel functions are

$$j_0(x) = \frac{\sin x}{x}, \qquad y_0(x) = -\frac{\cos x}{x},$$

$$j_1(x) = \frac{\sin x}{x^2} - \frac{\cos x}{x}, \qquad y_1(x) = -\frac{\cos x}{x^2} - \frac{\sin x}{x},$$

$$j_2(x) = \left(\frac{3}{x^2} - \frac{1}{x}\right)\sin x - \frac{3}{x^2}\cos x, \quad y_2(x) = \left(-\frac{3}{x^2} + \frac{1}{x}\right)\cos x - \frac{3}{x^2}\sin x.$$

Lemma 7.15. The spherical Bessel functions satisfy the following properties:

1. $j_0(0) = 1$, $j_n(0) = 0$ for $n \geqslant 1$.
2. $y_n(x) \to -\infty$ as $x \to 0^+$.
3. $j_n(x) = x^n \left(-\frac{1}{x}\frac{d}{dx}\right)^n \left(\frac{\sin x}{x}\right), \quad y_n(x) = x^n \left(-\frac{1}{x}\frac{d}{dx}\right)^n \left(\frac{\cos x}{x}\right).$
4. $\displaystyle\int_0^{\alpha_{nm}} j_n(\mu r)^2\, r^2\, dr = \frac{\pi}{4}\alpha_{nm}^2 J'_{n+\frac{1}{2}}(\alpha_{nm}\mu)^2$, where $j_n(\alpha_{nm}) = 0$.
5. $\displaystyle\frac{1}{x}\sin\sqrt{x^2 + 2xt} = \sum_{n=0}^{\infty} \frac{(-1)^n y_{n-1}(x)}{n!}\, t^n$ for $2|t| < |x|$.
6. $\displaystyle\frac{1}{x}\cos\sqrt{x^2 - 2xt} = \sum_{n=0}^{\infty} \frac{j_{n-1}(x)}{n!}\, t^n$ for $2|t| < |x|$.

See, for example, [28].

7.5 INTERIOR DIRICHLET PROBLEM FOR A SPHERE

In Example 7.5 we consider Laplace's equation defined in a spherical region. The interior Dirichlet problem for Laplace's equation on the sphere $B(a)$ models, among other things, the steady-state temperature distribution of a spherical ball with the boundary held at a given temperature.

Example 7.5. (*Potential in a Sphere*)
We want to solve the Dirichlet problem for Laplace's equation

$$\begin{aligned} \nabla^2 u &= 0 \quad \text{on } B(a), \\ u(a, \theta, \phi) &= f(\theta, \phi) \quad \text{on } \partial B(a), \end{aligned}$$

where $B(a) = \{\, (x, y, z) \mid x^2 + y^2 + z^2 \leqslant a^2 \,\}$. The appropriate coordinate system consists of spherical coordinates r, θ, and ϕ, and here $x = r\sin\theta\cos\phi$, $y = r\sin\theta\sin\phi$, and $z = r\cos\theta$. The sphere above can then be described as

$$B(a) = \{\, (r, \theta, \phi) \mid 0 \leqslant r \leqslant a,\ 0 \leqslant \theta \leqslant \pi,\ -\pi \leqslant \phi \leqslant \pi \,\}.$$

Analogous to Example 6.1, the boundaries of the rectangular parallelepiped in $\mathbb{R}^3$ with the spherical coordinate system are not real boundaries or physical boundaries. Nevertheless, as before, we treat the region as a rectangular parallelepiped and apply our rule of thumb for side conditions. Hence, we need to impose boundary conditions on these "nonphysical boundaries."

A formal statement of the problem is given below,

(i) The function $u(r, \theta, \phi)$ must satisfy Laplace's equation in spherical coordinates r, θ, ϕ; that is,

$$\nabla^2 u = \frac{1}{r^2}\frac{\partial}{\partial r}\left(r^2 \frac{\partial u}{\partial r}\right) + \frac{1}{r^2 \sin\theta}\frac{\partial}{\partial \theta}\left(\sin\theta \frac{\partial u}{\partial \theta}\right) + \frac{1}{r^2 \sin^2\theta}\frac{\partial^2 u}{\partial \phi^2} = 0 \tag{7.32}$$

for $(r, \theta, \phi) \in B(a)$.

(ii) To ensure that the solution is single-valued, $u(r, \theta, \phi)$ must satisfy the periodicity conditions at $\phi = \pm\pi$; that is,

$$u(r, \theta, \pi) = u(r, \theta, -\pi),$$

$$\frac{\partial u}{\partial \phi}(r, \theta, \pi) = \frac{\partial u}{\partial \phi}(r, \theta, -\pi)$$

for $0 \leqslant r \leqslant a$ and $0 \leqslant \theta \leqslant \pi$.

(iii) To ensure that the solution is continuous, $u(r, \theta, \phi)$ must satisfy boundedness conditions; that is,

$$|u(r, \theta, \phi)| \quad \text{bounded everywhere.}$$

(iv) Finally, the solution must satisfy the boundary condition at $r = a$; that is,

$$u(a, \theta, \phi) = f(\theta, \phi)$$

for $0 \leqslant \theta \leqslant \pi$ and $-\pi \leqslant \phi \leqslant \pi$.

Assumption (iii) is made to preclude the occurrence of singular solutions that may arise due to the singular nature of the transformation to spherical coordinates. Separating variables, we look for a solution of the form

$$u(r, \theta, \phi) = R(r) \cdot S(\theta) \cdot T(\phi),$$

and substituting this into Laplace's equation (7.32), we obtain

$$(r^2 R')' \cdot S \cdot T + \frac{R \cdot T}{\sin\theta}(\sin\theta\, S')' + \frac{R \cdot S}{\sin^2\theta} T'' = 0;$$

that is,

$$\frac{(r^2 R')'}{R} + \frac{(\sin\theta\, S')'}{S \sin\theta} + \frac{T''}{T \sin^2\theta} = 0,$$

so that

$$\frac{(r^2 R')'}{R} = -\frac{(\sin\theta\, S')'}{S \sin\theta} - \frac{T''}{T \sin^2\theta} = \lambda,$$

where λ is a separation constant. Thus, we have the two equations

$$(r^2 R')' - \lambda R = 0 \tag{7.33}$$

and

$$\frac{\sin\theta(\sin\theta\, S')'}{S} + \lambda \sin^2\theta = -\frac{T''}{T} = \nu,$$

where ν is another separation constant.
The second of these equations leads to

$$\sin\theta(\sin\theta\, S')' + (\lambda \sin^2\theta - \nu)S = 0 \tag{7.34}$$

and

$$T'' + \nu T = 0. \tag{7.35}$$

Applying the periodicity conditions, the boundary value problem for T is

$$\begin{aligned} &T'' + \nu T = 0, \quad -\pi < \phi < \pi, \\ &T(-\pi) = T(\pi), \\ &T'(-\pi) = T'(\pi), \end{aligned}$$

with eigenvalues and corresponding eigenfunctions

$$\nu_m = m^2 \qquad \text{and} \qquad T_m(\phi) = \alpha_m \cos m\phi + \beta_m \sin m\phi$$

for $m = 0, 1, 2, \ldots$.

If we let $x = \cos\theta$ and $v(x) = S(\theta)$, then equation (7.34) for S becomes

$$(1 - x^2)v'' - 2xv' + \left(\lambda - \frac{m^2}{1 - x^2}\right)v = 0,$$

which is the associated Legendre equation. Since we want bounded solutions, from Theorem 7.13, we must have $\lambda = n(n+1)$ for $n \geqslant 0$. Therefore, v satisfies

$$(1 - x^2)v'' - 2xv' + \left(n(n+1) - \frac{m^2}{1 - x^2}\right)v = 0,$$

with general solution

$$v(x) = A_n^m P_n^m(x) + B_n^m Q_n^m(x).$$

Again, since $|v(x)|$ must be bounded, $B_n^m = 0$ and

$$v(x) = A_n^m P_n^m(x),$$

and hence

$$S_n^m(\theta) = A_n^m P_n^m(\cos\theta)$$

for $n, m \geqslant 0$.

Finally, the equation for R is now

$$r^2 R'' + 2rR' - n(n+1)R = 0,$$

which is a Cauchy-Euler equation. Assuming a solution of the form $R(r) = r^\alpha$, we have the characteristic equation

$$\alpha^2 + \alpha - n(n+1) = 0$$

with roots $\alpha_1 = n$ and $\alpha_2 = -1 - n$. The solution is

$$R(r) = Ar^n + Br^{-1-n},$$

and since $|R(r)|$ must be bounded, we need $B = 0$, so that $R(r) = r^n$ for $n \geqslant 0$. Thus, for each m, $n \geqslant 0$, the function

$$u_n^m(r, \theta, \phi) = r^n P_n^m(\cos\theta)\big(\alpha_n^m \cos m\phi + \beta_n^m \sin m\phi\big)$$

satisfies Laplace's equation, the periodicity conditions, and the boundedness conditions.

Applying the superposition principle, we have

$$u(r, \theta, \phi) = \sum_{n=0}^{\infty} \sum_{m=0}^{n} r^n P_n^m(\cos\theta)\big(\alpha_n^m \cos m\phi + \beta_n^m \sin m\phi\big), \tag{7.36}$$

and all that remains is to apply the boundary condition

$$u(a, \theta, \phi) = f(\theta, \phi)$$

to get the solution to Dirichlet's problem in the sphere. Setting $r = a$ in equation (7.36), we have

$$f(\theta, \phi) = \sum_{n=0}^{\infty} \sum_{m=0}^{n} a^n P_n^m(\cos\theta)\big(\alpha_n^m \cos m\phi + \beta_n^m \sin m\phi\big).$$

The final solution is

$$u(r, \theta, \phi) = \sum_{n=0}^{\infty} \sum_{m=0}^{n} r^n P_n^m(\cos\theta)(\alpha_n^m \cos(m\phi) + \beta_n^m \sin(m\phi)),$$

where the coefficients are found using the orthogonality relations in Lemma 7.14, and

$$\alpha_n^m = (2 - \delta_{m0})\frac{2n+1}{4\pi a^n} \cdot \frac{(n-m)!}{(n+m)!} \int_{-\pi}^{\pi} \int_0^{\pi} f(\theta, \phi) P_n^m(\cos\theta) \cos(m\phi) \sin\theta \, d\theta \, d\phi,$$

$$\beta_n^m = (2 - \delta_{m0})\frac{2n+1}{4\pi a^n} \cdot \frac{(n-m)!}{(n+m)!} \int_{-\pi}^{\pi} \int_0^{\pi} f(\theta, \phi) P_n^m(\cos\theta) \sin(m\phi) \sin\theta \, d\theta \, d\phi,$$

where $0 \leqslant m \leqslant n$, and $n \geqslant 0$.

As a special case, when f is independent of ϕ, that is,

$$f(\theta, \phi) = f(\theta),$$

we get

$$u(r, \theta) = \sum_{n=0}^{\infty} A_n \left(\frac{r}{a}\right)^n P_n(\cos\theta),$$

where

$$A_n = \frac{2n+1}{2} \int_0^{\pi} f(\theta) P_n(\cos\theta) \sin\theta \, d\theta.$$

■

7.6 SUMMARY

It is safe to say that this is the most demanding chapter in the book. Spherical coordinates are typically used in dealing with problems that are rotationally symmetric in three dimensions. The most prominent example is the description of the orbits of the electron in the hydrogen atom; the solution of the Schrödinger equation for the energy levels of the electron, that is, the quantum numbers, follow the computations made in this chapter.

The Cartesian coordinates (x, y, z) of a point in three-dimensional space are transformed to radius $r > 0$ and two angular coordinates $\phi \in (-\pi, \pi)$ and $\theta \in (0, \pi)$ according to

$$x = r\cos\phi\sin\theta, \quad y = r\sin\phi\sin\theta, \quad z = r\cos\theta.$$

The Laplacian transforms into

$$\Delta u = \frac{1}{r^2}\frac{\partial}{\partial r}\left(r^2 \frac{\partial u}{\partial r}\right) + \frac{1}{r^2 \sin\theta}\frac{\partial}{\partial\theta}\left(\sin\theta\frac{\partial u}{\partial\theta}\right) + \frac{1}{r^2\sin^2\theta}\frac{\partial^2 u}{\partial\phi^2},$$

and we note that this operation is singular at $r = 0$, and also along the entire z-axis; that is, $\theta = 0$ and $\theta = \pi$.

The eigenvalue problem for the spherical Laplacian,

$$\Delta u = -\lambda u, \quad \text{on } \Omega$$

$$u = u_0, \quad \text{on } \partial\Omega$$

gives rise to the **Helmholtz equation** $\Delta u = -\lambda u$.

In the method of separation of variables for a time-dependent problem, we assume a solution of the form

$$u(r, \theta, \phi, t) = R(r) \cdot S(\theta) \cdot Q(\phi) \cdot T(t),$$

and we often see $S(\theta) \cdot Q(\phi)$ written as $Y(\theta, \phi)$. Separating variables, results in a radial problem and angular problems.

The radial equation is known as the **spherical Bessel equation**

$$\frac{d}{dr}\left(r^2 \frac{dR}{dr}\right) + (\mu^2 r^2 - \lambda)R = 0,$$

and the solutions are known as **spherical Bessel functions**. The spherical Bessel functions of the **first kind** are denoted by $j_n(r)$, while the spherical Bessel functions of the **second kind** are denoted by $y_n(r)$ and are given by

$$j_n(r) = \sqrt{\frac{\pi}{2r}}\, J_{n+\frac{1}{2}}(r), \qquad y_n(r) = \sqrt{\frac{\pi}{2r}}\, Y_{n+\frac{1}{2}}(r)$$

for $n \in \mathbb{Z}$. Here $J_{n+\frac{1}{2}}(r)$ and $Y_{n+\frac{1}{2}}(r)$ are Bessel functions of the first and second kind, respectively, of order $n + \frac{1}{2}$.

The angular equation is known as the **surface Helmholtz equation**

$$\frac{1}{\sin\theta}\frac{\partial}{\partial\theta}\left(\sin\theta \frac{\partial Y}{\partial\theta}\right) + \frac{1}{\sin^2\theta}\frac{\partial^2 Y}{\partial\phi^2} + \lambda Y = 0,$$

and the solutions are known as **spherical harmonics** or **surface harmonics**.

The equation for spherical harmonics can be separated further into equations for $S(\theta)$ and $Q(\phi)$. The Q equation becomes one of our standard examples,

$$Q'' + \nu Q = 0,$$

and is easily solved. The S equation can be transformed into **Legendre's equation** on $[-1, 1]$ by means of a change of independent variable $x = \cos\theta$ (note that this is not the x-coordinate from Cartesian coordinates):

$$(1 - x^2)v'' - 2xv' + \lambda v = 0.$$

The only bounded solutions are polynomials $P_n(x)$, called **Legendre polynomials** or **zonal harmonics of order** n, and their properties are summarized in this chapter. Notice that this is a three-dimensional problem; hence, eventually we end up working with three separation constants (λ, μ, ν).

At the end of the chapter, we obtain a generalized Fourier series as a series in spherical harmonics and spherical Bessel functions.

Following is a short table of the various types of Sturm-Liouville problems that we have encountered in the preceding two chapters.

Name	Equation	Solutions
Cauchy Euler eqn.	$x^2u'' + xu' - m^2u = 0$	x^m, x^{-m}
Cauchy Euler eqn.	$x^2u'' + 2xu' - m(m+1)u = 0$	x^m, $x^{-(m+1)}$
Bessel's eqn. of order m	$x^2u'' + xu' + (x^2 - m^2)u = 0$	$J_m(x), Y_m(x)$
Spherical Bessel's eqn. of order m	$x^2u'' + 2xu' + (x^2 - m(m+1))u = 0$	$j_m(x), y_m(x)$
Legendre's eqn. of degree n	$(1-x^2)v'' - 2xv' + n(n+1)v = 0$	$P_n(x), Q_n(x)$
Assoc. Legendre eqn. of degree n**, order** m	$(1-x^2)v'' - 2xv' + \left(n(n+1) - \frac{m^2}{1-x^2}\right)v = 0$	$P_n^m(x), Q_n^m(x)$

7.6.1 Problems and Notes

Problems from Part II:

Exercise	13.19	13.20	14.20	14.21	15.13	15.14
Notes						

CHAPTER 8

FOURIER TRANSFORMS

8.1 FOURIER INTEGRALS

In previous chapters, all of the partial differential equations we studied were defined on a finite domain $[a, b]$, and the differential operators we obtained by separating variables had a discrete spectrum. In this case, the solution could be represented as an infinite series of orthogonal eigenfunctions.

In this chapter we study partial differential equations defined on an infinite domain, either $(0, \infty)$, or $(-\infty, \infty)$, and usually the differential operators do not have a discrete spectrum. In this case, instead of getting a representation for solutions on a finite domain $[a, b]$ in terms of generalized Fourier series, we get a representation on $(0, \infty)$ or $(-\infty, \infty)$ in terms of improper integrals, called **Fourier integrals**.

8.1.1 Fourier Integral Representation

In this section we give a heuristic argument for the Fourier integral formula: Assuming that f is defined for all real numbers x, that both f and f' are piecewise continuous on every finite interval $[-a, a]$, and that f is absolutely integrable on $\mathbb{R}$, that is,

$$\int_{-\infty}^{\infty} |f(x)|\, dx = M < \infty,$$

f can be represented by its Fourier integral

$$\int_0^{\infty} [A(\omega)\cos\omega x + B(\omega)\sin\omega x]\, d\omega,$$

where

$$A(\omega) = \frac{1}{\pi}\int_{-\infty}^{\infty} f(x)\cos\omega x\, dx \qquad \text{and} \qquad B(\omega) = \frac{1}{\pi}\int_{-\infty}^{\infty} f(x)\sin\omega x\, dx$$

for $\omega > 0$. Note that since f is absolutely integrable, both of these integrals converge.

As with Fourier series, before we can determine the convergence of the integrals, we write

$$f(x) \sim \int_0^{\infty} [A(\omega)\cos\omega x + B(\omega)\sin\omega x]\, d\omega,$$

where $A(\omega)$ and $B(\omega)$ are as given above. This is called the **Fourier integral formula** for $f(x)$ on the interval $-\infty < x < \infty$. To show why this might be true, we assume that $f(x)$ is actually continuous on every finite interval and that $f'(x)$ is piecewise continuous. Then $f(x)$ has a Fourier series expansion on $[-a, a]$, and by Dirichlet's theorem we can write

$$f(x) = a_0 + \sum_{n=1}^{\infty}\left(a_n\cos\frac{n\pi x}{a} + b_n\sin\frac{n\pi x}{a}\right)$$

for $-a < x < a$, where

$$a_0 = \frac{1}{2a}\int_{-a}^{a} f(x)\, dx, \quad a_n = \frac{1}{a}\int_{-a}^{a} f(x)\cos\frac{n\pi x}{a}\, dx,$$

$$b_n = \frac{1}{a}\int_{-a}^{a} f(x)\sin\frac{n\pi x}{a}\, dx$$

for $n \geqslant 1$. Now we define $\omega_n = n\pi/a$, and rewrite the coefficients as

$$a_n = \frac{\pi}{a}A(\omega_n) \quad \text{and} \quad b_n = \frac{\pi}{a}B(\omega_n),$$

where

$$A(\omega_n) = \frac{1}{\pi}\int_{-a}^{a} f(x)\cos\omega_n x\, dx \quad \text{and} \quad B(\omega_n) = \frac{1}{\pi}\int_{-a}^{a} f(x)\sin\omega_n x\, dx$$

for $n \geqslant 1$.

The formula for the Fourier series becomes

$$f(x) = a_0 + \sum_{n=1}^{\infty} [A(\omega_n)\cos\omega_n x + B(\omega_n)\sin\omega_n x]\ \Delta\omega_n,$$

that is,

$$f(x) = \frac{1}{2a}\int_{-a}^{a} f(\xi)\,d\xi + \frac{1}{\pi}\sum_{n=1}^{\infty}\left[\left[\int_{-a}^{a} f(\xi)\cos\omega_n\xi\,d\xi\ \cos\omega_n x\right.\right.$$
$$\left. + \int_{-a}^{a} f(\xi)\sin\omega_n\xi\,d\xi\ \sin\omega_n x\right]\Delta\omega_n, \qquad -a < x < a \tag{8.1}$$

where $\Delta\omega_n = \omega_{n+1} - \omega_n = \pi/a$ for $n \geqslant 1$. Now, as $a \to \infty$ we have $\Delta\omega_n \to 0$ and $\omega_1 \to 0$, and since

$$|a_0| \leqslant \frac{1}{2a}\int_{-a}^{a} |f(x)|\,dx \leqslant \frac{1}{2a}\int_{-\infty}^{\infty} |f(x)|\,dx \leqslant \frac{M}{2a},$$

where M is a constant, $a_0 \to 0$ as $a \to \infty$. The series resembles a limit of Riemann sums:

$$f(x) = \int_0^{\infty} [A(\omega)\cos\omega x + B(\omega)\sin\omega x]\ d\omega, \tag{8.2}$$

where

$$A(\omega) = \frac{1}{\pi}\int_{-\infty}^{\infty} f(x)\cos\omega x\,dx; \quad \text{and} \quad B(\omega) = \frac{1}{\pi}\int_{-\infty}^{\infty} f(x)\sin\omega x\,dx$$

that is,

$$f(x) = \int_0^{\infty}\left[\cos\omega x\,\frac{1}{\pi}\int_{-\infty}^{\infty} f(\xi)\cos\omega\xi\,d\xi\right.$$
$$\left. + \sin\omega x\,\frac{1}{\pi}\int_{-\infty}^{\infty} f(\xi)\sin\omega\xi\,d\xi\right] d\omega \tag{8.3}$$

for $0 < \omega < \infty$. A rigorous proof of the Fourier integral formulas (8.2) and (8.3) requires a careful study of the limits involved, and will be given in Section 8.1.4.

In the case where f is not continuous everywhere on $\mathbb{R}$ but is piecewise continuous on every finite interval, then according to Dirichlet's theorem, we would expect that

$$\frac{f(x^+) + f(x^-)}{2} = \int_0^{\infty} [A(\omega)\cos\omega x + B(\omega)\sin\omega x]\ d\omega \tag{8.4}$$

for $-\infty < x < \infty$.

The formula (8.4) is called the **Fourier integral representation** of the function $f(x)$, while $A(\omega)$ and $B(\omega)$ are called the **Fourier integral coefficients** of f. Note that if f is continuous at x, (8.4) becomes (8.2). We have the following convergence theorem, similar to Dirichlet's theorem for Fourier series.

Theorem 8.1. (*Fourier Integral Theorem*)
If f and f' are piecewise continuous on every finite interval $[a, b] \subset \mathbb{R}$ and f is absolutely integrable on $\mathbb{R}$, that is,

$$\int_{-\infty}^{\infty} |f(x)|\, dx < \infty,$$

the Fourier integral

$$\int_0^{\infty} [A(\omega)\cos\omega x + B(\omega)\sin\omega x]\, d\omega,$$

where

$$A(\omega) = \frac{1}{\pi}\int_{-\infty}^{\infty} f(x)\cos\omega x\, dx \quad \text{and} \quad B(\omega) = \frac{1}{\pi}\int_{-\infty}^{\infty} f(x)\sin\omega x\, dx$$

for $0 < \omega < \infty$, converges to

$$\frac{f(x^+) + f(x^-)}{2}$$

at each point $x \in \mathbb{R}$.

8.1.2 Examples

Example 8.1. Given the rectangular pulse

$$f(x) = \begin{cases} 1, & \text{for } |x| < 1, \\ 0, & \text{for } |x| > 1, \end{cases}$$

and $f(1) = f(-1) = \frac{1}{2}$. Show that the Fourier integral formula for f is

$$f(x) = \frac{2}{\pi}\int_0^{\infty} \frac{\sin\omega\cos\omega x}{\omega}\, d\omega$$

for all $x \in \mathbb{R}$.

Solution. The graph of the rectangular pulse is shown in Figure 8.1. Clearly, the function f is piecewise smooth on $\mathbb{R}$, and

$$\int_{-\infty}^{\infty} |f(x)|\, dx = \int_{-1}^{1} 1\, dx = 2 < \infty,$$

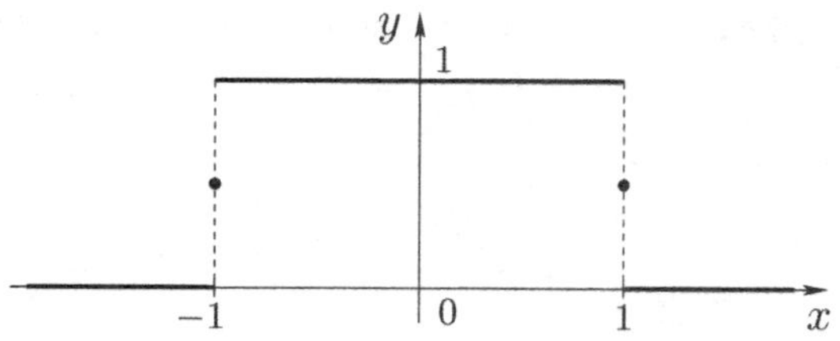

Figure 8.1 Rectangular pulse.

so the conditions for Dirichlet's theorem hold. Now,

$$A(\omega) = \frac{1}{\pi}\int_{-\infty}^{\infty} f(x)\cos\omega x\,dx = \frac{1}{\pi}\int_{-1}^{1}\cos\omega x\,dx = \frac{1}{\pi}\frac{\sin\omega x}{\omega}\bigg|_{-1}^{1} = \frac{2}{\pi}\frac{\sin\omega}{\omega}$$

for $0 < \omega < \infty$, and

$$B(\omega) = \frac{1}{\pi}\int_{-\infty}^{\infty} f(x)\sin\omega x\,dx = \frac{1}{\pi}\int_{-1}^{1}\sin\omega x\,dx = -\frac{1}{\pi\omega}\cos\omega x\bigg|_{-1}^{1} = 0$$

for $0 < \omega < \infty$. From Dirichlet's theorem we have

$$f(x) = \frac{2}{\pi}\int_0^{\infty}\frac{\sin\omega\cos\omega x}{\omega}\,d\omega$$

for all $x \in \mathbb{R}$, and hence

$$\int_0^{\infty}\frac{\sin\omega\cos\omega x}{\omega}\,d\omega = \begin{cases} \dfrac{\pi}{2}, & \text{for } 0 \leqslant x < 1, \\ \dfrac{\pi}{4}, & \text{for } x = 1, \\ 0, & \text{for } x > 1. \end{cases}$$

This integral is called **Dirichlet's discontinuous factor**.

In particular, for $x = 0$, since $f(0) = 1$, we have

$$\int_0^{\infty}\frac{\sin\omega}{\omega}\,d\omega = \frac{\pi}{2};$$

thus, the integral converges, even though, as we show later in Lemma 8.4,

$$\int_0^{\infty}\left|\frac{\sin\omega}{\omega}\right|\,d\omega = +\infty.$$

∎

Example 8.2. Find the Fourier integral representation of the function

$$f(x) = \begin{cases} e^{-x}, & \text{for } x > 0, \\ 0, & \text{for } x < 0, \end{cases}$$

and $f(0) = \frac{1}{2}$.

Solution. The graph of the function $f(x)$ is shown in Figure 8.2. Again, it is clear that the function f is piecewise smooth on $\mathbb{R}$, and

$$\int_{-\infty}^{\infty} |f(x)|\,dx = \int_{0}^{\infty} e^{-x}\,dx = -e^{-x}\Big|_{0}^{\infty} = 1 < \infty,$$

so the conditions for Dirichlet's theorem hold. Now

$$\begin{aligned} A(\omega) &= \frac{1}{\pi}\int_{-\infty}^{\infty} f(x)\cos\omega x\,dx \\ &= \frac{1}{\pi}\int_{0}^{\infty} e^{-x}\cos\omega x\,dx \\ &= \frac{1}{\pi}\left[\frac{e^{-x}\sin\omega x}{\omega}\Big|_{0}^{\infty} + \frac{1}{\omega}\int_{0}^{\infty} e^{-x}\sin\omega x\,dx\right] \\ &= \frac{1}{\pi\omega}\int_{0}^{\infty} e^{-x}\sin\omega x\,dx. \end{aligned}$$

Therefore,

$$A(\omega) = \frac{1}{\pi\omega}\left[-\frac{e^{-x}\cos\omega x}{\omega}\Big|_{0}^{\infty} - \frac{1}{\omega}\int_{0}^{\infty} e^{-x}\cos\omega x\,dx\right],$$

so that

$$A(\omega) = \frac{1}{\pi\omega^2} - \frac{1}{\omega^2}A(\omega),$$

that is,

$$\left(1 + \frac{1}{\omega^2}\right)A(\omega) = \frac{1}{\pi\omega^2},$$

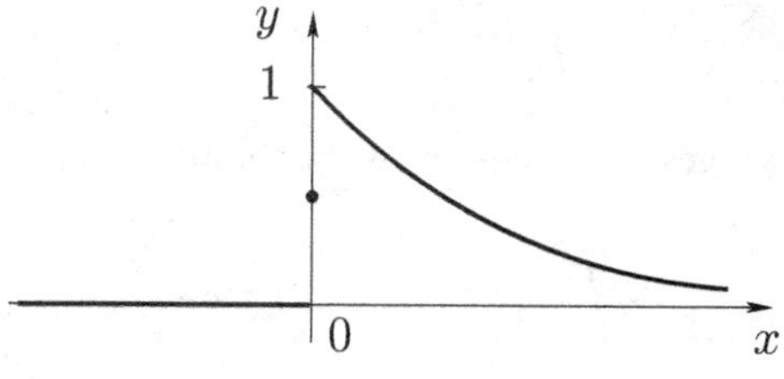

Figure 8.2 Exponential pulse.

and

$$A(\omega) = \frac{1}{\pi(1+\omega^2)}$$

for $0 < \omega < \infty$. Also,

$$\begin{aligned}
B(\omega) &= \frac{1}{\pi}\int_{-\infty}^{\infty} f(x)\sin\omega x\,dx = \frac{1}{\pi}\int_{0}^{\infty} e^{-x}\sin\omega x\,dx \\
&= \frac{1}{\pi}\left[-\frac{e^{-x}\cos\omega x}{\omega}\bigg|_0^{\infty} - \frac{1}{\omega}\int_0^{\infty} e^{-x}\cos\omega x\,dx\right] = \frac{1}{\pi\omega} - \frac{1}{\omega}A(\omega) \\
&= \frac{1}{\pi\omega} - \frac{1}{\pi\omega(1+\omega^2)} = \frac{1}{\pi\omega}\left(1 - \frac{1}{1+\omega^2}\right)
\end{aligned}$$

and

$$B(\omega) = \frac{\omega}{\pi(1+\omega^2)}$$

for $0 < \omega < \infty$. Thus,

$$f(x) = \int_0^{\infty}\left[A(\omega)\cos\omega x + B(\omega)\sin\omega x\right]d\omega = \frac{1}{\pi}\int_0^{\infty}\frac{\cos\omega x + \omega\sin\omega x}{1+\omega^2}\,d\omega$$

for all $x \in \mathbb{R}$. ∎

Example 8.3. Calculate the Fourier integral of the function

$$f(x) = e^{-|x|}, \quad -\infty < x < \infty.$$

Solution. Clearly, f is piecewise continuous on $(-\infty, \infty)$; is absolutely integrable, since

$$\int_{-\infty}^{\infty} e^{-|x|}\,dx = 2\int_0^{\infty} e^{-x}\,dx = 2 < \infty;$$

and hence f satisfies the hypotheses of Dirichlet's theorem. Computing the Fourier integral coefficients, we have

$$\begin{aligned}
A(\omega) &= \frac{1}{\pi}\int_{-\infty}^{\infty} e^{-|x|}\cos\omega x\,dx \\
&= \frac{2}{\pi}\int_0^{\infty} e^{-x}\cos\omega x\,dx \\
&= \frac{2}{\pi}\frac{e^{-x}}{1+\omega^2}(\omega\sin\omega x - \cos\omega x)\bigg|_0^{\infty} \\
&= \frac{2}{\pi(1+\omega^2)},
\end{aligned}$$

and since $f(x)$ is even, then

$$B(\omega) = \frac{1}{\pi}\int_{-\infty}^{\infty} e^{-|x|}\sin\omega x\,dx = 0.$$

Therefore,

$$e^{-|x|} = \frac{2}{\pi}\int_{0}^{\infty} \frac{\cos\omega x}{1+\omega^2}\,d\omega$$

for all $x \in (-\infty, \infty)$. ∎

8.1.3 Fourier Sine and Cosine Integral Representations

From Examples 8.1 and 8.3, the Fourier integral of an even function contained only cosine terms, just as with Fourier series; similarly, the Fourier integral of an odd function will contain only sine terms. If the function f is defined on the interval $(-\infty, \infty)$, the Fourier integral coefficients $A(\omega)$ and $B(\omega)$ are uniquely determined, and the Fourier integral representation for f on $\mathbb{R}$ may contain both sine and cosine terms. However, on infinite domains, the PDEs we solve using separation of variables typically have either no boundary conditions or boundary conditions specified on the subinterval $(0, \infty)$ and usually lead to integrals containing only sine terms or to integrals containing only cosine terms. If the function f is defined only on the interval $(0, \infty)$, we may extend it to the interval $(-\infty, \infty)$ either as an odd function f_{odd} or as an even function f_{even}. From the previous remarks, the Fourier integral for f_{odd} on the interval $(-\infty, \infty)$ contains only sine terms, while the Fourier integral for f_{even} on the interval $(-\infty, \infty)$ contains only cosine terms. Both f_{odd} and f_{even} agree with f on the interval $(0, \infty)$; hence, f has two different Fourier integral representations on the interval $(0, \infty)$.

Definition 8.2. Suppose that f is defined on $(0, \infty)$ and is absolutely integrable there, that is,

$$\int_{0}^{\infty} |f(x)|\,dx < \infty;$$

then:

(i) the **Fourier cosine integral** for f on $(0, \infty)$ is

$$f(x) \sim \int_{0}^{\infty} A(\omega)\cos\omega x\,d\omega,$$

where

$$A(\omega) = \frac{1}{\pi}\int_{-\infty}^{\infty} f_{\text{even}}(x)\cos\omega x\,dx = \frac{2}{\pi}\int_{0}^{\infty} f(x)\cos\omega x\,dx$$

for $0 < \omega < \infty$, and

(ii) the **Fourier sine integral** for f on $(0, \infty)$ is

$$f(x) \sim \int_0^\infty B(\omega) \sin \omega x \, d\omega,$$

where

$$B(\omega) = \frac{1}{\pi} \int_{-\infty}^{\infty} f_{\text{odd}}(x) \sin \omega x \, dx = \frac{2}{\pi} \int_0^\infty f(x) \sin \omega x \, dx$$

for $0 < \omega < \infty$.

We have a result similar to Dirichlet's convergence theorem for the Fourier sine and cosine integrals, and it follows immediately from the Fourier integral theorem.

Theorem 8.3. If f is defined on the interval $(0, \infty)$, f and f' are piecewise continuous on every finite subinterval of $(0, \infty)$, and f is absolutely integrable on $(0, \infty)$, then the following are true:

(i) For the Fourier cosine integral,

$$\int_0^\infty A(\omega) \cos \omega x_0 \, d\omega = \begin{cases} \dfrac{f(x_0^+) + f(x_0^-)}{2}, & \text{for } 0 < x_0 < \infty, \\ f(0^+), & \text{for } x_0 = 0, \end{cases}$$

where

$$A(\omega) = \frac{2}{\pi} \int_0^\infty f(x) \cos \omega x \, dx$$

for $0 < \omega < \infty$.

(ii) For the Fourier sine integral,

$$\int_0^\infty B(\omega) \sin \omega x_0 \, d\omega = \begin{cases} \dfrac{f(x_0^+) + f(x_0^-)}{2}, & \text{for } 0 < x_0 < \infty, \\ 0, & \text{for } x_0 = 0, \end{cases}$$

where

$$B(\omega) = \frac{2}{\pi} \int_0^\infty f(x) \sin \omega x \, dx$$

for $0 < \omega < \infty$.

Example 8.4. Find the Fourier cosine and sine integral formulas for

$$f(x) = \begin{cases} 1, & \text{for } 0 < x < 1, \\ 0, & \text{for } x > 1, \end{cases}$$

and $f(0) = 1, \; f(1) = \frac{1}{2}$.

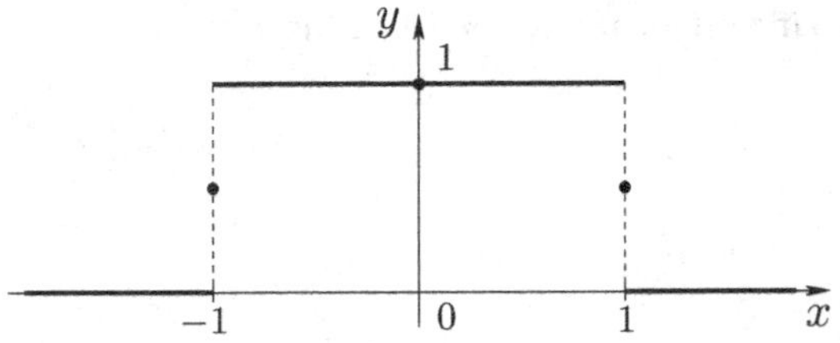

Figure 8.3 Even extension.

Solution.

- For the Fourier cosine integral, the graph of the even extension (Fig. 8.3) is the same as the graph in Example 8.1. Here,

$$A(\omega) = \frac{2}{\pi}\int_0^\infty f(x)\cos\omega x\,dx = \frac{2}{\pi}\int_0^1 \cos\omega x\,dx = \frac{2}{\pi}\frac{\sin\omega}{\omega};$$

thus, the Fourier cosine integral of f is

$$\int_0^\infty A(\omega)\cos\omega x\,d\omega = \frac{2}{\pi}\int_0^\infty \frac{\sin\omega}{\omega}\cos\omega x\,d\omega = \begin{cases} 1, & \text{for } 0 \leqslant x < 1, \\ \dfrac{1}{2}, & \text{for } x = 1, \\ 0, & \text{for } x > 1. \end{cases}$$

- For the Fourier sine integral, the graph of the odd extension of f is shown in Figure 8.4. Here,

$$B(\omega) = \frac{2}{\pi}\int_0^\infty f(x)\sin\omega x\,dx = \frac{2}{\pi}\int_0^1 \sin\omega x\,dx = \frac{2}{\pi\omega}(1-\cos\omega);$$

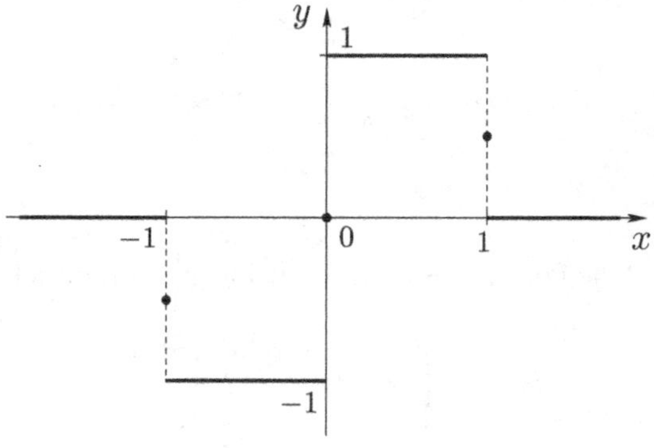

Figure 8.4 Odd extension.

thus, the Fourier sine integral of f is

$$\int_0^\infty B(\omega)\sin\omega x\,d\omega = \frac{2}{\pi}\int_0^\infty \frac{1-\cos\omega}{\omega}\sin\omega x\,d\omega;$$

that is,

$$\int_0^\infty B(\omega)\sin\omega x\,d\omega = \begin{cases} 1, & \text{for } 0 < x < 1, \\ \dfrac{1}{2}, & \text{for } x = 1, \\ 0, & \text{for } x = 0 \text{ and for } x > 1. \end{cases}$$

∎

8.1.4 Proof of Fourier's Theorem

We begin by giving an elementary proof of the following lemma.

Lemma 8.4. (*Dirichlet's Integral*)

(a) $\displaystyle\int_0^\infty \frac{\sin x}{x}\,dx = \frac{\pi}{2}.$

(b) $\displaystyle\int_0^\infty \left|\frac{\sin x}{x}\right| dx = +\infty.$

Proof. The following proof is outlined on page 397, Miscellaneous Exercise 39, in G. H. Hardy's *A Course of Pure Mathematics* [26].

(a) For each $n \geqslant 1$, define

$$u_n = \int_0^{\pi/2} \sin 2nx \cot x\,dx \qquad \text{and} \qquad v_n = \int_0^{\pi/2} \frac{\sin 2nx}{x}\,dx,$$

and consider the difference $u_{n+1} - u_n$,

$$\begin{aligned}
u_{n+1} - u_n &= \int_0^{\pi/2} [\sin 2(n+1)x - \sin 2nx] \cot x \, dx \\
&= 2\int_0^{\pi/2} \sin x \cos(2n+1)x \cot x \, dx \\
&= 2\int_0^{\pi/2} \cos x \cos(2n+1)x \, dx \\
&= \int_0^{\pi/2} [\cos(2n+2)x + \cos 2nx] \, dx \\
&= \frac{1}{2(n+1)} \sin 2(n+1)x \Big|_0^{\pi/2} + \frac{1}{2n} \sin 2nx \Big|_0^{\pi/2} \\
&= 0;
\end{aligned}$$

therefore, $u_{n+1} - u_n = 0$ for all $n \geqslant 1$. Thus, u_n is constant for $n \geqslant 1$, and

$$\begin{aligned}
u_1 &= \int_0^{\pi/2} \sin 2x \cot x \, dx = 2\int_0^{\pi/2} \cos^2 x \, dx \\
&= \int_0^{\pi/2} (1 + \cos 2x) \, dx = \frac{\pi}{2} + \frac{\sin 2x}{2} \Big|_0^{\pi/2} \\
&= \frac{\pi}{2},
\end{aligned}$$

and therefore $u_n = u_1 = \pi/2$ for all $n \geqslant 1$.

Making the substitution $t = 2nx$ in the integral $\int_0^{\pi/2} (\sin 2nx/x) \, dx$, then

$$v_n = \int_0^{\pi/2} \frac{\sin 2nx}{x} \, dx = \int_0^{n\pi} \frac{\sin t}{t} \, dt, \tag{8.5}$$

and integrating by parts, we have

$$\begin{aligned}
u_n - v_n &= \int_0^{\pi/2} \left(\cot x - \frac{1}{x} \right) \sin 2nx \, dx \\
&= -\frac{\cos 2nx}{2n} \left(\cot x - \frac{1}{x} \right) \Big|_0^{\pi/2} \\
&\qquad - \frac{1}{2n} \int_0^{\pi/2} \left(\csc^2 x - \frac{1}{x^2} \right) \cos 2nx \, dx,
\end{aligned}$$

and therefore

$$u_n - v_n = \frac{(-1)^n}{n\pi} + \lim_{p\to 0^+} \frac{\cos 2np}{2n}\left(\cot p - \frac{1}{p}\right) - \frac{1}{2n}\int_0^{\pi/2}\left(\csc^2 x - \frac{1}{x^2}\right)\cos 2nx\,dx. \quad (8.6)$$

Using L'Hospital's rule, we have

$$\lim_{p\to 0^+}\left(\cot p - \frac{1}{p}\right) = 0, \quad \text{and} \quad \lim_{x\to 0^+}\left(\csc^2 x - \frac{1}{x^2}\right) = \frac{1}{3}.$$

Therefore, we may redefine the integrand on the right-hand side of (8.6) to be continuous on the interval $[0, \pi/2]$, and hence bounded there. Thus, there exists a constant $M > 0$ such that

$$\left|\int_0^{\pi/2}\left(\csc^2 x - \frac{1}{x^2}\right)\cos 2nx\,dx\right| \leqslant \int_0^{\pi/2}\left|\left(\csc^2 x - \frac{1}{x^2}\right)\cos 2nx\right|dx \leqslant \frac{\pi M}{2}.$$

Letting $n \to \infty$ in (8.5) and (8.6), we obtain

$$\lim_{n\to\infty} v_n = \lim_{n\to\infty}(v_n - u_n + u_n) = \lim_{n\to\infty}(v_n - u_n) + \lim_{n\to\infty} u_n,$$

so that

$$\int_0^\infty \frac{\sin t}{t}\,dt = \frac{\pi}{2}.$$

(b) For each positive integer n, we have

$$\begin{aligned}\int_0^{(n+1)\pi}\left|\frac{\sin x}{x}\right|dx &= \sum_{k=0}^{n}\int_{k\pi}^{(k+1)\pi}\left|\frac{\sin x}{x}\right|dx \\ &= \sum_{k=0}^{n}\int_0^{\pi}\left|\frac{\sin(t+k\pi)}{t+k\pi}\right|dt = \sum_{k=0}^{n}\int_0^{\pi}\left|\frac{(-1)^k\sin t}{t+k\pi}\right|dt \\ &= \sum_{k=0}^{n}\int_0^{\pi}\frac{|\sin t|}{t+k\pi}\,dt = \sum_{k=0}^{n}\int_0^{\pi}\frac{\sin t}{t+k\pi}\,dt,\end{aligned}$$

where the last equality follows since $\sin t \geqslant 0$ for $0 \leqslant t \leqslant \pi$. For $0 < t < \pi$, we have $1/(t + k\pi) > 1/[\pi(k+1)]$, so that

$$\int_0^{(n+1)\pi}\left|\frac{\sin x}{x}\right|dx > \sum_{k=0}^{n}\frac{1}{\pi(k+1)}\int_0^{\pi}\sin t\,dt = \frac{2}{\pi}\sum_{k=0}^{n}\frac{1}{k+1}.$$

Now, for $x \geqslant 1$ we have

$$\log x = \int_1^x \frac{1}{t}\,dt \leqslant \int_1^x 1\,dt = x - 1,$$

and replacing x by $(k+2)/(k+1)$, we have

$$\log\left(\frac{k+2}{k+1}\right) \leqslant \frac{1}{k+1}$$

for all $k \geqslant 0$. Therefore,

$$\sum_{k=0}^{n} \frac{1}{k+1} \geqslant \sum_{k=0}^{n} (\log(k+2) - \log(k+1)) = \log(n+2),$$

so

$$\lim_{n\to\infty} \sum_{k=0}^{n} \frac{1}{k+1} \geqslant \lim_{n\to\infty} \log(n+2) = +\infty,$$

and this implies that

$$\int_0^\infty \left|\frac{\sin x}{x}\right| dx = +\infty.$$

□

The following two lemmas are key to the proof of the Fourier integral theorem.

Lemma 8.5. If f is piecewise continuous on an interval (a, b), then

$$\lim_{h\to 0} \frac{1}{2h} \int_{x-h}^{x+h} f(\xi)\,d\xi = \frac{f(x^+) + f(x^-)}{2}$$

for all $x \in (a, b)$.

Proof. If $x \in (a, b)$ and f is continuous at x, since f is piecewise continuous on (a, b), we can choose $h > 0$ so small that f is continuous on the interval $(x - 2h, x + 2h)$. Writing

$$\int_{x-h}^{x+h} f(\xi)\,d\xi = \int_a^{x+h} f(\xi)\,d\xi - \int_a^{x-h} f(\xi)\,d\xi,$$

from the fundamental theorem of calculus, we have

$$\frac{d}{dh}\left(\int_{x-h}^{x+h} f(\xi)\,d\xi\right) = f(x+h) - f(x-h)(-1) = f(x+h) + f(x-h).$$

From L'Hospital's rule, since $f(x^+)$ and $f(x^-)$ both exist, we have

$$\lim_{h\to 0} \frac{1}{2h} \int_{x-h}^{x+h} f(\xi)\,d\xi = \lim_{h\to 0} \frac{f(x+h) + f(x-h)}{2} = \frac{f(x^+) + f(x^-)}{2}.$$

If f has a jump discontinuity at x, again since f is piecewise continuous on (a, b), we can choose $h > 0$ so small that f is continuous on the interval $(x - 2h, x)$ and also on the interval $(x, x + 2h)$. Now the same argument given above works. □

Lemma 8.6.

$$\int_0^\infty \frac{\sin \omega\xi}{\omega}\, d\omega = \begin{cases} \dfrac{\pi}{2}, & \text{for } \xi > 0, \\ 0, & \text{for } \xi = 0, \\ -\dfrac{\pi}{2}, & \text{for } \xi < 0. \end{cases}$$

Proof. If $\xi = 0$, the result is obvious. If $\xi > 0$, making the change of variable $x = \omega\xi$, then

$$\int_0^\infty \frac{\sin \omega\xi}{\omega}\, d\omega = \int_0^\infty \frac{\sin x}{x}\, dx = \frac{\pi}{2}.$$

On the other hand, if $\xi < 0$, making the change of variable $x = -\omega\xi$, then

$$\int_0^\infty \frac{\sin \omega\xi}{\omega}\, d\omega = -\int_0^\infty \frac{\sin x}{x}\, dx = -\frac{\pi}{2}.$$

□

Now, for real numbers x and h, we define

$$K(x, h) = \frac{2}{\pi} \int_0^\infty \frac{\sin \omega h}{\omega} \cos \omega x \, d\omega,$$

and note that K is an even function of x. We have

$$\begin{aligned} K(x, h) &= \frac{1}{\pi} \int_0^\infty \frac{1}{\omega} \left[\sin \omega(x + h) - \sin \omega(x - h) \right] d\omega \\ &= \frac{1}{\pi} \left\{ \int_0^\infty \frac{\sin \omega(x + h)}{\omega}\, d\omega - \int_0^\infty \frac{\sin \omega(x - h)}{\omega}\, d\omega \right\}. \end{aligned}$$

Using Lemma 8.6, we have

$$K(x, h) = \frac{1}{\pi} \begin{cases} -\dfrac{\pi}{2} - \left(-\dfrac{\pi}{2} \right), & \text{for } x < -h, \\ \dfrac{\pi}{2} - \left(-\dfrac{\pi}{2} \right), & \text{for } -h < x < h, \\ \dfrac{\pi}{2} - \dfrac{\pi}{2}, & \text{for } x > h, \end{cases}$$

and therefore

$$K(x,h) = \frac{1}{\pi}\begin{cases} 0, & \text{for } x < -h, \\ \pi, & \text{for } -h < x < h, \\ 0, & \text{for } x > h; \end{cases}$$

that is,

$$K(x,h) = \begin{cases} 1, & \text{for } |x| < h, \\ 0, & \text{for } |x| > h. \end{cases}$$

Finally, we can prove Theorem 8.1.

Theorem 8.7. (*Fourier Integral Theorem*)
If f and f' are piecewise continuous on every interval (a,b), and f is absolutely integrable on $(-\infty,\infty)$, that is,

$$\int_{-\infty}^{\infty} |f(x)|\, dx < \infty,$$

the Fourier integral

$$\int_0^{\infty} [A(\omega)\cos\omega x + B(\omega)\sin\omega x]\, d\omega,$$

where

$$A(\omega) = \frac{1}{\pi}\int_{-\infty}^{\infty} f(x)\cos\omega x\, dx \qquad \text{and} \qquad B(\omega) = \frac{1}{\pi}\int_{-\infty}^{\infty} f(x)\sin\omega x\, dx$$

for $0 < \omega < \infty$, converges to

$$\frac{f(x^+) + f(x^-)}{2}$$

at each point $x \in \mathbb{R}$.

Proof. From Lemma 8.5 and the definition of $K(x,h)$, we have

$$\begin{aligned} \frac{f(x^+) + f(x^-)}{2} &= \lim_{h\to 0}\frac{1}{2h}\int_{x-h}^{x+h} f(\xi)\, d\xi \\ &= \lim_{h\to 0}\frac{1}{2h}\int_{-\infty}^{\infty} f(\xi)K(x-\xi,h)\, d\xi \\ &= \lim_{h\to 0}\frac{1}{2h}\int_{-\infty}^{\infty} f(\xi)\left\{\frac{2}{\pi}\int_0^{\infty}\frac{\sin\omega h}{\omega}\cos\omega(x-\xi)\, d\omega\right\} d\xi. \end{aligned}$$

Interchanging the limit and integration processes yields

$$\frac{f(x^+)+f(x^-)}{2} = \frac{1}{\pi}\int_{-\infty}^{\infty}\int_{0}^{\infty} f(\xi)\left(\lim_{h\to 0}\frac{\sin\omega h}{\omega h}\right)\cos\omega(x-\xi)\,d\omega\,d\xi$$

$$= \frac{1}{\pi}\int_{-\infty}^{\infty}\int_{0}^{\infty} f(\xi)\cos\omega(x-\xi)\,d\omega\,d\xi.$$

Finally, interchanging the order of integration, we have

$$\frac{f(x^+)+f(x^-)}{2} = \frac{1}{\pi}\int_{0}^{\infty}\int_{-\infty}^{\infty} f(\xi)[\cos\omega x\cos\omega\xi + \sin\omega x\sin\omega\xi]\,d\xi\,d\omega$$

$$= \frac{1}{\pi}\int_{0}^{\infty}\left\{\left(\int_{-\infty}^{\infty} f(\xi)\cos\omega\xi\,d\xi\right)\cos\omega x + \left(\int_{-\infty}^{\infty} f(\xi)\sin\omega\xi\,d\xi\right)\sin\omega x\right\}d\omega;$$

that is,

$$\frac{f(x^+)+f(x^-)}{2} = \int_{0}^{\infty}[A(\omega)\cos\omega x + B(\omega)\sin\omega x]d\omega,$$

from the definition of the Fourier integral coefficients. □

8.2 FOURIER TRANSFORMS

In this section we give the definitions and basic properties of the Fourier transform. Assuming that f is defined for all real numbers x, that f is continuous on $\mathbb{R}$, that f' is piecewise continuous on every finite interval (a, b), and that f is absolutely integrable on $\mathbb{R}$, that is,

$$\int_{-\infty}^{\infty}|f(x)|\,dx < \infty,$$

then by Dirichlet's theorem, f can be represented by the Fourier integral formula

$$f(x) = \int_{0}^{\infty}[A(\omega)\cos\omega x + B(\omega)\sin\omega x]\,d\omega, \tag{8.7}$$

where

$$A(\omega) = \frac{1}{\pi}\int_{-\infty}^{\infty} f(x)\cos\omega x\,dx \qquad \text{and} \qquad B(\omega) = \frac{1}{\pi}\int_{-\infty}^{\infty} f(x)\sin\omega x\,dx$$

for $\omega > 0$. Thus,

$$f(x) = \frac{1}{\pi}\int_{0}^{\infty}\left(\int_{-\infty}^{\infty}[f(\xi)\,(\cos\omega\xi\cos\omega x + \sin\omega\xi\sin\omega x)]\,d\xi\right)d\omega;$$

that is,

$$f(x) = \frac{1}{\pi}\int_0^\infty \left(\int_{-\infty}^\infty f(\xi)\cos\omega(\xi - x)\,d\xi\right)d\omega$$

for $-\infty < x < \infty$. Note that

$$\int_{-\infty}^\infty f(\xi)\cos\omega(\xi - x)\,d\xi$$

is an even function of ω, so that

$$\int_0^\infty \left(\int_{-\infty}^\infty f(\xi)\cos\omega(\xi - x)\,d\xi\right)d\omega = \frac{1}{2}\int_{-\infty}^\infty \left(\int_{-\infty}^\infty f(\xi)\cos\omega(\xi - x)\,d\xi\right)d\omega,$$

and thus,

$$f(x) = \frac{1}{2\pi}\int_{-\infty}^\infty \left(\int_{-\infty}^\infty f(\xi)\cos\omega(\xi - x)\,d\xi\right)d\omega. \tag{8.8}$$

Note also that

$$\int_{-\infty}^\infty f(\xi)\sin\omega(\xi - x)\,d\xi$$

is an odd function of ω, so that

$$\int_{-\infty}^\infty \left(\int_{-\infty}^\infty f(\xi)\sin\omega(\xi - x)\,d\xi\right)d\omega = 0,$$

and thus

$$f(x) = \frac{1}{2\pi}\int_{-\infty}^\infty \left(\int_{-\infty}^\infty f(\xi)\left[\cos\omega(\xi - x) + i\sin\omega(\xi - x)\right]\,d\xi\right)d\omega.$$

Therefore,

$$f(x) = \frac{1}{2\pi}\int_{-\infty}^\infty \left(\int_{-\infty}^\infty f(\xi)\,e^{i\omega(\xi - x)}\,d\xi\right)d\omega,$$

and we rewrite this last equality as

$$f(x) = \int_{-\infty}^\infty \underbrace{\left(\frac{1}{2\pi}\int_{-\infty}^\infty f(\xi)\,e^{i\omega\xi}\,d\xi\right)}_{\widehat{f}(\omega)} e^{-i\omega x}\,d\omega \tag{8.9}$$

for $-\infty < x < \infty$. Note that if f has a jump discontinuity at $x_0 \in \mathbb{R}$, then from Dirichlet's theorem we have

$$\frac{f(x_0^+) + f(x_0^-)}{2} = \int_{-\infty}^\infty \underbrace{\left(\frac{1}{2\pi}\int_{-\infty}^\infty f(\xi)\,e^{i\omega\xi}\,d\xi\right)}_{\widehat{f}(\omega)} e^{-i\omega x_0}\,d\omega. \tag{8.10}$$

Definition 8.8. If f is piecewise smooth on every finite interval (a, b) and

$$\int_{-\infty}^{\infty} |f(x)|\, dx < \infty,$$

the **Fourier transform** of $f(x)$ is

$$\widehat{f}(\omega) = \frac{1}{2\pi} \int_{-\infty}^{\infty} f(x)\, e^{i\omega x}\, dx \tag{8.11}$$

for $-\infty < x < \infty$.

Using (8.9) and (8.10), we can restate Theorem 8.1 in terms of Fourier transforms.

Theorem 8.9. If f and f' are piecewise continuous on every finite interval (a, b) and f is absolutely integrable on $(-\infty, \infty)$, then

$$\frac{f(x^+) + f(x^-)}{2} = \int_{-\infty}^{\infty} \widehat{f}(\omega)\, e^{-i\omega x}\, d\omega$$

for $-\infty < x < \infty$.

Definition 8.10. If f is piecewise smooth on every finite interval (a, b) and absolutely integrable on $(-\infty, \infty)$, then

$$\widehat{f}(\omega) = \frac{1}{2\pi} \int_{-\infty}^{\infty} f(x)\, e^{i\omega x}\, dx, \quad \omega \in \mathbb{R} \tag{8.12}$$

and

$$\frac{f(x^+) + f(x^-)}{2} = \int_{-\infty}^{\infty} \widehat{f}(\omega)\, e^{-i\omega x}\, d\omega, \quad x \in \mathbb{R} \tag{8.13}$$

form what is called a **Fourier transform pair.**

Definition 8.11. If f is piecewise smooth on every finite interval (a, b), absolutely integrable on $(-\infty, \infty)$ and f is continuous on $(-\infty, \infty)$, then

$$\widehat{f}(\omega) = \frac{1}{2\pi} \int_{-\infty}^{\infty} f(x)\, e^{i\omega x}\, dx, \quad \omega \in \mathbb{R} \tag{8.14}$$

and

$$f(x) = \int_{-\infty}^{\infty} \widehat{f}(\omega)\, e^{-i\omega x}\, d\omega, \quad x \in \mathbb{R} \tag{8.15}$$

and $f(x)$ is called the **inverse Fourier transform** of $\widehat{f}(\omega)$.

Notation: For the Fourier transform and the inverse Fourier transform we have:

- The Fourier transform of f is sometimes denoted by

$$\widehat{f}(\omega) = F(\omega) = \mathcal{F}\left[\, f(x)\right](\omega),$$

and

- the inverse Fourier transform of $\widehat{f}$ is sometimes denoted by

$$f(x) = \mathcal{F}^{-1}\left[\, F(\omega)\right](x) = \mathcal{F}^{-1}\left[\,\widehat{f}(\omega)\right](x) = \mathcal{F}^{-1}\left[\mathcal{F}\left[\, f(x)\right](\omega)\right](x)$$

or any combination thereof. The reader is cautioned that there is **no standard definition** of the Fourier transform and the inverse Fourier transform. Hence, in some textbooks you will find the Fourier transform and the inverse Fourier transform with different factors in front and/or different signs in the exponent. We will be consistent throughout this book and use the notation presented in the preceding definitions.

Example 8.5. In this example we find the Fourier transform pair for $f(x) = e^{-|x|}$.

Clearly, f satisfies the hypothesis of the Fourier integral theorem, and

$$\begin{aligned}
F(\omega) = \mathcal{F}\left[e^{-|x|}\right](\omega) &= \frac{1}{2\pi}\int_{-\infty}^{\infty} e^{-|x|}\, e^{i\omega x}\, dx \\
&= \frac{1}{2\pi}\left\{\int_{-\infty}^{0} e^{x(1+i\omega)}\, dx + \int_{0}^{\infty} e^{-x(1-i\omega)}\, dx\right\} \\
&= \frac{1}{2\pi}\left\{\frac{e^{x(1+i\omega)}}{1+i\omega}\bigg|_{-\infty}^{0} - \frac{e^{-x(1-i\omega)}}{1-i\omega}\bigg|_{0}^{\infty}\right\} \\
&= \frac{1}{2\pi}\left(\frac{1}{1+i\omega} + \frac{1}{1-i\omega}\right) \\
&= \frac{1}{\pi(1+\omega^2)}.
\end{aligned}$$

Therefore,

$$e^{-|x|} = \mathcal{F}^{-1}\left[\, F(\omega)\right](x) = \int_{-\infty}^{\infty} F(\omega) e^{-i\omega x}\, d\omega = \frac{1}{\pi}\int_{-\infty}^{\infty} \frac{e^{-i\omega x}}{1+\omega^2}\, d\omega.$$

∎

8.2.1 Operational Properties of the Fourier Transform

In the next few theorems, we show that the Fourier transform and the inverse Fourier transform are linear operators, and state the shift theorems and the dilation theorem for the Fourier transform. We will need these results in Chapter 9 when we use transform methods to solve PDEs,

Theorem 8.12. (*Linearity*)

(a) If f and g are piecewise smooth on every finite interval, and absolutely integrable on $\mathbb{R}$, then

$$\mathcal{F}\left[\, af + bg \,\right] = a\mathcal{F}\left[\, f \,\right] + b\mathcal{F}\left[\, g \,\right].$$

(b) If, in addition, f and g are continuous on $\mathbb{R}$, then

$$\mathcal{F}^{-1}\left[\, af + bg \,\right] = a\mathcal{F}^{-1}\left[\, f \,\right] + b\mathcal{F}^{-1}\left[\, g \,\right].$$

Proof. We will give the proof for the Fourier transform; the proof for the inverse transform is exactly the same.

$$\begin{aligned} \mathcal{F}\left[\, af(x) + bg(x) \,\right](\omega) &= \frac{1}{2\pi}\int_{-\infty}^{\infty} \left(af(x) + bg(x)\right) e^{i\omega x}\, dx \\ &= a \cdot \frac{1}{2\pi}\int_{-\infty}^{\infty} f(x)\, e^{i\omega x}\, dx + b \cdot \frac{1}{2\pi}\int_{-\infty}^{\infty} g(x)\, e^{i\omega x}\, dx \\ &= a\mathcal{F}\left[f(x)\right](\omega) + b\mathcal{F}\left[g(x)\right](\omega) \end{aligned}$$

for $-\infty < \omega < \infty$. □

Theorem 8.13. (*Shift Theorems*)
If f is piecewise continuous on every finite interval and is absolutely integrable on $\mathbb{R}$, and $a \neq 0$ is a real number, then

(a) $F\left[f(x-a)\right](\omega) = e^{ia\omega}\widehat{f}(\omega)$,

(b) $F\left[e^{-iax} f(x)\right](\omega) = \widehat{f}(\omega - a)$,

(c) $F\left[f(ax)\right](\omega) = (1/|a|)\,\widehat{f}\,(\omega/a)$

for $-\infty < \omega < \infty$.

Proof. Suppose that f is piecewise smooth on every finite interval and is absolutely integrable on $\mathbb{R}$.

(a) From the definition of the Fourier transform, we have

$$F\left[f(x-a)\right](\omega) = \frac{1}{2\pi}\int_{-\infty}^{\infty} f(x-a)\,e^{i\omega x}\,dx = \frac{1}{2\pi}\int_{-\infty}^{\infty} f(t)e^{i\omega(t+a)}\,dt$$

$$= e^{i\omega a}\frac{1}{2\pi}\int_{-\infty}^{\infty} f(x)e^{i\omega x}\,dx = e^{ia\omega}\,\widehat{f}(\omega).$$

(b) Again, from the definition we have

$$F\left[e^{-iax}f(x)\right](\omega) = \frac{1}{2\pi}\int_{-\infty}^{\infty} e^{-iax}\,f(x)\,e^{i\omega x}\,dx$$

$$= \frac{1}{2\pi}\int_{-\infty}^{\infty} f(x)\,e^{i(\omega-a)x}\,dx$$

$$= \widehat{f}(\omega - a).$$

We leave the proof of (c) as an exercise. □

Now we indicate the relationship between the Fourier transform operator and the operations of differentiation and integration.

Theorem 8.14. (*Transforms of Derivatives*)

(a) If f is piecewise smooth, f and f' are integrable, and $f(x) \to 0$ as $|x| \to \infty$, then

$$\mathcal{F}\left[f'(x)\right] = -i\omega\,\mathcal{F}\left[f(x)\right].$$

(b) If f' is piecewise smooth, f'' is integrable, and $f'(x) \to 0$ as $|x| \to \infty$, then

$$\mathcal{F}\left[f''(x)\right] = (-i\omega)\,\mathcal{F}\left[f'(x)\right] = (-i\omega)^2\,\mathcal{F}\left[f(x)\right].$$

(c) In general, if f and $f^{(k)}$, for $k = 1, 2, \ldots, n-1$, are piecewise smooth and approach 0 as $|x| \to \infty$, and $f, f', \ldots, f^{(n)}$ are integrable, then

$$\mathcal{F}\left[f^{(n)}(x)\right] = (-i\omega)^n\,\mathcal{F}\left[f(x)\right].$$

Proof. We will prove part (a); the proofs of (b) and (c) are left as an exercise.

(a) Integrate by parts and use the fact that $f(x) \to 0$ as $|x| \to \infty$; then

$$\mathcal{F}\left[f'(x)\right](\omega) = \frac{1}{2\pi}\int_{-\infty}^{\infty} f'(x)\,e^{i\omega x}\,dx$$

$$= \frac{1}{2\pi}\left[f(x)\,e^{i\omega x}\Big|_{-\infty}^{\infty} - \int_{-\infty}^{\infty}(i\omega)\,f(x)\,e^{i\omega x}\,dx\right],$$

and therefore

$$\mathcal{F}\left[f'(x)\right](\omega) = (-i\omega)\,\frac{1}{2\pi}\int_{-\infty}^{\infty} f(x)\, e^{i\omega x}\, dx = -i\omega\, \mathcal{F}\left[f(x)\right].$$

□

Theorem 8.15. (*Transform of an Integral*)
If f is continuous on $(-\infty, \infty)$, f' is piecewise continuous on every finite interval (a, b), and f is absolutely integrable on $(-\infty, \infty)$, then

$$\mathcal{F}\left[\int_0^x f(s)\, ds\right](\omega) = -\frac{1}{i\omega}\mathcal{F}[f(x)](\omega)$$

for $-\infty < \omega < \infty$.

Proof. Using Theorem 8.14, from the fundamental theorem of calculus we have

$$\frac{d}{dx}\int_0^x f(s)\, ds = f(x),$$

and the result follows. □

Theorem 8.16. (*Derivatives of Transforms*)
Suppose that f is piecewise smooth on every finite interval (a, b).

(a) If $f(x)$ and $xf(x)$ are absolutely integrable on $(-\infty, \infty)$, then

$$\mathcal{F}\left[\, x f(x)\,\right](\omega) = -i\,\frac{d}{d\omega}\mathcal{F}\left[f(x)\right](\omega).$$

(b) If $f(x)$ and $x^n f(x)$ are integrable, then

$$\mathcal{F}\left[\, x^n f(x)\,\right](\omega) = (-i)^n\,\frac{d^n}{d\omega^n}\mathcal{F}\left[f(x)\right](\omega).$$

Proof. We give the proof for part (a) and leave the proof for part (b) as an exercise.

(a) We have

$$\begin{aligned}\frac{d}{d\omega}\left(\frac{1}{2\pi}\int_{-\infty}^{\infty} f(x)\, e^{i\omega x}\, dx\right) &= \frac{1}{2\pi}\int_{-\infty}^{\infty}\frac{d}{d\omega}\left(e^{i\omega x}\right)\, f(x)\, dx\\ &= \frac{i}{2\pi}\int_{-\infty}^{\infty} x f(x)\, e^{i\omega x}\, dx\\ &= i\,\mathcal{F}\left[x f(x)\right](\omega),\end{aligned}$$

and therefore

$$-i\,\frac{d}{d\omega}\left(\mathcal{F}\left[\, f(x)\,\right](\omega)\right) = \mathcal{F}\left[\, x f(x)\,\right](\omega)$$

for $-\infty < \omega < \infty$.

□

8.2.2 Fourier Sine and Cosine Transforms

In this section we give the definitions and basic properties of the Fourier sine and cosine transforms. Given a function f defined on the semi-infinite domain $[0, \infty)$, we can extend f to $\mathbb{R}$ as an odd function f_{odd}, or as an even function f_{even}, and hence f can have two Fourier transforms:

$$B(\omega) = \frac{2}{\pi} \int_0^\infty f(\xi) \sin \omega\xi \, d\xi$$

and

$$A(\omega) = \frac{2}{\pi} \int_0^\infty f(\xi) \cos \omega\xi \, d\xi,$$

called the **Fourier sine transform** and the **Fourier cosine transform**, respectively.

• Fourier Sine Transform

Let f_{odd} be the odd extension of f to $(-\infty, \infty)$, as in Figure 8.5. The Fourier transform of f_{odd} is

$$\begin{aligned}
\widehat{f_{\text{odd}}}(\omega) &= \mathcal{F}\left[\, f_{\text{odd}}(x) \,\right](\omega) \\
&= \frac{1}{2\pi} \int_{-\infty}^{\infty} f_{\text{odd}}(x)\, e^{i\omega x} \, dx \\
&= \frac{1}{2\pi} \int_{-\infty}^{\infty} f_{\text{odd}}(x) \left[\cos \omega x + i \sin \omega x\right] dx,
\end{aligned}$$

and since f_{odd} is an odd function, then

$$\widehat{f_{\text{odd}}}(\omega) = \frac{i}{2\pi} \int_{-\infty}^{\infty} f_{\text{odd}}(x) \sin \omega x \, dx = \frac{2i}{2\pi} \int_0^\infty f_{\text{odd}}(x) \sin \omega x \, dx.$$

Now, $f_{\text{odd}} = f$ on $[0, \infty)$, and hence

$$\widehat{f_{\text{odd}}}(\omega) = \frac{i}{\pi} \int_0^\infty f(\xi) \sin \omega\xi \, d\xi$$

for $-\infty < \omega < \infty$. Note that this is an odd function of ω.

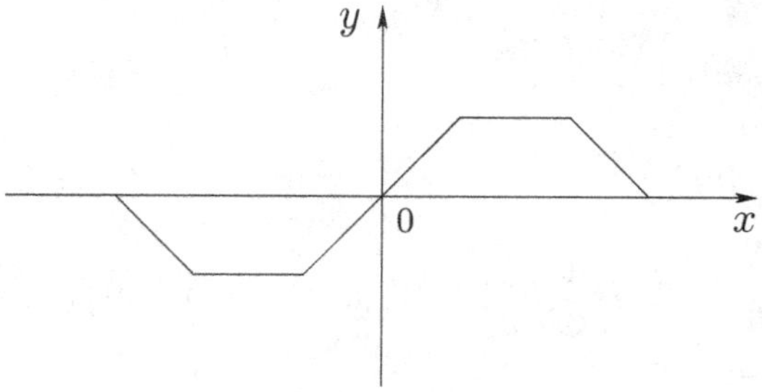

Figure 8.5 Odd extension.

The inverse Fourier transform of $\widehat{f}_{\text{odd}}$ is

$$
\begin{aligned}
f_{\text{odd}}(x) &= \mathcal{F}^{-1}\left[\,\widehat{f}_{\text{odd}}(\omega)\,\right](x) \\
&= \int_{-\infty}^{\infty} \widehat{f}_{\text{odd}}(\omega)\, e^{-i\omega x}\, d\omega \\
&= \int_{-\infty}^{\infty} \widehat{f}_{\text{odd}}(\omega)\, (\cos\omega x - i\sin\omega x)\; d\omega \\
&= -i\int_{-\infty}^{\infty} \widehat{f}_{\text{odd}}(\omega)\sin\omega x\, d\omega \\
&= -2i\int_{0}^{\infty} \widehat{f}_{\text{odd}}(\omega)\sin\omega x\, d\omega.
\end{aligned}
$$

However, $f_{\text{odd}}(x) = f(x)$ for $x \in [0,\infty)$, so that

$$
\begin{aligned}
f(x) &= -2i\int_{0}^{\infty} \widehat{f}_{\text{odd}}(\omega)\sin\omega x\, d\omega \\
&= -2i\int_{0}^{\infty}\left(\frac{i}{\pi}\int_{0}^{\infty} f(\xi)\sin\omega\xi\, d\xi\right)\sin\omega x\, d\omega \\
&= \int_{0}^{\infty}\left(\frac{2}{\pi}\int_{0}^{\infty} f(\xi)\sin\omega\xi\, d\xi\right)\sin\omega x\, d\omega.
\end{aligned}
$$

Therefore, if $f\,:\,[0,\infty)\longrightarrow\mathbb{R}$ is extended as an odd function, we have

$$
f(x) = \int_{0}^{\infty}\underbrace{\left(\frac{2}{\pi}\int_{0}^{\infty} f(\xi)\sin\omega\xi\, d\xi\right)}_{B(\omega)}\sin\omega x\, d\omega \tag{8.16}
$$

for $0 < x < \infty$; note that (8.16) is just the Fourier sine integral formula.

• Fourier Cosine Transform

Let f_{even} be the even extension of f to $(-\infty,\infty)$, as in Figure 8.6. The Fourier transform of f_{even} is

$$
\begin{aligned}
\widehat{f}_{\text{even}}(\omega) &= \mathcal{F}\left[\, f_{\text{even}}(x)\,\right](\omega) \\
&= \frac{1}{2\pi}\int_{-\infty}^{\infty} f_{\text{even}}(x)\, e^{i\omega x}\, dx \\
&= \frac{1}{2\pi}\int_{-\infty}^{\infty} f_{\text{even}}(x)\left[\cos\omega x + i\sin\omega x\right]\, dx,
\end{aligned}
$$

and since f_{even} is an even function, then

$$
\widehat{f}_{\text{even}}(\omega) = \frac{1}{2\pi}\int_{-\infty}^{\infty} f_{\text{even}}(x)\cos\omega x\, dx = \frac{1}{\pi}\int_{0}^{\infty} f_{\text{even}}(x)\cos\omega x\, dx,
$$

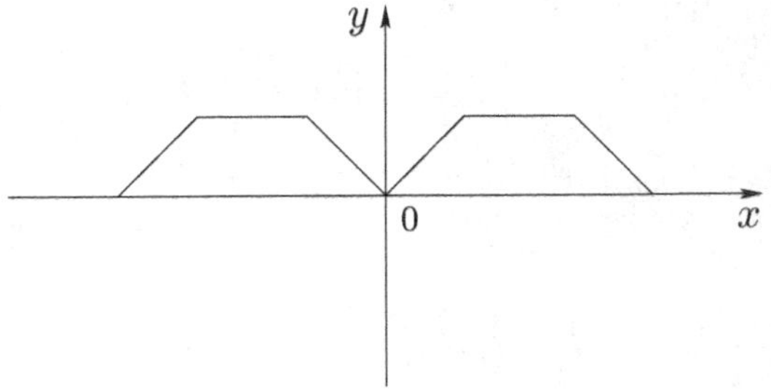

Figure 8.6 Even extension.

Now, $f_{\text{even}} = f$ on $[0, \infty)$, and hence

$$\widehat{f}_{\text{even}}(\omega) = \frac{1}{\pi} \int_0^{\infty} f(\xi) \cos \omega\xi \, d\xi$$

for $-\infty < \omega < \infty$. Note that this is an <u>even</u> function of ω. The inverse Fourier transform of $\widehat{f}_{\text{even}}$ is

$$\begin{aligned} f_{\text{even}}(x) &= \mathcal{F}^{-1}\left[\widehat{f}_{\text{even}}(\omega)\right](x) \\ &= \int_{-\infty}^{\infty} \widehat{f}_{\text{even}}(\omega)\, e^{-i\omega x}\, d\omega \\ &= \int_{-\infty}^{\infty} \widehat{f}_{\text{even}}(\omega)\, (\cos \omega x - i \sin \omega x)\, d\omega \\ &= \int_{-\infty}^{\infty} \widehat{f}_{\text{even}}(\omega) \cos \omega x \, d\omega \\ &= 2 \int_0^{\infty} \widehat{f}_{\text{even}}(\omega) \cos \omega x \, d\omega. \end{aligned}$$

However, $f_{\text{even}}(x) = f(x)$ for $x \in [0, \infty)$, so that

$$\begin{aligned} f(x) &= 2 \int_0^{\infty} \widehat{f}_{\text{even}}(\omega) \cos \omega x \, d\omega \\ &= 2 \int_0^{\infty} \left(\frac{1}{\pi} \int_0^{\infty} f(\xi) \cos \omega\xi \, d\xi \right) \cos \omega x \, d\omega \\ &= \int_0^{\infty} \left(\frac{2}{\pi} \int_0^{\infty} f(\xi) \cos \omega\xi \, d\xi \right) \cos \omega x \, d\omega. \end{aligned}$$

Therefore, if $f : [0,\infty) \longrightarrow \mathbb{R}$ is extended as an even function, we have

$$f(x) = \int_0^\infty \underbrace{\left(\frac{2}{\pi}\int_0^\infty f(\xi)\cos\omega\xi\,d\xi\right)}_{A(\omega)} \cos\omega x\,d\omega \tag{8.17}$$

for $0 < x < \infty$; and note that (8.17) is just the Fourier cosine integral formula.

Definition 8.17. Let $f : [0,\infty) \longrightarrow \mathbb{R}$ be continuous and absolutely integrable on $(0,\infty)$, and let f' be piecewise continuous on every finite interval $(a,b) \subset (0,\infty)$. Then the sine and cosine transform and inverse transform are given by:

(i) The **Fourier sine transform** of $f(x)$ is

$$\mathcal{S}\,[\,f(x)\,]\,(\omega) = \frac{2}{\pi}\int_0^\infty f(x)\sin\omega x\,dx,$$

and the **inverse sine transform** of $g(\omega)$ is

$$\mathcal{S}^{-1}\,[\,g(\omega)\,]\,(x) = \int_0^\infty g(\omega)\sin\omega x\,d\omega.$$

(ii) The **Fourier cosine transform** of $f(x)$ is

$$\mathcal{C}\,[\,f(x)\,]\,(\omega) = \frac{2}{\pi}\int_0^\infty f(x)\cos\omega x\,dx,$$

and the **inverse cosine transform** of $g(\omega)$ is

$$\mathcal{C}^{-1}\,[\,g(\omega)\,]\,(x) = \int_0^\infty g(\omega)\cos\omega x\,d\omega.$$

Note: From the definitions above and Dirichlet's theorem, we have

(i) $\mathcal{S}^{-1}\left[\mathcal{S}\left[f(x)\right](\omega)\right](x) = f(x)$, that is,

$$f(x) = \int_0^\infty \left(\frac{2}{\pi}\int_0^\infty f(\xi)\sin\omega\xi\,d\xi\right)\sin\omega x\,d\omega$$

for $0 < x < \infty$.

(ii) $\mathcal{C}^{-1}\left[\mathcal{C}\left[f(x)\right](\omega)\right](x) = f(x)$, that is,

$$f(x) = \int_0^\infty \left(\frac{2}{\pi}\int_0^\infty f(\xi)\cos\omega\xi\,d\xi\right)\cos\omega x\,d\omega$$

for $0 < x < \infty$.

8.2.3 Operational Properties of the Fourier Sine and Cosine Transforms

The linearity of the sine and cosine transforms is evident; we only give the results on transforms of derivatives here.

Theorem 8.18. (*Transforms of Derivatives*)
If f is piecewise smooth, f and f' are integrable on $[0, \infty)$, and $\lim_{x\to\infty} f(x) \to 0$, then:

(a) For the Fourier sine transform, we have

$$\mathcal{S}\left[f'(x)\right](\omega) = -\omega\, \mathcal{C}\left[f(x)\right](\omega)$$

and if f'' is integrable on $[0, \infty)$ and $\lim_{x\to\infty} f'(x) \to 0$ also, then

$$\mathcal{S}\left[f''(x)\right](\omega) = \frac{2\omega}{\pi}\, f(0) - \omega^2\, \mathcal{S}\left[f(x)\right](\omega).$$

(b) For the Fourier cosine transform, we have

$$\mathcal{C}\left[f'(x)\right](\omega) = -\frac{2}{\pi}\, f(0) + \omega\, \mathcal{S}\left[f'(x)\right](\omega)$$

and if f'' is integrable on $[0, \infty)$ and $\lim_{x\to\infty} f'(x) \to 0$ also, then

$$\mathcal{C}\left[f''(x)\right](\omega) = -\frac{2}{\pi}\, f'(0) - \omega^2\, \mathcal{C}\left[f(x)\right](\omega).$$

Proof. We give the proof of the first part of (a); proofs of the remaining results are left as an exercise.

(a) Integrating by parts gives

$$\begin{aligned}
\mathcal{S}\left[f'(x)\right](\omega) &= \frac{2}{\pi}\int_0^\infty f'(x)\sin\omega x\, dx \\
&= \frac{2}{\pi}\, f(x)\sin\omega x\Big|_0^\infty - \frac{2}{\pi}\int_0^\infty f(x)\,\omega\,\cos\omega x\, dx,
\end{aligned}$$

and since $f(x) \to 0$ as $x \to \infty$, then

$$\begin{aligned}
\mathcal{S}\left[f'(x)\right](\omega) &= -\frac{2}{\pi}\int_0^\infty f(x)\,\omega\,\cos\omega x\, dx \\
&= -\omega\, C\left[f(x)\right](\omega).
\end{aligned}$$

□

Note: From Theorem 8.18, it is obvious that the following are true:

(i) If the Fourier sine transform is used to solve a PDE, then $f(0)$; that is, $u(0,t)$ must be known, for example, a Dirichlet boundary condition at $x = 0$.

(ii) If the Fourier cosine transform is used to solve a PDE, then $f'(0)$; that is, $\partial u(0,t)/\partial x$, must be known, for example, a Neumann boundary condition at $x = 0$.

8.2.4 Fourier Transforms and Convolutions

When solving PDEs using Fourier transforms, one is faced with the problem of having to find the inverse Fourier transform of a product of the form $F(\omega) \cdot G(\omega)$, where F and G are, respectively, the Fourier transforms of the functions f and g. Unfortunately, the inverse Fourier transform of the product is not the product of the inverse Fourier transforms. To find the inverse Fourier transform of a product, we need to introduce the convolution of two functions.

Definition 8.19. (*Convolution Product*)
If f and g are defined on all of $\mathbb{R}$, and are integrable over $\mathbb{R}$, the **convolution** of f and g is given by

$$f * g\,(x) = \int_{-\infty}^{\infty} f(x-t)\,g(t)\,dt$$

for $-\infty < x < \infty$.

The following lemma lists some of the properties satisfied by the convolution.

Lemma 8.20. (*Properties of the Convolution*)
Suppose that $a \in \mathbb{R}$ and that f, g, h are integrable over $\mathbb{R}$, then

(a) $f * g = g * f$,

(b) $a\,(f * g) = (af) * g = f * (ag)$,

(c) $(f * g) * h = f * (g * h)$,

(d) $f * (g + h) = f * g + f * h$.

Proof. We will prove (a) and leave the rest as an exercise.

(a) If f and g are integrable over $\mathbb{R}$, then

$$\begin{aligned} g * f\,(x) &= \int_{-\infty}^{\infty} g(x-t)\,f(t)\,dt \underset{s=x-t}{=} -\int_{\infty}^{-\infty} g(s)\,f(x-s)\,ds \\ &= \int_{-\infty}^{\infty} f(x-s)\,g(s)\,ds \\ &= f * g\,(x), \end{aligned}$$

for all $x \in \mathbb{R}$.

□

Example 8.6. (*Convolution with a Sine*)
Let f be an even integrable function on $\mathbb{R}$, and let $g(x) = \sin ax$ for $x \in \mathbb{R}$, where $a > 0$ is constant; then

$$f * g(x) = 2\pi \sin ax \, \widehat{f}(a),$$

where $\widehat{f}$ is the Fourier transform of f.
Solution. We have

$$\begin{aligned} f * g(x) &= \int_{-\infty}^{\infty} f(t) \sin a(x - t) \, dt \\ &= \int_{-\infty}^{\infty} f(t) [\sin ax \cos at - \cos ax \sin at] \, dt \\ &= \sin ax \int_{-\infty}^{\infty} f(t) \cos at \, dt - \cos ax \int_{-\infty}^{\infty} f(t) \sin at \, dt, \end{aligned}$$

and since f is even,

$$f * g(x) = \sin ax \int_{-\infty}^{\infty} f(t) \cos at \, dt.$$

And, again, since f is even,

$$\begin{aligned} f * g(x) &= \sin ax \int_{-\infty}^{\infty} f(t) \cos at \, dt + i \sin ax \int_{-\infty}^{\infty} f(t) \sin at \, dt \\ &= \sin ax \int_{-\infty}^{\infty} f(t) (\cos at + i \sin at) \, dt \\ &= \sin ax \int_{-\infty}^{\infty} f(t) \, e^{iat} \, dt \\ &= 2\pi \sin ax \frac{1}{2\pi} \int_{-\infty}^{\infty} f(t) \, e^{iat} \, dt \\ &= 2\pi \sin ax \, \widehat{f}(a). \end{aligned}$$

∎

Theorem 8.21. (*Convolution Theorem*)
If f and g are integrable and satisfy the hypotheses of Dirichlet's theorem, that is, the Fourier integral theorem, then

(a) $\mathcal{F}\left[\frac{1}{2\pi} f * g\right] = \widehat{f} \cdot \widehat{g}$.

(b) If, in addition, f and g are continuous, then $f * g = 2\pi \mathcal{F}^{-1}\left[\widehat{f} \cdot \widehat{g}\right]$.

Proof. For part (a), we have

$$\begin{aligned}\mathcal{F}[f * g](\omega) &= \frac{1}{2\pi}\int_{-\infty}^{\infty} f * g(x)\, e^{i\omega x}\, dx \\ &= \frac{1}{2\pi}\int_{-\infty}^{\infty}\left(\int_{-\infty}^{\infty} f(x-t)\, g(t)\, dt\right) e^{i\omega x}\, dx \\ &= \frac{1}{2\pi}\int_{-\infty}^{\infty} g(t)\left(\int_{-\infty}^{\infty} f(x-t)\, e^{i\omega x}\, dx\right) dt,\end{aligned}$$

and letting $u = x - t$ yields

$$\begin{aligned}\mathcal{F}[f * g](\omega) &= \frac{1}{2\pi}\int_{-\infty}^{\infty} g(t)\left(\int_{-\infty}^{\infty} f(u)\, e^{i\omega(u+t)} du\right) dt \\ &= \frac{1}{2\pi}\left(\int_{-\infty}^{\infty} g(t)\, e^{i\omega t}\, dt\right)\left(\int_{-\infty}^{\infty} f(u)\, e^{i\omega u}\, du\right) \\ &= 2\pi \widehat{f}(\omega)\, \widehat{g}(\omega).\end{aligned}$$

Therefore,

$$\mathcal{F}\left[\frac{1}{2\pi} f * g\right](\omega) = \widehat{f}(\omega)\, \widehat{g}(\omega)$$

for $-\infty < \omega < \infty$. The proof of part (b) follows from the definition of the inverse Fourier transform. □

Example 8.7. Find the Fourier transform of the function

$$g(x) = \begin{cases} 1 - \dfrac{|x|}{2}, & \text{for } |x| < 2, \\ 0, & \text{for } |x| \geqslant 2. \end{cases}$$

Solution. The graph of g is shown in Figure 8.7. Clearly, g and g' are piecewise

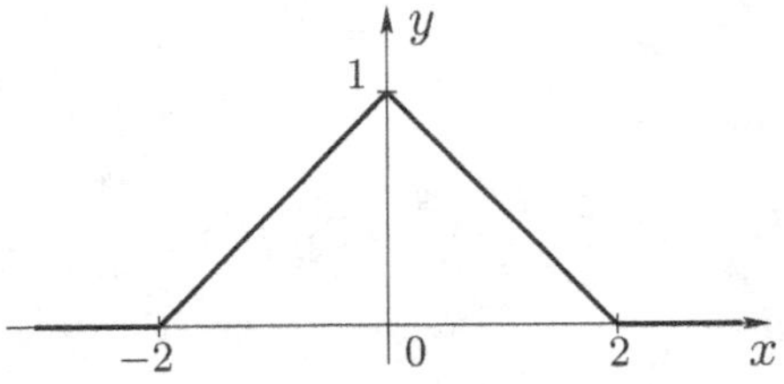

Figure 8.7 Tent function $g(x)$.

continuous on $\mathbb{R}$, and

$$\int_0^\infty |g(x)|\,dx = \int_{-2}^{2}\left|1-\frac{|x|}{2}\right|\,dx = 2\int_0^2\left(1-\frac{x}{2}\right)\,dx = 2 < \infty;$$

thus, g satisfies the hypotheses of the Fourier integral theorem (i.e., Dirichlet's theorem), and therefore, for $\omega \neq 0$,

$$\begin{aligned}
\widehat{g}(\omega) &= \frac{1}{2\pi}\int_{-\infty}^{\infty} g(x)\,e^{i\omega x}\,dx \\
&= \frac{1}{2\pi}\int_{-2}^{2}\left(1-\frac{|x|}{2}\right)\,e^{i\omega x}\,dx \\
&= \frac{1}{2\pi}\int_{-2}^{2}\left(1-\frac{|x|}{2}\right)\,(\cos\omega x + i\sin\omega x)\,dx \\
&= \frac{1}{2\pi}\int_{-2}^{2}\left(1-\frac{|x|}{2}\right)\,\cos\omega x\,dx,
\end{aligned}$$

since $g(x)$ is an even function. Therefore, for $\omega \neq 0$, we have

$$\begin{aligned}
\widehat{g}(\omega) &= \frac{1}{\pi}\int_0^2\left(1-\frac{x}{2}\right)\,\cos\omega x\,dx \\
&= \frac{1}{\pi}\int_0^2 \cos\omega x\,dx - \frac{1}{2\pi}\int_0^2 x\cos\omega x\,dx \\
&= \frac{1}{\pi\omega}\sin\omega x\Big|_0^2 - \frac{1}{2\pi}\left[\frac{x\sin\omega x}{\omega}\Big|_0^2 - \frac{1}{\omega}\int_0^2 \sin\omega x\,dx\right] \\
&= \frac{\sin 2\omega}{\pi\omega} - \frac{1}{2\pi}\frac{2\sin 2\omega}{\omega} - \frac{1}{2\pi\omega^2}\cos\omega x\Big|_0^2 \\
&= \frac{1}{2\pi\omega^2}\left(1-\cos 2\omega\right),
\end{aligned}$$

and thus,

$$\widehat{g}(\omega) = \frac{\sin^2\omega}{\pi\omega^2}.$$

For $\omega = 0$, we have

$$\widehat{g}(0) = \frac{1}{2\pi}\int_{-\infty}^{\infty} g(x)\,dx = \frac{1}{2\pi}\int_{-2}^{2}\left(1-\frac{|x|}{2}\right)\,dx = \frac{1}{\pi}\int_0^2\left(1-\frac{x}{2}\right)\,dx = \frac{1}{\pi}.$$

Note that

$$\lim_{\omega\to 0}\widehat{g}(\omega) = \frac{1}{\pi}\lim_{\omega\to 0}\left(\frac{\sin\omega}{\omega}\right)^2 = \frac{1}{\pi},$$

and $\widehat{g}$ is continuous at all $\omega \in \mathbb{R}$. ∎

In fact, the next result shows that this is true in general (we omit the proof).

Theorem 8.22. If the function $f : \mathbb{R} \longrightarrow \mathbb{R}$ is piecewise smooth on every finite interval and is absolutely integrable on $\mathbb{R}$, that is,

$$\int_{-\infty}^{\infty} |f(t)|\, dt < \infty,$$

then the Fourier transform

$$\widehat{f}(\omega) = \frac{1}{2\pi} \int_{-\infty}^{\infty} f(t) e^{i\omega t}\, dt$$

is uniformly continuous on $\mathbb{R}$.

Example 8.8. Let $f(x)$ be the rectangular pulse

$$f(x) = \begin{cases} 1, & \text{for } |x| < 1, \\ 0, & \text{for } |x| > 1, \end{cases}$$

and $f(-1) = f(1) = \frac{1}{2}$, whose graph is shown in Figure 8.8 Let $h(x)$ be the convolution of f with itself, that is,

$$h(x) = \int_{-\infty}^{\infty} f(x-t) f(t)\, dt.$$

Find the Fourier transform of $h(x)$, and use the convolution theorem to identify $h(x)$.

Solution. First we find the Fourier transform of $f(x)$; for $\omega \neq 0$ we have

$$\begin{aligned}
\widehat{f}(\omega) &= \frac{1}{2\pi} \int_{-\infty}^{\infty} f(x)\, e^{i\omega x}\, dx = \frac{1}{2\pi} \int_{-1}^{1} (\cos \omega x + i \sin \omega x)\, dx \\
&= \frac{1}{2\pi} \int_{-1}^{1} \cos \omega x\, dx = \frac{1}{\pi} \int_{0}^{1} \cos \omega x\, dx \\
&= \left. \frac{\sin \omega x}{\pi \omega} \right|_0^1 = \frac{\sin \omega}{\pi \omega},
\end{aligned}$$

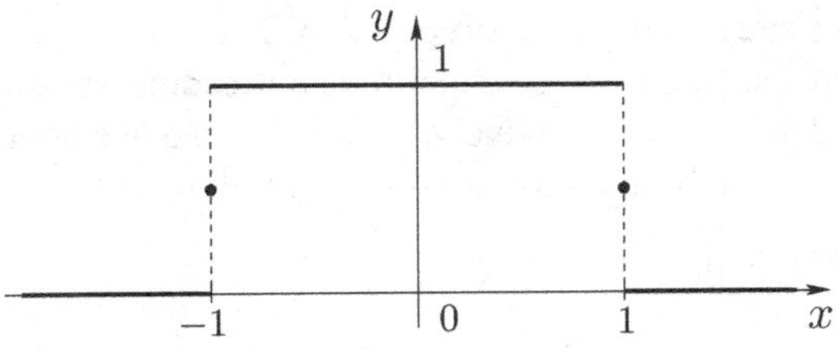

Figure 8.8 Rectangular pulse.

while for $\omega = 0$, we have

$$\widehat{f}(0) = \frac{1}{\pi}.$$

Now, the Fourier transform of $h(x) = (f * f)(x)$ is given by

$$\widehat{h}(\omega) = 2\pi\, \widehat{f}(\omega) \cdot \widehat{f}(\omega) = \frac{2\sin^2\omega}{\pi\omega^2}$$

for $\omega \neq 0$, and $\widehat{h}(0) = 2/\pi$. Therefore,

$$h(x) = (f{*}f)(x) = \mathcal{F}^{-1}\left[\frac{2\sin^2\omega}{\pi\omega^2}\right](x) = 2\,\mathcal{F}^{-1}\left[\frac{\sin^2\omega}{\pi\omega^2}\right](x) = 2\,\mathcal{F}^{-1}\left[\widehat{g}(\omega)\right](x),$$

where $\widehat{g}(\omega)$ was found in Example 8.7. Thus,

$$h(x) = \int_{-\infty}^{\infty} f(x-t)\, f(t)\, dt = \begin{cases} 2\left(1 - \dfrac{|x|}{2}\right), & \text{for } |x| < 2, \\ 0, & \text{for } |x| \geqslant 2. \end{cases}$$

Note: This can be used to evaluate certain improper integrals. For example, from the above we have

$$\mathcal{F}^{-1}\left[\frac{\sin^2\omega}{\pi\omega^2}\right](x) = \int_{-\infty}^{\infty} \frac{\sin^2\omega}{\pi\omega^2}\, e^{-i\omega x}\, d\omega = \begin{cases} 1 - \dfrac{|x|}{2}, & \text{for } |x| < 2, \\ 0, & \text{for } |x| \geqslant 2, \end{cases}$$

and therefore

$$\int_0^{\infty} \frac{\sin^2\omega}{\omega^2} \cos\omega x\, d\omega = \begin{cases} \dfrac{\pi}{2}\left(1 - \dfrac{|x|}{2}\right), & \text{for } |x| < 2, \\ 0, & \text{for } |x| \geqslant 2. \end{cases}$$

∎

8.2.5 Fourier Transform of a Gaussian Function

In the next chapter we solve the problem of heat conduction in an unbounded region, first for an infinite rod, where there are no boundary conditions, and next for a semi-infinite rod, where there is one boundary condition. In the first case we use the Fourier transform. In the second case we use either the Fourier sine transform or the Fourier cosine transform, depending, respectively, on whether the one boundary condition is a Dirichlet condition or a Neumann condition. We will need the following integrals.

Lemma 8.23. If $b > 0$, then

$$\int_0^{\infty} e^{-\omega^2 b}\, d\omega = \sqrt{\frac{\pi}{4b}}.$$

Proof. We write

$$I = \int_0^\infty e^{-\omega^2 b}\, d\omega,$$

so that

$$I^2 = \left(\int_0^\infty e^{-x^2 b}\, dx\right) \cdot \left(\int_0^\infty e^{-y^2 b}\, dy\right) = \int_0^\infty \int_0^\infty e^{-(x^2+y^2)b}\, dx\, dy.$$

Introducing polar coordinates ρ and θ, we have

$$\begin{aligned} I^2 &= \int_0^{\pi/2} \int_0^\infty e^{-\rho^2 b} \rho\, d\rho\, d\theta = \frac{\pi}{2}\left(-\frac{1}{2b}\right) \int_0^\infty e^{-\rho^2 b}(-2\rho b)\, d\rho \\ &= -\frac{\pi}{4b}\, e^{-\rho^2 b}\Big|_0^\infty = \frac{\pi}{4b} \end{aligned}$$

and

$$I = \sqrt{\frac{\pi}{4b}}.$$

□

Lemma 8.24. If $b > 0$, then

$$\int_0^\infty e^{-\omega^2 b} \cos \omega r\, d\omega = \sqrt{\frac{\pi}{4b}}\, e^{-\frac{r^2}{4b}}$$

for all real numbers r.

Proof. Let $b > 0$ and define the function g by

$$g(r) = \int_0^\infty e^{-\omega^2 b} \cos \omega r\, d\omega,$$

for $-\infty < r < \infty$. Since the improper integral converges uniformly on every finite interval and the integral

$$\int_0^\infty \frac{\partial}{\partial r}\left(e^{-\omega^2 b} \cos \omega r\right) d\omega = -\int_0^\infty \omega\, e^{-\omega^2 b} \sin \omega r\, d\omega$$

converges uniformly on every finite interval, g is differentiable. Differentiating under the integral sign, and then integrating by parts, we have

$$
\begin{aligned}
g'(r) &= -\int_0^\infty \omega\, e^{-\omega^2 b} \sin \omega r \, d\omega \\
&= \frac{1}{2b}\int_0^\infty (-2\omega b)\, e^{-\omega^2 b} \underbrace{\sin \omega r}_{u} \, d\omega \\
&= \frac{1}{2b}\left[e^{-\omega^2 b} \sin \omega r \bigg|_0^\infty - \int_0^\infty e^{-\omega^2 b} r \cos \omega r \, d\omega \right] \\
&= -\frac{r}{2b}\int_0^\infty e^{-\omega^2 b} \cos \omega r \, d\omega \\
&= -\frac{r}{2b}\, g(r),
\end{aligned}
$$

and $g(r)$ satisfies the first-order linear differential equation

$$
g'(r) + \frac{r}{2b}\, g(r) = 0.
$$

Multiplying by the integrating factor $e^{\frac{r^2}{4b}}$, we have

$$
\frac{d}{dr}\left(e^{\frac{r^2}{4b}} g(r) \right) = 0,
$$

so that

$$
g(r) = C \cdot e^{-\frac{r^2}{4b}}
$$

for $-\infty < r < \infty$, where C is constant. The constant of integration is given by

$$
C = g(0) = \int_0^\infty e^{-\omega^2 b} \, d\omega,
$$

and from Lemma 8.23, we have

$$
g(r) = \int_0^\infty e^{-\omega^2 b} \cos \omega r \, d\omega = \sqrt{\frac{\pi}{4b}}\, e^{-\frac{r^2}{4b}}
$$

for $-\infty < r < \infty$. □

Now we prove the important result that the Fourier transform of a Gaussian function is again a Gaussian function (to within a multiplicative constant).

Theorem 8.25. (*Fourier Transform of a Gaussian Function*)
Let $a > 0$ be an arbitrary constant; then

$$
\mathcal{F}\left[e^{-ax^2} \right](\omega) = \frac{1}{\sqrt{4\pi a}}\, e^{-\frac{\omega^2}{4a}}.
$$

Proof. Let $\widehat{f}(\omega)$ be the Fourier transform of the function

$$f(x) = e^{-ax^2}, \quad -\infty < x < \infty,$$

that is,

$$\widehat{f}(\omega) = \frac{1}{2\pi} \int_{-\infty}^{\infty} e^{-ax^2} e^{i\omega x}\, dx$$

for $-\infty < \omega < \infty$; then

$$\begin{aligned}
\widehat{f}(\omega) &= \frac{1}{2\pi} \int_{-\infty}^{\infty} e^{-ax^2} \big(\cos\omega x + i \sin\omega x\big)\, dx \\
&= \frac{1}{2\pi} \int_{-\infty}^{\infty} e^{-ax^2} \cos\omega x\, dx + \frac{i}{2\pi} \int_{-\infty}^{\infty} e^{-ax^2} \sin\omega x\, dx
\end{aligned}$$

for $-\infty < \omega < \infty$. Since e^{-ax^2} is an even function, then from Lemma 8.24 we have

$$\widehat{f}(\omega) = \frac{1}{2\pi} \int_{-\infty}^{\infty} e^{-ax^2} \cos\omega x\, dx = \frac{1}{\pi} \int_{0}^{\infty} e^{-ax^2} \cos\omega x\, dx = \frac{1}{\sqrt{4\pi a}}\, e^{-\frac{\omega^2}{4a}}$$

for $-\infty < \omega < \infty$. □

Corollary 8.26. Let $a > 0$ be an arbitrary constant; then

$$\mathcal{F}^{-1}\big[e^{-a\omega^2}\big](x) = \sqrt{\frac{\pi}{a}}\, e^{-\frac{x^2}{4a}} = 2\pi \cdot \frac{1}{\sqrt{4\pi a}}\, e^{-\frac{x^2}{4a}}$$

for $-\infty < x < \infty$.

Proof. This follows from Theorems 8.7 and 8.25. □

8.3 SUMMARY

In Chapter 2 on Fourier series, we defined the Fourier series for functions in $PWC(-\ell, \ell)$, and the Fourier sine series and Fourier cosine series for functions in $PWC(0, \ell)$. Thus, Fourier series are used on symmetric domains $(-\ell, \ell)$, while Fourier sine series and Fourier cosine series are used on the domain $[0, \ell)$. The situation is similar here: The **Fourier transform** is applied to problems on $(-\infty, \infty)$, while the **Fourier sine transform** and **Fourier cosine transform** are applied to problems on $(0, \infty)$. The Fourier transform can be written as an integral transform

$$\mathcal{F}[f(x)](\omega) = F(\omega) = \widehat{f}(\omega) = \frac{1}{2\pi} \int_{-\infty}^{\infty} f(x) e^{i\omega x} dx$$

and the inverse Fourier transform as

$$\mathcal{F}^{-1}[\widehat{f}(\omega)](x) = f(x) = \int_{-\infty}^{\infty} \widehat{f}(\omega)e^{-i\omega x}\,d\omega.$$

In using the Fourier transform, we enter a new world, the ω-world. In physical applications, the ω-world is also called the *frequency domain*, and indeed, local peaks of the Fourier transform $\widehat{f}(\omega)$ correspond to dominant frequencies of the original function.

The Fourier sine and cosine transforms apply to functions defined on $(0, \infty)$ and are closely related to the Fourier transform, and as such, they have very similar properties. We give several detailed examples in the text.

8.3.1 Problems and Notes

Problems from Part II:

Exercise	16.1	16.2	16.3	16.4	16.5	16.8
Notes						
Exercise	16.9	16.10	16.11	16.13	16.14	
Notes						

Final exam questions:

Exercise	19.1	19.10
Notes		

You should now be able to do **Final Exams 1 and 3.**

CHAPTER 9

FOURIER TRANSFORM METHODS IN PDEs

In this chapter we give examples of solving the classical partial differential equations

$$\frac{\partial u}{\partial t} = k\frac{\partial^2 u}{\partial x^2}, \qquad \frac{\partial^2 u}{\partial t^2} = c^2\frac{\partial^2 u}{\partial x^2}, \qquad \frac{\partial^2 u}{\partial x^2} + \frac{\partial^2 u}{\partial y^2} = 0,$$

that is, the heat equation, the wave equation, and Laplace's equation, defined on infinite domains. We will use Fourier transforms, and in most of the examples we will be taking the Fourier transform with respect to the spatial variable x, converting the PDE and any initial conditions into an ODE for the transform $\widehat{u}(\omega, t)$. We then solve the problem in terms of the transform variable ω and take the inverse Fourier transform to get the solution to the original problem.

A way to understand the duality between a function $f(x)$ and its Fourier transform $\widehat{f}(\omega)$ is to consider a different world, the ω-**world**, as opposed to the x-**world**. The ω-world has a different set of rules, and things that are complicated in the x-world may be much simpler in the ω-world. We move between the two worlds through the use of the Fourier transform and the inverse Fourier transform.

For example, differentiation in the x-world corresponds to multiplication by $-i\omega$ in the ω-world, and multiplication by x in the x-world corresponds to $-i(d/d\omega)$ in the ω-world. Convolution in the x-world corresponds to multiplication in the ω-world.

Partial Differential Equations: Theory and Completely Solved Problems, First Edition. By T. Hillen, I. E. Leonard, H. van Roessel

In fact, we can use the operational properties of the Fourier transform as a list of the correspondences.

Thus, to solve PDEs, we **transform** the problem from the x-world to the ω-world, where we obtain a simple ODE. We **solve** this ODE in the ω-world, and then **invert** this solution by transforming the solution back to the x-world using the inverse Fourier transform. We then have the solution to the original problem in the x-world.

This technique, that is,

TRANSFORM, SOLVE, INVERT

is typical of many branches of applied mathematics.

For the heat equation and wave equation above, we define

$$\widehat{u}(\omega,t)=\mathcal{F}\big[\,u(x,t)\,\big](\omega)=\frac{1}{2\pi}\int_{-\infty}^{\infty}u(x,t)\,e^{i\omega x}\,dx,$$

and recall the following operational properties of the Fourier transform:

1. $\mathcal{F}\left[\dfrac{\partial u}{\partial t}(x,t)\right](\omega)=\dfrac{d}{dt}\widehat{u}(\omega,t).$

2. $\mathcal{F}\left[\dfrac{\partial^n u}{\partial t^n}(x,t)\right](\omega)=\dfrac{d^n}{dt^n}\widehat{u}(\omega,t).$

3. $\mathcal{F}\left[\dfrac{\partial u}{\partial x}(x,t)\right](\omega)=(-i\omega)\widehat{u}(\omega,t).$

4. $\mathcal{F}\left[\dfrac{\partial^n u}{\partial x^n}(x,t)\right](\omega)=(-i\omega)^n\widehat{u}(\omega,t).$

Here ω is treated as a parameter and we have total derivatives on the right-hand side.

9.1 THE WAVE EQUATION

9.1.1 d'Alembert's Solution to the One-Dimensional Wave Equation

We use the Fourier transform to solve the following boundary value–initial value problem for the wave equation, and again we obtain d'Alembert's solution. The wave equation on an infinite domain reads

$$\frac{\partial^2 u}{\partial t^2}=c^2\frac{\partial^2 u}{\partial x^2},\qquad -\infty<x<\infty,\quad t>0,$$

$$u(x,0)=f(x),$$

$$\frac{\partial u}{\partial t}(x,0)=g(x),$$

where $\int_{-\infty}^{\infty} |f(x)|\, dx < \infty$ and $\int_{-\infty}^{\infty} |g(x)|\, dx < \infty$.

To find the solution of this problem, we consider two problems, each with one zero initial condition:

$$\begin{aligned} &\frac{\partial^2 v}{\partial t^2} = c^2 \frac{\partial^2 v}{\partial x^2}, \quad -\infty < x < \infty, \quad t > 0, \\ &v(x,0) = f(x), \\ &\frac{\partial v}{\partial t}(x,0) = 0 \end{aligned} \tag{I}$$

and

$$\begin{aligned} &\frac{\partial^2 w}{\partial t^2} = c^2 \frac{\partial^2 w}{\partial x^2}, \quad -\infty < x < \infty, \quad t > 0, \\ &w(x,0) = 0, \\ &\frac{\partial w}{\partial t}(x,0) = g(x), \end{aligned} \tag{II}$$

and then use the superposition principle to get the solution to the original problem, $u = v + w$.

- Problem I: In the first problem, we let

$$\widehat{v}(\omega,t) = \mathcal{F}\big[v(x,t)\big](\omega) = \frac{1}{2\pi}\int_{-\infty}^{\infty} v(x,t)\, e^{i\omega x}\, dx$$

and

$$\widehat{f}(\omega) = \mathcal{F}\big[f(x)\big](\omega) = \frac{1}{2\pi}\int_{-\infty}^{\infty} f(x)\, e^{i\omega x}\, dx.$$

Taking the Fourier transform of both sides of the wave equation, we have the following initial value problem:

$$\begin{aligned} &\frac{d^2\widehat{v}(\omega,t)}{dt^2} + c^2\omega^2\widehat{v}(\omega,t) = 0, \quad t > 0, \\ &\widehat{v}(\omega,0) = \widehat{f}(\omega), \\ &\frac{d\widehat{v}}{dt}(\omega,0) = 0. \end{aligned}$$

This is a simple ODE in the ω-world, and we can solve it to obtain

$$\widehat{v}(\omega,t) = \widehat{f}(\omega)\cos\omega ct.$$

To get back to the x-world, we take the inverse transform, and we have

$$\begin{aligned} v(x,t) &= \int_{-\infty}^{\infty} \widehat{f}(\omega)\cos\omega ct\, e^{-i\omega x}\,d\omega \\ &= \int_{-\infty}^{\infty} \widehat{f}(\omega)\left(\frac{e^{i\omega ct}+e^{-i\omega ct}}{2}\right) e^{-i\omega x}\,d\omega \\ &= \frac{1}{2}\left[\int_{-\infty}^{\infty} \widehat{f}(\omega)e^{-i\omega(x-ct)}\,d\omega + \int_{-\infty}^{\infty} \widehat{f}(\omega)e^{-i\omega(x+ct)}\,d\omega\right] \\ &= \frac{1}{2}\left[f(x-ct)+f(x+ct)\right] \qquad (9.1) \end{aligned}$$

so that

$$v(x,t) = \frac{1}{2}\left[f(x-ct)+f(x+ct)\right], \qquad (9.2)$$

for $-\infty < x < \infty,\ t > 0$.

• Problem II: In the second problem, we let

$$\widehat{w}(\omega,t) = \mathcal{F}\big[w(x,t)\big](\omega) = \frac{1}{2\pi}\int_{-\infty}^{\infty} w(x,t)\,e^{i\omega x}\,dx$$

and

$$\widehat{g}(\omega) = \mathcal{F}\big[g(x)\big](\omega) = \frac{1}{2\pi}\int_{-\infty}^{\infty} g(x)\,e^{i\omega x}\,dx.$$

Taking the Fourier transform of both sides of the wave equation, we have the following initial value problem:

$$\begin{aligned} &\frac{d^2\widehat{w}(\omega,t)}{dt^2} + c^2\omega^2\widehat{w}(\omega,t) = 0, \quad t>0, \\ &\widehat{w}(\omega,0) = 0, \\ &\frac{d\widehat{w}}{dt}(\omega,0) = \widehat{g}(\omega), \end{aligned}$$

with solution

$$\widehat{w}(\omega,t) = \frac{1}{\omega c}\,\widehat{g}(\omega)\sin\omega ct.$$

Taking the inverse transform, we have

$$\begin{aligned}
w(x,t) &= \frac{1}{c}\int_{-\infty}^{\infty} \frac{\widehat{g}(\omega)}{\omega} \sin \omega ct\, e^{-i\omega x}\, d\omega \\
&= \frac{1}{2c}\int_{-\infty}^{\infty} \frac{\widehat{g}(\omega)}{i\omega} \left(e^{i\omega ct} - e^{-i\omega ct}\right) e^{-i\omega x}\, d\omega \\
&= \frac{1}{2c}\left[\int_{-\infty}^{\infty} \frac{\widehat{g}(\omega)}{i\omega} e^{-i\omega(x-ct)}\, d\omega - \int_{-\infty}^{\infty} \frac{\widehat{g}(\omega)}{i\omega} e^{-i\omega(x+ct)}\, d\omega\right] \\
&= -\frac{1}{2c}\int_{0}^{x-ct} g(s)\, ds + \frac{1}{2c}\int_{0}^{x+ct} g(s)\, ds \\
&= \frac{1}{2c}\int_{x-ct}^{x+ct} g(s)\, ds, \qquad (9.3)
\end{aligned}$$

so that

$$w(x,t) = \frac{1}{2c}\int_{x-ct}^{x+ct} g(s)\, ds \qquad (9.4)$$

for $-\infty < x < \infty,\ t > 0$. From the superposition principle, the general solution to the wave equation is

$$u(x,t) = v(x,t) + w(x,t) = \frac{1}{2}\left[f(x-ct) + f(x+ct)\right] + \frac{1}{2c}\int_{x-ct}^{x+ct} g(s)\, ds$$

for $-\infty < x < \infty,\ t > 0$.

Note: We used the fact that

$$\mathcal{F}\left[\int_0^x g(s)\, ds\right](\omega) = -\frac{1}{i\omega}\widehat{g}(\omega)$$

for $-\infty < \omega < \infty$.

Thus, we have recovered **d'Alembert's solution** to the one-dimensional wave equation. This derivation of d'Alembert's solution required that both of the initial conditions have a Fourier transform. However, as can be checked by differentiating, d'Alembert's solution is valid for any twice continuously differentiable functions f and g, even if their integrals are unbounded.

Example 9.1. We solve the wave problem

$$\frac{\partial^2 u}{\partial t^2} = 25\frac{\partial^2 u}{\partial x^2}, \qquad -\infty < x < \infty, \quad t > 0,$$

$$u(x,0) = \begin{cases} 1, & \text{for} \quad x < 0, \\ 0, & \text{for} \quad x > 0, \end{cases}$$

$$\frac{\partial u}{\partial t}(x,0) = 0,$$

to get

$$u(x,t) = \frac{1}{2}\left[u(x-5t,0) + u(x+5t,0)\right]$$

for $-\infty < x < \infty$ and $t > 0$. In Figure 9.1 we sketch the solution as a function of x for various values of the time t. The solution can be written as

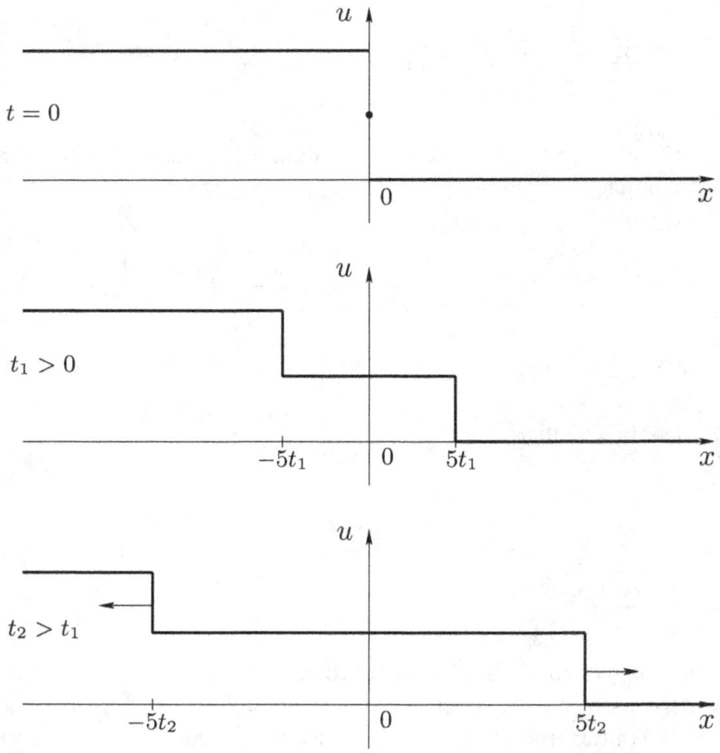

Figure 9.1 Wave equation solution for various times.

$$u(x,t) = \begin{cases} 1, & \text{if } \; x < -5t, \\ \dfrac{1}{2}, & \text{if } \; -5t < x < 5t, \\ 0, & \text{if } \; x > 5t \end{cases}$$

for $-\infty < x < \infty$ and $t > 0$. ∎

Example 9.2. If

$$f(x) = 1 \qquad \text{and} \qquad g(x) = \cos 2\pi x$$

for $-\infty < x < \infty$, neither f nor g has a Fourier transform, since neither of them is absolutely integrable. However, d'Alembert's solution,

$$u(x,t) = 1 + \frac{1}{4}\int_{x-2t}^{x+2t} \cos 2\pi s\, ds = 1 + \frac{1}{8\pi}\left[\sin 2\pi(x+2t) - \sin 2\pi(x-2t)\right]$$

for $-\infty < x < \infty$ and $t > 0$ is a solution to the wave equation

$$\begin{aligned} &\frac{\partial^2 u}{\partial t^2} = 4\frac{\partial^2 u}{\partial x^2}, \qquad -\infty < x < \infty, \quad t > 0, \\ &u(x,0) = 1, \\ &\frac{\partial u}{\partial t}(x,0) = \cos 2\pi x. \end{aligned}$$

∎

9.2 THE HEAT EQUATION

In this section we solve the heat conduction problem in unbounded regions. For example, in an infinite rod, a semi-infinite rod, a half-plane, and a semi-infinite strip.

9.2.1 Heat Flow in an Infinite Rod

Example 9.3. Find a bounded solution to the initial value problem

$$\begin{aligned} &\frac{\partial u}{\partial t} = k\frac{\partial^2 u}{\partial x^2}, \qquad -\infty < x < \infty, \quad t > 0, \\ &u(x,0) = e^{-|x|}, \\ &|u(x,t)| \text{ bounded as } x \to \pm\infty, \end{aligned}$$

where $k > 0$ is a constant. Here we require the additional assumption that the solution be bounded for all $x \in \mathbb{R}$.

Solution. We use Fourier transforms. A separation of variables solution is given later. Let $\widehat{u}(\omega, t)$ be the Fourier transform of $u(x, t)$; then

$$\widehat{u}(\omega, t) = \frac{1}{2\pi} \int_{-\infty}^{\infty} u(x, t) e^{i\omega x}\, dx,$$

and for $f(x) = e^{-|x|}$, from Example 8.3 we have

$$\widehat{f}(\omega) = \widehat{u}(\omega, 0) = \frac{1}{2\pi} \int_{-\infty}^{\infty} e^{-|x|}\, e^{i\omega x}\, dx = \frac{1}{\pi(1+\omega^2)}$$

for $-\infty < \omega < \infty$. Taking the Fourier transform of both sides of the heat equation, we have

$$\frac{d\widehat{u}(\omega, t)}{dt} = -k\omega^2 \widehat{u}(\omega, t), \quad t > 0,$$

$$\widehat{u}(\omega, 0) = \widehat{f}(\omega),$$

with solution

$$\widehat{u}(\omega, t) = \widehat{f}(\omega)\, e^{-\omega^2 kt} = \frac{1}{\pi(1+\omega^2)}\, e^{-\omega^2 kt}.$$

Therefore, taking the inverse Fourier transform, we have

$$u(x, t) = \int_{-\infty}^{\infty} \widehat{u}(\omega, t)\, e^{-i\omega x}\, d\omega = \frac{1}{\pi} \int_{-\infty}^{\infty} \frac{e^{-i\omega x - k\omega^2 t}}{1+\omega^2}\, d\omega$$

for $-\infty < x < \infty$ and $t > 0$. ∎

9.2.2 Fundamental Solution to the Heat Equation

Now we study the heat equation for an infinite rod, and for an arbitrary initial condition, obtain an explicit representation of the solution called the **fundamental solution to the heat equation**. The situation is similar to d'Alembert's solution to the wave equation. The fundamental solution to the heat equation allows us to solve the general heat equation with a single integration. The problem to be solved is

$$\frac{\partial u}{\partial t} = k \frac{\partial^2 u}{\partial x^2}, \quad -\infty < x < \infty, \quad t > 0,$$

$$u(x, 0) = f(x),$$

$$|u(x, t)| \text{ bounded as } x \to \pm\infty.$$

Let $\widehat{u}(\omega, t)$ and $\widehat{f}(\omega)$ be the Fourier transforms of $u(x, t)$ and $f(x)$, respectively; then

$$\widehat{u}(\omega, t) = \frac{1}{2\pi} \int_{-\infty}^{\infty} u(x, t) e^{i\omega x}\, dx \quad \text{and} \quad u(x, t) = \int_{-\infty}^{\infty} \widehat{u}(\omega, t) e^{-i\omega x}\, d\omega,$$

while

$$\widehat{f}(\omega) = \frac{1}{2\pi}\int_{-\infty}^{\infty} f(x)e^{i\omega x}\,dx \quad \text{and} \quad f(x) = \int_{-\infty}^{\infty} \widehat{f}(\omega)e^{-i\omega x}\,d\omega.$$

Taking the Fourier transform of the heat equation and the initial condition, we have

$$\frac{d\widehat{u}}{dt} + k\omega^2\widehat{u} = 0,$$
$$\widehat{u}(\omega, 0) = \widehat{f}(\omega),$$

with solution

$$\widehat{u}(\omega, t) = \widehat{f}(\omega)e^{-k\omega^2 t},$$

and taking the inverse Fourier transform, the solution is

$$u(x,t) = \int_{-\infty}^{\infty} \widehat{f}(\omega)e^{-k\omega^2 t}e^{-i\omega x}\,d\omega. \tag{9.5}$$

Replacing $\widehat{f}(\omega)$ by the integral and interchanging the order of integration, we have

$$\begin{aligned} u(x,t) &= \int_{-\infty}^{\infty}\left(\frac{1}{2\pi}\int_{-\infty}^{\infty} f(\xi)e^{i\omega\xi}\,d\xi\right)e^{-k\omega^2 t}e^{-i\omega x}\,d\omega \\ &= \frac{1}{2\pi}\int_{-\infty}^{\infty} f(\xi)\left(\int_{-\infty}^{\infty} e^{i\omega(\xi - x)}e^{-k\omega^2 t}\,d\omega\right)d\xi; \end{aligned}$$

that is,

$$\begin{aligned} u(x,t) &= \frac{1}{2\pi}\int_{-\infty}^{\infty} f(\xi)\left(\int_{-\infty}^{\infty}\left[\cos\omega(\xi - x) + i\sin\omega(\xi - x)\right]e^{-k\omega^2 t}\,d\omega\right)d\xi, \\ &= \frac{1}{\pi}\int_{-\infty}^{\infty} f(\xi)\left(\int_{0}^{\infty}\cos\omega(\xi - x)\,e^{-k\omega^2 t}\,d\omega\right)d\xi, \end{aligned} \tag{9.6}$$

since $\sin\omega(\xi - x)$ is an odd function of ω and $\cos\omega(\xi - x)$ is an even function of ω.

Now, since $kt > 0$, from Lemma 8.24 we have

$$\int_{0}^{\infty}\cos\omega(\xi - x)\,e^{-\omega^2 kt}\,d\omega = \sqrt{\frac{\pi}{4kt}}\,e^{-\frac{(\xi - x)^2}{4kt}},$$

and the explicit representation of the solution is given by

$$u(x,t) = \frac{1}{\sqrt{4k\pi t}}\int_{-\infty}^{\infty} f(\xi)\,e^{-\frac{(\xi - x)^2}{4kt}}\,d\xi \tag{9.7}$$

for $-\infty < x < \infty,\ t > 0$. We can also solve this problem using the method of separation of variables, and this will be done in Part II.

The advantage of this form of the solution is that it requires evaluation of only a single integral instead of three separate integrals, and it makes sense for many functions f for which the Fourier transform is undefined: for example, any bounded continuous function. Also, it does not require any smoothness in order to satisfy the initial value problem, and the condition

$$\int_{-\infty}^{\infty} |f(x)|\, dx < \infty$$

is not needed for this form of the solution to be valid. The function

$$G(x,t) = \frac{1}{\sqrt{4\pi kt}}\, e^{-\frac{x^2}{4kt}}$$

in the integrand is called the **fundamental solution** of the heat equation or the **Gauss-Weierstrass kernel** for the heat equation.

As the next theorem shows, we can write the solution to the heat equation as a convolution.

Theorem 9.1. The solution $u(x,t)$ of the linear heat equation

$$\frac{\partial u}{\partial t} = k\frac{\partial^2 u}{\partial x^2}, \qquad -\infty < x < \infty, \quad t > 0,$$

$$u(x,0) = f(x),$$

$$|u(x,t)| \text{ bounded as } x \to \pm\infty$$

can be written as

$$u(x,t) = f(x) * G(x,t) = \int_{-\infty}^{\infty} f(\xi)G(\xi - x, t)\, d\xi,$$

where $G(x,t)$ is the Gauss-Weierstrass kernel.

9.2.3 Error Function

Another helpful tool in solving the heat equation on the real line $\mathbb{R}$ is the **error function**

$$\text{erf}(x) = \frac{2}{\sqrt{\pi}} \int_0^x e^{-t^2}\, dt,$$

which is twice the area under a normal distribution with mean 0 and variance $\frac{1}{2}$.

Lemma 9.2. The error function is a monotone increasing function which satisfies

$$\lim_{x\to-\infty} \text{erf}(x) = -1 \qquad \text{and} \qquad \lim_{x\to\infty} \text{erf}(x) = 1.$$

Proof. The function e^{-t^2} is continuous on $\mathbb{R}$, and from the fundamental theorem of calculus, the error function is differentiable at each $x \in \mathbb{R}$. In fact,

$$\frac{d}{dx}\Big(\text{erf}(x)\Big) = \frac{2}{\sqrt{\pi}}\, e^{-x^2} > 0$$

for all $x \in \mathbb{R}$, hence the error function is monotone increasing on $\mathbb{R}$. Also, from Lemma 8.23 with $b = 1$, we have

$$\lim_{x\to\infty} \text{erf}(x) = \frac{2}{\sqrt{\pi}} \int_0^\infty e^{-t^2}\, dt = \frac{2}{\sqrt{\pi}} \frac{\sqrt{\pi}}{2} = 1.$$

Similarly, $\lim_{x\to-\infty} \text{erf}(x) = -1$. □

We illustrate the use of the error function with an example:

Example 9.4. Use convolutions, the error function, and operational properties of the Fourier transform to solve the boundary value–initial value problem

$$\frac{\partial u}{\partial t} = \frac{1}{100}\frac{\partial^2 u}{\partial x^2}, \qquad -\infty < x < \infty, \quad t > 0,$$

$$u(x,0) = \begin{cases} 100, & \text{if } -2 < x < 0, \\ 50, & \text{if } \quad 0 < x < 1, \\ 0, & \text{otherwise.} \end{cases}$$

Solution. Transforming the heat equation and the initial conditions, we get the solution to the transformed problem

$$\widehat{u}(\omega, t) = \widehat{f}(\omega)\, e^{-\omega^2 t/100}.$$

Since this is the product of two Fourier transforms, we know that it is $1/2\pi$ times the Fourier transform of a convolution, and using the convolution theorem, the solution can be written as

$$u(x,t) = \frac{1}{2\pi}\mathcal{F}^{-1}\left[e^{-\omega^2 t/100}\right] * f(x).$$

Computing the inverse Fourier transform of $e^{-\omega^2 t/100}$

$$\mathcal{F}^{-1}\left[e^{-\omega^2 t/100}\right](x) = \int_{-\infty}^{\infty} e^{-\omega^2 t/100} e^{-i\omega x}\, d\omega,$$

we let

$$\frac{\sigma^2}{4} = \frac{\omega^2 t}{100},$$

so that

$$\omega = \frac{\sigma\sqrt{25}}{\sqrt{t}} \qquad \text{and} \qquad d\omega = \frac{\sqrt{25}}{\sqrt{t}}\, d\sigma$$

and

$$\begin{aligned}\mathcal{F}^{-1}\left[e^{-\omega^2 t/100}\right](x) &= \frac{\sqrt{25}}{\sqrt{t}}\sqrt{4\pi}\int_{-\infty}^{\infty}\frac{1}{\sqrt{4\pi}}\,e^{-\sigma^2/4}\,e^{-i\sigma\frac{25x}{\sqrt{t}}}\,d\sigma \\ &= \frac{\sqrt{25}}{\sqrt{t}}\,\sqrt{4\pi}\,e^{-\frac{25x^2}{t}} \\ &= 10\sqrt{\frac{\pi}{t}}\,e^{-\frac{25x^2}{t}}.\end{aligned}$$

Therefore, the solution is

$$u(x,t) = \frac{1}{2\pi}\left[10\sqrt{\frac{\pi}{t}}\,e^{-\frac{25x^2}{t}}\right] * f(x) = \frac{5}{\sqrt{\pi t}}\int_{-\infty}^{\infty} f(s)\,e^{-\frac{25(x-s)^2}{t}}\,ds;$$

that is,

$$u(x,t) = \frac{5}{\sqrt{\pi t}}\left[\int_{-2}^{0} 100\,e^{-\frac{25(x-s)^2}{t}}\,ds + \int_{0}^{1} 50\,e^{-\frac{25(x-s)^2}{t}}\,ds\right].$$

We can write the solution

$$u(x,t) = \frac{5}{\sqrt{\pi t}}\left[\int_{-2}^{0} 100\,e^{-\frac{25(x-s)^2}{t}}\,ds + \int_{0}^{1} 50\,e^{-\frac{25(x-s)^2}{t}}\,ds\right]$$

in terms of the error function

$$\operatorname{erf}(w) = \frac{2}{\sqrt{\pi}}\int_0^w e^{-z^2}\,dz,$$

by letting $z = 5(x-s)/\sqrt{t}$, so that $dz = -\left(5/\sqrt{t}\right)ds$, and

$$u(x,t) = \frac{100}{\sqrt{\pi}}\int_{5x/\sqrt{t}}^{5(x+2)/\sqrt{t}} e^{-z^2}\,dz + \frac{50}{\sqrt{\pi}}\int_{5(x-1)/\sqrt{t}}^{5x/\sqrt{t}} e^{-z^2}\,dz;$$

that is,

$$u(x,t) = 50\left[\operatorname{erf}\left(\frac{5(x+2)}{\sqrt{t}}\right) - \operatorname{erf}\left(\frac{5x}{\sqrt{t}}\right)\right] + 25\left[\operatorname{erf}\left(\frac{5x}{\sqrt{t}}\right) - \operatorname{erf}\left(\frac{5(x-1)}{\sqrt{t}}\right)\right].$$

∎

9.2.4 Heat Flow in a Semi-infinite Rod: Dirichlet Condition

Here we study the heat equation on a semi-infinite domain, $0 < x < \infty$, with a homogeneous Dirichlet boundary condition at $x = 0$:

$$\frac{\partial u}{\partial t} = k\frac{\partial^2 u}{\partial x^2}, \quad 0 < x < \infty, \quad t > 0,$$

$$u(0,t) = 0,$$

$$u(x,0) = f(x),$$

$$|u(x,t)| \text{ bounded as } x \to \infty.$$

We give two solutions to this problem: one using Fourier sine transforms and the other using separation of variables. In this case, the solution using Fourier transforms is more difficult and uses a technique called the **method of images**.

• Using Fourier transforms, the homogeneous Dirichlet boundary condition suggests a Fourier sine transform, so we define

$$\widehat{u}(\omega,t) = S\left[u(x,t)\right](\omega) = \frac{2}{\pi}\int_0^\infty u(x,t)\sin\omega x\,dx \tag{9.8}$$

and

$$\widehat{f}(\omega) = S\left[f(x)\right](\omega) = \frac{2}{\pi}\int_0^\infty f(x)\sin\omega x\,dx; \tag{9.9}$$

then

$$S\left[\frac{\partial u}{\partial t}\right](\omega) = \frac{d\widehat{u}}{dt}(\omega,t)$$

and

$$S\left[\frac{\partial^2 u}{\partial x^2}\right](\omega) = \frac{2\omega}{\pi}u(0,t) - \omega^2\,\widehat{u}(\omega,t),$$

and from the boundary condition, $u(0,t) = 0$, so that

$$S\left[\frac{\partial^2 u}{\partial x^2}\right](\omega) = -\omega^2\,\widehat{u}(\omega,t).$$

After taking the Fourier sine transform of both sides of the partial differential equation,

$$\frac{\partial u}{\partial t} = k\frac{\partial^2 u}{\partial x^2},$$

the transform $\widehat{u}(\omega,t)$ satisfies the initial value problem

$$\frac{d\widehat{u}}{dt}(\omega,t) + k\omega^2\,\widehat{u}(\omega,t) = 0,$$

$$\widehat{u}(\omega,0) = \widehat{f}(\omega),$$

with solution

$$\widehat{u}(\omega,t)=\widehat{u}(\omega,0)\,e^{-\omega^2kt}=\widehat{f}(\omega)\,e^{-\omega^2kt}$$

for $0<\omega<\infty$ and $t>0$. Therefore,

$$u(x,t)=\int_0^\infty \widehat{f}(\omega)\,e^{-\omega^2kt}\sin\omega x\,d\omega \tag{9.10}$$

for $0<x<\infty$ and $t>0$.

To get an explicit representation of the solution in (9.10) in the x-world, we need to evaluate the inverse sine transform, and this is where the *method of images* is used. Note that the Fourier sine transform $\widehat{f}(\omega)$ in equation (9.9) actually makes sense for all ω, that is, $-\infty<\omega<\infty$, and, in fact, each of the functions $\widehat{f}(\omega)$ and $\sin\omega x$ in the integrand is an odd function of ω, so we can write

$$\int_0^\infty \widehat{f}(\omega)\,e^{-\omega^2kt}\sin\omega x\,d\omega=\frac{1}{2}\int_{-\infty}^\infty \widehat{f}(\omega)\,e^{-\omega^2kt}\sin\omega x\,d\omega. \tag{9.11}$$

Since $\cos\omega x$ is an even function of ω, then

$$\int_{-\infty}^\infty \widehat{f}(\omega)\,e^{-\omega^2kt}\cos\omega x\,d\omega=0,$$

and we can write the solution $u(x,t)$ as

$$\begin{aligned}u(x,t)&=-\frac{1}{2i}\int_{-\infty}^\infty \widehat{f}(\omega)\,e^{-\omega^2kt}(\cos\omega x-i\sin\omega x)\,d\omega\\&=\int_{-\infty}^\infty\left(-\frac{\widehat{f}(\omega)}{2i}\right)e^{-\omega^2kt}e^{-i\omega x}\,d\omega;\end{aligned}$$

that is,

$$u(x,t)=\mathcal{F}^{-1}\left[-\frac{\widehat{f}(\omega)}{2i}\,e^{-\omega^2kt}\right](x). \tag{9.12}$$

Let f_{odd} be the odd extension of $f(x)$ to $(-\infty,\infty)$; then

$$\begin{aligned}\frac{\widehat{f}(\omega)}{2}&=\frac{1}{2}\frac{2}{\pi}\int_0^\infty f(x)\sin\omega x\,dx\\&=\frac{1}{2\pi}\int_{-\infty}^\infty f_{\text{odd}}(x)\sin\omega x\,dx\\&=-i\frac{1}{2\pi}\int_{-\infty}^\infty f_{\text{odd}}(x)(\cos\omega x+i\sin\omega x)\,dx\\&=-i\frac{1}{2\pi}\int_{-\infty}^\infty f_{\text{odd}}(x)e^{i\omega x}\,dx\\&=-i\mathcal{F}\left[f_{\text{odd}}(x)\right](\omega),\end{aligned}$$

so that

$$-\frac{\widehat{f}(\omega)}{2i} = \mathcal{F}\left[f_{\text{odd}}(x)\right](\omega). \tag{9.13}$$

From (9.12) and (9.13) it follows that $u(x,t)$ is the solution to the heat flow problem where the *image of the initial data on* $(0,\infty)$ is the extension of the initial data to all of $(-\infty,\infty)$ as an odd function; thus,

$$\begin{aligned}
&\frac{\partial u}{\partial t} = k\frac{\partial^2 u}{\partial x^2}, \quad -\infty < x < \infty, \quad t > 0, \\
&u(0,t) = 0, \\
&u(x,0) = f_{\text{odd}}(x).
\end{aligned}$$

Therefore,

$$u(x,t) = f_{\text{odd}}(x) * G(x,t),$$

where $G(x,t)$ is the heat kernel or Gaussian kernel,

$$G(x,t) = \frac{1}{\sqrt{4\pi kt}}\, e^{-x^2/4kt}.$$

The solution is then

$$u(x,t) = f_{\text{odd}} * G(x,t) = \frac{1}{\sqrt{4\pi kt}} \int_{-\infty}^{\infty} f_{\text{odd}}(s) e^{-(x-s)^2/4kt}\, ds$$

for $-\infty < x < \infty,\ t > 0$. We write this as

$$u(x,t) = \frac{1}{\sqrt{4\pi kt}} \left[\int_{-\infty}^{0} f_{\text{odd}}(s) e^{-(x-s)^2/4kt}\, ds + \int_{0}^{\infty} f_{\text{odd}}(s) e^{-(x-s)^2/4kt}\, ds\right],$$

and replacing s by $-s$ in the first integral, we have

$$u(x,t) = \frac{1}{\sqrt{4\pi kt}} \left[-\int_{\infty}^{0} f_{\text{odd}}(-s) e^{-(x+s)^2/4kt}\, ds + \int_{0}^{\infty} f_{\text{odd}}(s) e^{-(x-s)^2/4kt}\, ds\right].$$

Using the fact that f_{odd} is an odd function, we have

$$u(x,t) = \frac{1}{\sqrt{4\pi kt}} \left[-\int_{0}^{\infty} f_{\text{odd}}(s) e^{-(x+s)^2/4kt}\, ds + \int_{0}^{\infty} f_{\text{odd}}(s) e^{-(x-s)^2/4kt}\, ds\right];$$

that is,

$$u(x,t) = \frac{1}{\sqrt{4\pi kt}} \left[\int_{0}^{\infty} f(s) e^{-(x-s)^2/4kt}\, ds - \int_{0}^{\infty} f(s) e^{-(x+s)^2/4kt}\, ds\right].$$

The solution to the heat equation on the semi-infinite interval $(0, \infty)$ is therefore

$$u(x,t) = \frac{1}{\sqrt{4\pi kt}} \int_0^\infty f(s) \left[e^{-(x-s)^2/4kt} - e^{-(x+s)^2/4kt} \right] ds \tag{9.14}$$

for $0 < x < \infty,\ t > 0$.

In this example we see that, indeed, solution (9.10) of the ODE in the ω-world is very simple. However, we pay a price in finding the inverse sine transform before arriving at the final solution (9.14) in the x-world.

• Using separation of variables, we assume that $u(x,t) = X(x) \cdot T(t)$, obtaining two ordinary differential equations

$$\begin{aligned} &X''(x) + \lambda X(x) = 0, \quad 0 < x < \infty, \qquad T'(t) + \lambda k T(t) = 0, \quad t > 0, \\ &X(0) = 0, \\ &|X(x)| \text{ bounded}, \ 0 < x < \infty, \end{aligned}$$

and in order to satisfy the boundedness condition we need $\lambda \geqslant 0$, say $\lambda = \omega^2$, where $\omega \geqslant 0$. The general solution to the spatial equation is

$$X_\omega(x) = A(\omega) \cos \omega x + B(\omega) \sin \omega x,$$

and applying the boundary condition $X(0) = 0$, we have $A(\omega) = 0$, so the solution is

$$X_\omega(x) = B(\omega) \sin \omega x$$

for $0 < x < \infty$. The corresponding solution to the time equation is

$$T_\omega(t) = e^{-\omega^2 kt}.$$

For each $\omega \geqslant 0$, the product solution

$$u_\omega(x,t) = X_\omega(x) \cdot T_\omega(t) = B(\omega) \sin \omega x \, e^{-\omega^2 kt}$$

satisfies the heat equation, the boundary condition, and the boundedness condition. Thus, we have a continuum of eigenvalues, and we apply the superposition principle, where the solution is an integral over the eigenvalues $\omega \geqslant 0$, rather than a sum. Therefore, we write

$$u(x,t) = \int_0^\infty B(\omega) \sin \omega x \, e^{-\omega^2 kt} \, d\omega \tag{9.15}$$

and use the initial condition to determine the coefficient $B(\omega)$. Setting $t = 0$ in (9.15), we have

$$f(x) = u(x,0) = \int_0^\infty B(\omega) \sin \omega x \, d\omega,$$

for $0 < x < \infty$, which is the Fourier sine integral representation for $f(x)$, and therefore if f satisfies the hypotheses in Dirichlet's theorem, then

$$B(\omega) = \frac{2}{\pi} \int_0^\infty f(x) \sin \omega x \, dx \tag{9.16}$$

for $0 < \omega < \infty$. Substituting the formula for $B(\omega)$ into the solution (9.15), we have

$$u(x,t) = \int_0^\infty \left(\frac{2}{\pi} \int_0^\infty f(\xi) \sin \omega \xi \, d\xi \right) \sin \omega x \, e^{-\omega^2 kt} \, d\omega;$$

that is,

$$u(x,t) = \frac{1}{\pi} \int_0^\infty \int_0^\infty f(\xi) \left[\cos \omega(\xi - x) - \cos \omega(\xi + x) \right] e^{-\omega^2 kt} \, d\xi \, d\omega.$$

Interchanging the order of integration, we have

$$u(x,t) = \frac{1}{\pi} \int_0^\infty f(\xi) \left(\int_0^\infty \left[\cos \omega(\xi - x) - \cos \omega(\xi + x) \right] e^{-\omega^2 kt} \, d\omega \right) d\xi;$$

that is,

$$\begin{aligned} u(x,t) &= \frac{1}{\pi} \int_0^\infty f(\xi) \left(\int_0^\infty \cos \omega(\xi - x) \, e^{-\omega^2 kt} \, d\omega \right) d\xi \\ &\quad - \frac{1}{\pi} \int_0^\infty f(\xi) \left(\int_0^\infty \cos \omega(\xi + x) \, e^{-\omega^2 kt} \, d\omega \right) d\xi \end{aligned} \tag{9.17}$$

for $0 < x < \infty, \ t > 0$, and we can evaluate the inner integrals exactly.

Since $kt > 0$, then from Lemma 8.24 we have

$$\int_0^\infty \cos \omega(\xi - x) \, e^{-\omega^2 kt} \, d\omega = \frac{1}{2} \sqrt{\frac{\pi}{kt}} \, e^{-\frac{(\xi - x)^2}{4kt}}$$

and

$$\int_0^\infty \cos \omega(\xi + x) \, e^{-\omega^2 kt} \, d\omega = \frac{1}{2} \sqrt{\frac{\pi}{kt}} \, e^{-\frac{(\xi + x)^2}{4kt}},$$

and solution (9.17) is given by

$$\begin{aligned} u(x,t) &= \frac{1}{\sqrt{4k\pi t}} \int_0^\infty f(\xi) \, e^{-\frac{(\xi - x)^2}{4kt}} \, d\xi \\ &\quad - \frac{1}{\sqrt{4k\pi t}} \int_0^\infty f(\xi) \, e^{-\frac{(\xi + x)^2}{4kt}} \, d\xi \end{aligned} \tag{9.18}$$

for $0 < x < \infty, \ t > 0$. This is the same as the solution we obtained using the Fourier sine transform, but it involved much less work.

We can simplify the solution somewhat if in the first integral we make the substitution

$$\xi = x + 2\sigma\sqrt{kt},$$

and in the second integral we make the substitution

$$\xi = -x + 2\sigma\sqrt{kt}.$$

Then solution (9.18) can be written

$$u(x,t) = \frac{1}{\sqrt{\pi}} \int_{-x/\sqrt{4kt}}^{\infty} f\left(x + 2\sigma\sqrt{kt}\right) e^{-\sigma^2}\, d\sigma - \frac{1}{\sqrt{\pi}} \int_{x/\sqrt{4kt}}^{\infty} f\left(-x + 2\sigma\sqrt{kt}\right) e^{-\sigma^2}\, d\sigma \tag{9.19}$$

for $0 < x < \infty,\ t > 0$.

Note: If the initial temperature distribution is constant, say

$$f(x) = u_0,$$

the solution becomes

$$u(x,t) = \frac{u_0}{\sqrt{\pi}} \left(\int_{-x/\sqrt{4kt}}^{\infty} e^{-\sigma^2}\, d\sigma - \int_{x/\sqrt{4kt}}^{\infty} e^{-\sigma^2}\, d\sigma \right).$$

In terms of the **error function**

$$\text{erf}\,(x) = \frac{2}{\sqrt{\pi}} \int_0^x e^{-\sigma^2}\, d\sigma, \quad 0 < x < \infty,$$

the solution to the heat conduction problem for the semi-infinite rod above is given by

$$u(x,t) = u_0\, \text{erf}\left(\frac{x}{\sqrt{4kt}}\right)$$

for $0 < x < \infty,\ t > 0$. Note that

$$\lim_{x\to\infty} \text{erf}\left(\frac{x}{\sqrt{4kt}}\right) = 1,$$

so that

$$\lim_{x\to\infty} u(x,t) = u_0$$

for all $t > 0$.

9.2.5 Heat Flow in a Semi-infinite Rod: Neumann Condition

In this section we solve the problem of heat conduction on a semi-infinite domain, $0 < x < \infty$, with a homogeneous Neumann boundary condition at $x = 0$:

$$\frac{\partial u}{\partial t} = k\frac{\partial^2 u}{\partial x^2}, \quad 0 < x < \infty, \quad t > 0,$$

$$\frac{\partial u}{\partial x}(0,t) = 0,$$

$$u(x,0) = f(x),$$

$$|u(x,t)| \text{ bounded as } x \to \infty.$$

Here we assume that the initial temperature distribution is given by $u(x,0) = f(x)$, and the boundary $x = 0$ is insulated so that $\partial u(0,t)/\partial x = 0$.

• Using Fourier transforms, the homogeneous Neumann condition suggests a Fourier cosine transform, so we define

$$\widehat{u}(\omega,t) = C\left[u(x,t)\right](\omega) = \frac{2}{\pi}\int_0^\infty u(x,t)\cos\omega x\, dx \tag{9.20}$$

and

$$\widehat{f}(\omega) = C\left[f(x)\right](\omega) = \frac{2}{\pi}\int_0^\infty f(x)\cos\omega x\, dx. \tag{9.21}$$

Now, using the operational properties of the Fourier cosine transform, we have

$$C\left[\frac{\partial u}{\partial t}\right](\omega) = \frac{d\widehat{u}}{dt}(\omega,t)$$

and

$$C\left[\frac{\partial^2 u}{\partial x^2}\right](\omega) = -\frac{2}{\pi}\frac{\partial u}{\partial x}(0,t) - \omega^2\widehat{u}(\omega,t),$$

and from the boundary condition, $\partial u(0,t)/\partial x = 0$, we have

$$C\left[\frac{\partial^2 u}{\partial x^2}\right](\omega) = -\omega^2\,\widehat{u}(\omega,t).$$

Taking the Fourier cosine transform of both sides of the partial differential equation, the Fourier cosine transform $\widehat{u}(\omega,t)$ satisfies the initial value problem

$$\frac{d\widehat{u}}{dt}(\omega,t) + k\omega^2\,\widehat{u}(\omega,t) = 0,$$

$$\widehat{u}(\omega,0) = \widehat{f}(\omega),$$

with solution

$$\widehat{u}(\omega,t) = \widehat{u}(\omega,0)\, e^{-\omega^2 kt} = \widehat{f}(\omega)\, e^{-\omega^2 kt}$$

for $0 < \omega < \infty$ and $t > 0$. Therefore,

$$u(x,t) = \int_0^\infty \widehat{f}(\omega)\, e^{-\omega^2 kt} \cos \omega x \, d\omega \tag{9.22}$$

for $0 < x < \infty$ and $t > 0$. To get an explicit representation of the solution in (9.22) in the x-world, we need to evaluate the inverse cosine transform, and again we use the method of images. Again, we note that the Fourier cosine transform $\widehat{f}(\omega)$ in equation (9.21) actually makes sense for all ω, with $-\infty < \omega < \infty$, and that each of the functions $\widehat{f}(\omega)$, $e^{-\omega^2 kt}$, and $\cos \omega x$ in the integrand is an even function of ω, so we can write

$$\int_0^\infty \widehat{f}(\omega)\, e^{-\omega^2 kt} \cos \omega x \, d\omega = \frac{1}{2} \int_{-\infty}^\infty \widehat{f}(\omega)\, e^{-\omega^2 kt} \cos \omega x \, d\omega.$$

Since $\sin \omega x$ is an odd function of ω, then

$$\int_{-\infty}^\infty \widehat{f}(\omega)\, e^{-\omega^2 kt} \sin \omega x \, d\omega = 0,$$

and we can write the solution $u(x,t)$ as

$$\begin{aligned} u(x,t) &= \frac{1}{2} \int_{-\infty}^\infty \widehat{f}(\omega)\, e^{-\omega^2 kt} (\cos \omega x - i \sin \omega x) \, d\omega \\ &= \int_{-\infty}^\infty \frac{\widehat{f}(\omega)}{2}\, e^{-\omega^2 kt} e^{-i\omega x} \, d\omega; \end{aligned}$$

that is,

$$u(x,t) = \mathcal{F}^{-1} \left[\frac{\widehat{f}(\omega)}{2}\, e^{-\omega^2 kt} \right] (x). \tag{9.23}$$

Let f_{even} be the even extension of $f(x)$ to $(-\infty, \infty)$; then

$$\begin{aligned} \frac{\widehat{f}(\omega)}{2} &= \frac{1}{2} \frac{2}{\pi} \int_0^\infty f(x) \cos \omega x \, dx \\ &= \frac{1}{2\pi} \int_{-\infty}^\infty f_{\text{even}}(x) \cos \omega x \, dx \\ &= \frac{1}{2\pi} \int_{-\infty}^\infty f_{\text{even}}(x) (\cos \omega x + i \sin \omega x) \, dx \\ &= \frac{1}{2\pi} \int_{-\infty}^\infty f_{\text{even}}(x)\, e^{i\omega x} \, dx \\ &= \mathcal{F} \left[f_{\text{even}}(x) \right] (\omega), \end{aligned}$$

so that

$$\frac{\widehat{f}(\omega)}{2} = \mathcal{F}\left[f_{\text{even}}(x)\right](\omega). \tag{9.24}$$

From (9.23) and (9.24) it follows that $u(x,t)$ is the solution to the heat flow problem on the infinite interval $-\infty < x < \infty$; that is,

$$\frac{\partial u}{\partial t} = k\frac{\partial^2 u}{\partial x^2}, \quad -\infty < x < \infty, \quad t > 0,$$

$$\frac{\partial u}{\partial x}(0,t) = 0,$$

$$u(x,0) = f_{\text{even}}(x).$$

Therefore,

$$u(x,t) = f_{\text{even}}(x) * G(x,t),$$

where $G(x,t)$ is the heat kernel or Gaussian kernel,

$$G(x,t) = \frac{1}{\sqrt{4\pi kt}}\, e^{-x^2/4kt}.$$

The solution is then

$$\begin{aligned} u(x,t) &= f_{\text{even}} * G(x,t) \\ &= \frac{1}{\sqrt{4\pi kt}} \int_{-\infty}^{\infty} f_{\text{even}}(s)\, e^{-(x-s)^2/4kt}\, ds \\ &= \frac{1}{\sqrt{4\pi kt}} \int_{0}^{\infty} f(s) \left(e^{-(x+s)^2/4kt} + e^{-(x-s)^2/4kt}\right) ds, \end{aligned}$$

so that

$$u(x,t) = \frac{1}{\sqrt{4\pi kt}} \int_{0}^{\infty} f(s) \left(e^{-(x-s)^2/4kt} + e^{-(x+s)^2/4kt}\right) ds$$

for $0 < x < \infty$, $t > 0$.

9.3 LAPLACE'S EQUATION

9.3.1 Laplace's Equation in a Half-Plane

Solving the problem of steady-state heat conduction in a half-plane involves finding a function $u(x,y)$ defined in the upper half-plane, $-\infty < x < \infty,\ y > 0$, that is bounded and harmonic and satisfies the boundary and boundedness conditions

$$\nabla^2 u = \frac{\partial^2 u}{\partial x^2} + \frac{\partial^2 u}{\partial y^2} = 0, \quad -\infty < x < \infty, \quad y > 0,$$

$$u(x,0) = f(x),$$

$$|u(x,y)| \text{ bounded as } x \to \pm\infty.$$

• We use the Fourier transform with respect to the x-variable to solve the problem. Let

$$\widehat{u}(\omega, y) = \frac{1}{2\pi}\int_{-\infty}^{\infty} u(x, y)e^{i\omega x}\,dx$$

and

$$\widehat{f}(\omega) = \frac{1}{2\pi}\int_{-\infty}^{\infty} f(x)e^{i\omega x}\,dx,$$

then taking the Fourier transform of Laplace's equation and the boundary condition at $y = 0$, we have

$$\frac{d^2\widehat{u}}{dy^2}(\omega, y) - \omega^2\widehat{u}(\omega, y) = 0, \quad y > 0,$$

$$\widehat{u}(\omega, 0) = \widehat{f}(\omega).$$

Since the solution must be bounded, for $\omega > 0$ the solution is taken to be $e^{-\omega y}$, while for $\omega < 0$ we take $e^{\omega y}$. Thus, the general solution can be written as

$$\widehat{u}(\omega, y) = C(\omega)e^{-|\omega|y}.$$

Setting $y = 0$, we have $C(\omega) = \widehat{f}(\omega)$, so

$$\widehat{u}(\omega, y) = \widehat{f}(\omega)e^{-|\omega|y} \tag{9.25}$$

for $0 < y < \infty$, for all ω, which is the solution in the ω-world.

To obtain the solution in the x-world, we apply the inverse Fourier transform to get

$$u(x, y) = \mathcal{F}^{-1}\left[\widehat{u}(\omega, y)\right] = \int_{-\infty}^{\infty} \widehat{f}(\omega)e^{-|\omega|y}e^{-i\omega x}\,d\omega,$$

so that

$$u(x, y) = \frac{1}{2\pi}\int_{-\infty}^{\infty}\left(\int_{-\infty}^{\infty} f(s)e^{i\omega s}\,ds\right)e^{-|\omega|y}e^{-i\omega x}\,d\omega \tag{9.26}$$

for $-\infty < x < \infty$, and $y > 0$.

We can get another representation for the solution by interchanging the order of integration:

$$u(x, y) = \frac{1}{2\pi}\int_{-\infty}^{\infty} f(s)\underbrace{\left(\int_{-\infty}^{\infty} e^{-|\omega|y}e^{-i\omega(x-s)}\,d\omega\right)}_{g(x-s,y)}ds,$$

and evaluating the inner integral directly. For $y > 0$, we let

$$g(x, y) = \int_{-\infty}^{\infty} e^{-|\omega|y}e^{-i\omega x}\,d\omega,$$

so that

$$\begin{aligned}
g(x,y) &= \int_{-\infty}^{0} e^{-|\omega|y}e^{-i\omega x}\,d\omega + \int_{0}^{\infty} e^{-|\omega|y}e^{-i\omega x}\,d\omega \\
&= -\int_{\infty}^{0} e^{-|\lambda|y}e^{i\lambda x}\,d\lambda + \int_{0}^{\infty} e^{-|\omega|y}e^{-i\omega x}\,d\omega \\
&= \int_{0}^{\infty} e^{-\lambda y}\left(e^{i\lambda x}+e^{-i\lambda x}\right)d\lambda \\
&= 2\int_{0}^{\infty} e^{-\lambda y}\cos\lambda x\,d\lambda,
\end{aligned}$$

and integrating by parts twice, we have

$$g(x,y) = \frac{2y}{x^2+y^2}$$

for $-\infty < x < \infty$, $y > 0$. Therefore, the solution to Laplace's equation in the half-plane $-\infty < x < \infty$, $y > 0$, is given by

$$u(x,y) = \frac{y}{\pi}\int_{-\infty}^{\infty} \frac{f(t)}{(x-t)^2+y^2}\,dt \tag{9.27}$$

which is the **Poisson integral formula for the half-plane**.

Note: Now that we have this explicit representation of the solution, we can show that (9.27) is valid for any piecewise smooth function $f(x)$, $-\infty < x < \infty$, for which

$$|f(x)| \leqslant M, \quad -\infty < x < \infty,$$

for some constant $M > 0$. It is not necessary to assume that $f(x)$ is absolutely integrable; that is, it is not necessary to assume that

$$\int_{-\infty}^{\infty} |f(x)|\,dx < \infty.$$

In fact, it can be shown that (9.27) is a bounded harmonic function on the half-plane and that

$$\lim_{y\to 0+} u(x,y) = \frac{1}{2}\left[f(x^+)+f(x^-)\right]$$

for $-\infty < x < \infty$ (see Pinsky [39], p. 333).

• Using separation of variables, we write $u(x,y) = X(x)\cdot Y(y)$ and obtain the two ordinary differential equations

$$\begin{aligned}
&X'' + \lambda^2 X = 0, && Y'' - \lambda^2 Y = 0,\\
&|X(x)| \quad \text{bounded}, && |Y(y)| \quad \text{bounded},\\
&\text{for} \quad -\infty < x < \infty, && \text{for} \quad 0 < y < \infty.
\end{aligned}$$

The general solutions to these two equations are

$$X(x) = c_1 \cos \lambda x + c_2 \sin \lambda x,$$

$$Y(y) = d_1 e^{\lambda y} + d_2 e^{-\lambda y},$$

and the first of these is bounded automatically. Since the solution to the second equation must also be bounded, we choose

$$Y(y) = \begin{cases} e^{-\lambda y}, & \text{for } \lambda > 0, \\ e^{\lambda y}, & \text{for } \lambda < 0; \end{cases}$$

that is,

$$Y(y) = e^{-|\lambda| y}$$

for $-\infty < \lambda < \infty$. Thus, for any $\lambda \neq 0$, the function

$$u(x, y, \lambda) = e^{-|\lambda| y} \left(A(\lambda) \cos \lambda x + B(\lambda) \sin \lambda x \right)$$

satisfies Laplace's equation and is bounded in the upper half-plane.

Using the superposition principle, we have

$$u(x, y) = \int_0^\infty e^{-|\lambda| y} \left(A(\lambda) \cos \lambda x + B(\lambda) \sin \lambda x \right) d\lambda. \tag{9.28}$$

Now we choose $A(\lambda)$ and $B(\lambda)$ to satisfy the boundary condition

$$f(x) = u(x, 0) = \int_0^\infty \left(A(\lambda) \cos \lambda x + B(\lambda) \sin \lambda x \right) d\lambda,$$

so that

$$A(\lambda) = \frac{1}{\pi} \int_{-\infty}^\infty f(x) \cos \lambda x \, dx, \tag{9.29}$$

$$B(\lambda) = \frac{1}{\pi} \int_{-\infty}^\infty f(x) \sin \lambda x \, dx, \tag{9.30}$$

for $0 < \lambda < \infty$.

Example 9.5. Solve the problem

$$\nabla^2 u = \frac{\partial^2 u}{\partial x^2} + \frac{\partial^2 u}{\partial y^2} = 0, \quad -\infty < x < \infty, \quad y > 0,$$

$$u(x, 0) = f(x),$$

$$|u(x, y)| \text{ bounded as } x \to \pm\infty$$

if the temperature along the bottom of the half-plane $y = 0$ is given by

$$f(x) = \begin{cases} T_0, & \text{for } |x| < a, \\ 0, & \text{for } |x| > a \end{cases}$$

(see Andrews, [3], p. 366).

Solution. From the separation of variables solution given in (9.28), (9.29), and (9.30), we have

$$A(\lambda) = \frac{T_0}{\pi} \int_{-a}^{a} \cos \lambda x \, dx = \frac{2T_0}{\pi} \frac{\sin \lambda a}{\lambda}$$

and

$$B(\lambda) = \frac{T_0}{\pi} \int_{-a}^{a} \sin \lambda x \, dx = 0$$

for $\lambda > 0$, and the solution is

$$u(x, y) = \frac{2T_0}{\pi} \int_0^\infty e^{-\lambda y} \frac{\sin \lambda a \cos \lambda x}{\lambda} \, d\lambda.$$

Alternatively, from the Poisson integral formula we have

$$\begin{aligned} u(x, y) &= \frac{yT_0}{\pi} \int_{-a}^{a} \frac{dt}{(x - t)^2 + y^2} \\ &= \frac{T_0}{\pi} \left[\tan^{-1} \left(\frac{x + a}{y} \right) - \tan^{-1} \left(\frac{x - a}{y} \right) \right]. \end{aligned}$$

Using the trigonometric identity

$$\tan(A - B) = \frac{\tan A - \tan B}{1 + \tan A \tan B},$$

we can rewrite the solution in the form

$$u(x, y) = \frac{T_0}{\pi} \tan^{-1} \left(\frac{2ay}{x^2 + y^2 - a^2} \right).$$

Curves in the upper half-plane for which the steady-state temperature is constant are called **isotherms**. For this particular problem, the isotherms are defined by the family of circular arcs

$$x^2 + y^2 - ky = a^2,$$

with centers on the y-axis and endpoints on the x-axis at $x = \pm a$ (Fig. 9.2). The use of the Poisson integral formula made it easy to identify these curves.

∎

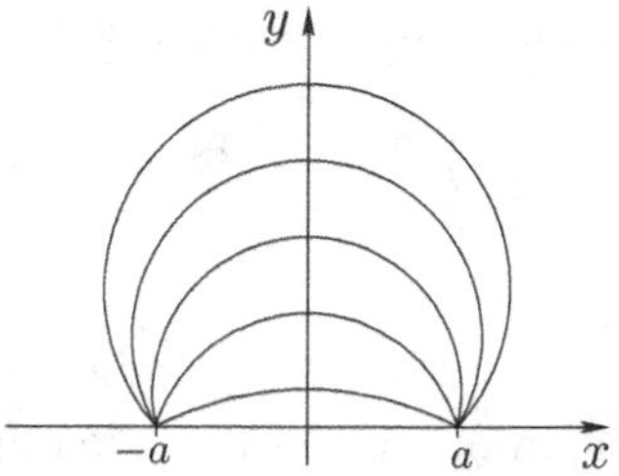

Figure 9.2 Isotherms.

9.3.2 Laplace's Equation in a Semi-infinite Strip

Solving the problem of steady-state heat conduction in a semi-infinite strip with Dirichlet boundary conditions involves finding the function $u(x, y)$ that is bounded and harmonic in the vertical semi-infinite strip $0 < x < a,\ y > 0$ and satisfies the following problem:

$$\nabla^2 u = \frac{\partial^2 u}{\partial x^2} + \frac{\partial^2 u}{\partial y^2} = 0, \quad 0 < x < a, \quad y > 0,$$

$$u(0, y) = g_1(y),$$

$$u(a, y) = g_2(y),$$

$$u(x, 0) = f(x),$$

$$|u(x, y)| \text{ bounded as } y \to \infty.$$

Using the superposition principle, we write $u(x, y) = v(x, y) + w(x, y)$, where v and w satisfy Laplace's equation and the boundary conditions shown below.

Problem I:

$$\frac{\partial^2 v}{\partial x^2} + \frac{\partial^2 v}{\partial y^2} = 0, \quad 0 < x < a,\ y > 0,$$

$$v(0, y) = 0,$$

$$v(a, y) = 0,$$

$$v(x, 0) = f(x),$$

$$|v(x, y)| \text{ bounded as } y \to \infty.$$

Problem II:

$$\frac{\partial^2 w}{\partial x^2} + \frac{\partial^2 w}{\partial y^2} = 0, \quad 0 < x < a,\ y > 0,$$

$$w(0, y) = g_1(y),$$

$$w(a, y) = g_2(y),$$

$$w(x, 0) = 0,$$

$$|w(x, y)| \text{ bounded as } y \to \infty.$$

• Clearly, the solution to Problem I is

$$v(x,y) = \sum_{n=1}^{\infty} a_n e^{-\frac{n\pi}{a}y} \sin\frac{n\pi}{a}x, \quad 0 < x < a, \tag{9.31}$$

where

$$a_n = \frac{2}{a}\int_0^a f(x)\sin\frac{n\pi}{a}x\,dx \tag{9.32}$$

for $n \geqslant 1$.

• We use separation of variables to solve Problem II. Assuming a solution of the form $w(x,y) = X(x)\cdot Y(y)$, we obtain two ordinary differential equations,

$$X'' - \lambda X = 0, \quad 0 < x < a, \qquad Y'' + \lambda Y = 0, \quad y > 0,$$

$$Y(0) = 0,$$

$$|Y(y)| \text{ bounded as } y \to \infty.$$

We solve the singular Sturm-Liouville problem for Y first; again we consider three cases.

Case 1: If $\lambda = 0$, the equation becomes $Y'' = 0$, with general solution

$$Y(y) = Ay + B,$$

and applying the boundary condition $Y(0) = B = 0$, and then the boundedness condition implies that $A = 0$, so $\lambda = 0$ is not an eigenvalue.

Case 2: If $\lambda < 0$, say $\lambda = -\mu^2$, where $\mu \neq 0$, the equation becomes

$$Y'' - \mu^2 Y = 0,$$

with general solution

$$Y(y) = A\cosh\mu y + B\sinh\mu y.$$

Applying the boundary condition $Y(0) = A = 0$, and then the boundedness condition implies that $B = 0$ also, and there are no negative eigenvalues.

Case 3: If $\lambda > 0$, say $\lambda = \mu^2$, where $\mu \neq 0$, the equation becomes

$$Y'' + \mu^2 Y = 0,$$

with general solution

$$Y(y) = A\cos\mu y + B\sin\mu y.$$

Applying the boundary condition $Y(0) = A = 0$, and then the solution is

$$Y(y) = B\sin\mu y,$$

which satisfies the boundedness condition also.

Note that this problem has a continuous spectrum; the eigenvalues are

$$\lambda = \mu^2,$$

where $\mu > 0$, and the corresponding eigenfunctions are

$$Y_\mu(y) = \sin \mu y, \quad y > 0$$

for $\mu > 0$. The general solution to the X-equation is given by

$$X_\mu(x) = A(\mu) \sinh \mu x + B(\mu) \sinh \mu(a - x),$$

and from the superposition principle,

$$w(x, y) = \int_0^\infty [A(\mu) \sinh \mu x + B(\mu) \sinh \mu(a - x)] \sin \mu y \, d\mu$$

satisfies Laplace's equation, the boundary condition on the face $y = 0$, and the boundedness condition.

We use the boundary conditions on the faces $x = 0$ and $x = a$ to determine the coefficients $A(\mu)$ and $B(\mu)$. We want

$$g_1(y) = w(0, y) = \int_0^\infty B(\mu) \sinh \mu a \sin \mu y \, d\mu$$

and

$$g_2(y) = w(a, y) = \int_0^\infty A(\mu) \sinh \mu a \sin \mu y \, d\mu,$$

so $A(\mu) \sinh \mu a$ and $B(\mu) \sinh \mu a$ are the Fourier sine integral coefficients of g_2 and g_1, respectively. Therefore, the solution to Problem II is

$$w(x, y) = \int_0^\infty [A(\mu) \sinh \mu x + B(\mu) \sinh \mu(a - x)] \sin \mu y \, d\mu, \tag{9.33}$$

where

$$A(\mu) = \frac{2}{\pi \sinh \mu a} \int_0^\infty g_2(y) \sin \mu y \, dy \tag{9.34}$$

and

$$B(\mu) = \frac{2}{\pi \sinh \mu a} \int_0^\infty g_1(y) \sin \mu y \, dy \tag{9.35}$$

for $0 < \mu < \infty$. Adding the solutions to Problems I and II, the temperature distribution in the semi-infinite strip is

$$u(x, y) = \sum_{n=1}^\infty a_n e^{-\frac{n\pi}{a} y} \sin \frac{n\pi}{a} x$$

$$+ \int_0^\infty \left[A_1(\mu) \frac{\sinh \mu x}{\sinh \mu a} + B_1(\mu) \frac{\sinh \mu(a - x)}{\sinh \mu a} \right] \sin \mu y \, d\mu, \tag{9.36}$$

where

$$a_n = \frac{2}{a}\int_0^a f(x)\sin\frac{n\pi}{a}x\,dx \tag{9.37}$$

and

$$B_1(\mu) = \frac{2}{\pi}\int_0^\infty g_1(y)\sin\mu y\,dy \tag{9.38}$$

and

$$A_1(\mu) = \frac{2}{\pi}\int_0^\infty g_2(y)\sin\mu y\,dy. \tag{9.39}$$

- Alternatively, to solve Problem II using Fourier transforms, we want

$$\begin{aligned}
&\frac{\partial^2 w}{\partial x^2} + \frac{\partial^2 w}{\partial y^2} = 0, \quad 0 < x < a, \quad y > 0,\\
&w(x,0) = 0,\\
&w(0,y) = g_1(y),\\
&w(a,y) = g_2(y),\\
&|w(x,y)| \text{ bounded as } y \to \infty,
\end{aligned}$$

and the homogeneous Dirichlet boundary condition at $y = 0$ indicates that we want to use the Fourier sine transform to solve this problem.

Let

$$\widehat{w}(x,\omega) = \frac{2}{\pi}\int_0^\infty w(x,y)\sin\omega y\,dy;$$

then taking the Fourier sine transform of the partial differential equation, we have

$$\frac{d^2\widehat{w}}{dx^2}(x,\omega) + w(x,0) - \omega^2\widehat{w}(x,\omega) = 0.$$

Since $w(x,0) = 0$, then $\widehat{w}(x,\omega)$ satisfies the initial value problem

$$\begin{aligned}
&\frac{d^2\widehat{w}}{dx^2} - \omega^2\widehat{w} = 0, \quad 0 < x < a,\\
&\widehat{w}(0,\omega) = \widehat{g}_1(\omega),\\
&\widehat{w}(a,\omega) = \widehat{g}_2(\omega),
\end{aligned}$$

where

$$\widehat{g}_1(\omega) = \frac{2}{\pi}\int_0^\infty g_1(\omega)\sin\omega y\,dy \qquad \text{and} \qquad \widehat{g}_2(\omega) = \frac{2}{\pi}\int_0^\infty g_2(\omega)\sin\omega y\,dy.$$

The general solution to the ordinary differential equation for $\widehat{w}$ is

$$\widehat{w}(x,\omega) = A_1(\omega)\sinh\omega x + B_1(\omega)\sinh\omega(a-x), \quad 0 < x < a,$$

and applying the initial conditions, we have

$$B_1(\omega) = \frac{\widehat{g}_1(\omega)}{\sinh\omega a} \qquad \text{and} \qquad A_1(\omega) = \frac{\widehat{g}_2(\omega)}{\sinh\omega a}.$$

Taking the inverse sine transform, the solution is given by

$$\begin{aligned} w(x,y) &= \int_0^\infty \widehat{w}(x,\omega)\sin\omega y\,d\omega \\ &= \int_0^\infty \left[\widehat{g}_2(\omega)\frac{\sinh\omega x}{\sinh\omega a} + \widehat{g}_1(\omega)\frac{\sinh\omega(a-x)}{\sinh\omega a}\right]\sin\omega y\,d\omega, \end{aligned}$$

where

$$\widehat{g}_1(\omega) = \frac{2}{\pi}\int_0^\infty g_1(\omega)\sin\omega y\,dy \qquad \text{and} \qquad \widehat{g}_2(\omega) = \frac{2}{\pi}\int_0^\infty g_2(\omega)\sin\omega y\,dy.$$

This is exactly the same solution as that obtained above using separation of variables.

9.4 SUMMARY

In this chapter we used the Fourier transform to solve PDEs. Given a PDE, we transform the original problem in the x-world into a problem in the ω-world and obtain an ODE that is easily solved. We then transform the solution that we obtained in the ω-world back to the x-world. As we have seen, finding the inverse Fourier transform is often very difficult, and this is where a table of Fourier transforms and their inverses comes in handy. However, before using the table, be sure that the definition of the Fourier transform and inverse Fourier transform are the same as those given in this book. As a final note, some of the problems on an infinite domain can also be solved using the method of separation of variables.

The Fourier transform method presented us with two special general solution methods.

• The linear heat equation

$$\begin{aligned} &\frac{\partial u}{\partial t} = k\frac{\partial^2 u}{\partial x^2}, \quad -\infty < x < \infty, \quad t > 0, \\ &u(x,0) = f(x) \end{aligned}$$

can be solved by using a **convolution with the Gauss-Weierstrass kernel**:

$$u(x,t) = G * f(x) = \int_{-\infty}^{\infty} f(s)\,G(x-s,t)\,ds,$$

where

$$G(x,t) = \frac{1}{\sqrt{4\pi kt}} e^{-x^2/4kt}.$$

- The linear wave equation

$$\frac{\partial^2 u}{\partial t^2} = c^2 \frac{\partial^2 u}{\partial x^2}, \quad -\infty < x < \infty, \quad t > 0,$$

$$u(x,0) = f(x),$$

$$u_t(x,0) = g(x)$$

can be solved using **d'Alembert's solution**:

$$u(x,t) = \frac{1}{2}\left[f(x-ct) + f(x+ct)\right] + \frac{1}{2c}\int_{x-ct}^{x+ct} g(s)\,ds.$$

Another important function in this context is the **error function**

$$\text{erf}(z) = \frac{2}{\pi}\int_0^z e^{-x^2}\,dx,$$

and we can express solutions to the heat equation in terms of this function.

9.4.1 Problems and Notes

Problems from Part II:

Exercise	14.6	14.7	14.10	16.6	16.7	16.12
Notes						
Exercise	16.15	16.16	16.17			
Notes						

Final exam questions:

Exercise	19.18
Notes	

You should now be able to do **Final Exam 4**.

CHAPTER 10

METHOD OF CHARACTERISTICS

10.1 INTRODUCTION TO THE METHOD OF CHARACTERISTICS

The **method of characteristics**, which we develop now, is quite different from the methods we have studied so far. The method of characteristics is mathematically rather simple; in fact, this material could have been covered much earlier. It involves an interesting logical twist, and to explain it, we introduce the idea of an **anchor point**, a concept we find very useful but is not used in other textbooks.

We begin with a simple first-order partial differential equation,

$$\frac{\partial u}{\partial t} + c\frac{\partial u}{\partial x} = 0, \quad -\infty < x < \infty, \quad t > 0, \tag{10.1}$$

$$u(x, 0) = f(x).$$

Notice that no boundary condition is needed, since we are working on the entire real line. As seen earlier, a solution of a partial differential equation for a function of two variables, $u(x, t)$, can be understood as a surface over the xt-plane. To parametrize this surface, we use **characteristic curves**, shown in Figure 10.1. We want to find curves $x(t)$ in the xt-plane such that the PDE can be reduced to an ODE along these

Partial Differential Equations: Theory and Completely Solved Problems, First Edition. By T. Hillen, I. E. Leonard, H. van Roessel

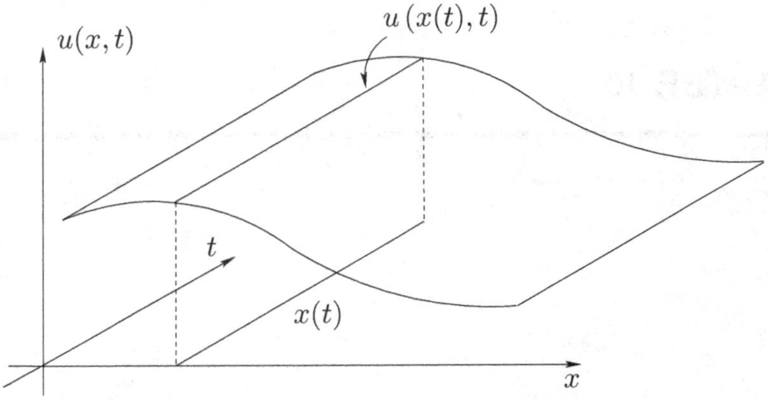

Figure 10.1 Schematic of a typical characteristic curve $x(t)$ and the part $u(x(t), t)$ of the solution surface u which lies directly above it.

curves. These ODEs can be solved and pieced back together to find the solution of the PDE.

We assume that the curve $x(t)$ is given and study u along this curve, $u(x(t), t)$. From the chain rule, we obtain

$$\frac{d}{dt}\Big[u(x(t), t)\Big] = \frac{\partial u}{\partial x}(x(t), t) \cdot \frac{dx(t)}{dt} + \frac{\partial u}{\partial t}(x(t), t).$$

Note the clear distinction between a total derivative d/dt and a partial derivative $\partial/\partial t$. Now, if we compare the last expression with the original PDE, we see that these expressions are identical if

$$\frac{dx(t)}{dt} = c.$$

Indeed, for this choice of $x = x(t)$, we have

$$\frac{d}{dt}\Big[u(x(t), t)\Big] = c\frac{\partial u}{\partial x}(x(t), t) + \frac{\partial u}{\partial t}(x(t), t) = 0;$$

hence, the solution is constant along the curve $x = x(t)$. Thus, to make all this happen, that is, to apply the idea of characteristic curves, we have to solve two ODEs, which are called the **characteristic ODEs**:

$$\frac{dx(t)}{dt} = c, \tag{10.2}$$

$$\frac{d}{dt}\Big[u(x(t), t)\Big] = 0. \tag{10.3}$$

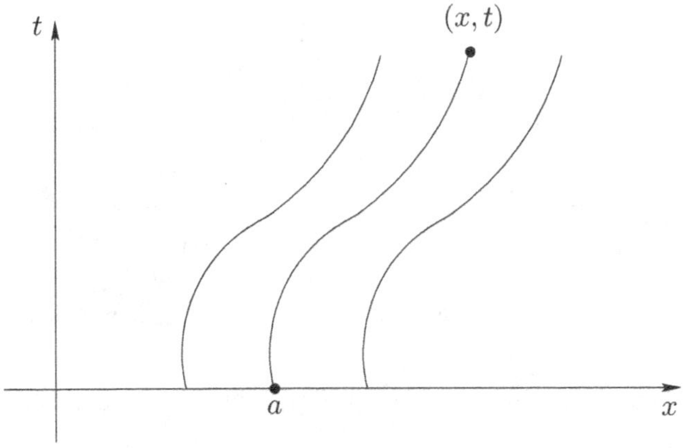

Figure 10.2 Schematic of a typical characteristic curve $x(t)$ and the anchor point. Notice that, in general, characteristic curves are not necessarily straight lines.

We solve the first of these characteristic ODEs to get

$$x(t) = ct + a,$$

where a is an unknown constant.

We observe that at $t = 0$, we have $x(0) = a$, and hence this constant a is the point where the characteristic curve starts. We call it the **anchor point** (Fig. 10.2). If a point (x, t) is given, we can always find the corresponding anchor point as

$$a = x - ct.$$

Now we solve the second characteristic ODE to get

$$u(x(t), t) = k,$$

where k is also an arbitrary constant.

If u is constant along the characteristic, it must have the same value that it has at the point where the characteristic starts, that is, at the anchor point. Hence,

$$u\big(x(t), t\big) = u\big(x(0), 0\big) = f\big(x(0)\big) = f(a) = f\big(x(t) - ct\big).$$

Here we used the initial condition $u(x, 0) = f(x)$. Therefore, given a point (x, t), there is a unique characteristic curve passing through (x, t), with anchor point at $a = x - ct$, and the solution at (x, t) is

$$u(x, t) = f(a) = f(x - ct). \tag{10.4}$$

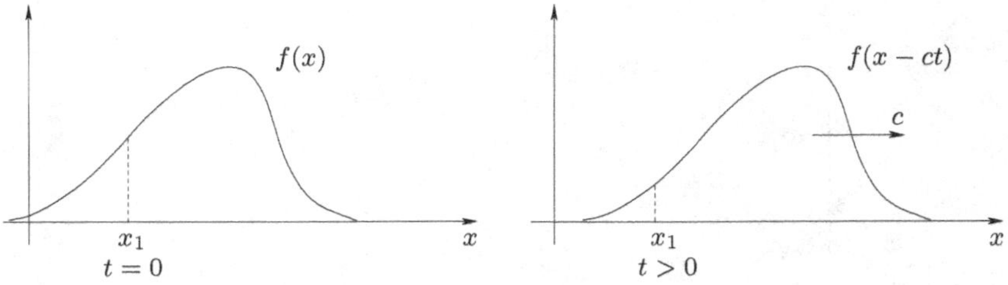

Figure 10.3 Schematic of a traveling wave that moves to the right with constant speed c.

This is a translation of the initial data to the right with velocity c, as shown in Figure 10.3.

Example 10.1. Solve the following PDE for $u(x,t)$:

$$\frac{\partial u}{\partial t} + 5\frac{\partial u}{\partial x} = 0, \quad -\infty < x < \infty, \quad t > 0,$$

$$u(x,0) = e^{-x^2},$$

using the method of characteristics.

Solution. The first characteristic equation is

$$\frac{dx(t)}{dt} = 5,$$

with solution

$$x(t) = 5t + a,$$

so that

$$a = x(t) - 5t.$$

The second characteristic equation has the solution

$$u(x(t),t) = f(a) = e^{-a^2} = e^{-(x(t)-5t)^2}.$$

Thus, the solution is

$$u(x,t) = e^{-(x-5t)^2}.$$

∎

Now we extend the method to allow for linear source or sink terms.

Example 10.2. Solve the following PDE for $u(x,t)$ on $-\infty < x < \infty$

$$\frac{\partial u}{\partial t} + \alpha\frac{\partial u}{\partial x} + \beta u = 0,$$

$$u(x,0) = f(x),$$

using the method of characteristics.

Solution. Again we look for solutions of the form $u = u(x(t), t)$. From the chain rule we get

$$\frac{d}{dt}\Big[u\big(x(t), t\big)\Big] = \frac{\partial u}{\partial x} \cdot \frac{dx}{dt} + \frac{\partial u}{\partial t}.$$

Hence, the characteristic equations become

$$\frac{dx(t)}{dt} = \alpha,$$

$$\frac{d}{dt}\Big[u\big(x(t), t\big)\Big] = -\beta u(x(t), t).$$

The solution to the first characteristic equation is

$$x(t) = \alpha t + a.$$

Thus, the anchor point for the characteristic passing through the point (x, t) is

$$a = x - \alpha t.$$

The solution to the second characteristic equation is

$$u\big(x(t), t\big) = u(x(0), 0)e^{-\beta t},$$

and the solution along the characteristic $x = x(t)$ is

$$u\big(x(t), t\big) = u(a, 0)e^{-\beta t} = f\big(x(t) - \alpha t\big)e^{-\beta t}.$$

Therefore, given a point (x, t), the anchor point for the unique characteristic passing through the point is

$$a = x - \alpha t,$$

and the solution at the point (x, t) is

$$u(x, t) = f(x - \alpha t)e^{-\beta t}.$$

∎

10.2 GEOMETRIC INTERPRETATION

In Section 10.1 we considered some simple homogeneous first-order PDEs. Now we extend the previous concepts to more general PDEs of the form

$$A(x, y)\frac{\partial u}{\partial x} + B(x, y)\frac{\partial u}{\partial y} + C_1(x, y)u = C_0(x, y), \tag{10.5}$$

where we assume that the coefficient functions are sufficiently smooth. It is convenient to rewrite this PDE in the form

$$A(x,y)\frac{\partial u}{\partial x} + B(x,y)\frac{\partial u}{\partial y} = C(x,y,u), \tag{10.6}$$

where $C(x,y,u) = C_0(x,y) - C_1(x,y)u$. In applications this PDE is usually accompanied by an auxiliary condition (called an initial condition if one of the independent variables is time) of the form

$$u(x,y) = f(x,y) \tag{10.7}$$

for $(x,y) \in \Gamma_a$, where Γ_a is a curve of anchor points (Fig. 10.1) and f is a given function.

It turns out that there is a nice geometric interpretation for this PDE. Suppose that $u(x,y)$ is the solution to (10.6) subject to the auxiliary condition equation (10.7). We can think of $z = u(x,y)$ as a surface S in $\mathbb{R}^3$. Denote by Γ the curve on the surface $z = u(x,y)$ whose projection onto the xy-plane is Γ_a (Fig. 10.1). The curve Γ is called the **initial curve**. If we parameterize Γ_a using the anchor points,

$$\Gamma_a : \begin{cases} x = x_0(a), \\ y = y_0(a), \end{cases} \tag{10.8}$$

the initial curve Γ is given by

$$\Gamma : \begin{cases} x = x_0(a), \\ y = y_0(a), \\ z = z_0(a) = f(x_0(a), y_0(a)). \end{cases} \tag{10.9}$$

Recall from elementary calculus that a normal vector to the surface $z = u(x,y)$ is given by

$$\mathbf{N} = \left(\frac{\partial u}{\partial x}, \frac{\partial u}{\partial y}, -1\right).$$

If equation (10.6) is rewritten as

$$A\frac{\partial u}{\partial x} + B\frac{\partial u}{\partial y} - C = 0,$$

this PDE says that the vector field

$$\mathbf{F} = \big(A(x,y), B(x,y), C(x,y,u)\big)$$

is perpendicular to the normal vector; that is,

$$\mathbf{F} \cdot \mathbf{N} = (A, B, C) \cdot \left(\frac{\partial u}{\partial x}, \frac{\partial u}{\partial y}, -1 \right) = A \frac{\partial u}{\partial x} + B \frac{\partial u}{\partial y} - C = 0.$$

Therefore, the vector field $\mathbf{F}$ is everywhere tangent to the solution surface $z = u(x, y)$. In other words, the solution $u(x, y)$ to the PDE (10.6), when considered as a surface S in $\mathbb{R}^3$, is made up of the integral curves of the vector field $\mathbf{F}$.

Integral curves

$$\mathbf{r}(s) = (x(s), y(s), z(s))$$

of $\mathbf{F}$ which start from the initial curve Γ satisfy the following vector ODE:

$$\begin{aligned} \frac{d\mathbf{r}}{ds} &= \mathbf{F}, \\ \mathbf{r}(0) &= (x_0(a), y_0(a), z_0(a)), \end{aligned}$$

or, in component form, the system of ODEs

$$\begin{aligned} \frac{dx}{ds} &= A(x, y), \\ x(0) &= x_0(a), \end{aligned} \tag{10.10}$$

$$\begin{aligned} \frac{dy}{ds} &= B(x, y), \\ y(0) &= y_0(a), \end{aligned} \tag{10.11}$$

$$\begin{aligned} \frac{dz}{ds} &= C(x, y, z), \\ z(0) &= z_0(a). \end{aligned} \tag{10.12}$$

The system of ODEs (10.10), (10.11), and (10.12) are called the **characteristic equations** for the PDE given in Equation (10.6). The solutions to the characteristic equations are called the **characteristic curves** for the PDE.

Solution procedure

1. Solve the first two characteristic equations (10.10), and (10.11) to get x and y in terms of the characteristic variable s and the anchor point a:

$$x = X(s, a),$$

$$y = Y(s, a).$$

2. Insert the solution from the previous step into equation (10.12), and solve the resulting equation for z:

$$z = Z(s, a).$$

3. Write the characteristic variable s and anchor point a in terms of the original independent variables x and y; that is, invert

$$x = X(s, a),$$
$$y = Y(s, a)$$

to get

$$s = S(x, y),$$
$$a = \Lambda(x, y).$$

4. Write the solution for z in terms of x and y to get the solution to the original PDE:

$$u(x, y) = Z\big(S(x, y), \Lambda(x, y)\big).$$

Remarks on the solution procedure

1. (a) The first two characteristic equations do not involve z and so can be solved independent of the third characteristic equation. However, if $A(x, y)$ and $B(x, y)$ are complicated functions, these equations are nonlinear and can be very difficult to solve.

   (b) Normally, the solution to an ODE is a function of only one variable, the independent variable, which in this case is the characteristic variable s. But clearly, the solution also depends on the initial condition, and in this case, the initial condition depends on the anchor point a. Thus, we consider the solution of these characteristic equations as functions of both s and a.

2. Recall that $C(x, y, z) = C_0(x, y) - C_1(x, y)z$, so that equation (10.12) can be written as

$$\frac{dz}{ds} + C_1(x, y)z = C_0(x, y).$$

   This is a first-order linear ODE in z, and once x and y are known in terms of the characteristic variable s, it can be solved easily by means of an integrating factor.

3. We can think of $\{s, a\}$ and $\{x, y\}$ as two different coordinate systems for our solution surface S. Solving the characteristic equations gives us the solution surface S in terms of $\{s, a\}$; however, we would like to express our solution $z = u(x, y)$ in terms of our original independent variables x and y. This leads us to attempt to find the inverse transformation. However, the functions

$X(s,a)$ and $Y(s,a)$ are often quite complicated, so that finding the inverse transformation can be rather difficult.

4. Once the first three steps have been completed, the solution is known explicitly.

5. If we apply this method to time-dependent problems of the form

$$\frac{\partial u}{\partial t} + B(x,t)\frac{\partial u}{\partial x} = C(x,t),$$

$$u(x,0) = f(x),$$

we have $A(x,t) = 1$, and we can easily solve the first characteristic equation (10.10), getting $x = t$. Note that you have to make a mental shift from (y,x) to (x,t), Hence, we can chose the time t as a parameter along the characteristic curves $x(t)$; then the second and third characteristic equations (10.11) and (10.12) become

$$\frac{dx}{dt} = B(x,t),$$

$$\frac{du}{dt} = C(x,t).$$

In this context, the two equations above are called the **characteristic equations**. Hence, the method from Section 10.1 is a simplified version of this more general geometric method.

To summarize: steps 1 and 3 can be difficult, while steps 2 and 4 are easy. We illustrate the solution procedure outlined above with several examples.

Example 10.3. Solve the following PDE:

$$\frac{\partial u}{\partial t} + c\frac{\partial u}{\partial x} = 0, \quad -\infty < x < \infty, \quad t > 0,$$

$$u(x,0) = f(x),$$

where c is a constant.

Solution. This is the example we solved at the beginning of the chapter. We want to apply the solution method outlined above, but in this example the independent variables are $\{x,t\}$ rather than $\{x,y\}$. From the auxiliary condition, in this case the initial condition $u(x,0) = f(x)$, we see that the curve of anchor points, Γ_a, is just the x-axis. Thus,

$$\Gamma_a = \{\,(x,t)\,|\, x \in \mathbb{R},\ t = 0\,\},$$

which can be parameterized as follows:

$$\Gamma_a : \begin{cases} x = x_0(a) = a, \\ t = t_0(a) = 0 \end{cases}$$

for $-\infty < a < \infty$.

This leads us to the following parametric representation of the initial curve Γ:

$$\Gamma : \begin{cases} x = x_0(a) = a, \\ t = t_0(a) = 0, \\ z = z_0(a) = f(a), \end{cases}$$

for $-\infty < a < \infty$. For this example the characteristic equations are

$$\frac{dx}{ds} = c, \qquad x(0) = a, \tag{10.13}$$

$$\frac{dt}{ds} = 1, \qquad t(0) = 0, \tag{10.14}$$

$$\frac{dz}{ds} = 0, \qquad z(0) = f(a). \tag{10.15}$$

We now proceed to solve the problem.

Step 1: Solve the first two characteristic equations.
In this case these ODEs are easily solved by direct integration, yielding

$$x = X(s,a) = cs + a,$$

$$t = T(s,a) = s.$$

Step 2: Solve the third characteristic equation.
Again, this is easily done, giving

$$z = Z(s,a) = f(a).$$

Step 3: Invert the transformation.
We invert

$$x = cs + a,$$

$$t = s,$$

to get

$$s = S(x,t) = t,$$

$$a = \Lambda(x,t) = x - ct.$$

Step 4: Write down the final solution.

$$u(x,t) = Z(S(x,t), \Lambda(x,t)) = f(\Lambda(x,t)) = f(x-ct),$$

which is the same solution as found earlier: namely, equation (10.4). ∎

Example 10.4. Solve the following PDE:

$$x\frac{\partial u}{\partial x} + y\frac{\partial u}{\partial y} = u, \quad -\infty < x < \infty, \quad y > 0,$$

$$u(x, x^2) = 1 + x^2$$

using the method of characteristics.

Solution. In this problem, the side condition is not an initial condition at $x = 0$ but, rather, a condition along the parabola $y = x^2$. Hence, the curve of anchor points Γ_a and the initial curve Γ can be parameterized as follows:

$$\Gamma_a : \begin{cases} x = a, \\ y = a^2 \end{cases}$$

for $-\infty < a < \infty$, and

$$\Gamma \ : \begin{cases} x = a, \\ y = a^2, \\ z = 1 + a^2 \end{cases}$$

for $-\infty < a < \infty$. For this example the characteristic equations are

$$\frac{dx}{ds} = x, \tag{10.16}$$

$$x(0) = a, \tag{10.17}$$

$$\frac{dy}{ds} = y, \tag{10.18}$$

$$y(0) = a^2, \tag{10.19}$$

$$\frac{dz}{ds} = z, \tag{10.20}$$

$$z(0) = 1 + a^2. \tag{10.21}$$

Step 1: The first two characteristic equations can be integrated to obtain

$$x = X(s,a) = ae^s,$$

$$y = Y(s,a) = a^2e^s.$$

Step 2: The third characteristic equation can be solved to obtain

$$z = Z(s,a) = (1+a^2)e^s.$$

Step 3: We can invert the transformation

$$x = ae^s,$$

$$y = a^2e^s,$$

to obtain

$$s = S(x,y) = \log\frac{x^2}{y},$$

$$a = \Lambda(x,y) = \frac{y}{x}.$$

Step 4: We write the final solution as

$$u(x,y) = Z\left(\log\frac{x^2}{y}, \frac{y}{x}\right)$$

$$= \left(1+\left(\frac{y}{x}\right)^2\right)\frac{x^2}{y},$$

that is,

$$u(x,y) = \frac{x^2+y^2}{y}.$$

∎

Example 10.5. Solve the following PDE for $u(x,y)$:

$$\frac{\partial u}{\partial x} + 2x\frac{\partial u}{\partial y} = y, \quad x>0, \quad -\infty<y<\infty,$$

$$u(0,y) = 1+y^2$$

using the method of characteristics.

Solution. From the auxiliary condition we see that the curve of anchor points, Γ_a, is just the y-axis, that is, the line $x=0$. Thus, the curves Γ_a and Γ can be parameterized as follows:

$$\Gamma_a : \begin{cases} x = 0, \\ y = a \end{cases}$$

and

$$\Gamma : \begin{cases} x = 0, \\ y = a, \\ z = 1 + a^2 \end{cases}$$

for $-\infty < a < \infty$. For this example the characteristic equations are

$$\frac{dx}{ds} = 1, \tag{10.22}$$

$$x(0) = 0, \tag{10.23}$$

$$\frac{dy}{ds} = 2x, \tag{10.24}$$

$$y(0) = a, \tag{10.25}$$

$$\frac{dz}{ds} = y, \tag{10.26}$$

$$z(0) = 1 + a^2. \tag{10.27}$$

We now proceed to solve the problem.

Step 1: Solve the first two characteristic equations.

The first characteristic equation can be integrated to obtain $x = s$. Inserting this into the second characteristic equation yields

$$\frac{dy}{ds} = 2s.$$

Integrating this equation and applying the initial condition results in

$$y = s^2 + a.$$

Thus,

$$x = X(s, a) = s,$$
$$y = Y(s, a) = s^2 + a.$$

Step 2: Solve the third characteristic equation.

Using the results of step 1, the third characteristic equation becomes

$$\frac{dz}{ds} = s^2 + a,$$

and integration yields

$$z = \frac{s^3}{3} + as + c,$$

where c is a constant of integration. The initial condition implies that

$$c = 1 + a^2,$$

and the result is

$$z = Z(s, a) = \frac{s^3}{3} + as + 1 + a^2.$$

Step 3: Invert the transformation.

We invert

$$x = s,$$
$$y = s^2 + a$$

to get

$$s = x,$$
$$a = y - x^2.$$

Step 4: Write the final solution.

The solution is

$$u(x, y) = Z(x, y - x^2) = \frac{x^3}{3} + (y - x^2)x + 1 + (y - x^2)^2;$$

that is,

$$u(x, y) = x^4 - \frac{2}{3}x^3 - 2x^2y + xy + y^2 + 1.$$

∎

10.3 D'ALEMBERT'S SOLUTION

The method of characteristics can also be used to derive d'Alembert's solution to the one-dimensional wave equation on the entire real line $-\infty < x < \infty$:

$$\frac{\partial^2 u}{\partial t^2} = c^2 \frac{\partial^2 u}{\partial x^2}, \quad -\infty < x < \infty, \quad t > 0,$$
$$u(x, 0) = f(x),$$
$$\frac{\partial u}{\partial t}(x, 0) = g(x).$$

We assume that the solution is twice continuously differentiable and is such that

$$\frac{\partial^2 u}{\partial x\,\partial t} = \frac{\partial^2 u}{\partial t\,\partial x}.$$

We take a more abstract view and perform algebraic manipulations of the differential operators. We use the factorization of the binomial $a^2 - b^2$ as

$$a^2 - b^2 = (a+b)(a-b) = (a-b)(a+b)$$

to factor the wave operator

$$\frac{\partial^2}{\partial t^2} - c^2\frac{\partial^2}{\partial x^2}.$$

We can write the wave equation as

$$\frac{\partial^2 u}{\partial t^2} - c^2\frac{\partial^2 u}{\partial x^2} = \left(\frac{\partial}{\partial t} + c\frac{\partial}{\partial x}\right)\left(\frac{\partial}{\partial t} - c\frac{\partial}{\partial x}\right)u = 0. \tag{10.28}$$

Now we introduce a new variable:

$$v(x,t) = \left(\frac{\partial}{\partial t} - c\frac{\partial}{\partial x}\right)u(x,t) = \frac{\partial u}{\partial t}(x,t) - c\frac{\partial u}{\partial x}(x,t).$$

At $t = 0$ we have

$$v(x,0) = \frac{\partial u}{\partial t}(x,0) - c\frac{\partial u}{\partial x}(x,0) = g(x) - cf'(x).$$

We can rewrite the second-order wave equation as a pair of first-order linear PDEs to be solved sequentially:

$$\frac{\partial v}{\partial t} + c\frac{\partial v}{\partial x} = 0, \tag{10.29}$$

$$v(x,0) = F(x)$$

and

$$\frac{\partial u}{\partial t} - c\frac{\partial u}{\partial x} = v, \tag{10.30}$$

$$u(x,0) = f(x),$$

where $F(x) = g(x) - c\,f'(x)$. The first of these is, in fact, the simple first-order equation (10.1), which we studied at the beginning of this chapter, and its solution had the form

$$v(x,t) = F(x - ct). \tag{10.31}$$

Now we put this solution for v into equation (10.30) to get

$$\frac{\partial u}{\partial t} - c\frac{\partial u}{\partial x} = F(x - ct), \tag{10.32}$$

$$u(x, 0) = f(x).$$

where $F(x) = g(x) - c\,f'(x)$, and we solve problem (10.32) using the method of characteristics.

From the auxiliary condition $u(x, 0) = f(x)$, we see that the curve of anchor points, Γ_a, is just the x-axis, as was the case in the example at the beginning of this chapter. Thus, the curve Γ_a and the initial curve Γ can be parameterized as follows:

$$\Gamma_a : \begin{cases} x = x_0(a) = a, \\ t = t_0(a) = 0 \end{cases}$$

and

$$\Gamma \ : \begin{cases} x = x_0(a) = a, \\ t = t_0(a) = 0, \\ z = z_0(a) = f(a) \end{cases}$$

for $-\infty < a < \infty$.

Since the coefficient $A(x, t) = 1$, we can use the simpler approach mentioned in the *Remarks on the solution procedure* and choose $s = t$ as the parameter. In this case the remaining characteristic equations are

$$\frac{dx}{dt} = -c, \tag{10.33}$$

$$x(0) = a, \tag{10.34}$$

$$\frac{dz}{dt} = F(x - ct), \tag{10.35}$$

$$z(0) = f(a). \tag{10.36}$$

We now proceed to solve the problem.
Step 1: Solve the first two characteristic equations.

In this case this ODE is easily solved by direct integration, yielding

$$x = X(t, a) = -ct + a.$$

Step 2: Solve the third characteristic equation.

Using the results of step 1, the third characteristic equation becomes

$$\frac{dz}{dt} = F(-ct - a - ct) = F(a - 2ct),$$

and integration yields

$$z = \int_0^t F(a - 2c\xi)\, d\xi + K,$$

where K is a constant of integration. The initial condition implies that $K = f(a)$, so we obtain

$$z = Z(t,a) = \int_0^t F(a - 2c\xi)\, d\xi + f(a),$$

and recalling the definition of F in terms of f and g, $F(x) = g(x) - c\,f'(x)$, and making the substitution $\mu = a - 2c\,\xi$, the integral of F becomes

$$\begin{aligned}
\int_0^t F(x - 2c\,\xi)\, d\xi &= -\frac{1}{2c}\int_a^{a-2ct} F(\mu)\, d\mu \\
&= -\frac{1}{2c}\int_a^{a-2ct} \left[g(\mu) - cf'(\mu)\right] d\mu \\
&= -\frac{1}{2c}\int_a^{a-2ct} g(\mu)\, d\mu + \frac{1}{2} f(\mu)\Big|_a^{a-2ct} \\
&= \frac{1}{2c}\int_{a-2ct}^{a} g(\mu)\, d\mu + \frac{1}{2}\left[f(a - 2ct) - f(a)\right].
\end{aligned}$$

The final result is

$$Z(t,a) = \frac{1}{2}\left[f(a) + f(a - 2ct)\right] + \frac{1}{2c}\int_{a-2ct}^{a} g(\mu)\, d\mu.$$

Step 3: Invert the transformation.

We invert $x = -ct + a$ to get $a = \Lambda(x,t) = x + ct$.

Step 4: Write the final solution.

We have

$$\begin{aligned}
u(x,t) &= Z\big(t, \Lambda(x,t)\big) = Z(t, x + ct) \\
&= \frac{1}{2}[f(x + ct) + f(x + ct - 2ct)] + \frac{1}{2c}\int_{x+ct-2ct}^{x+ct} g(\mu)\, d\mu,
\end{aligned}$$

and simplifying this expression, we recover d'Alembert's solution:

$$u(x,t) = \frac{1}{2}[f(x + ct) + f(x - ct)] + \frac{1}{2c}\int_{x-ct}^{x+ct} g(\mu)\, d\mu.$$

10.4 EXTENSION TO QUASILINEAR EQUATIONS

Up to this point we have considered only linear equations. However, the equations often encountered in applications are, at least mildly, nonlinear. The extension to quasilinear PDEs which are nonlinear in u, but still linear in u_x and u_y is quite straightforward. We rewrite equation (10.6) as

$$A(x, y, u)\frac{\partial u}{\partial x} + B(x, y, u)\frac{\partial u}{\partial y} = C(x, y, u), \tag{10.37}$$

but now allow the coefficient functions A and B to depend on the unknown function u in addition to the independent variables x and y. In addition, we no longer restrict C to be linear in u.

The characteristic equations are

$$\frac{dx}{ds} = A(x, y, z), \tag{10.38}$$

$$x(0) = x_0(a), \tag{10.39}$$

$$\frac{dy}{ds} = B(x, y, z), \tag{10.40}$$

$$y(0) = y_0(a), \tag{10.41}$$

$$\frac{dz}{ds} = C(x, y, z), \tag{10.42}$$

$$z(0) = z_0(a). \tag{10.43}$$

What is different here is that the first two characteristic equations are no longer independent of z, and now the complete system must be solved. However, once this is done, we proceed with steps 3 and 4 as before to obtain the solution.

Example 10.6. (*A Quasilinear Example*)
Solve the first-order quasilinear equation

$$(y - u)\frac{\partial u}{\partial x} + (u - x)\frac{\partial u}{\partial y} = x - y, \quad -\infty < x, y < \infty,$$

$$u\left(x, \frac{1}{x}\right) = 0.$$

Solution. The side condition is given along the hyperbola

$$\Gamma_a = \left\{ \left(a, \frac{1}{a}\right) \,\middle|\, a \neq 0 \right\}.$$

The characteristic equations are

$$\frac{dx}{ds} = y - z,$$

$$\frac{dy}{ds} = z - x,$$

$$\frac{dz}{ds} = x - y,$$

and since all equations are coupled, they have to be solved simultaneously.

Under closer inspection, we find two conserved quantities:

• First,

$$\begin{aligned}\frac{d}{ds}\Big(x(s) + y(s) + z(s)\Big) &= \frac{dx}{ds}(s) + \frac{dy}{ds}(s) + \frac{dz}{ds}(s)\\ &= \Big(y(s) - z(s)\Big) + \Big(z(s) - x(s)\Big) + \Big(x(s) - y(s)\Big)\\ &= 0,\end{aligned}$$

and hence there is a constant A such that

$$x(s) + y(s) + z(s) = A$$

along the characteristics.

• Next,

$$\begin{aligned}\frac{d}{ds}\Big(x(s)^2 + y(s)^2 + z(s)^2\Big) &= 2\Big(x(s)x'(s) + y(s)y'(s) + z(s)z'(s)\Big)\\ &= 2\Big(x(s)\big[y(s) - z(s)\big] + y(s)\big[z(s) - x(s)\big] + z(s)\big[x(s) - y(s)\big]\Big)\\ &= 0,\end{aligned}$$

and hence there is a constant B such that

$$x(s)^2 + y(s)^2 + z(s)^2 = B$$

along the characteristics.

All we know so far is that the characteristics lie on the intersection of a plane and a sphere in $\mathbb{R}^3$, so they are circles, or points. We still cannot solve explicitly for $x(s)$, $y(s)$, and $z(s)$, but we can use the side conditions to compute the unknown constants A and B as follows.

On Γ_a we have $u(a, 1/a) = 0$; hence,

$$a + \frac{1}{a} + 0 = \frac{a^2 + 1}{a} = A$$

and

$$a^2 + \frac{1}{a^2} + 0^2 = \frac{a^4 + 1}{a^2} = B,$$

so that

$$A^2 = \frac{a^4 + 2a^2 + 1}{a^2} = B + 2.$$

We can use this relation directly, whereby we circumvent finding explicit expressions for s and a. Since $A^2 = B + 2$, then

$$(x + y + z)^2 = x^2 + y^2 + z^2 + 2,$$

and therefore

$$xy + xz + yz = 1.$$

Thus, solving for $z = u(x, y)$, we have

$$u(x, y) = \frac{1 - xy}{x + y},$$

the solution to our problem.

∎

10.5 SUMMARY

The **method of characteristics** is quite simple, and hence this chapter could have come much earlier in the book. However, herein, this method applies primarily to first-order equations. Thus, it stands alone, separate from all the other chapters. Nevertheless, using the method of characteristics, we were able to obtain a third derivation of **d'Alembert's formula** for the solution to the boundary value–initial value problem for the one-dimensional wave equation by factoring the wave operator into a product of two linear first-order differential operators.

The important idea in the method is to find characteristic curves $x(s)$, such that the PDE becomes an ODE along those curves. Once those curves $x(s)$ are found, we need to find the anchor point a, which defines the initial condition for the ODE on the characteristic curve. Finally, we solve this ODE, using the anchor point for the initial condition.

This method needs to be practiced; hence, we recommend that you do as many of the following problems as possible.

10.5.1 Problems and Notes

Problems from Part II:

Exercise	17.1	17.2	17.3	17.4	17.5	17.6
Notes						
Exercise	17.7	17.8	17.9	17.10	17.11	17.12
Notes						
Exercise	17.13	14.6	14.7	14.10		
Notes						

Final exam questions:

Exercise	19.8
Notes	

You should now be able to do **Final Exam 2**.

PART II

EXPLICITLY SOLVED PROBLEMS

CHAPTER 11

FOURIER SERIES PROBLEMS

Exercise 11.1. ✗
Suppose that f is T-periodic and let F be an antiderivative of f; for example,

$$F(x) = \int_0^x f(t)\,dt, \quad -\infty < x < \infty.$$

Show that F is T-periodic if and only if the integral of f over any interval of length T is 0.

Solution. Note that

$$F(x+T) = \int_0^{x+T} f(t)\,dt = \int_0^x f(t)\,dt + \int_x^{x+T} f(t)\,dt = F(x) + \int_x^{x+T} f(t)\,dt$$

for all $x \in \mathbb{R}$. Therefore, $F(x+T) = F(x)$ for all $x \in \mathbb{R}$ if and only if

$$\int_x^{x+T} f(t)\,dt = 0$$

Partial Differential Equations: Theory and Completely Solved Problems, First Edition. By T. Hillen, I. E. Leonard, H. van Roessel

for all $x \in \mathbb{R}$. This holds if and only if the integral of f over *any* interval of length T is 0. Since f is T-periodic, F is T-periodic if and only if

$$\int_0^T f(t)\,dt = 0.$$

Exercise 11.2. XXX

Let $f(x) = x - 2\left[(x+1)/2\right]$, (where [ ] denotes the floor function, i.e., the greatest integer less than or equal to the argument), and consider the function

$$h(x) = |f(x)| = \left|x - 2\left[\frac{x+1}{2}\right]\right|.$$

(a) Show that h is 2-periodic.

(b) Plot the graph of h.

(c) Generalize (a) by finding a closed formula that describes the $2a$-periodic triangular wave

$$g(x) = |x| \quad \text{if} \quad -a < x < a$$

and

$$g(x + 2a) = g(x) \quad \text{otherwise.}$$

Solution. Note that if we can show that

$$f(x) = x - 2\left[\frac{x+1}{2}\right]$$

is 2-periodic, then for any $x \in \mathbb{R}$ we have

$$h(x+2) = |f(x+2)| = |f(x)| = h(x)$$

for all $x \in \mathbb{R}$, so that h is also 2-periodic.

(a) Now

$$\begin{aligned} f(x+2) &= x + 2 - 2\left[\frac{(x+2)+1}{2}\right] \\ &= x + 2 - 2\left[\frac{x+1}{2} + 1\right] \\ &= x + 2 - 2\left(\left[\frac{x+1}{2}\right] + 1\right) \\ &= x - 2\left[\frac{x+1}{2}\right] \\ &= f(x) \end{aligned}$$

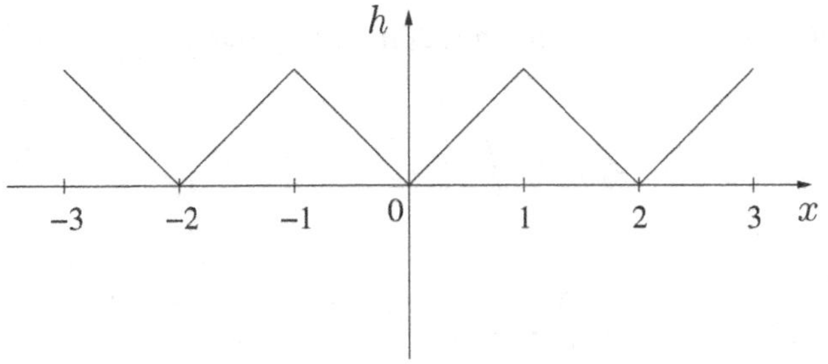

Figure 11.1 Triangular wave.

and f is 2-periodic, and from the remark above $h = |f|$ is also 2-periodic.

(b) Since $f(x) = x$ for $-1 < x < 1$, then $h(x) = |x|$ for $-1 < x < 1$, and the graph of h is shown in Figure 11.1.

(c) To find a $2a$-periodic triangular wave, we use the $2a$-periodic function

$$f(x) = x - 2a\left[\frac{x+a}{2a}\right],$$

and note that $f(x) = x$ on the interval $-a < x < a$. We leave it to you to check, exactly as in part (a), that this is $2a$-periodic and that $f(x) = x$ for $-a < x < a$. Therefore,

$$g(x) = \left|x - 2a\left[\frac{x+a}{2a}\right]\right|$$

is a $2a$-periodic triangular wave equal to $|x|$ on the interval $-a < x < a$.

Exercise 11.3. ✗
Evaluate

$$\int_0^a \cos\frac{n\pi x}{a}\cos\frac{m\pi x}{a}\,dx$$

for $n \geqslant 0$, $m \geqslant 0$. Use the trigonometric identity

$$\cos A \cos B = \frac{1}{2}\left[\cos(A+B) + \cos(A-B)\right]$$

(consider $A - B = 0$ and $A + B = 0$ separately).

Solution. If $n > m \geqslant 0$, then from the addition formula for the cosine, we have

$$\begin{aligned}
\int_0^a \cos\frac{n\pi x}{a}\cos\frac{m\pi x}{a}\,dx &= \frac{1}{2}\int_0^a \left[\cos\frac{(n-m)\pi x}{a} + \cos\frac{(n+m)\pi x}{a}\right] dx \\
&= \frac{a}{2\pi(n-m)}\sin\frac{(n-m)\pi x}{a}\bigg|_0^a \\
&\quad + \frac{a}{2\pi(n+m)}\sin\frac{(n+m)\pi x}{a}\bigg|_0^a \\
&= \frac{a}{2\pi(n-m)}\sin(n-m)\pi \\
&\quad + \frac{a}{2\pi(n+m)}\sin(n+m)\pi \\
&= 0.
\end{aligned}$$

If $n = m > 0$, then

$$\begin{aligned}
\int_0^a \cos\frac{n\pi x}{a}\cos\frac{m\pi x}{a}\,dx = \int_0^a \cos^2\frac{m\pi x}{a}\,dx &= \frac{1}{2}\int_0^a \left(1 + \cos\frac{2m\pi x}{a}\right) dx \\
&= \frac{a}{2} + \frac{a}{4m\pi}\sin\frac{2m\pi x}{a}\bigg|_0^a = \frac{a}{2}.
\end{aligned}$$

If $n = m = 0$, then

$$\int_0^a 1\cdot 1\,dx = a.$$

Exercise 11.4. ✗

Evaluate

$$\int_0^a \sin\frac{n\pi x}{a}\sin\frac{m\pi x}{a}\,dx$$

for $n > 0$, $m > 0$ and consider $n = m$ separately. Use the trignometric identity:

$$\sin A \sin B = \frac{1}{2}\left[\cos(A-B) - \cos(A+B)\right].$$

Solution. If m and n are positive integers with $m \neq n$, then

$$\begin{aligned}
\int_0^a \sin\frac{n\pi x}{a}\sin\frac{m\pi x}{a}\,dx &= \int_0^a \frac{1}{2}\left[\cos\frac{(n-m)\pi x}{a} - \cos\frac{(n+m)\pi x}{a}\right] dx \\
&= \frac{a}{2\pi(n-m)}\sin\frac{(n-m)\pi x}{a}\Bigg|_0^a \\
&\qquad - \frac{a}{2\pi(n+m)}\sin\frac{(n+m)\pi x}{a}\Bigg|_0^a \\
&= \frac{a}{2\pi(n-m)}\left[\sin(n-m)\pi - \sin 0\right] \\
&\qquad - \frac{a}{2\pi(n+m)}\left[\sin(n+m)\pi - \sin 0\right] \\
&= 0,
\end{aligned}$$

while if $n = m > 0$, then

$$\begin{aligned}
\int_0^a \sin\frac{n\pi x}{a}\sin\frac{n\pi x}{a}\,dx &= \int_0^a \sin^2\frac{n\pi x}{a}\,dx \\
&= \int_0^a \frac{1}{2}\left(1 - \cos\frac{2n\pi x}{a}\right) dx \\
&= \frac{1}{2}\cdot x\Bigg|_0^a - \frac{a}{4n\pi}\cdot\sin\frac{2n\pi x}{a}\Bigg|_0^a \\
&= \frac{a}{2} - \frac{a}{4n\pi}(\sin 2n\pi - \sin 0) \\
&= \frac{a}{2}.
\end{aligned}$$

Therefore,

$$\int_0^a \sin\frac{n\pi x}{a}\sin\frac{m\pi x}{a}\,dx = \begin{cases} \dfrac{a}{2}, & \text{for} \quad m = n > 0, \\ 0, & \text{for} \quad m > 0,\ n > 0,\ m \neq n. \end{cases}$$

Exercise 11.5. XX

Compute the Fourier series of the 2π-periodic function f given by

$$f(x) = \begin{cases} 1, & \text{if} \quad 0 < x < \pi/2, \\ 0, & \text{if} \quad \pi/2 < |x| < \pi, \\ -1, & \text{if} \quad -\pi/2 < x < 0. \end{cases}$$

For which values of x does the Fourier series converge to f? Sketch the graph of the Fourier series.

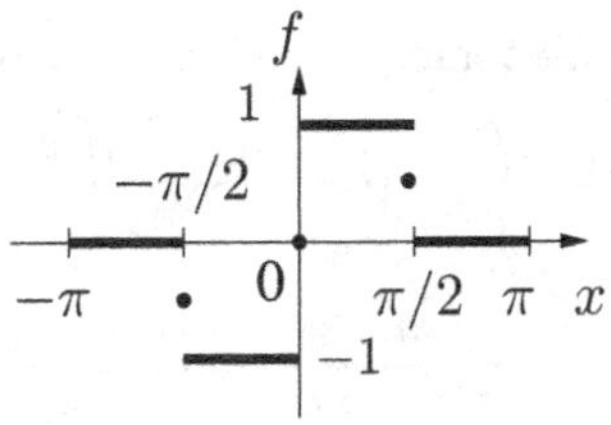

Figure 11.2 Step function.

Solution. Note that f is an odd function on the interval $-\pi < x < \pi$, so that

$$a_0 = \frac{1}{2\pi}\int_{-\pi}^{\pi} f(x)\,dx = 0$$

and

$$a_n = \frac{1}{\pi}\int_{-\pi}^{\pi} f(x)\cos nx\,dx = 0$$

for $n = 1, 2, \ldots$. We calculate the b_n,

$$\begin{aligned}
b_n &= \frac{1}{\pi}\int_{-\pi}^{\pi} f(x)\sin nx\,dx \\
&= \frac{1}{\pi}\int_{-\pi/2}^{0} (-1)\sin nx\,dx + \frac{1}{\pi}\int_{0}^{\pi/2} \sin nx\,dx \\
&= \frac{2}{\pi}\int_{0}^{\pi/2} \sin nx\,dx \\
&= \frac{2}{\pi}\left(-\frac{1}{n}\cos nx\right)\Big|_{0}^{\pi/2} \\
&= \frac{2}{n\pi}\left(1 - \cos\frac{n\pi}{2}\right)
\end{aligned}$$

and the Fourier series is

$$\frac{2}{\pi}\sum_{n=1}^{\infty} \frac{1 - \cos\dfrac{n\pi}{2}}{n}\sin nx.$$

The graph of the Fourier series of f on the interval $-\pi < x < \pi$ is shown Figure 11.2. Note that the original function f is piecewise smooth and has only a finite jump discontinuity at $x = 0$ and $x = \pm\pi/2$. Thus, from the Fourier series representation theorem, the Fourier series of f will converge to 0 at all points $x = n\pi$, $n = 0, \pm 1, \pm 2, \pm 3, \ldots$, and for all points $(2n+1)\pi/2$, the Fourier series of f will converge to $(-1)^n/2$, $n = 0, \pm 1, \pm 2, \pm 3, \ldots$. The rest of the graph of the Fourier series can be obtained by translating this graph by an integer multiple of 2π in the x-direction.

Exercise 11.6. XXX
Compute the Fourier series of the 2π-periodic function given on $-\pi \leqslant x \leqslant \pi$ by $f(x) = |\cos x|$. For which values of x does the Fourier series converge to f?

Solution. Note that f is even, since

$$f(-x) = |\cos(-x)| = |\cos x| = f(x)$$

for all $x \in \mathbb{R}$; therefore, $b_n = 0$ for all $n \geqslant 1$, and we only need to compute a_n for $n \geqslant 0$. Now

$$\begin{aligned} a_0 &= \frac{1}{2\pi}\int_{-\pi}^{\pi} |\cos x|\, dx = \frac{1}{\pi}\int_0^{\pi} |\cos x|\, dx = \frac{1}{\pi}\int_0^{\pi/2} \cos x\, dx - \frac{1}{\pi}\int_{\pi/2}^{\pi} \cos x\, dx \\ &= \frac{1}{\pi}\sin x\Big|_0^{\pi/2} - \frac{1}{\pi}\sin x\Big|_{\pi/2}^{\pi} = \frac{1}{\pi} - (-1)\frac{1}{\pi} = \frac{2}{\pi}, \end{aligned}$$

and for $n \geqslant 1$, since $\cos nx$ is also an even function, we have

$$\begin{aligned} a_n &= \frac{1}{\pi}\int_{-\pi}^{\pi} |\cos x| \cos nx\, dx = \frac{2}{\pi}\int_0^{\pi} |\cos x| \cos nx\, dx \\ &= \frac{2}{\pi}\int_0^{\pi/2} \cos x \cos nx\, dx - \frac{2}{\pi}\int_{\pi/2}^{\pi} \cos x \cos nx\, dx. \end{aligned}$$

If $n = 1$, then

$$\begin{aligned} a_1 &= \frac{2}{\pi}\int_0^{\pi/2} \cos^2 x\, dx - \frac{2}{\pi}\int_{\pi/2}^{\pi} \cos^2 x\, dx \\ &= \frac{2}{\pi}\int_0^{\pi/2} \left(\frac{1}{2} + \frac{1}{2}\cos 2x\right) dx - \frac{2}{\pi}\int_{\pi/2}^{\pi} \left(\frac{1}{2} + \frac{1}{2}\cos 2x\right) dx \\ &= \frac{1}{\pi}\left[\frac{\pi}{2} - \left(\pi - \frac{\pi}{2}\right)\right] = \frac{1}{\pi}\left(\frac{\pi}{2} - \frac{\pi}{2}\right) = 0. \end{aligned}$$

Now

$$2\cos x \cos nx = \cos(n+1)x + \cos(n-1)x,$$

so that for $n \neq 1$ we have

$$\begin{aligned} \frac{2}{\pi}\int_0^{\pi/2} \cos x \cos nx\, dx &= \frac{1}{\pi}\int_0^{\pi/2} (\cos(n+1)x + \cos(n-1)x)\, dx \\ &= \frac{1}{\pi}\left[\frac{\sin(n+1)\pi/2}{n+1} + \frac{\sin(n-1)\pi/2}{n-1}\right] \end{aligned}$$

and

$$\frac{2}{\pi}\int_{\pi/2}^{\pi}\cos x\cos nx\,dx = \frac{1}{\pi}\int_{\pi/2}^{\pi}(\cos(n+1)x+\cos(n-1)x)\,dx$$
$$= -\frac{1}{\pi}\left[\frac{\sin(n+1)\pi/2}{n+1}+\frac{\sin(n-1)\pi/2}{n-1}\right].$$

For $n \neq 1$, we have

$$a_n = \frac{2}{\pi}\left[\frac{\sin(n+1)\pi/2}{n+1}+\frac{\sin(n-1)\pi/2}{n-1}\right],$$

and if n is odd, then $a_n = 0$. However, if n is even, say $n = 2k$, with $k \neq 0$, then

$$\begin{aligned} a_{2k} &= \frac{2}{\pi}\left[\frac{\sin(2k+1)\pi/2}{2k+1}+\frac{\sin(2k-1)\pi/2}{2k-1}\right] \\ &= \frac{2}{\pi}\left[\frac{(-1)^k}{2k+1}-\frac{(-1)^k}{2k-1}\right] \\ &= \frac{4}{\pi}\frac{(-1)^{k+1}}{4k^2-1} \end{aligned}$$

and the Fourier series is

$$a_0 + \sum_{k=1}^{\infty} a_{2k}\cos 2kx = \frac{2}{\pi} + \frac{4}{\pi}\sum_{k=1}^{\infty}\frac{(-1)^{k+1}}{4k^2-1}\cos 2kx.$$

Since $f(-\pi) = f(\pi)$, the piecewise smooth 2π-periodic function with

$$f(x) = |\cos x|, \quad -\pi \leqslant x \leqslant \pi$$

is continuous at each $x \in \mathbb{R}$, and therefore the Fourier series converges to $f(x)$ for each $x \in \mathbb{R}$.

Exercise 11.7. ✗✗

Consider the parabola $f(x) = x^2$ on $-a \leqslant x \leqslant a$ and show that the Fourier series of f is given by

$$\frac{a^2}{3} - \frac{4a^2}{\pi^2}\left[\cos(\pi x/a) - \frac{1}{2^2}\cos(2\pi x/a) + \frac{1}{3^2}\cos(3\pi x/a) - + \cdots\right].$$

Find its values at the points of discontinuity of f.

Solution. Note that since $f(a) = a^2 = (-a)^2 = f(-a)$, the piecewise smooth $2a$-periodic function is continuous everywhere and so has no points of discontinuity. Also, since f is an even function, then $b_n = 0$ for all $n \geqslant 1$, and the Fourier series for f has only cosine terms:

$$a_0 + \sum_{n=1}^{\infty} a_n \cos(n\pi x/a)$$

where

$$a_0 = \frac{1}{a}\int_0^a f(x)\,dx$$

and

$$a_n = \frac{2}{a}\int_0^a f(x)\cos(n\pi x/a)\,dx$$

for $n \geqslant 1$.

To calculate the coefficients a_n, we have

$$a_0 = \frac{1}{a}\int_0^a x^2\,dx = \frac{1}{a}\frac{x^3}{3}\Big|_0^a = \frac{a^2}{3}.$$

For $n \geqslant 1$, we integrate by parts twice to get

$$\begin{aligned} a_n &= \frac{2}{a}\int_0^a x^2\cos(n\pi x/a)\,dx = \frac{2}{a}\left[\frac{a}{n\pi}x^2\sin(n\pi x/a)\Big|_0^a \right. \\ &\qquad \left. - \frac{2a}{n\pi}\int_0^a x\sin(n\pi x/a)\,dx\right] \\ &= -\frac{4}{n\pi}\int_0^a x\sin(n\pi x/a)\,dx = -\frac{4}{n\pi}\left[-\frac{a}{n\pi}x\cos(n\pi x/a)\Big|_0^a \right. \\ &\qquad \left. + \frac{a}{n\pi}\int_0^a \cos(n\pi x/a)\,dx\right] \\ &= \frac{4a^2}{n^2\pi^2}(-1)^n \end{aligned}$$

for $n = 1, 2, 3, \ldots$. The Fourier series of f is

$$\frac{a^2}{3} - \frac{4a^2}{\pi^2}\left[\cos(\pi x/a) - \frac{1}{2^2}\cos(2\pi x/a) + \frac{1}{3^2}\cos(3\pi x/a) - + \cdots\right],$$

and since $\overline{f}$ is piecewise smooth and continuous everywhere, the Fourier series given above converges to $\overline{f}(x)$ for each $x \in \mathbb{R}$, where $\overline{f}$ is the $2a$-periodic extension of f.

Exercise 11.8. ✗✗

Consider the $2a$-periodic function f that is given on the interval $-a < x < a$ by $f(x) = x$. Show that the Fourier series of f is given by

$$\frac{2a}{\pi}\sum_{n=1}^{\infty}\frac{(-1)^{n+1}}{n}\sin(n\pi x/a)$$

by differentiating the Fourier series in Exercise 11.7 term by term. Justify your work.

Solution. Since the $2a$-periodic function $F(x)$ in Exercise 11.7 is piecewise smooth and continuous everywhere, the Fourier series converges to the function everywhere, and

$$F(x) = \frac{a^2}{3} - \frac{4a^2}{\pi^2} \sum_{n=1}^{\infty} \frac{(-1)^{n+1}}{n^2} \cos(n\pi x/a),$$

where $F(x) = x^2$ for $-a < x < a$. Since this function also has a piecewise smooth derivative and

$$F'(x) = 2x = 2 \cdot f(x)$$

for $-a < x < a$, the coefficients in the Fourier series of $F'(x)$ can be obtained by differentiating the series above term by term. Therefore, the Fourier series of $F'(x)$ is given by

$$\frac{4a^2}{\pi^2} \sum_{n=1}^{\infty} \frac{(-1)^{n+1} n\pi}{n^2 a} \sin(n\pi x/a) = \frac{4a}{\pi} \sum_{n=1}^{\infty} \frac{(-1)^{n+1}}{n} \sin(n\pi x/a),$$

and the Fourier series of $f(x)$ is

$$\frac{2a}{\pi} \sum_{n=1}^{\infty} \frac{(-1)^{n+1}}{n} \sin(n\pi x/a),$$

which converges to $f(x)$ for $-a < x < a$ and to 0 for $x = \pm a$.

Exercise 11.9. XXX
Obtain the expansion

$$e^{ax} = \frac{\sinh \pi a}{\pi} \sum_{n=-\infty}^{\infty} \frac{(-1)^n}{a^2 + n^2}(a \cos nx - n \sin nx),$$

valid for all real numbers $a \neq 0$ and all $-\pi < x < \pi$.

Solution. Note that the solution can be obtained using the standard Fourier series of e^{ax}. However, it turns out that it is easier to use the (equivalent) complex form of the Fourier series, which we do here. If $f(x)$ is a 2π-periodic piecewise smooth function, the complex form of the Fourier series of $f(x)$ is

$$\sum_{n=-\infty}^{\infty} c_n e^{inx},$$

where the Fourier coefficients are given by

$$c_n = \frac{1}{2\pi} \int_{-\pi}^{\pi} f(t) e^{-int}\, dt.$$

Here the Nth partial sum,

$$S_N(x) = \sum_{n=-N}^{N} c_n e^{inx},$$

is the same as the usual partial sum (check this).

Now, if $f(x) = e^{ax}$ for $-\pi < x < \pi$, then

$$\begin{aligned} c_n &= \frac{1}{2\pi}\int_{-\pi}^{\pi} e^{ax}e^{-inx}\,dx = \frac{1}{2\pi}\frac{e^{(a-in)x}}{a-in}\bigg|_{-\pi}^{\pi} \\ &= \frac{1}{2\pi}\frac{e^{(a-in)\pi} - e^{-(a-in)\pi}}{a-in} = \frac{1}{2\pi}\frac{e^{\pi a}(-1)^n - e^{-\pi a}(-1)^n}{a-in} \\ &= \frac{(-1)^n \sinh \pi a}{\pi(a-in)} = \frac{(-1)^n(a+in)\sinh \pi a}{\pi(a^2+n^2)}, \end{aligned}$$

and the Fourier series of f is

$$\begin{aligned} &\frac{\sinh \pi a}{\pi}\sum_{n=-\infty}^{\infty}\frac{(-1)^n(a+in)}{(a^2+n^2)}e^{inx} \\ &\qquad = \frac{\sinh \pi a}{\pi}\sum_{n=-\infty}^{\infty}\frac{(-1)^n}{a^2+n^2}(a\cos nx - n\sin nx), \end{aligned}$$

where we used the fact that

$$\begin{aligned} (a+in)e^{inx} &= (a+in)(\cos nx + i\sin nx) \\ &= (a\cos nx - n\sin nx) + i(a\sin nx + n\cos nx) \end{aligned}$$

and the fact that the Fourier series of a real-valued function is real valued, so that

$$\sum_{n=-\infty}^{\infty}\frac{(-1)^n}{a^2+n^2}(a\sin nx + n\cos nx) = 0.$$

Since the function f is piecewise smooth and is continuous for $-\pi < x < \pi$, we have

$$e^{ax} = \frac{\sinh \pi a}{\pi}\sum_{n=-\infty}^{\infty}\frac{(-1)^n}{a^2+n^2}(a\cos nx - n\sin nx)$$

for $-\pi < x < \pi$.

Exercise 11.10. ✗✗✗

For any complex number $z \in \mathbb{C}$ with $z \neq 1$, show the identity

$$1 + z + z^2 + \cdots + z^n = \frac{1 - z^{n+1}}{1 - z}$$

and then use it to derive *Lagrange's trigonometric identity*:

$$1 + \cos\theta + \cos 2\theta + \cdots + \cos n\theta = \frac{1}{2} + \frac{\sin\left[(2n+1)\theta/2\right]}{2\,\sin(\theta/2)} \qquad (0 < \theta < 2\pi).$$

Use the fact that $\cos n\theta = \text{Re}\left(e^{in\theta}\right)$.

Solution. If $z \neq 1$, then

$$\begin{aligned}(1 - z)(1 + z + z^2 + \cdots + z^n) &= 1 + z + z^2 + \cdots + z^n \\ &\qquad - (z + z^2 + \cdots + z^{n+1}) \\ &= 1 - z^{n+1},\end{aligned}$$

so that

$$1 + z + z^2 + \cdots + z^n = \begin{cases} \dfrac{1 - z^{n+1}}{1 - z}, & \text{if } z \neq 1 \\ n + 1, & \text{if } z = 1. \end{cases}$$

Taking $z = e^{i\theta}$, where $0 < \theta < 2\pi$, then $z \neq 1$, so that

$$\sum_{k=0}^{n} e^{ki\theta} = \frac{1 - e^{(n+1)i\theta}}{1 - e^{i\theta}},$$

and

$$\begin{aligned}\sum_{k=0}^{n} e^{ki\theta} &= \frac{1 - e^{(n+1)i\theta}}{-e^{i\theta/2}\left(e^{i\theta/2} - e^{-i\theta/2}\right)} \\ &= \frac{-e^{-i\theta/2}\left(1 - e^{(n+1)i\theta}\right)}{2i\sin(\theta/2)} \\ &= \frac{i\left(e^{-i\theta/2} - e^{(n+\frac{1}{2})i\theta}\right)}{2\sin(\theta/2)} \\ &= \frac{1}{2} + \frac{\sin\left(n + \frac{1}{2}\right)\theta}{2\sin(\theta/2)} \\ &\qquad + \frac{i}{2\sin(\theta/2)}\left(\cos(\theta/2) - \cos\left(n + \tfrac{1}{2}\right)\theta\right).\end{aligned}$$

Equating real and imaginary parts, we have

$$1 + \cos\theta + \cos 2\theta + \cdots + \cos n\theta = \frac{1}{2} + \frac{\sin\left(n + \frac{1}{2}\right)\theta}{2\sin(\theta/2)}$$

for $0 < \theta < 2\pi$, and as an added bonus,

$$\sin\theta + \sin 2\theta + \cdots + \sin n\theta = \frac{1}{2}\cot(\theta/2) - \frac{\cos\left(n + \frac{1}{2}\right)\theta}{2\sin(\theta/2)}$$

for $0 < \theta < 2\pi$.

Exercise 11.11. XX

Let $f(x) = \cosh x$, $-\pi \leqslant x \leqslant \pi$, $f(x + 2\pi) = f(x)$.

(a) Find the Fourier series of f.

Note: $\cosh x = \frac{1}{2}(e^x + e^{-x})$.

(b) For which values of $x \in [-\pi, \pi]$ does the Fourier series of f converge to $f(x)$?

(c) Evaluate the infinite series $\sum_{n=1}^{\infty} \dfrac{1}{n^2 + 1}$ using part (b).

Solution.

(a) Writing $f(x) = a_0 + \sum_{n=1}^{\infty}(a_n \cos nx + b_n \sin nx)$, the coefficients in the Fourier series are computed as follows: Since $f(x) = \cosh x$ is an even function on the interval $[-\pi, \pi]$, then $b_n = 0$ for all $n \geqslant 1$, and

$$a_0 = \frac{1}{2\pi}\int_{-\pi}^{\pi} \cosh x \, dx = \frac{1}{2\pi} \sinh x \Big|_{-\pi}^{\pi} = \frac{\sinh \pi}{\pi},$$

while for $n \geqslant 1$,

$$\begin{aligned}
a_n &= \frac{1}{\pi}\int_{-\pi}^{\pi} \cosh x \cos nx \, dx = \frac{1}{2\pi}\int_{-\pi}^{\pi} \left(e^x + e^{-x}\right)\cos nx \, dx \\
&= \frac{1}{2\pi}\int_{-\pi}^{\pi} e^x \cos nx \, dx + \frac{1}{2\pi}\int_{-\pi}^{\pi} e^{-x} \cos nx \, dx \\
&= \frac{1}{2\pi}\left\{\frac{e^x(\cos nx + n \sin nx)}{n^2 + 1}\right\}\Bigg|_{-\pi}^{\pi} \\
&\qquad + \frac{1}{2\pi}\left\{\frac{e^{-x}(-\cos nx + n \sin nx)}{n^2 + 1}\right\}\Bigg|_{-\pi}^{\pi} \\
&= \frac{2(-1)^n}{n^2 + 1}\,\frac{\sinh \pi}{\pi},
\end{aligned}$$

where we integrated by parts twice to get

$$\int e^{x} \cos nx \, dx = \frac{e^{x}(\cos nx + n \sin nx)}{n^2+1}$$

and

$$\int e^{-x} \cos nx \, dx = \frac{e^{-x}(-\cos nx + n \sin nx)}{n^2+1}.$$

The Fourier series for f is therefore

$$\cosh x \sim 2\frac{\sinh \pi}{\pi}\left\{\frac{1}{2} + \sum_{n=1}^{\infty} \frac{(-1)^n}{n^2+1} \cos nx\right\}$$

for $-\infty < x < \infty$.

(b) From Dirichlet's theorem, the Fourier series converges to $f(x)$ for all x in the interval $[-\pi, \pi]$, since f is piecewise smooth on $[-\pi, \pi]$, continuous at each $x \in [-\pi, \pi]$, and $f(-\pi^+) = f(\pi^-)$.

(c) Again, from Dirichlet's theorem, the Fourier series converges to $\cosh \pi$ at $x = \pi$, so

$$\cosh \pi = 2\frac{\sinh \pi}{\pi}\left\{\frac{1}{2} + \sum_{n=1}^{\infty} \frac{(-1)^n}{n^2+1} \cos n\pi\right\};$$

therefore,

$$\sum_{n=1}^{\infty} \frac{1}{n^2+1} = \frac{\pi \coth \pi - 1}{2}.$$

Exercise 11.12. ✗✗

Sketch the Fourier series of $f(x)$ and determine the Fourier coefficients for the following functions defined on the interval $-a \leqslant x \leqslant a$.

(a) $f(x) = \begin{cases} 1, & \text{for } |x| < a/2, \\ 0, & \text{for } |x| > a/2. \end{cases}$

(b) $f(x) = \begin{cases} 1, & \text{if } 0 < x < a, \\ 0, & \text{if } -a < x < 0. \end{cases}$

Solution.

(a) From Dirichlet's theorem the Fourier series of $f(x)$ converges to

$$\frac{1}{2}\left[f(x^+) + f(x^-)\right].$$

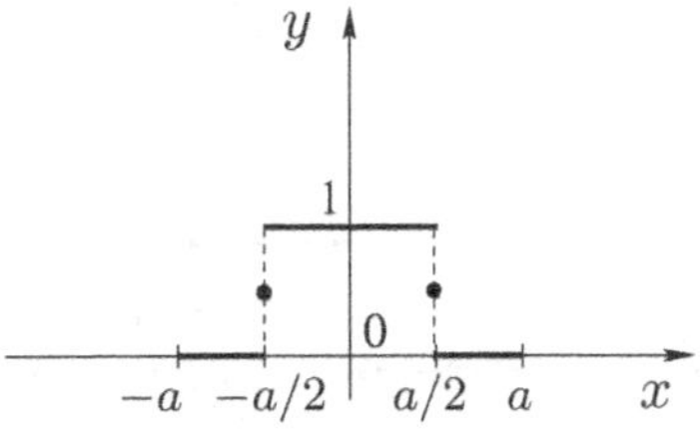

Figure 11.3

The graph of the Fourier series of $f(x)$ on the interval $-a \leqslant x \leqslant a$ is shown in Figure 11.3 Since $f(x)$ is an even piecewise smooth function on the interval $[-a, a]$, it has a Fourier series representation of the form

$$f(x) \sim a_0 + \sum_{n=1}^{\infty} a_n \cos \frac{n\pi x}{a}$$

where

$$a_0 = \frac{1}{a} \int_0^a f(x)\, dx = \frac{1}{a} \int_0^{a/2} 1\, dx = \frac{1}{2}$$

and

$$\begin{aligned} a_n &= \frac{2}{a} \int_0^a f(x) \cos \frac{n\pi x}{a}\, dx = \frac{2}{a} \int_0^{a/2} \cos \frac{n\pi x}{a}\, dx \\ &= \frac{2}{n\pi} \sin \frac{n\pi x}{a} \bigg|_0^{a/2} = \frac{2}{n\pi} \sin \frac{n\pi}{2} \end{aligned}$$

for $n \geqslant 1$; that is,

$$a_n = \begin{cases} \dfrac{2(-1)^k}{\pi(2k+1)}, & \text{if } n = 2k+1 \text{ is odd,} \\ 0, & \text{if } n = 2k \text{ is even.} \end{cases}$$

Hence, the Fourier series of $f(x)$ is

$$\frac{1}{2} + \frac{2}{\pi} \sum_{k=0}^{\infty} \frac{(-1)^k}{(2k+1)} \cos \frac{(2k+1)\pi x}{a}.$$

(b) Again, from Dirichlet's theorem the Fourier series of $f(x)$ converges to

$$\frac{1}{2}\left[f(x^+) + f(x^-)\right].$$

The graph of the Fourier series of $f(x)$ on the interval $-a \leqslant x \leqslant a$ is shown in Figure 11.4. We could compute the Fourier series directly, but instead we

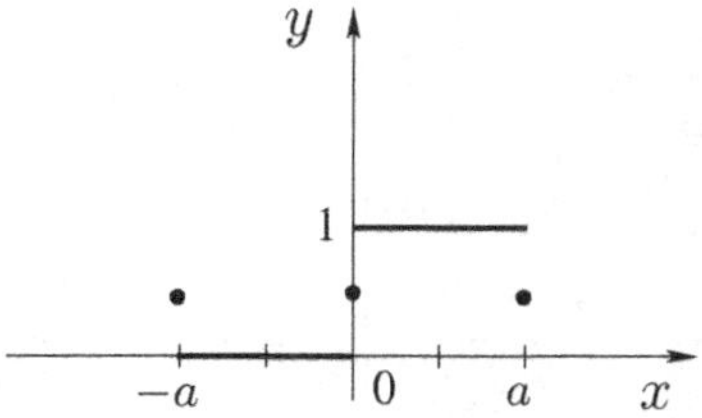

Figure 11.4

observe that $f(x) - \frac{1}{2}$ is an odd piecewise smooth function on the interval $[-a, a]$. Hence, it has a Fourier series representation of the form

$$f(x) - \frac{1}{2} \sim \sum_{n=1}^{\infty} b_n \sin \frac{n\pi x}{a},$$

where

$$\begin{aligned} b_n &= \frac{2}{a} \int_0^a \left(f(x) - \frac{1}{2} \right) \sin \frac{n\pi x}{a}\, dx = \frac{1}{a} \int_0^a \sin \frac{n\pi x}{a}\, dx \\ &= -\frac{1}{n\pi} \cos \frac{n\pi x}{a} \bigg|_0^a = -\frac{1}{n\pi} (\cos n\pi - 1) \\ &= \frac{1}{n\pi} [1 - (-1)^n] \end{aligned}$$

for $n \geqslant 1$; that is,

$$b_n = \begin{cases} \dfrac{2}{\pi(2k+1)}, & \text{if } n = 2k+1 \text{ is odd,} \\ 0, & \text{if } n = 2k \text{ is even.} \end{cases}$$

Hence, the Fourier series of $f(x)$ is

$$f(x) \sim \frac{1}{2} + \frac{2}{\pi} \sum_{k=0}^{\infty} \frac{1}{2k+1} \sin \frac{(2k+1)\pi x}{a}.$$

Exercise 11.13. ✗✗

Show that finding the Fourier series operation is a linear operation; that is, the Fourier series of

$$c_1 f(x) + c_2 g(x)$$

is the sum of c_1 times the Fourier series of $f(x)$ plus c_2 times the Fourier series of $g(x)$.

Solution. Suppose that the Fourier series of f and g are given by

$$f(x) \sim A_0 + \sum_{n=1}^{\infty} \left(A_n \cos \frac{n\pi x}{a} + B_n \sin \frac{n\pi x}{a} \right)$$

and

$$g(x) \sim C_0 + \sum_{n=1}^{\infty} \left(C_n \cos \frac{n\pi x}{a} + D_n \sin \frac{n\pi x}{a} \right),$$

where

$$A_0 = \frac{1}{2a} \int_{-a}^{a} f(x)\, dx, \quad A_n = \frac{1}{a} \int_{-a}^{a} f(x) \cos \frac{n\pi x}{a}\, dx,$$

$$B_n = \frac{1}{a} \int_{-a}^{a} f(x) \sin \frac{n\pi x}{a}\, dx$$

for $n \geqslant 1$, and

$$C_0 = \frac{1}{2a} \int_{-a}^{a} g(x)\, dx, \quad C_n = \frac{1}{a} \int_{-a}^{a} g(x) \cos \frac{n\pi x}{a}\, dx,$$

$$D_n = \frac{1}{a} \int_{-a}^{a} g(x) \sin \frac{n\pi x}{a}\, dx$$

for $n \geqslant 1$. If c_1 and c_2 are scalars, and the Fourier series of $c_1 f + c_2 g$ is given by

$$\left[c_1 f + c_2 g\right](x) \sim E_0 + \sum_{n=1}^{\infty} \left(E_n \cos \frac{n\pi x}{a} + F_n \sin \frac{n\pi x}{a} \right),$$

then

$$E_0 = \frac{1}{2a} \int_{-a}^{a} \left[c_1 f(x) + c_2 g(x)\right]\, dx = \frac{c_1}{2a} \int_{-a}^{a} f(x)\, dx + \frac{c_2}{2a} \int_{-a}^{a} g(x)\, dx;$$

that is,

$$E_0 = c_1 A_0 + c_2 C_0.$$

Also,

$$\begin{aligned} E_n &= \frac{1}{a} \int_{-a}^{a} \left[c_1 f(x) + c_2 g(x)\right] \cos \frac{n\pi x}{a}\, dx \\ &= \frac{c_1}{a} \int_{-a}^{a} f(x) \cos \frac{n\pi x}{a}\, dx + \frac{c_2}{a} \int_{-a}^{a} g(x) \cos \frac{n\pi x}{a}\, dx, \end{aligned}$$

so that

$$E_n = c_1 A_n + c_2 C_n$$

for $n \geqslant 1$. Similarly,

$$F_n = \frac{1}{a}\int_{-a}^{a} [c_1 f(x) + c_2 g(x)] \sin\frac{n\pi x}{a}\, dx$$
$$= \frac{c_1}{a}\int_{-a}^{a} f(x)\sin\frac{n\pi x}{a}\, dx + \frac{c_2}{a}\int_{-a}^{a} g(x)\sin\frac{n\pi x}{a}\, dx$$

and

$$F_n = c_1 B_n + c_2 D_n$$

for $n \geqslant 1$. Therefore, the Fourier series for $c_1 f + c_2 g$ is

$$\big[c_1 f + c_2 g\big](x) \sim c_1 A_0 + c_2 C_0 + \sum_{n=1}^{\infty}\left[(c_1 A_n + c_2 C_n)\cos\frac{n\pi x}{a} + (c_1 B_n + c_2 D_n)\sin\frac{n\pi x}{a}\right];$$

that is,

$$\big[c_1 f + c_2 g\big](x) \sim c_1\left[A_0 + \sum_{n=1}^{\infty}\left(A_n\cos\frac{n\pi x}{a} + B_n\sin\frac{n\pi x}{a}\right)\right] + c_2\left[C_0 + \sum_{n=1}^{\infty}\left(C_n\cos\frac{n\pi x}{a} + D_n\sin\frac{n\pi x}{a}\right)\right]$$
$$\sim c_1 f(x) + c_2 g(x).$$

Exercise 11.14. ✗
Show that e^x is the sum of an even function and an odd function.

Solution. We can write

$$e^x = \frac{e^x + e^{-x}}{2} + \frac{e^x - e^{-x}}{2} = \cosh x + \sinh x,$$

and $\cosh x$ is an even function while $\sinh x$ is an odd function. In general, if $f(x)$ is an arbitrary function, we can write

$$f(x) = f_{\text{even}}(x) + f_{\text{odd}}(x),$$

where

$$f_{\text{even}}(x) = \frac{f(x) + f(-x)}{2}$$

is even, and

$$f_{\text{odd}}(x) = \frac{f(x) - f(-x)}{2}$$

is odd.

Exercise 11.15. XX
Given the function

$$f(x) = \cos\frac{\pi x}{a}, \quad 0 \leqslant x < a,$$

find the Fourier sine series for f.

Solution. Writing

$$f(x) = \cos\frac{\pi x}{a} \sim \sum_{n=1}^{\infty} b_n \sin\frac{n\pi x}{a},$$

the coefficients b_n in the Fourier sine series are computed as follows:

$$\begin{aligned}
b_n &= \frac{2}{a}\int_0^a \cos\frac{\pi x}{a}\sin\frac{n\pi x}{a}\,dx = \frac{1}{a}\int_0^a \Big(\sin\frac{(n+1)\pi x}{a} + \sin\frac{(n-1)\pi x}{a}\Big)\,dx \\
&= \frac{1}{\pi}\left(-\frac{1}{n+1}\cos\frac{(n+1)\pi x}{a}\bigg|_0^a\right) + \frac{1}{\pi}\left(-\frac{1}{n-1}\cos\frac{(n-1)\pi x}{a}\bigg|_0^a\right) \\
&= \frac{1}{\pi(n+1)}\left((-1)^n + 1\right) + \frac{1}{\pi(n-1)}\left((-1)^n + 1\right) \\
&= \frac{1+(-1)^n}{\pi}\left(\frac{1}{n+1} + \frac{1}{n-1}\right) = \frac{1+(-1)^n}{\pi}\,\frac{2n}{n^2-1}.
\end{aligned}$$

Therefore,

$$b_n = \begin{cases} \dfrac{4n}{\pi(n^2-1)}, & \text{if } n \text{ is even}, \\ 0, & \text{if } n \text{ is odd}, \quad n \geqslant 3. \end{cases}$$

For $n = 1$,

$$b_1 = \frac{2}{a}\int_0^a \sin\frac{\pi x}{a}\cos\frac{\pi x}{a}\,dx = \frac{1}{\pi}\sin^2\frac{\pi x}{a}\bigg|_0^a = 0.$$

The Fourier sine series for f is therefore

$$\cos\frac{\pi x}{a} \sim \frac{8}{\pi}\sum_{n=1}^{\infty}\frac{n}{4n^2-1}\sin\frac{2n\pi x}{a}$$

for $0 \leqslant x < a$.

Exercise 11.16. ✗✗✗

For each of the following functions, find the Fourier series of $f(x)$, the Fourier sine series of $f(x)$ on $[0, a]$, and the Fourier cosine series of $f(x)$ on $[0, a]$, and sketch the appropriate extensions of the functions and all the series involved.

(a) $f(x) = \begin{cases} x, & -a < x < 0, \\ 1 + x, & 0 < x < a. \end{cases}$

(b) $f(x) = \begin{cases} 2, & -a < x < 0, \\ e^{-x}, & 0 < x < a. \end{cases}$

Solution.

(a) *Fourier Series*: The graphs of $f(x)$ and the Fourier series of $f(x)$ are shown in Figure 11.5 The Fourier series representation of $f(x)$ is

$$f(x) \sim a_0 + \sum_{n=1}^{\infty} \left(a_n \cos \frac{n\pi x}{a} + b_n \sin \frac{n\pi x}{a} \right),$$

where

$$a_0 = \frac{1}{2a} \int_{-a}^{a} f(x)\, dx = \frac{1}{2a} \int_{-a}^{a} x\, dx + \frac{1}{2a} \int_{0}^{a} 1\, dx = \frac{1}{2a}\, a = \frac{1}{2},$$

since the function x is an odd function on $[-a, a]$.

For $n \geqslant 1$,

$$a_n = \frac{1}{a} \int_{-a}^{a} f(x) \cos \frac{n\pi x}{a}\, dx = \frac{1}{a} \int_{-a}^{a} x \cos \frac{n\pi x}{a}\, dx + \frac{1}{a} \int_{0}^{a} \cos \frac{n\pi x}{a}\, dx$$

$$= \frac{1}{n\pi} \sin \frac{n\pi x}{a} \bigg|_0^a = 0,$$

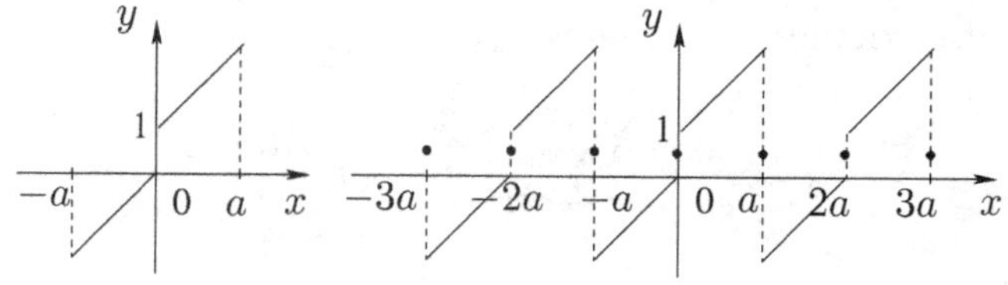

Figure 11.5 Fourier series.

since $x\cos(n\pi x/a)$ is an odd function on $[-a, a]$, and

$$b_n = \frac{1}{a}\int_{-a}^{a} f(x)\sin\frac{n\pi x}{a}\,dx = \frac{1}{a}\int_{-a}^{a} x\sin\frac{n\pi x}{a}\,dx + \frac{1}{a}\int_{0}^{a}\sin\frac{n\pi x}{a}\,dx$$

$$= \frac{2}{a}\int_{0}^{a} x\sin\frac{n\pi x}{a}\,dx + \frac{1}{a}\int_{0}^{a}\sin\frac{n\pi x}{a}\,dx$$

$$= \frac{2}{a}\left[-\frac{a}{n\pi}x\cos\frac{n\pi x}{a}\bigg|_0^a + \frac{a}{n\pi}\int_0^a \cos\frac{n\pi x}{a}\,dx\right] + \frac{1}{a}\left(-\frac{a}{n\pi}\cos\frac{n\pi x}{a}\right)\bigg|_0^a$$

$$= \frac{1}{n\pi}\left[1-(-1)^n\right] - \frac{2a(-1)^n}{n\pi}.$$

Hence, the Fourier series of $f(x)$ is

$$\frac{1}{2} + \frac{1}{\pi}\sum_{n=1}^{\infty}\frac{[1-(-1)^n] - 2a(-1)^n}{n}\sin\frac{n\pi x}{a}$$

which is a sine series plus a constant term.

Fourier Sine Series: The graphs of the odd extension of $f(x)$ to the interval $[-a, a]$ and the Fourier sine series of $f(x)$ are shown in Figure 11.6. The Fourier sine series representation of $f(x)$ is

$$f(x) \sim \sum_{n=1}^{\infty} b_n \sin\frac{n\pi x}{a},$$

where

$$b_n = \frac{2}{a}\int_0^a f(x)\sin\frac{n\pi x}{a}\,dx = \frac{2}{a}\int_0^a (1+x)\sin\frac{n\pi x}{a}\,dx$$

$$= \frac{2}{a}\left[-\frac{a}{n\pi}(1+x)\cos\frac{n\pi x}{a}\bigg|_0^a + \frac{a}{n\pi}\int_0^a\cos\frac{n\pi x}{a}\,dx\right]$$

$$= \frac{2}{n\pi}\left[1-(-1)^n\right] - \frac{2a(-1)^n}{n\pi}.$$

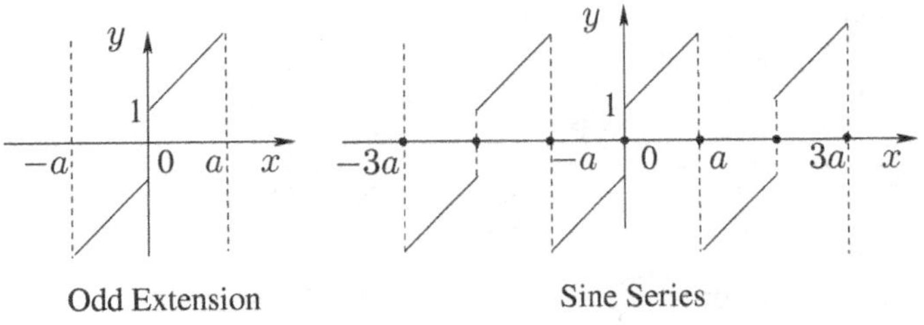

Figure 11.6 Fourier sine series.

Hence, the Fourier sine series of $f(x)$ is

$$\frac{2}{\pi}\sum_{n=1}^{\infty}\frac{[1-(-1)^n]-a(-1)^n}{n}\sin\frac{n\pi x}{a}.$$

Fourier Cosine Series: The graphs of the even extension of $f(x)$ to the interval $[-a, a]$ and the Fourier cosine series of $f(x)$ are shown below. The Fourier cosine series representation of $f(x)$ is

$$f(x) \sim a_0 + \sum_{n=1}^{\infty} a_n \cos\frac{n\pi x}{a},$$

where

$$a_0 = \frac{1}{a}\int_0^a f(x)\,dx = \frac{1}{a}\int_0^a (1+x)\,dx = \frac{1}{a}\left[x\Big|_0^a + \frac{x^2}{2}\Big|_0^a\right] = 1 + \frac{a}{2},$$

and for $n \geqslant 1$,

$$\begin{aligned} a_n &= \frac{2}{a}\int_0^a f(x)\cos\frac{n\pi x}{a}\,dx = \frac{2}{a}\int_0^a (1+x)\cos\frac{n\pi x}{a}\,dx \\ &= \frac{2}{a}\left[\frac{a}{n\pi}(1+x)\sin\frac{n\pi x}{a}\Big|_0^a - \frac{a}{n\pi}\int_0^a \sin\frac{n\pi x}{a}\,dx\right] \\ &= \frac{2a}{n^2\pi^2}\cos\frac{n\pi x}{a}\Big|_0^a = \frac{2a}{n^2\pi^2}\left[(-1)^n - 1\right]; \end{aligned}$$

that is,

$$a_n = \begin{cases} -\dfrac{4a}{\pi^2(2k+1)^2}, & \text{if } n = 2k+1 \text{ is odd,} \\ 0, & \text{if } n = 2k \text{ is even.} \end{cases}$$

Hence, the Fourier cosine series of $f(x)$ is

$$1 + \frac{a}{2} - \frac{4a}{\pi^2}\sum_{k=0}^{\infty}\frac{1}{(2k+1)^2}\cos\frac{(2k+1)\pi x}{a}.$$

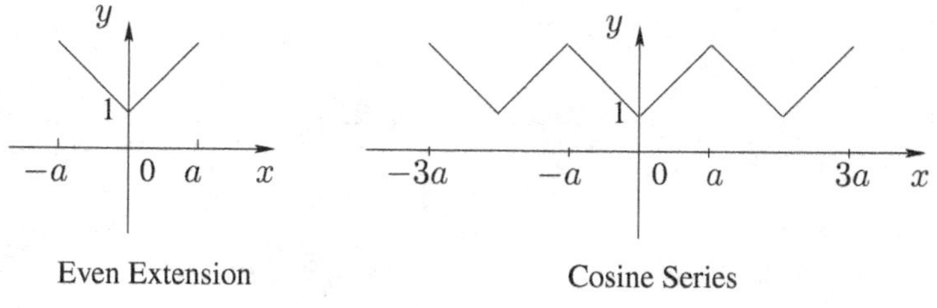

Figure 11.7 Fourier cosine series

(b) *Fourier Series*: The graphs of $f(x)$ and the Fourier series of $f(x)$ are shown below. The Fourier series representation of $f(x)$ is

$$f(x) \sim a_0 + \sum_{n=1}^{\infty} \left(a_n \cos \frac{n\pi x}{a} + b_n \sin \frac{n\pi x}{a} \right),$$

where

$$a_0 = \frac{1}{2a} \int_{-a}^{a} f(x)\, dx = \frac{1}{2a} \int_{-a}^{0} 2\, dx + \frac{1}{2a} \int_{0}^{a} e^{-x}\, dx$$
$$= \frac{1}{2a} \left(2a + 1 - e^{-a} \right).$$

Using integration by parts twice, we obtain

$$\int e^{\alpha x} \cos \beta x\, dx = \frac{e^{\alpha x}}{\alpha^2 + \beta^2} \left(\beta \sin \beta x + \alpha \cos \beta x \right),$$

so that

$$a_n = \frac{1}{a} \int_{-a}^{a} f(x) \cos \frac{n\pi x}{a}\, dx = \frac{1}{a} \int_{-a}^{0} 2 \cos \frac{n\pi x}{a}\, dx$$
$$+ \frac{1}{a} \int_{0}^{a} e^{-x} \cos \frac{n\pi x}{a}\, dx$$
$$= \frac{2}{n\pi} \sin \frac{n\pi x}{a} \Big|_{-a}^{0} + \frac{n\pi}{a^2 + n^2\pi^2} e^{-x} \sin \frac{n\pi x}{a} \Big|_{0}^{a}$$
$$- \frac{a}{a^2 + n^2\pi^2} e^{-x} \cos \frac{n\pi x}{a} \Big|_{0}^{a}$$
$$= \frac{a}{a^2 + n^2\pi^2} \left[1 - e^{-a}(-1)^n \right]$$

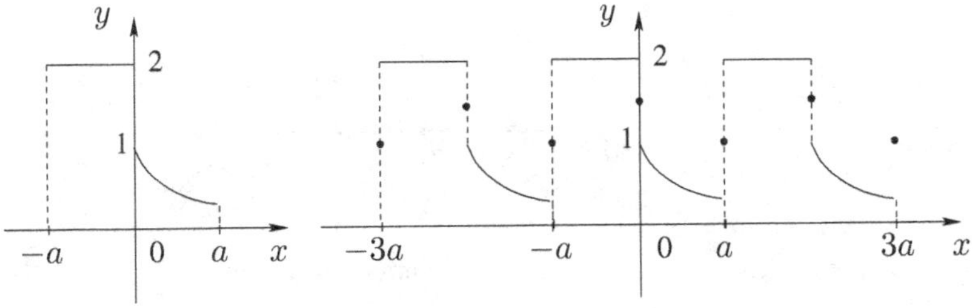

Figure 11.8 Fourier series.

for $n \geqslant 1$. Similarly, we have

$$\int e^{\alpha x} \sin \beta x \, dx = \frac{e^{\alpha x}}{\alpha^2 + \beta^2} (\alpha \sin \beta x - \beta \cos \beta x),$$

so that

$$b_n = \frac{1}{a} \int_{-a}^{a} f(x) \sin \frac{n\pi x}{a} \, dx = \frac{1}{a} \int_{-a}^{0} 2 \sin \frac{n\pi x}{a} \, dx + \frac{1}{a} \int_{0}^{a} e^{-x} \sin \frac{n\pi x}{a} \, dx$$

and

$$\begin{aligned} b_n &= -\frac{2}{n\pi} \cos \frac{n\pi x}{a} \bigg|_{-a}^{0} - \frac{a}{a^2 + n^2\pi^2} e^{-x} \sin \frac{n\pi x}{a} \bigg|_{0}^{a} \\ &\qquad - \frac{n\pi}{a^2 + n^2\pi^2} e^{-x} \cos \frac{n\pi x}{a} \bigg|_{0}^{a} \\ &= -\frac{2}{n\pi} [1 - (-1)^n] + \frac{n\pi}{a^2 + n^2\pi^2} \left[1 - e^{-a}(-1)^n\right] \end{aligned}$$

for $n \geqslant 1$. Hence the Fourier series of $f(x)$ is

$$\frac{2a + 1 - e^{-a}}{2a} + \sum_{n=1}^{\infty} \left(a_n \cos \frac{n\pi x}{a} + b_n \sin \frac{n\pi x}{a} \right),$$

where a_n and b_n are as computed above.

Fourier Sine Series: The graphs of the odd extension of $f(x)$ to the interval $[-a, a]$ and the Fourier sine series of $f(x)$ are shown in Figure 11.9. The Fourier sine series representation of $f(x)$ is

$$f(x) \sim \sum_{n=1}^{\infty} b_n \sin \frac{n\pi x}{a},$$

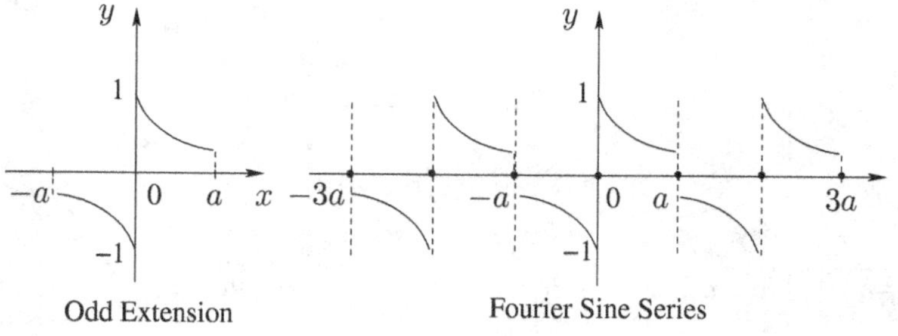

Figure 11.9 Fourier sine series.

where

$$b_n = \frac{2}{a}\int_0^a f(x)\sin\frac{n\pi x}{a}\,dx = \frac{2}{a}\int_0^a e^{-x}\sin\frac{n\pi x}{a}\,dx$$

$$= -\frac{2a}{a^2+n^2\pi^2}e^{-x}\sin\frac{n\pi x}{a}\bigg|_0^a - \frac{2n\pi}{a^2+n^2\pi^2}e^{-x}\cos\frac{n\pi x}{a}\bigg|_0^a$$

$$= \frac{2n\pi}{a^2+n^2\pi^2}\left[1-e^{-a}(-1)^n\right]$$

for $n \geqslant 1$, and the Fourier sine series of $f(x)$ is

$$2\pi\sum_{n=1}^{\infty}\frac{1-e^{-a}(-1)^n}{a^2+n^2\pi^2}\,n\sin\frac{n\pi x}{a}.$$

Fourier Cosine Series: The graphs of the even extension of $f(x)$ to the interval $[-a, a]$ and the Fourier cosine series of $f(x)$ are shown in Figure 11.10. The Fourier cosine series representation of $f(x)$ is

$$f(x) \sim a_0 + \sum_{n=1}^{\infty} a_n \cos\frac{n\pi x}{a},$$

where

$$a_0 = \frac{1}{a}\int_0^a f(x)\,dx = \frac{1}{a}\int_0^a e^{-x}\,dx = -\frac{1}{a}e^{-x}\bigg|_0^a = \frac{1}{a}\left(1-e^{-a}\right).$$

Since

$$\int e^{\alpha x}\cos\beta x\,dx = \frac{e^{\alpha x}}{\alpha^2+\beta^2}\left(\beta\sin\beta x+\alpha\cos\beta x\right),$$

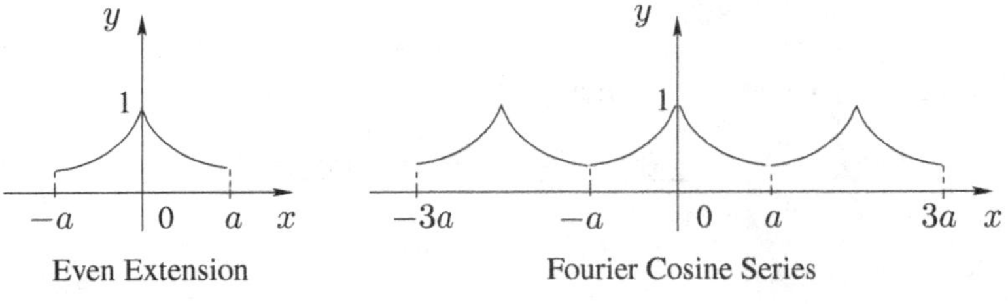

Figure 11.10 Fourier cosine series.

then

$$\begin{aligned}a_n &= \frac{2}{a}\int_0^a f(x)\cos\frac{n\pi x}{a}\,dx = \frac{2}{a}\int_0^a e^{-x}\cos\frac{n\pi x}{a}\,dx \\ &= \frac{2n\pi}{a^2+n^2\pi^2}e^{-x}\sin\frac{n\pi x}{a}\bigg|_0^a - \frac{2a}{a^2+n^2\pi^2}e^{-x}\cos\frac{n\pi x}{a}\bigg|_0^a \\ &= \frac{2a}{a^2+n^2\pi^2}\left[1-e^{-a}(-1)^n\right]\end{aligned}$$

for $n \geqslant 1$, and the Fourier cosine series of $f(x)$ is

$$\frac{1-e^{-a}}{a} + 2a\sum_{n=1}^{\infty}\frac{1-e^{-a}(-1)^n}{a^2+n^2\pi^2}\cos\frac{n\pi x}{a}.$$

Exercise 11.17. ✗✗✗✗

Consider the integral $\displaystyle\int_0^1 \frac{dx}{1+x^2}$.

(a) Evaluate the integral explicitly.

(b) Use the Taylor series of $1/(1+x^2)$ (a geometric series) to obtain an infinite series for the integral.

(c) Equate part (a) to part (b) to derive a formula for π.

Solution.

(a) Since

$$\frac{d}{dx}\left(\tan^{-1}x\right) = \frac{1}{1+x^2},$$

we have

$$\int_0^1 \frac{1}{1+x^2}\,dx = \tan^{-1}x\bigg|_0^1 = \tan^{-1}1 - \tan^{-1}0 = \frac{\pi}{4}.$$

(b) Recall that the geometric series

$$\frac{1}{1+t^2} = 1 - t^2 + t^4 - t^6 + \cdots,$$

that is,

$$\frac{1}{1+t^2} = \sum_{n=0}^{\infty}(-1)^n t^{2n},$$

converges for all $-1 < t < 1$. Integrating from 0 to x, where $|x| < 1$, we get

$$\begin{aligned}\int_0^x \frac{1}{1+t^2}\,dt &= \int_0^x \left(\sum_{n=0}^{\infty}(-1)^n t^{2n}\right)\,dt \\ &= \sum_{n=0}^{\infty}(-1)^n \int_0^x t^{2n}\,dt \\ &= \sum_{n=0}^{\infty}(-1)^n \frac{t^{2n+1}}{2n+1}\bigg|_0^x \\ &= \sum_{n=0}^{\infty}\frac{(-1)^n}{2n+1}x^{2n+1},\end{aligned}$$

and therefore

$$\tan^{-1} x = \sum_{n=0}^{\infty}\frac{(-1)^n}{2n+1}x^{2n+1} = x - \frac{1}{3}x^3 + \frac{1}{5}x^5 - \frac{1}{7}x^7 + \cdots$$

for $-1 < x < 1$, and this is *Gregory's series* for $\tan^{-1} x$, discovered by James Gregory about 1670.

Letting $x \to 1^-$, a theorem of Abel (see [4], p. 245) tells us that

$$\begin{aligned}\frac{\pi}{4} &= \int_0^1 \frac{1}{1+t^2}\,dt = \lim_{x\to 1^-}\int_0^x \frac{1}{1+t^2}\,dt = \lim_{x\to 1^-}\sum_{n=0}^{\infty}\frac{(-1)^n}{2n+1}x^{2n+1} \\ &= \sum_{n=0}^{\infty}\frac{(-1)^n}{2n+1}\lim_{x\to 1^-}x^{2n+1} = \sum_{n=0}^{\infty}\frac{(-1)^n}{2n+1},\end{aligned}$$

so that

$$\frac{\pi}{4} = 1 - \frac{1}{3} + \frac{1}{5} - \frac{1}{7} + \frac{1}{9} - \frac{1}{11} + \cdots,$$

and this is *Leibniz's formula* for $\pi/4$, discovered by Leibniz in 1673.

(c) From part (b) we have

$$\pi = 4\sum_{n=0}^{\infty}\frac{(-1)^n}{2n+1}.$$

The convergence is very slow, however.

Another proof of Leibniz's formula which doesn't require integrating an infinite series term by term is given below.

$$
\begin{aligned}
1-\frac{1}{3}+\frac{1}{5}-\frac{1}{7}+\cdots+\frac{(-1)^{n-1}}{2n-1} &= \int_0^1 \left(1-x^2+x^4-x^6+\cdots+(-1)^{n-1}x^{2n-2}\right)\,dx \\
&= \int_0^1 \frac{1-x^{2n}}{1+x^2}\,dx \\
&= \int_0^1 \frac{1}{1+x^2}\,dx - \int_0^1 \frac{x^{2n}}{1+x^2}\,dx,
\end{aligned}
$$

and therefore,

$$
\begin{aligned}
\left|\frac{\pi}{4}-\left(1-\frac{1}{3}+\frac{1}{5}-\frac{1}{7}+\cdots+\frac{(-1)^{n-1}}{2n-1}\right)\right| &= \int_0^1 \frac{x^{2n}}{1+x^2}\,dx \\
&\leqslant \int_0^1 x^{2n}\,dx \\
&= \frac{1}{2n+1} \longrightarrow 0
\end{aligned}
$$

as $n \to \infty$.

Exercise 11.18. XXX

Assume that the function $f(x)$ is continuous on $[-a, a]$.

(a) Under what conditions does $f(x)$ equal its Fourier series for all $x \in [-a, a]$?

(b) Under what conditions does $f(x)$ equal its Fourier sine series for all $x \in [0, a]$?

(c) Under what conditions does $f(x)$ equal its Fourier cosine series for all $x \in [0, a]$?

Hint: What does the Fourier series converge to at the endpoints of the interval?

(a) From Dirichlet's theorem, we know that for any x_0 with $-a < x_0 < a$, the Fourier series of f converges to $f(x_0)$, since f is continuous at x_0. We also know that at the endpoints $x = -a$ and $x = a$, the Fourier series converges to

$$\frac{1}{2}\left[f(a^-)+f(-a^+)\right],$$

and if f is continuous at the endpoints, that is, continuous from the left at $x = a$ and continuous from the right at $x = -a$, the Fourier series converges to

$$\frac{f(a) + f(-a)}{2}$$

at $x = a$ and at $x = -a$, so that the Fourier series converges to $f(x)$ for all $x \in [-a, a]$ if and only if $f(a) = f(-a)$.

(b) Again, from Dirichlet's theorem, if $0 < x_0 < a$, the Fourier sine series of f converges to $f(x_0)$ since f is continuous at x_0. If f_{odd} is the odd extension of f to $[-a, a]$, then at $x = 0$, the Fourier sine series of f converges to

$$\begin{aligned}\frac{1}{2}\left[f_{\text{odd}}(0^-) + f_{\text{odd}}(0^+)\right] &= \frac{1}{2}\left[-f(0) + f(0)\right] \\ &= 0,\end{aligned}$$

and the Fourier sine series converges to f at $x = 0$ if and only if $f(0) = 0$.

If f_{odd} is the odd extension of f to $[-a, a]$, then at $x = a$, the Fourier sine series of f converges to

$$\begin{aligned}\frac{1}{2}\left[f_{\text{odd}}(a^-) + f_{\text{odd}}(-a^+)\right] &= \frac{1}{2}\left[f_{\text{odd}}(a) + f_{\text{odd}}(-a)\right] \\ &= \frac{1}{2}\left[f(a) - f(a)\right] = 0,\end{aligned}$$

and the Fourier sine series converges to f at $x = a$ if and only if $f(a) = 0$.

(c) From Dirichlet's theorem, if $0 < x_0 < a$, the Fourier cosine series of f converges to $f(x_0)$ since f is continuous at x_0. If f_{even} is the even extension of f to $[-a, a]$, then at $x = 0$, the Fourier cosine series of f converges to

$$\frac{1}{2}\left[f_{\text{even}}(0^-) + f_{\text{even}}(0^+)\right] = \frac{1}{2}\left[f(0) + f(0)\right] = f(0),$$

and the Fourier cosine series of f converges to f at $x = 0$ if and only if f is continuous from the right at $x = 0$.

If f_{even} is the even extension of f to $[-a, a]$, then at $x = a$, the Fourier cosine series of f converges to

$$\begin{aligned}\frac{1}{2}\left[f_{\text{even}}(a^-) + f_{\text{even}}(-a^+)\right] &= \frac{1}{2}\left[f_{\text{even}}(a) + f_{\text{even}}(-a)\right] = \frac{1}{2}\left[f(a) + f(a)\right] \\ &= f(a),\end{aligned}$$

and the Fourier cosine series of f converges to f at $x = a$ if and only if f is continuous from the left at $x = a$.

Exercise 11.19. XXX

Let f be the periodic function shown below, each segment of which is a semicircle of radius π.

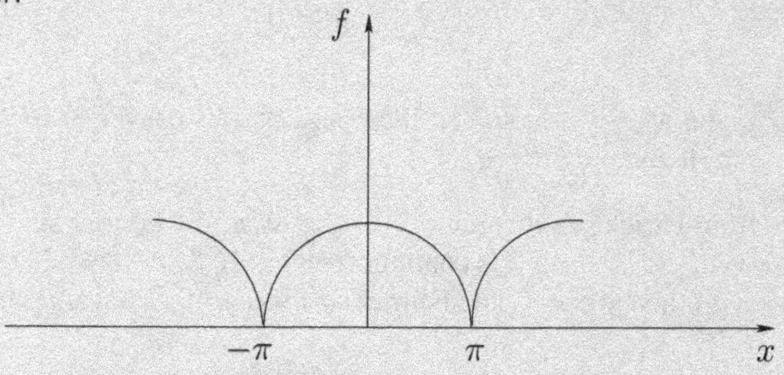

Show that its Fourier series is

$$f(x) = \frac{\pi^2}{4} + \frac{\pi^2}{2}\sum_{n=1}^{\infty}[J_0(n\pi) + J_2(n\pi)]\cos nx,$$

where the J_n's denote Bessel functions of the first kind.

Hint: You may use any of the following formulas:

$$\cos(a\sin\theta) = J_0(a) + 2J_2(a)\cos 2\theta + 2J_4(a)\cos 4\theta + \cdots,$$

$$\sin(a\sin\theta) = 2J_1(a)\sin\theta + 2J_3(a)\sin 3\theta + 2J_5(a)\sin 5\theta + \cdots,$$

$$\cos(a\cos\theta) = J_0(a) - 2J_2(a)\cos 2\theta + 2J_4(a)\cos 4\theta - \cdots,$$

$$\sin(a\cos\theta) = 2J_1(a)\cos\theta - 2J_3(a)\cos 3\theta + 2J_5(a)\cos 5\theta - \cdots,$$

from Corollary 6.7.

Solution. Since $f(x)$ is an even function on the interval $[-\pi, \pi]$, the Fourier series for $f(x)$ is given by

$$f(x) \sim a_0 + \sum_{n=1}^{\infty} a_n \cos nx,$$

where

$$a_0 = \frac{1}{\pi}\int_0^{\pi}\sqrt{\pi^2 - x^2}\,dx = \frac{1}{\pi}\cdot\frac{\pi^3}{4} = \frac{\pi^2}{4},$$

since the integral is just the area of a quarter circle of radius π, and

$$a_n = \frac{2}{\pi}\int_0^{\pi}\sqrt{\pi^2 - x^2}\cos nx\,dx$$

for $n \geqslant 1$.

To compute a_n, we make the substitution $x = \pi \cos\theta$, so that $dx = -\pi \sin\theta \, d\theta$, so that

$$\begin{aligned} a_n &= \frac{2}{\pi} \int_0^{\pi} \sqrt{\pi^2 - x^2} \cos nx \, dx \\ &= \frac{2}{\pi} \int_{\pi/2}^{0} \sqrt{\pi^2 - \pi^2 \cos^2\theta} \cos(n\pi \cos\theta)(-\pi \sin\theta) \, d\theta \\ &= 2\pi \int_0^{\pi/2} \sqrt{\sin^2\theta} \cos(n\pi\cos\theta) \sin\theta \, d\theta \\ &= 2\pi \int_0^{\pi/2} \cos(n\pi\cos\theta) \sin^2\theta \, d\theta \\ &= 2\pi \int_0^{\pi/2} \cos(n\pi\cos\theta) \frac{1}{2}(1 - \cos 2\theta) \, d\theta, \end{aligned}$$

and thus

$$a_n = \pi \int_0^{\pi/2} \cos(n\pi\cos\theta) \, d\theta - \pi \int_0^{\pi/2} \cos(n\pi\cos\theta) \cos 2\theta \, d\theta,$$

Now,

$$\begin{aligned} \int_0^{\pi/2} \cos(n\pi\cos\theta)\cos 2\theta \, d\theta &= \int_0^{\pi/2} \Big[J_0(n\pi)\cos 2\theta - 2J_2(n\pi)\cos^2 2\theta \\ &\qquad + 2J_4(n\pi)\cos 4\theta \cos 2\theta - + \cdots \Big] \, d\theta \\ &= -2J_2(n\pi) \cdot \frac{1}{2} \cdot \frac{\pi}{2}, \end{aligned}$$

and

$$\int_0^{\pi/2} \cos(n\pi\cos\theta)\cos 2\theta \, d\theta = -\frac{\pi}{2} J_2(n\pi)$$

from the orthogonality of the family

$$\{1, \cos 2\theta, \cos 4\theta, \dots\}$$

on the interval $[0, \pi/2]$.

Similarly,

$$\begin{aligned} \int_0^{\pi/2} \cos(n\pi\cos\theta) \, d\theta &= \int_0^{\pi/2} \Big[J_0(n\pi) - 2J_2(n\pi)\cos 2\theta \\ &\qquad + 2J_4(n\pi)\cos 4\theta - + \cdots \Big] \, d\theta \\ &= \frac{\pi}{2} J_0(n\pi), \end{aligned}$$

so that

$$\int_0^{\pi/2} \cos(n\pi \cos\theta)\, d\theta = \frac{\pi}{2} J_0(n\pi).$$

Therefore, since f is continuous everywhere, from Dirichlet's theorem

$$f(x) = \frac{\pi^2}{4} + \frac{\pi^2}{2} \sum_{n=1}^{\infty} \left[J_0(n\pi) + J_2(n\pi) \right] \cos nx$$

for all real numbers x.

CHAPTER 12

STURM-LIOUVILLE PROBLEMS

Exercise 12.1. XXX

Given the differential equation

$$\frac{d^2\phi}{dx^2} + \lambda\phi = 0,$$

determine the eigenvalues λ and corresponding eigenfunctions if ϕ satisfies the following boundary conditions.

(a) $\phi(a) = 0$ and $\phi(b) = 0$.

(b) $\frac{d\phi}{dx}(0) = 0$ and $\frac{d\phi}{dx}(a) = 0$.

(c) $\phi(0) = 0$ and $\frac{d\phi}{dx}(a) = 0$.

(d) $\frac{d\phi}{dx}(0) = 0$ and $\phi(a) = 0$.

(e) $\phi(0) = 0$ and $\frac{d\phi}{dx}(a) + \phi(a) = 0$.

Analyze three cases: $\lambda > 0,\ \lambda = 0,\ \lambda < 0$. You may assume that the eigenvalues are real.

Partial Differential Equations: Theory and Completely Solved Problems, First Edition. By T. Hillen, I. E. Leonard, H. van Roessel

Solution.

(a) If ϕ satisfies the boundary value problem

$$\begin{aligned}
&\phi''(x) + \lambda\phi(x) = 0, \quad a < x < b,\\
&\phi(a) = 0,\\
&\phi(b) = 0,
\end{aligned}$$

we have to consider three cases.

Case 1: If $\lambda = 0$, the equation is

$$\phi''(x) = 0,$$

with general solution

$$\phi(x) = Ax + B, \quad a < x < b.$$

The first boundary condition gives $\phi(a) = Aa + B = 0$, so that $B = -aA$, and the solution is now

$$\phi(x) = A(x - a), \quad a < x < b.$$

The second boundary condition gives $\phi(b) = A(b - a) = 0$, so that $A = 0$. In this case the only solution is the trivial solution $\phi(x) = 0$ for $a < x < b$.

Case 2: If $\lambda < 0$, then $\lambda = -\mu^2$ where $\mu \neq 0$, and the equation is

$$\phi''(x) - \mu^2\phi(x) = 0,$$

with general solution

$$\phi(x) = Ae^{\mu x} + Be^{-\mu x}, \quad a < x < b.$$

The first and second boundary conditions give the following homogeneous system of linear equations for A and B :

$$\begin{aligned}
Ae^{\mu a} + Be^{-\mu a} &= 0,\\
Ae^{\mu b} + Be^{-\mu b} &= 0,
\end{aligned}$$

and the determinant of the coefficient matrix is

$$\begin{vmatrix} e^{\mu a} & e^{-\mu a} \\ e^{\mu b} & e^{-\mu b} \end{vmatrix} = e^{\mu(a-b)} - e^{-\mu(a-b)} = 2\sinh \mu(a - b) \neq 0,$$

since $a \neq b$. Therefore this system of equations has a unique solution $A = 0$ and $B = 0$, and in this case the only solution to the boundary value problem is the trivial solution $\phi(x) = 0$ for $a < x < b$.

Case 3: If $\lambda > 0$, then $\lambda = \mu^2$ where $\mu \neq 0$, and the equation is

$$\phi''(x) + \mu^2 \phi(x) = 0,$$

with general solution

$$\phi(x) = A\cos \mu x + B \sin \mu x.$$

The first boundary condition gives

$$\phi(a) = A\cos \mu a + B \sin \mu a = 0,$$

while the second boundary condition gives

$$\phi(b) = A\cos \mu b + B \sin \mu b = 0.$$

We have the following homogeneous system of linear equations for A and B

$$\begin{aligned} A\cos \mu a + B \sin \mu a &= 0, \\ A\cos \mu b + B \sin \mu b &= 0, \end{aligned}$$

and the determinant of the coefficient matrix is

$$\begin{vmatrix} \cos \mu a & \sin \mu a \\ \cos \mu b & \sin \mu b \end{vmatrix} = \cos \mu a \sin \mu b - \sin \mu a \cos \mu b = \sin \mu(b-a).$$

The boundary value problem has the trivial solution if and only if this determinant is nonzero, so we get a nontrivial solution if and only if $\sin \mu(b-a) = 0$, that is, if and only if $\mu(b-a) = n\pi$ for some integer n. The eigenvalues are

$$\lambda_n = \left(\frac{n\pi}{b-a}\right)^2,$$

with corresponding eigenfunctions

$$\phi_n(x) = \sin n\pi \left(\frac{x-a}{b-a}\right), \quad a < x < b,$$

for $n \geqslant 1$.

(b) If ϕ satisfies the boundary value problem

$$\begin{aligned} &\phi''(x) + \lambda \phi(x) = 0, \quad 0 < x < a, \\ &\phi'(0) = 0, \\ &\phi'(a) = 0, \end{aligned}$$

we have to consider three cases.

Case 1: If $\lambda = 0$, the equation is

$$\phi''(x) = 0,$$

with general solution

$$\phi(x) = Ax + B, \quad 0 < x < a.$$

The first boundary condition gives $\phi'(0) = A = 0$, and the solution is now

$$\phi(x) = B, \quad 0 < x < a.$$

The second boundary condition, $\phi'(a) = 0$, is now satisfied automatically. In this case we have a nontrivial solution, $\phi(x) = B$ for $0 < x < a$.

Case 2: If $\lambda < 0$, then $\lambda = -\mu^2$ where $\mu \neq 0$, and the equation is

$$\phi''(x) - \mu^2\phi(x) = 0,$$

with general solution

$$\phi(x) = A\cosh \mu x + B\sinh \mu x, \quad 0 < x < a.$$

The first boundary condition gives $\phi'(0) = \mu B = 0$, so that $B = 0$, and the solution is now

$$\phi(x) = A\cosh \mu x, \quad 0 < x < a.$$

The second boundary condition gives $\phi'(a) = \mu A \sinh \mu a = 0$, so that $A = 0$, since $\sinh \mu a \neq 0$. In this case the only solution is the trivial solution $\phi(x) = 0$ for $0 < x < a$.

Case 3: If $\lambda > 0$, then $\lambda = \mu^2$ where $\mu \neq 0$, and the equation is

$$\phi''(x) + \mu^2\phi(x) = 0,$$

with general solution

$$\phi(x) = A\cos \mu x + B\sin \mu x.$$

The first boundary condition gives $\phi'(0) = \mu B = 0$, so that $B = 0$, and the solution is now

$$\phi(x) = A\cos \mu x, \quad 0 < x < a.$$

The second boundary condition gives $\phi'(a) = -\mu A \sin \mu a = 0$, and if $A = 0$, then again we get the trivial solution. In this case we have a nontrivial solution if and only if $\mu a = n\pi$ for some integer n; that is, $\mu = n\pi/a$ for some integer n. The eigenvalues are $\lambda_n = (n\pi/a)^2$, with corresponding eigenfunctions

$$\phi_n(x) = \cos\frac{n\pi x}{a}, \quad 0 < x < a$$

for $n \geqslant 0$. Note that this includes the constant solution obtained for $\lambda_0 = 0$.

(c) If ϕ satisfies the boundary value problem

$$\begin{aligned}
&\phi''(x) + \lambda\phi(x) = 0, \quad 0 < x < a,\\
&\phi(0) = 0,\\
&\phi'(a) = 0,
\end{aligned}$$

we have to consider three cases.

Case 1: If $\lambda = 0$, the equation is

$$\phi''(x) = 0,$$

with general solution

$$\phi(x) = Ax + B, \quad 0 < x < a.$$

The first boundary condition gives $\phi(0) = B = 0$, and the solution is now

$$\phi(x) = Ax, \quad 0 < x < a.$$

The second boundary condition gives $\phi'(a) = A = 0$. In this case we have only the trivial solution $\phi(x) = 0$ for $0 < x < a$.

Case 2: If $\lambda < 0$, then $\lambda = -\mu^2$ where $\mu \neq 0$, and the equation is

$$\phi''(x) - \mu^2\phi(x) = 0,$$

with general solution

$$\phi(x) = A\cosh\mu x + B\sinh\mu x, \quad 0 < x < a.$$

The first boundary condition gives $\phi(0) = A = 0$, and the solution is now

$$\phi(x) = B\sinh\mu x, \quad 0 < x < a.$$

The second boundary condition gives $\phi'(a) = \mu B\cosh\mu a = 0$, so that $B = 0$, since $\mu\cosh\mu a \neq 0$. In this case the only solution is the trivial solution $\phi(x) = 0$ for $0 < x < a$.

Case 3: If $\lambda > 0$, then $\lambda = \mu^2$ where $\mu \neq 0$, and the equation is

$$\phi''(x) + \mu^2\phi(x) = 0,$$

with general solution

$$\phi(x) = A\cos\mu x + B\sin\mu x, \quad 0 < x < a.$$

The first boundary condition gives $\phi(0) = A = 0$, and the solution is now

$$\phi(x) = B\sin\mu x, \quad 0 < x < a.$$

The second boundary condition gives $\phi'(a) = \mu B\cos\mu a = 0$, and if $B = 0$, again we get the trivial solution. In this case we have a nontrivial solution if and only if

$$\mu a = \frac{(2n-1)\pi}{2}$$

for some integer $n \geqslant 1$; that is,

$$\mu = \frac{(2n-1)\pi}{2a}$$

for some integer $n \geqslant 1$. The eigenvalues are

$$\lambda_n = \left(\frac{(2n-1)\pi}{2a}\right)^2,$$

with corresponding eigenfunctions

$$\phi_n(x) = \sin\frac{(2n-1)\pi x}{2a}, \quad 0 < x < a$$

for $n \geqslant 1$.

(d) If ϕ satisfies the boundary value problem

$$\begin{aligned} &\phi''(x) + \lambda\phi(x) = 0, \quad 0 < x < a, \\ &\phi'(0) = 0, \\ &\phi(a) = 0, \end{aligned}$$

we have to consider three cases.

Case 1: If $\lambda = 0$, the equation is

$$\phi''(x) = 0,$$

with general solution

$$\phi(x) = Ax + B, \quad 0 < x < a.$$

The first boundary condition gives $\phi'(0) = A = 0$, and the solution is now

$$\phi(x) = B, \quad 0 < x < a.$$

The second boundary condition gives $\phi(a) = B = 0$. In this case we have only the trivial solution $\phi(x) = 0$ for $0 < x < a$.

Case 2: If $\lambda < 0$, then $\lambda = -\mu^2$ where $\mu \neq 0$, and the equation is

$$\phi''(x) - \mu^2\phi(x) = 0,$$

with general solution

$$\phi(x) = A\cosh\mu x + B\sinh\mu x, \quad 0 < x < a.$$

The first boundary condition gives $\phi'(0) = \mu B = 0$, so that $B = 0$, and the solution is now

$$\phi(x) = A\cosh\mu x, \quad 0 < x < a.$$

The second boundary condition gives $\phi(a) = A\cosh\mu a = 0$, so that $A = 0$, since $\cosh\mu a \neq 0$. In this case the only solution is the trivial solution $\phi(x) = 0$ for $0 < x < a$.

Case 3: If $\lambda > 0$, then $\lambda = \mu^2$ where $\mu \neq 0$, and the equation is

$$\phi''(x) + \mu^2\phi(x) = 0,$$

with general solution

$$\phi(x) = A\cos\mu x + B\sin\mu x, \quad 0 < x < a.$$

The first boundary condition gives $\phi'(0) = \mu B = 0$, so that $B = 0$, and the solution is now

$$\phi(x) = A\cos\mu x, \quad 0 < x < a.$$

The second boundary condition gives $\phi(a) = A\cos\mu a = 0$, and if $A = 0$, then again we get the trivial solution. In this case we have a nontrivial solution if and only if

$$\mu a = \frac{(2n-1)\pi}{2}$$

for some integer $n \geqslant 1$; that is,

$$\mu = \frac{(2n-1)\pi}{2a}$$

for some integer $n \geqslant 1$. The eigenvalues are

$$\lambda_n = \left(\frac{(2n-1)\pi}{2a}\right)^2,$$

with corresponding eigenfunctions

$$\phi_n(x) = \cos\frac{(2n-1)\pi x}{2a}, \quad 0 < x < a$$

for $n \geqslant 1$.

(e) If ϕ satisfies the boundary value problem

$$\begin{aligned}&\phi''(x) + \lambda\phi(x) = 0, \quad 0 < x < a,\\ &\phi(0) = 0,\\ &\phi'(a) + \phi(a) = 0,\end{aligned}$$

we have to consider three cases.

Case 1: If $\lambda = 0$, the equation is

$$\phi''(x) = 0,$$

with general solution

$$\phi(x) = Ax + B, \quad 0 < x < a.$$

The first boundary condition gives $\phi(0) = B = 0$, and the solution is now

$$\phi(x) = Ax, \quad 0 < x < a.$$

The second boundary condition gives $\phi'(a) + \phi(a) = A(1 + a) = 0$, so that $A = 0$. In this case the only solution is the trivial solution $\phi(x) = 0$ for $0 < x < a$.

Case 2: If $\lambda < 0$, then $\lambda = -\mu^2$ where $\mu \neq 0$, and the equation is

$$\phi''(x) - \mu^2\phi(x) = 0,$$

with general solution

$$\phi(x) = A\cosh\mu x + B\sinh\mu x, \quad 0 < x < a.$$

The first boundary condition gives $\phi(0) = A = 0$, and the solution is now

$$\phi(x) = B\sinh\mu x.$$

The second boundary condition gives

$$\phi'(a) + \phi(a) = \mu B\cosh\mu a + B\sinh\mu a = B\cosh\mu a(\mu + \tanh\mu a) = 0.$$

If we look at the graphs of $y = \tanh\mu a$ and $y = -\mu$ in Figure 12.1 we see that $\mu + \tanh\mu a = 0$ if and only if $\mu = 0$. Therefore, since $\mu \neq 0$, and $\cosh\mu a \neq 0$, the second boundary condition implies that $B = 0$. In this case the only solution is the trivial solution $\phi(x) = 0$ for $0 < x < a$.

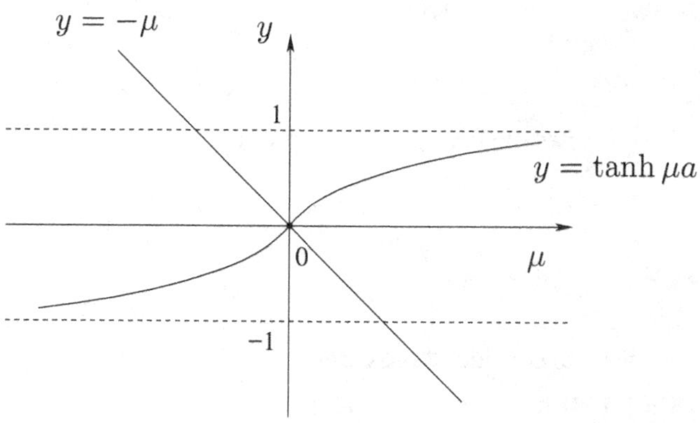

Figure 12.1 Intersection of graphs of $y = \tanh\mu a$ and $y = -\mu$

Case 3: If $\lambda > 0$, then $\lambda = \mu^2$ where $\mu \neq 0$, and the equation is

$$\phi''(x) + \mu^2\phi(x) = 0,$$

with general solution

$$\phi(x) = A\cos\mu x + B\sin\mu x, \quad 0 < x < a.$$

The first boundary condition gives $\phi(0) = A = 0$, and the solution is now

$$\phi(x) = B\sin\mu x.$$

The second boundary condition gives

$$\phi'(a) + \phi(a) = \mu B\cos\mu a + B\sin\mu a = B(\mu\cos\mu a + \sin\mu a) = 0,$$

and we have a nontrivial solution if and only if

$$\mu\cos\mu a + \sin\mu a = 0,$$

that is, if and only if

$$\tan\mu a = -\mu.$$

From the graphs of $y = \tan\mu a$ and $y = -\mu$ in Figure 12.2 we see that there are an infinite number of solutions μ_n. The eigenvalues are

$$\lambda_n = \mu_n^2,$$

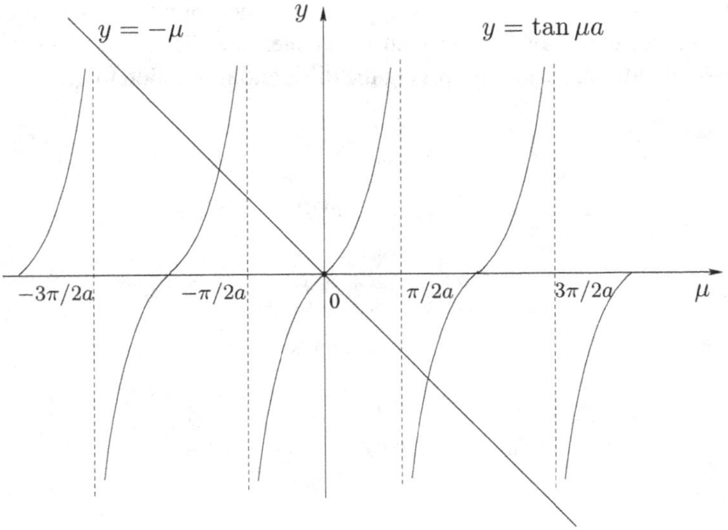

Figure 12.2 Intersection of graphs of $y = \tan\mu a$ and $y = -\mu$

where $\tan \mu_n a = -\mu_n$, and the corresponding eigenfunctions are

$$\phi_n(x) = \sin \mu_n x, \quad 0 < x < a$$

for $n \geqslant 1$.

Note: Since $\cot \mu_n a = -1/\mu_n$, and $\mu_n \to \infty$ as $n \to \infty$, the zeros of the equation $\tan \mu_n a = -\mu_n$ approach the zeros of $\cos \mu_n a$, and for *large* n, μ_n is given approximately by

$$\mu_n \approx \frac{(2n-1)\pi}{2a}.$$

Exercise 12.2. XX

Use the **energy method** to show that there are no negative eigenvalues for the Neumann problem

$$\frac{d^2\phi}{dx^2} + \lambda\phi = 0, \quad 0 < x < a$$

$$\frac{d\phi}{dx}(0) = 0$$

$$\frac{d\phi}{dx}(a) = 0.$$

This means: Multiply the equation by ϕ, integrate and solve for λ. Does the expression for λ look familiar?

Solution. Suppose that λ is an eigenvalue of the boundary value problem with corresponding eigenfunction ϕ; then ϕ satisfies the differential equation and the boundary conditions, and multiplying the differential equation by ϕ, we have

$$\phi \frac{d^2\phi}{dx^2} + \lambda\phi^2 = 0$$

for $0 < x < a$. Integrating over the interval $[0, a]$, we have

$$\int_0^a \phi \frac{d^2\phi}{dx^2}\, dx + \lambda \int_0^a \phi^2\, dx = 0,$$

and integrating the first integral by parts yields

$$\phi \frac{d\phi}{dx}\bigg|_0^a - \int_0^a \left(\frac{d\phi}{dx}\right)^2 dx + \lambda \int_0^a \phi^2\, dx = 0;$$

that is,

$$\phi(a)\frac{d\phi}{dx}(a) - \phi(0)\frac{d\phi}{dx}(0) - \int_0^a \left(\frac{d\phi}{dx}\right)^2 dx + \lambda \int_0^a \phi^2\, dx = 0,$$

and since $d\phi(0)/dx = d\phi(a)/dx = 0$, then

$$\lambda \int_0^a \phi^2\, dx - \int_0^a \left(\frac{d\phi}{dx}\right)^2 dx = 0;$$

that is,

$$\lambda = \frac{\displaystyle\int_0^a \left(\frac{d\phi}{dx}\right)^2 dx}{\displaystyle\int_0^a \phi^2\, dx} \geqslant 0,$$

and there are no negative eigenvalues for this boundary value problem. Note that the expression for λ is, in fact, the Rayleigh quotient for the Sturm-Liouville eigenvalue problem above .

Exercise 12.3. XXX

Solve the initial value problem

$$\begin{aligned} y'' + 9y &= F(t), \\ y(0) &= 0, \\ y'(0) &= 0, \end{aligned}$$

where $F(t)$ is the 2π-periodic input function given by its Fourier series,

$$F(t) = \sum_{n=1}^{\infty} \left[\frac{\cos nt}{n^2} + (-1)^n \frac{\sin nt}{n}\right].$$

Solution. Since the differential equation is a linear equation with constant coefficients, the general solution to the nonhomogeneous equation is given by

$$y(t) = y_h(t) + y_p(t),$$

where $y_h(t)$ is the general solution to the corresponding homogeneous equation and $y_p(t)$ is *any* particular solution to the nonhomogeneous equation. The solution to the homogeneous equation is

$$y_h(t) = c_1 \cos 3t + c_2 \sin 3t,$$

where c_1 and c_2 are arbitrary constants. To find a particular solution to the nonhomogeneous equation, we solve the equation

$$y''(t) + 9y(t) = A_n \cos nt + B_n \sin nt$$

for $n \geqslant 0$, where A_n and B_n are the Fourier coefficients of the driving force $F(t)$; that is, $A_0 = B_0 = 0$,

$$A_n = \frac{1}{n^2}, \qquad \text{and} \qquad B_n = \frac{(-1)^n}{n}$$

for $n \geqslant 1$.

Note that for $n \neq 3$, from the method of undetermined coefficients, the nth normal mode of vibration is

$$y_n(t) = a_n \cos nt + b_n \sin nt,$$

where the constants a_n and b_n are determined from the Fourier coefficients of $F(t)$ to be

$$a_0 = 0, \quad a_n = \frac{1}{n^2(9-n^2)}, \quad b_n = \frac{(-1)^n}{n(9-n^2)}$$

for $n \geqslant 1,\ n \neq 3$. For $n = 3$, the term in the driving force has the same frequency as the natural frequency of the system, and we have to solve the nonhomogeneous equation

$$y_3''(t) + 9y_3(t) = A_3 \cos 3t + B_3 \sin 3t.$$

In this case the method of undetermined coefficients suggests a solution of the form

$$y_3(t) = t(a_3 \cos 3t + b_3 \sin 3t).$$

In order to determine the constants a_3 and b_3, we substitute this expression into the differential equation

$$y_3'' + 9y_3 = A_3 \cos 3t + B_3 \sin 3t$$

to obtain

$$a_3 = -\frac{B_3}{6} \quad \text{and} \quad b_3 = \frac{A_3}{6}.$$

The particular solution to the nonhomogeneous equation can then be written as

$$y_p(t) = \sum_{\substack{n=1 \\ n\neq 3}}^{\infty} \left(\frac{1}{n^2(9-n^2)} \cos nt + \frac{(-1)^n}{n(9-n^2)} \sin nt \right) + \frac{t}{6}\left(\frac{1}{3} \cos 3t + \frac{1}{3^2} \sin 3t \right),$$

and the general solution to the nonhomogeneous equation is

$$y(t) = c_1 \cos 3t + c_2 \sin 3t + \sum_{\substack{n=1 \\ n\neq 3}}^{\infty} \left(\frac{1}{n^2(9-n^2)} \cos nt + \frac{(-1)^n}{n(9-n^2)} \sin nt \right) + \frac{t}{6}\left(\frac{1}{3} \cos 3t + \frac{1}{9} \sin 3t \right)$$

and the constants c_1 and c_2 can now be evaluated using the initial conditions

$$y(0) = y'(0) = 0.$$

Applying the initial conditions, we find that

$$c_1 = -\sum_{\substack{n=1\\ n\neq 3}}^{\infty} \frac{1}{n^2(9-n^2)} \qquad \text{and} \qquad c_2 = -\frac{1}{3^2\cdot 6} - \frac{1}{3}\sum_{\substack{n=1\\ n\neq 3}}^{\infty} \frac{(-1)^n}{9-n^2}.$$

Exercise 12.4. ✗✗✗✗

Find the eigenvalues and eigenfunctions of the regular Sturm-Liouville problem

$$\begin{aligned} &(x^3X')' + \lambda x X = 0, \quad 1 < x < e,\\ &X(1) = 0,\\ &X(e) = 0. \end{aligned}$$

Solution. Let $X = Y/x$, then $X' = Y'/x - Y/x^2$, so that $x^3X' = x^2Y' - xY$, and

$$(x^3X')' = (x\,x\,Y')' - Y - x\,Y' = x\,Y' + x(x\,Y')' - Y - x\,Y' = x(x\,Y')' - Y.$$

Therefore, the original boundary value problem is equivalent to the following problem:

$$\begin{aligned} &(x\,Y')' + \frac{\mu}{x}Y = 0, \quad 1 < x < e,\\ &Y(1) = 0,\\ &Y(e) = 0, \end{aligned}$$

where $\mu = \lambda - 1$. Now let $x = e^t$ and $\hat{Y}(t) = Y(e^t)$; then

$$\frac{dY}{dx} = \frac{d\hat{Y}}{dt}\frac{dt}{dx} = \frac{1}{x}\frac{d\hat{Y}}{dt},$$

that is,

$$x\frac{dY}{dx} = \frac{d\hat{Y}}{dt}.$$

Also,

$$\frac{d}{dx}\left(x\frac{dY}{dx}\right) = \frac{d}{dt}\left(\frac{d\hat{Y}}{dt}\right)\frac{dt}{dx} = \frac{1}{x}\frac{d^2\hat{Y}}{dt^2} = e^{-t}\frac{d^2\hat{Y}}{dt^2},$$

and therefore

$$\frac{d}{dx}\left(x\frac{dY}{dx}\right) + \frac{\mu}{x}Y = 0 \quad \text{for} \quad 1 < x < e$$

if and only if

$$e^{-t}\frac{d^2\hat{Y}}{dt^2} + e^{-t}\mu\hat{Y} = 0 \quad \text{for} \quad 0 < t < 1.$$

Thus, we have the equivalent regular Sturm-Liouville problem

$$\frac{d^2\hat{Y}}{dt^2} + \mu\hat{Y} = 0, \quad 0 < t < 1,$$

$$\hat{Y}(0) = 0,$$

$$\hat{Y}(1) = 0,$$

with eigenvalues $\mu_n = n^2\pi^2$ and eigenfunctions

$$\hat{Y}_n(t) = \sin n\pi t, \quad 0 < t < 1$$

for $n = 1, 2, 3, \ldots$. The eigenvalues for the original problem are therefore

$$\lambda_n = 1 + n^2\pi^2,$$

and the corresponding eigenfunctions are

$$X_n(x) = \frac{1}{x}\sin(n\pi \log x), \quad 1 < x < e$$

for $n = 1, 2, 3, \ldots$.

Exercise 12.5. XX

Solve the eigenvalue problem

$$\frac{d^2\phi}{dx^2} + \lambda\phi = 0$$

subject to the periodicity conditions

$$\phi(0) = \phi(2\pi) \quad \text{and} \quad \frac{d\phi}{dx}(0) = \frac{d\phi}{dx}(2\pi).$$

Solution. Again, we consider three cases.

Case 1: If $\lambda = 0$, the equation is $\phi'' = 0$ with general solution $\phi(x) = Ax + B$. From the first periodicity condition $\phi(0) = \phi(2\pi)$ we have

$$\phi(0) = A \cdot 0 + B = A \cdot 2\pi + B = \phi(2\pi),$$

so that $2\pi A = 0$, and $A = 0$. The solution is now

$$\phi(x) = B, \quad 0 < x < 2\pi.$$

The second periodicity condition $\phi'(0) = \phi'(2\pi)$ holds automatically, since

$$\phi'(0) = 0 = \phi'(2\pi).$$

Therefore, $\lambda_0 = 0$ is an eigenvalue with corresponding eigenfunction

$$\phi_0(x) = 1, \quad 0 < x < 2\pi.$$

Case 2: If $\lambda < 0$, then $\lambda = -\mu^2$ where $\mu \neq 0$. The differential equation is

$$\phi'' - \mu^2\phi = 0,$$

with general solution

$$\phi(x) = A\cosh \mu x + B\sinh \mu x, \quad 0 < x < 2\pi.$$

From the first periodicity condition

$$\phi(0) = A = A\cosh 2\pi\mu + B\sinh 2\pi\mu = \phi(2\pi),$$

while from the second periodicity condition

$$\phi'(0) = \mu B = \mu A\sinh 2\pi\mu + \mu B\cosh 2\pi\mu = \phi'(2\pi).$$

We have the homogeneous system of linear equations for A and B:

$$\begin{aligned} (\cosh 2\pi\mu - 1)\, A + \sinh 2\pi\mu\, B &= 0, \\ \sinh 2\pi\mu\, A + (\cosh 2\pi\mu - 1)\, B &= 0, \end{aligned}$$

and the determinant of the coefficient matrix is

$$\begin{vmatrix} \cosh 2\pi\mu - 1 & \sinh 2\pi\mu \\ \sinh 2\pi\mu & \cosh 2\pi\mu - 1 \end{vmatrix} = 2(1 - \cosh 2\pi\mu) = -4\sinh^2 \pi\mu \neq 0.$$

Since $\pi\mu \neq 0$, this system has only the trivial solution $A = B = 0$. In this case the boundary value problem has only the trivial solution $\phi(x) = 0$ for $0 < x < 2\pi$.
Case 3: If $\lambda > 0$, then $\lambda = \mu^2$ where $\mu \neq 0$, and the differential equation is

$$\phi'' + \mu^2\phi = 0,$$

with general solution

$$\phi(x) = A\cos \mu x + B\sin \mu x, \quad 0 < x < 2\pi.$$

From the first periodicity condition

$$\phi(0) = A = A\cos 2\pi\mu + B\sin 2\pi\mu = \phi(2\pi),$$

while from the second periodicity condition

$$\phi'(0) = \mu B = -\mu A\sin 2\pi\mu + \mu B\cos 2\pi\mu = \phi'(2\pi).$$

We have the homogeneous system of linear equations for A and B:

$$\begin{aligned} (1 - \cos 2\pi\mu)\, A - \sin 2\pi\mu\, B &= 0, \\ \sin 2\pi\mu\, A + (1 - \cosh 2\pi\mu)\, B &= 0, \end{aligned}$$

and the determinant of the coefficient matrix is

$$\begin{vmatrix} 1-\cos 2\pi\mu & -\sin 2\pi\mu \\ \sin 2\pi\mu & 1-\cos 2\pi\mu \end{vmatrix} = 2(1-\cos 2\pi\mu) = 4\sin^2\pi\mu,$$

and this system has a nontrivial solution if and only if this determinant is zero, that is, if and only if $\sin^2\pi\mu = 0$, that is if and only if $\pi\mu = n\pi$ for some integer n. In this case the boundary value problem has a nontrivial solution if and only if $\mu = n$ for some integer n. The eigenvalues are

$$\lambda_n = \mu_n^2 = n^2$$

for $n \geqslant 1$. For these eigenvalues the coefficient matrix becomes the zero matrix, and both coefficients are undetermined. Hence, for each $n \geqslant 1$, we have two linearly independent eigenfunctions:

$$\phi_{1,n}(x) = \cos nx \qquad \text{and} \qquad \phi_{2,n}(x) = \sin nx$$

for $n \geqslant 1$. Thus, this eigenvalue problem has eigenvalues

$$\lambda_n = n^2$$

and corresponding eigenfunctions

$$\{\cos nx,\ \sin nx\}$$

for $n \geqslant 0$.

Exercise 12.6. ✗
Assume that $f(x)$ is an even function and $g(x)$ is an odd function. Show that the set of functions $\{f(x), g(x)\}$ is orthogonal with respect to the weight function

$$w(x) = 1$$

on any symmetric interval $[-a, a]$ containing 0.

Solution. We have

$$\begin{aligned}
\int_{-a}^{a} f(x)g(x)\,dx &= \underbrace{\int_{-a}^{0} f(x)g(x)\,dx}_{t=-x} + \int_{0}^{a} f(x)g(x)\,dx \\
&= \int_{0}^{a} f(-t)g(-t)\,dt + \int_{0}^{a} f(x)g(x)\,dx \\
&= -\int_{0}^{a} f(t)g(t)\,dt + \int_{0}^{a} f(x)g(x)\,dx \\
&= 0,
\end{aligned}$$

and therefore f and g are orthogonal on the symmetric interval $[-a, a]$ with respect to the weight function $w(x) = 1$.

Exercise 12.7. XX

Show that the set of three Laguerre polynomials $\left\{1,\, 1-x,\, \frac{1}{2}(2-4x+x^2)\right\}$ is orthogonal with respect to the weight function

$$w(x) = e^{-x}$$

on the interval $[0, \infty)$.

Solution. This problem can be solved by pairwise integration of the functions with the weight function e^{-x}. Since this is quite tedious, we give a more elegant method using the gamma function. Recall that for $n \geqslant 0$ we have

$$\Gamma(n+1) = \int_0^\infty x^n e^{-x}\, dx = n!,$$

and therefore

$$\langle 1, 1-x\rangle = \int_0^\infty (1-x)e^{-x}\, dx = 0! - 1! = 0,$$

$$\left\langle 1, \frac{1}{2}(2-4x+x^2)\right\rangle = 0! - 2\cdot 1! + \frac{1}{2}\cdot 2! = 1 - 2 + 1 = 0$$

and finally,

$$\begin{aligned}\left\langle 1-x, \frac{1}{2}(2-4x+x^2)\right\rangle &= \left\langle 1, \frac{1}{2}(2-4x+x^2)\right\rangle - \left\langle x, \frac{1}{2}(2-4x+x^2)\right\rangle \\ &= 0 - \left(1! - 2\cdot 2! + \frac{3!}{2}\right) \\ &= -1 + 4 - 3 \\ &= 0,\end{aligned}$$

and the set of Laguerre polynomials $\left\{1,\, 1-x,\, \frac{1}{2}(2-4x+x^2)\right\}$ forms an orthogonal set on the interval $[0, \infty)$ with respect to the weight function $w(x) = e^{-x}$.

Exercise 12.8. XX

Is the set of functions $\left\{\frac{1}{2}(2-4x+x^2),\, -12x+8x^3\right\}$ orthogonal with respect to the weight function

$$w(x) = e^{-x}$$

on the interval $[0, \infty)$?

Solution. These functions are **not** orthogonal with respect to the weight function $w(x) = e^{-x}$ on the interval $[0, \infty)$; in fact,

$$
\begin{aligned}
< 8x^3 - 12x, \frac{1}{2}(x^2 - 4x + 2) > &= 2 < 2x^3 - 3x, x^2 - 4x + 2 > \\
&= 2 \int_0^\infty (2x^3 - 3x)(x^2 - 4x + 2)e^{-x}\, dx \\
&= 2 \int_0^\infty (2x^5 - 8x^4 + x^3 + 12x^2 - 6x)e^{-x}\, dx \\
&= 2\left[2 \cdot 5! - 8 \cdot 4! + 3! + 12 \cdot 2! - 6 \cdot 1!\right] \\
&= 2 \cdot 72 = 144.
\end{aligned}
$$

Exercise 12.9.
Given the boundary value problem

$$
\begin{aligned}
&(1 - x^2)y'' - 2xy' + (1 + \lambda x)y = 0 \\
&y(-1) = 0 \\
&y(1) = 0
\end{aligned}
$$

on the interval $-1 < x < 1$, put the equation in Sturm-Liouville form and decide whether the problem is regular or singular.

Solution. We can rewrite the boundary value problem in the form

$$
\begin{aligned}
&\left((1 - x^2)y'\right)' + (1 + \lambda x)y = 0 \\
&y(-1) = 0 \\
&y(1) = 0.
\end{aligned}
$$

Here $p(x) = 1 - x^2,\ p'(x) = -2x,\ q(x) = 1,\ \sigma(x) = x$ are all continuous on the interval $[-1, 1]$. Also, $\alpha_1 = \alpha_2 = 1$ and $\beta_1 = \beta_2 = 0$. However, $p(x) = 0$ at the endpoints of the interval $[-1, 1]$, and $\sigma(0) = 0$, so this is a **singular** Sturm-Liouville problem.

Exercise 12.10.
Find the eigenvalues and eigenfunctions of the periodic eigenvalue problem

$$
\begin{aligned}
&y'' + \lambda y = 0, \\
&y(-\pi) = y(\pi), \\
&y'(-\pi) = y'(\pi).
\end{aligned}
$$

Solution.
Case 1: If $\lambda = 0$, the equation $y'' = 0$ has general solution $y(x) = Ax + B$ with $y' = A$. The first periodicity condition gives

$$-A\pi + B = A\pi + B,$$

so that $A = 0$. The second periodicity condition is then satisfied automatically, so there is one nontrivial solution in this case. The eigenvalue is $\lambda = 0$ with corresponding eigenfunction $y_0 = 1$.

Case 2: If $\lambda < 0$, say $\lambda = -\mu^2$ where $\mu \neq 0$, then the differential equation becomes $y'' - \mu^2 y = 0$, and has general solution $y(x) = A\cosh\mu x + B\sinh\mu x$ with $y' = \mu A\sinh\mu x + \mu B\cosh\mu x$. The first periodicity condition gives

$$A\cosh\mu\pi - B\sinh\mu\pi = A\cosh\mu\pi + B\sinh\mu\pi$$

since $\cosh\mu x$ is an even function and $\sinh\mu x$ is an odd function. We have $2B\sinh\mu\pi = 0$, and since $\sinh\mu\pi \neq 0$, then $B = 0$. The solution is then $y = A\cosh\mu x$, and the second periodicity condition gives

$$-\mu A\sinh\mu\pi = \mu A\sinh\mu\pi,$$

so that $2\mu A\sinh\mu\pi = 0$, and since $\mu \neq 0$, then $\sinh\mu\pi \neq 0$, so we must have $A = 0$. Therefore, there are no nontrivial solutions in this case.

Case 3: If $\lambda > 0$, say $\lambda = \mu^2$ where $\mu \neq 0$, the differential equation becomes $y'' + \mu^2 y = 0$ and has general solution

$$y(x) = A\cos\mu x + B\sin\mu x, \quad \text{with} \quad y'(x) = -A\mu\sin\mu x + B\mu\cos\mu x.$$

Applying the first periodicity condition, we have

$$y(-\pi) = A\cos\mu\pi - B\sin\mu\pi = A\cos\mu\pi + B\sin\mu\pi = y(\pi),$$

so that $2B\sin\mu\pi = 0$. Applying the second periodicity condition, we have

$$y'(-\pi) = A\mu\sin\mu\pi + B\mu\cos\mu\pi = -A\mu\sin\mu\pi + B\mu\cos\mu\pi = y'(\pi),$$

so that $2A\sin\mu\pi = 0$. Therefore, the following equations must hold simultaneously:

$$\begin{aligned} A\sin\mu\pi &= 0, \\ B\sin\mu\pi &= 0. \end{aligned}$$

To get a nontrivial solution, we must have either $A \neq 0$ or $B \neq 0$, and if the equations hold, we must have $\sin\mu\pi = 0$. Therefore, μ must be an integer, so that the eigenvalues are

$$\lambda_n = \mu_n^2 = n^2$$

for $n = 1, 2, 3, \ldots$, and the eigenfunctions corresponding to these eigenvalues are $\sin nx$ and $\cos nx$ for $n = 1, 2, 3, \ldots$. The full set of orthogonal eigenfunctions for the periodic eigenvalue problem above is

$$\{\, 1,\ \cos nx,\ \sin nx;\ n = 1, 2, 3, \ldots \,\}$$

for $-\pi \leqslant x \leqslant \pi$.

Exercise 12.11. ✗✗✗
Find the eigenvalues and eigenfunctions of the Sturm-Liouville problem

$$\begin{aligned} &y'' + \lambda y = 0, \\ &y(0) + y'(0) = 0, \\ &y(2\pi) = 0. \end{aligned}$$

Solution.
Case 1: If $\lambda = 0$, the equation $y'' = 0$ has general solution $y(x) = Ax + B$ with $y' = A$. The first boundary condition gives

$$B + A = 0,$$

so that $A = -B$. The second boundary condition gives

$$2\pi A + B = 0,$$

so that $(2\pi - 1)A = 0$ and $A = -B = 0$, so there are no nontrivial solutions in this case.

Case 2: If $\lambda < 0$, say $\lambda = -\mu^2$ where $\mu \neq 0$, the differential equation becomes $y'' - \mu^2 y = 0$ and has general solution $y(x) = A\cosh\mu x + B\sinh\mu x$ with $y' = \mu A\sinh\mu x + \mu B\cosh\mu x$. The first boundary condition gives

$$A + \mu B = 0,$$

so that $A = -\mu B$. The second boundary condition gives

$$A\cosh 2\pi\mu + B\sinh 2\pi\mu = 0,$$

and since $\cosh 2\pi\mu \neq 0$, then

$$B(\tanh 2\pi\mu - \mu) = 0,$$

and to get nontrivial solutions we need

$$\tanh 2\pi\mu = \mu.$$

The graphs of $f(\mu) = \tanh 2\pi\mu$ and $g(\mu) = \mu$ intersect at the origin, $\mu = 0$, and since

$$\lim_{\mu\to\infty} \tanh 2\pi\mu = 1 \qquad \text{and} \qquad \lim_{\mu\to-\infty} \tanh 2\pi\mu = -1,$$

and

$$f'(0) = 2\pi > 1 = g'(0),$$

they intersect again in exactly two more points, $\mu = \pm\mu_0$, where μ_0 is the positive root of the equation $\tanh 2\pi\mu = \mu$. There is one nontrivial solution in this case, with eigenvalue $\lambda = -(\mu_0)^2$, and the corresponding eigenfunction is

$$\sinh \mu_0 x - \mu_0 \cosh \mu_0 x.$$

Case 3: If $\lambda > 0$, say $\lambda = \mu^2$ where $\mu \neq 0$, the differential equation becomes $y'' + \mu^2 y = 0$ and has general solution

$$y(x) = A\cos \mu x + B \sin \mu x,$$

with

$$y' = -\mu A \sin \mu x + \mu B \cos \mu x.$$

The first boundary condition gives

$$y(0) + y'(0) = A + \mu B = 0$$

so that $A = -\mu B$. The second boundary condition gives

$$y(2\pi) = A\cos 2\pi\mu + B\sin 2\pi\mu = 0,$$

so

$$B\left[\sin 2\pi\mu - \mu\cos 2\pi\mu\right] = 0,$$

and the eigenvalues are $\lambda_n = \mu_n^2$, where μ_n is the nth positive root of the equation $\tan 2\pi\mu = \mu$ (which has an infinite number of solutions μ_n, $n = 1, 2, 3, \ldots$). The corresponding eigenfunctions are

$$y_n = \sin \mu_n x - \mu_n \cos \mu_n x$$

for $n = 1, 2, 3, \ldots$.

Exercise 12.12. XXX

Show that the boundary value problem

$$\begin{aligned} y'' - \lambda y &= 0, \\ y(0) + y'(0) &= 0, \\ y(1) + y'(1) &= 0 \end{aligned}$$

has one positive eigenvalue. Does this contradict the theorem below?

Theorem. The eigenvalues of a regular Sturm-Liouville problem are all real and form an increasing sequence

$$\lambda_1 < \lambda_2 < \lambda_3 < \cdots,$$

where $\lambda_n \to \infty$ as $n \to \infty$.

Solution.

Case 1: If $\lambda = 0$, the differential equation $y'' = 0$ has general solution $y = Ax+B$, with $y' = A$. Applying the first boundary condition, we have

$$B + A = 0,$$

so that $B = -A$. Applying the second boundary condition, we have

$$A + B + A = 0,$$

so that $B = -2A$, and therefore $B = 2B$ and $B = A = 0$. Therefore, there are no nontrivial solutions in this case.

Case 2: If $\lambda < 0$, say $\lambda = -\mu^2$ where $\mu \neq 0$, the differential equation becomes $y'' + \mu^2 y = 0$ and has general solution

$$y = A\cos\mu x + B\sin\mu x \qquad \text{with} \qquad y' = -\mu A\sin\mu x + \mu B\cos\mu x.$$

The first boundary condition gives

$$y(0) + y'(0) = A + \mu B = 0,$$

so that $A = -\mu B$. The second boundary condition gives

$$y(1) + y'(1) = A\cos\mu + B\sin\mu - \mu A\sin\mu + \mu B\cos\mu = 0;$$

that is,

$$(\cos\mu - \mu\sin\mu)A + (\sin\mu + \mu\cos\mu)B = 0.$$

The system of linear equations for A and B,

$$\begin{aligned} &A + \mu B = 0, \\ &(\cos\mu - \mu\sin\mu)A + (\sin\mu + \mu\cos\mu)B = 0, \end{aligned}$$

has nontrivial solutions if and only if the determinant of the corresponding coefficient matrix is zero, that is,

$$(1 + \mu^2) \sin \mu = 0;$$

that is, if and only if $\sin \mu = 0$. The eigenvalues are $\lambda_n = -\mu_n^2 = -n^2\pi^2$ with corresponding eigenfunctions

$$y_n = \sin n\pi x - n \cos n\pi x$$

for $n = 1, 2, 3, \ldots$.

Case 3: If $\lambda > 0$, say $\lambda = \mu^2$, the differential equation becomes $y'' - \mu^2 y = 0$ and has general solution

$$y = A \cosh \mu x + B \sinh \mu x \qquad \text{with} \qquad y' = \mu A \sinh \mu x + \mu B \cosh \mu x.$$

The first boundary condition gives

$$y(0) + y'(0) = A + \mu B = 0$$

The second boundary condition gives

$$y(1) + y'(1) = A \cosh \mu + B \sinh \mu + \mu A \sinh \mu + \mu B \cosh \mu = 0;$$

that is,

$$(\cosh \mu + \mu \sinh \mu)A + (\sinh \mu + \mu \cosh \mu)B = 0.$$

The system of linear equations for A and B,

$$\begin{aligned} &A + \mu B = 0, \\ &(\cosh \mu + \mu \sinh \mu)A + (\sinh \mu + \mu \cosh \mu)B = 0 \end{aligned}$$

has nontrivial solutions if and only if the determinant of the corresponding coefficient matrix is zero, that is,

$$(1 - \mu^2) \sinh \mu = 0,$$

and since $\sinh \mu \neq 0$, the system has nontrivial solutions if and only if $1 - \mu^2 = 0$, that is, if and only if $\mu = \pm 1$. Therefore, there is only one positive eigenvalue,

$$\lambda = (\pm 1)^2 = 1,$$

with corresponding eigenfunction

$$y = \sinh x - \cosh x.$$

Note: Here the weight function is $\sigma(x) = -1 < 0$, and the problem is not a regular Sturm-Liouville problem, so this does not contradict the theorem since the theorem does not apply. We can, however, redefine the eigenvalue as $\tilde{\lambda} = -\lambda$; then the problem becomes a regular Sturm-Liouville problem and the theorem does apply. According to the computations above, we get for $\tilde{\lambda}$ the eigenvalues -1 and $n^2\pi^2$ for $n = 1, 2, \ldots$.

Exercise 12.13. XX

Consider the non-Sturm-Liouville differential equation

$$\frac{d^2\phi}{dx^2} + \alpha(x)\frac{d\phi}{dx} + \left[\lambda\beta(x) + \gamma(x)\right]\phi = 0.$$

Multiply this equation by $H(x)$. Determine $H(x)$ such that the equation may be reduced to the standard Sturm-Liouville form:

$$\frac{d}{dx}\left[p(x)\frac{d\phi}{dx}\right] + \left[\lambda\sigma(x) + q(x)\right]\phi = 0.$$

Given $\alpha(x)$, $\beta(x)$, and $\gamma(x)$, what are $p(x)$, $\sigma(x)$, and $q(x)$?

Solution. Multiplying the differential equation by $H(x)$, we have

$$H\,\frac{d^2\phi}{dx^2} + \alpha\,H\,\frac{d\phi}{dx} + \lambda\,\beta\,H\,\phi + \gamma\,H\,\phi = 0,$$

and we want to determine H so that the first two terms are an exact derivative; that is,

$$\frac{d}{dx}\left[p(x)\frac{d\phi}{dx}\right] = H\,\frac{d^2\phi}{dx^2} + \alpha\,H\,\frac{d\phi}{dx},$$

so

$$p(x)\,\frac{d^2\phi}{dx^2} + \frac{dp(x)}{dx}\frac{d\phi}{dx} = H\,\frac{d^2\phi}{dx^2} + \alpha\,H\,\frac{d\phi}{dx}.$$

Thus, we want $p(x) = H(x)$ and $p'(x) = \alpha(x)\,H(x)$, so that $H(x)$ satisfies the differential equation $H'(x) = \alpha(x)\,H(x)$.

If we take

$$p(x) = H(x) = \exp\left(\int \alpha(x)\,dx\right),$$

the differential equation is in Sturm-Liouville form,

$$\frac{d}{dx}\left[p(x)\frac{d\phi}{dx}\right] + \left[\lambda\sigma(x) + q(x)\right]\phi = 0.$$

where

$$p(x) = \exp\left(\int \alpha(x)\,dx\right),$$

$$q(x) = \gamma(x)\,\exp\left(\int \alpha(x)\,dx\right),$$

$$\sigma(x) = \beta(x)\,\exp\left(\int \alpha(x)\,dx\right).$$

Note that $p(x) > 0$ and $\sigma(x) > 0$ provided that $\beta(x) > 0$.

Exercise 12.14. ✗✗

For the Sturm-Liouville eigenvalue problem,

$$\frac{d^2\phi}{dx^2} + \lambda\phi = 0, \quad 0 < x < a,$$

$$\frac{d\phi}{dx}(0) = 0,$$

$$\phi(a) = 0,$$

verify the following general properties:

(a) There are an infinite number of eigenvalues with a smallest but no largest.

(b) The nth eigenfunction has $n - 1$ zeros in the open interval $0 < x < a$.

(c) The eigenfunctions are complete and orthogonal.

(d) What does the Rayleigh quotient say concerning negative and zero eigenvalues ?

Solution.

(a) Assuming that the eigenvalues are real, we have to consider the three cases when $\lambda = 0$, $\lambda < 0$, and $\lambda > 0$.

Case 1: If $\lambda = 0$, the general solution to the differential equation $\phi''(x) = 0$ is $\phi(x) = Ax + B$, with $\phi'(x) = A$, and applying the first boundary condition $\phi'(0) = 0$, we have $A = 0$, and the solution is $\phi(x) = B$ for $0 < x < a$. Applying the second boundary condition $\phi(a) = 0$, we have $B = 0$, and the only solution in this case is the trivial solution $\phi(x) = 0$ for $0 < x < a$. Therefore, $\lambda = 0$ is not an eigenvalue.

Case 2: If $\lambda < 0$, then $\lambda = -\mu^2$ where $\mu \neq 0$, and the general solution to the differential equation $\phi'' - \mu^2\phi = 0$ is

$$\phi(x) = A\cosh\mu x + B\sinh\mu x \quad \text{with} \quad \phi'(x) = \mu A\sinh\mu x + \mu B\cosh\mu x.$$

Applying the first boundary condition $\phi'(0) = \mu B = 0$ implies that $B = 0$, and the solution is now

$$\phi(x) = A\cosh\mu x.$$

Applying the second boundary condition $\phi(a) = 0$ implies that $A\cosh\mu a = 0$, so that $A = 0$, and in this case we have only the trivial solution $\phi(x) = 0$ for $0 < x < a$.

Case 3: If $\lambda > 0$, then $\lambda = \mu^2$ where $\mu \neq 0$, and the general solution to the differential equation $\phi'' + \mu^2\phi = 0$ is

$$\phi(x) = A\cos\mu x + B\sin\mu x \quad \text{with} \quad \phi'(x) = -\mu A\sin\mu x + \mu B\cos\mu x.$$

Applying the first boundary condition $\phi'(0) = 0$ implies that $\mu B = 0$, so that $B = 0$, and the solution is now

$$\phi(x) = A\cos\mu x.$$

Applying the second boundary condition $\phi(a) = 0$ implies that $A\cos\mu a = 0$, and if $A = 0$, we get only the trivial solution. The boundary value problem has a nontrivial solution if and only if $\cos\mu a = 0$, that is, if and only if $\mu a = (n - \frac{1}{2})\pi$ for some integer $n \geqslant 1$, and therefore the eigenvalues are

$$\lambda_n = \left(\frac{(2n-1)\pi}{2a}\right)^2$$

with corresponding eigenfunctions

$$\phi_n(x) = \cos\frac{(2n-1)\pi x}{2a}$$

for $n = 1, 2, \ldots$. The eigenvalues are therefore ordered as

$$0 < \lambda_1 < \lambda_2 < \cdots < \lambda_n < \cdots,$$

and there are an infinite number of eigenvalues, the smallest one being $\lambda_1 = \pi^2/4a^2$, but there is no largest eigenvalue.

(b) For $n \geqslant 1$, the eigenfunction ϕ_n is given by

$$\phi_n(x) = \cos\frac{(2n-1)\pi x}{2a}$$

for $0 < x < a$. Note that

$$\phi_n(0) = 1 \qquad \text{while} \qquad \phi_n(a) = \cos\frac{(2n-1)\pi}{2} = 0,$$

and all the zeros of ϕ_n occur in the interval $(0, a]$. Also, $\phi_n(x) = 0$ exactly when

$$\frac{(2n-1)\pi x}{2a} = \frac{(2k-1)\pi}{2}$$

for $1 \leqslant k \leqslant n$, that is,

$$x = \left(\frac{2k-1}{2n-1}\right)a$$

for $1 \leqslant k \leqslant n$, and the eigenfunction $\phi_n(x) = \cos[(2n-1)\pi x/2a]$ has exactly n zeros in the interval $(0, a]$; that is, $\phi_n(x)$ has exactly $n-1$ zeros in the open interval $(0, a)$.

(c) From Dirichlet's theorem we know that every f in the linear space of all piecewise smooth functions on $[0, a]$ has a Fourier series expansion in terms of the eigenfunctions; that is, the eigenfunctions form a complete set in the linear space $PWS\,[0,a]$. The eigenfunctions form what is usually called a **Schauder basis** for the linear space $PWS\,[0,a]$.

Finally, we note that

$$\int_0^a \phi_m(x)\phi_n(x)\,dx = \int_0^a \cos\frac{(2m-1)\pi x}{2a}\cos\frac{(2n-1)\pi x}{2a}\,dx = 0$$

for $m, n \geqslant 1$ with $m \neq n$, and the set of eigenfunctions forms an orthogonal set.

(d) Using the boundary conditions

$$\phi'(0) = 0 \qquad \text{and} \qquad \phi(a) = 0$$

for the regular Sturm-Liouville problem above, we can write the eigenvalues in terms of the corresponding eigenfunctions as

$$\lambda_n = R(\phi_n) = \frac{\displaystyle\int_0^a \phi_n'(x)^2\,dx}{\displaystyle\int_0^a \phi_n(x)^2\,dx},$$

and clearly, $\lambda_n \geqslant 0$. If $\lambda_0 = 0$ is an eigenvalue,

$$\lambda_0 = R(\phi_0) = \frac{\displaystyle\int_0^a \phi_0'(x)^2\,dx}{\displaystyle\int_0^a \phi_0(x)^2\,dx} = 0,$$

and then $\phi_0'(x) = 0$ for $0 \leqslant x \leqslant a$, and $\phi_0(x)$ is a constant, and then $\phi_0(a) = 0$ implies that $\phi_0(x) = 0$ for $0 < x < a$, which is a contradiction, and therefore $\lambda_0 = 0$ is **not** an eigenvalue.

Exercise 12.15. ✗✗

Show that $\lambda > 0$ for the eigenvalue problem

$$\frac{d^2\phi}{dx^2} + \left(\lambda - x^2\right)\phi = 0, \quad 0 < x < 1,$$

with

$$\frac{d\phi}{dx}(0) = 0, \qquad \text{and} \qquad \frac{d\phi}{dx}(1) = 0.$$

Solution. This is a regular Sturm-Liouville problem with

$$p(x) = 1, \qquad q(x) = -x^2 \leqslant 0, \qquad \text{and} \qquad \sigma(x) = 1$$

for $0 \leqslant x \leqslant 1$, and from the boundary conditions

$$\left[- p(x)\phi(x)\phi'(x)\right] \Big|_0^1 = 0,$$

the Rayleigh quotient reduces to

$$\lambda = R(\phi) = \frac{\displaystyle\int_0^1 \left[\phi'(x)^2 + x^2\phi(x)^2\right] dx}{\displaystyle\int_0^1 \phi(x)^2\, dx} \geqslant 0,$$

and all of the eigenvalues are nonnegative. If $\lambda = 0$ is an eigenvalue and ϕ_0 is the corresponding eigenfunction (and is thus not identically zero on the interval $[0, 1]$),

$$0 = R(\phi_0) = \frac{\displaystyle\int_0^1 \left[\phi_0'(x)^2 + x^2\phi_0(x)^2\right] dx}{\displaystyle\int_0^1 \phi_0(x)^2\, dx},$$

assuming that ϕ_0 and ϕ_0' are continuous on the interval $[0, 1]$. This implies that

$$\phi_0'(x)^2 = 0 \qquad \text{and} \qquad x^2\phi_0(x)^2 = 0$$

for all $x \in [0, 1]$, and this implies that $\phi_0(x) = 0$ for all $x \in [0, 1]$, which is a contradiction. Therefore, $\lambda_0 = 0$ is **not** an eigenvalue.

Exercise 12.16. XX

Give an example of an eigenvalue problem where there is more than one eigenfunction corresponding to an eigenvalue.

Solution. Consider the boundary value problem with periodicity conditions

$$\frac{d^2\phi}{dx^2} + \lambda\phi = 0, \qquad -\pi < x < \pi,$$

$$\phi(-\pi) = \phi(\pi), \qquad \frac{d\phi}{dx}(-\pi) = \frac{d\phi}{dx}(\pi).$$

The eigenvalues are $\lambda_n = n^2$ with corresponding eigenfunctions

$$\phi_n(x) = \cos nx \qquad \text{and} \qquad \psi_n(x) = \sin nx$$

for $n \geqslant 0$.

Therefore,there are two linearly independent eigenfunctions for each eigenvalue λ_n for $n \geqslant 1$. For $\lambda_0 = 0$, there is only one eigenfunction, $\phi_0(x) = 1$ for $-\pi \leqslant x \leqslant \pi$. Note that the periodicity conditions given here are not boundary conditions of Sturm-Liouville type; they are *mixed boundary conditions* in the sense that each contains the function or its derivative evaluated at *both* endpoints of the interval.

Exercise 12.17. XXX

Consider a fourth-order linear differential operator,

$$L = \frac{d^4}{dx^4}.$$

(a) Show that $u\,L(v) - v\,L(u)$ is an exact differential.

(b) Evaluate

$$\int_0^1 [u\,L(v) - v\,L(u)]\;dx$$

in terms of the boundary data for any functions u and v.

(c) Show that

$$\int_0^1 [u\,L(v) - v\,L(u)]\;dx = 0$$

if u and v are any two functions satisfying the boundary conditions

$$\phi(0) = 0, \qquad \phi(1) = 0,$$

$$\frac{d\phi}{dx}(0) = 0, \qquad \frac{d^2\phi}{dx^2}(1) = 0.$$

(d) Give another example of boundary conditions such that

$$\int_0^1 [u\,L(v) - v\,L(u)]\;dx = 0.$$

(e) For the eigenvalue problem below using the boundary conditions from (c),

$$\frac{d^4\phi}{dx^4} + \lambda e^x \phi = 0, \quad 0 < x < 1,$$

$$\phi(0) = 0, \qquad \phi(1) = 0,$$

$$\frac{d\phi}{dx}(0) = 0, \qquad \frac{d^2\phi}{dx^2}(1) = 0,$$

show that the eigenfunctions corresponding to distinct eigenvalues are orthogonal. What is the weight function ?

Solution.

(a) We consider

$$uv^{(4)} = (uv''')' - u'v''' = (uv''')' - (u'v'')' + u''v'', \tag{12.1}$$

and by symmetry,

$$vu^{(4)} = (vu''')' - (v'u'')' + v''u'', \tag{12.2}$$

and subtracting (12.2) from (12.1), we have

$$uL(v) - vL(u) = (uv''' - vu''' - u'v'' + v'u'')',$$

and $uL(v) - vL(u)$ is an exact differential.

(b) We have

$$\begin{aligned}\int_0^1 [uL(v) - vL(u)]\,dx = \left[uv''' - vu''' - u'v'' + v'u''\right]\Big|_0^1 \\ = \quad u(1)v'''(1) - v(1)u'''(1) \\ - u'(1)v''(1) + v'(1)u''(1) \\ - u(0)v'''(0) + v(0)u'''(0) \\ + u'(0)v''(0) - v'(0)u''(0).\end{aligned}$$

(c) Here u and v are any two functions satisfying the boundary conditions

$$\phi(0) = 0, \qquad \phi(1) = 0,$$

$$\phi'(0) = 0, \qquad \phi''(1) = 0.$$

From part (b) each of the first four terms contains either $u(1),\ v(1),\ u''(1)$, or $v''(1)$, each of which is 0, while each of the last four terms contains either $u(0),\ v(0),\ u'(0)$, or $v'(0)$, each of which is also 0.

(d) Another set of boundary conditions for which

$$\int_0^L [uL(v) - vL(u)]\,dx = 0$$

is given by

$$\phi'(0) = 0, \qquad \phi'(1) = 0,$$

$$\phi'''(0) = 0, \qquad \phi'''(1) = 0.$$

(e) Let (ϕ_n, λ_n) and (ϕ_m, λ_m) be distinct eigenpairs satisfying the boundary value problem

$$\begin{aligned}
&\frac{d^4\phi}{dx^4} + \lambda e^x \phi = 0, \quad 0 < x < 1,\\
&\phi(0) = 0, \quad \phi(1) = 0,\\
&\phi'(0) = 0, \quad \phi''(1) = 0;
\end{aligned}$$

then we have

$$\begin{aligned}
0 &= \int_0^1 \left[\phi_n L(\phi_m) - \phi_m L(\phi_n)\right] dx\\
&= \int_0^1 \left[\phi_n \left(-\lambda_m e^x \phi_m\right) - \phi_m \left(-\lambda_n e^x \phi_n\right)\right] dx = (\lambda_n - \lambda_m) \int_0^1 \phi_n \phi_m e^x \, dx
\end{aligned}$$

and if $\lambda_n \neq \lambda_m$, then

$$\int_0^1 \phi_n \phi_m e^x \, dx = 0$$

and ϕ_n and ϕ_m are orthogonal on the interval $[0, 1]$ with respect to the weight function

$$\sigma(x) = e^x$$

for $x \in [0, 1]$.

Exercise 12.18.
Let $u(x) = J_0(\alpha x)$ and $v(x) = J_0(\beta x)$, where J_0 is the Bessel function of the first kind of order zero.

(a) Show that $xu'' + u' + \alpha^2 xu = 0$ and $xv'' + v' + \beta^2 xv = 0$.

(b) Show that $[x\,(u'v - v'u)]' = (\beta^2 - \alpha^2)xuv$.

(c) Show that

$$(\beta^2 - \alpha^2) \int x J_0(\alpha x) J_0(\beta x)\, dx = x\left[\alpha J_0'(\alpha x) J_0(\beta x) - \beta J_0'(\beta x) J_0(\alpha x)\right].$$

This is one of a set of formulas called *Lommel's integrals.*

(d) Show that if α and β are distinct zeros of $J_0'(z)$, then

$$\int_0^1 x J_0(\alpha x) J_0(\beta x)\, dx = 0$$

so that $J_0(\alpha x)$ and $J_0(\beta x)$ are orthogonal on the interval $[0, 1]$ with respect to the weight function $\sigma(x) = x$.

Solution.

(a) Since $u(x) = J_0(\alpha x)$ and $v(x) = J_0(\beta x)$ are solutions to Bessel's equation of order zero, then

$$(xu'(x))' + \alpha^2 xu(x) = 0 \qquad \text{and} \qquad (xv'(x))' + \beta^2 xv(x) = 0;$$

that is,

$$xu''(x) + u'(x) + \alpha^2 xu(x) = 0 \quad \text{and} \quad xv''(x) + v'(x) + \beta^2 xv(x) = 0$$

for $0 \leqslant x \leqslant 1$.

(b) From part (a), we have

$$xu''(x)v(x) + u'(x)v(x) + \alpha^2 xu(x)v(x) = 0$$

and

$$xv''(x)u(x) + v'(x)u(x) + \beta^2 xv(x)u(x) = 0,$$

and subtracting the second equation from the first, we have

$$\begin{aligned} xu''(x)v(x) + u'(x)v(x) - (xv''(x)u(x) + v'(x)u(x)) \\ - (\beta^2 - \alpha^2)xu(x)v(x) = 0; \end{aligned}$$

that is,

$$(\beta^2 - \alpha^2)xu(x)v(x) = \big(xu'(x)v(x) - xv'(x)u(x)\big)'$$

for $0 \leqslant x \leqslant 1$.

(c) Integrating this last expression, we have an indefinite integral

$$\begin{aligned} (\beta^2 - \alpha^2)\int xu(x)v(x)\,dx &= \int \big(xu'(x)v(x) - xv'(x)u(x)\big)'\,dx \\ &= \big(xu'(x)v(x) - xv'(x)u(x)\big); \end{aligned}$$

that is,

$$(\beta^2 - \alpha^2)\int xJ_0(\alpha x)J_0(\beta x)\,dx = x\left[\alpha J_0'(\alpha x)J_0(\beta x) - \beta J_0'(\beta x)J_0(\alpha x)\right].$$

(d) Now, if α and β are distinct zeros of $J_0'(z)$, then

$$(\beta^2 - \alpha^2)\int_0^1 xJ_0(\alpha x)J_0(\beta x)\,dx = \alpha J_0'(\alpha)J_0(\beta) - \beta J_0'(\beta)J_0(\alpha) = 0,$$

and since $\alpha \neq \beta$, then

$$\int_0^1 xJ_0(\alpha x)J_0(\beta x)\,dx = 0.$$

Exercise 12.19. XXXX

Consider the boundary value problem

$$\frac{d^2\phi}{dx^2} + \lambda\phi = 0, \quad 0 < x < 1,$$

$$\phi(0) - \frac{d\phi}{dx}(0) = 0,$$

$$\phi(1) + \frac{d\phi}{dx}(1) = 0.$$

(a) Using the Rayleigh quotient, show that $\lambda \geqslant 0$. Why is $\lambda > 0$?

(b) Prove that eigenfunctions corresponding to different eigenvalues are orthogonal.

(c) Show that

$$\tan\sqrt{\lambda} = \frac{2\sqrt{\lambda}}{\lambda - 1}.$$

Determine the eigenvalues graphically. Estimate the large eigenvalues.

(d) Solve

$$\frac{\partial u}{\partial t} = k\frac{\partial^2 u}{\partial x^2},$$

with

$$u(0,t) - \frac{\partial u}{\partial x}(0,t) = 0,$$

$$u(1,t) + \frac{\partial u}{\partial x}(1,t) = 0,$$

$$u(x,0) = f(x).$$

You may call the relevant eigenfunctions $\phi_n(x)$ and assume that they are known.

Solution.

(a) We use the Rayleigh quotient to show that $\lambda > 0$ for all eigenvalues λ. Let λ be an eigenvalue of the Sturm-Liouville problem above, and let $\phi(x)$ be the corresponding eigenfunction; then

$$-p(x)\phi(x)\phi'(x)\Big|_0^1 = -\phi(1)\phi'(1) + \phi(0)\phi'(0) = \phi(1)^2 + \phi(0)^2 \geqslant 0,$$

and since $q(x) = 0 \leqslant 0$ for all $0 \leqslant x \leqslant 1$, then

$$\lambda = \frac{\phi(0)^2 + \phi(1)^2 + \displaystyle\int_0^1 \phi'(x)^2\,dx}{\displaystyle\int_0^1 \phi(x)^2\,dx} \geqslant 0$$

since $p(x) = \sigma(x) = 1$ for $0 \leqslant x \leqslant 1$. If $\lambda = 0$, then

$$\phi(0)^2 + \phi(1)^2 + \int_0^1 \phi'(x)^2\,dx = 0,$$

so that $\phi(0) = \phi(1) = 0$ and $\displaystyle\int_0^1 \phi'(x)^2\,dx = 0$, and since ϕ' is continuous on $[0,1]$, then $\phi(x)$ is constant on $[0,1]$, so that $\phi(x) = \phi(0) = 0$ for all $0 \leqslant x \leqslant 1$. Therefore, $\lambda = 0$ is **not** an eigenvalue of this boundary value problem, and all the eigenvalues satisfy $\lambda > 0$.

(b) If λ_1 and λ_2 are distinct eigenvalues of this boundary value problem with corresponding eigenfunctions ϕ_1 and ϕ_2, respectively, then

$$\phi_1'' + \lambda_1\phi_1 = 0 \qquad \text{and} \qquad \phi_2'' + \lambda_2\phi_2 = 0,$$

so that

$$\phi_2\phi_1'' - \phi_1\phi_2'' + (\lambda_1 - \lambda_2)\phi_1\phi_2 = 0,$$

that is,

$$(\phi_2\phi_1' - \phi_1\phi_2')' + (\lambda_1 - \lambda_2)\phi_1\phi_2 = 0,$$

and integrating, we have

$$(\phi_2\phi_1' - \phi_1\phi_2')\Big|_0^1 + (\lambda_1 - \lambda_2)\int_0^1 \phi_1\phi_2\,dx = 0.$$

However,

$$\begin{aligned}(\phi_2\phi_1' - \phi_1\phi_2')\Big|_0^1 &= \phi_2(1)\phi_1'(1) - \phi_1(1)\phi_2'(1) - \phi_2(0)\phi_1'(0) + \phi_1(0)\phi_2'(0)\\ &= -\phi_2(1)\phi_1(1) + \phi_1(1)\phi_2(1)\\ &\qquad - \phi_2(0)\phi_1(0) + \phi_1(0)\phi_2(0) = 0,\end{aligned}$$

so that

$$(\lambda_1 - \lambda_2)\int_0^1 \phi_1\phi_2\,dx = 0.$$

Since $\lambda_1 \neq \lambda_2$, then

$$\int_0^1 \phi_1(x)\phi_2(x)\,dx = 0;$$

that is, ϕ_1 and ϕ_2 are orthogonal on the interval $[0,1]$.

(c) If $\lambda > 0$, then $\lambda = \mu^2$ where $\mu \neq 0$, and the differential equation is

$$\phi'' + \mu^2 \phi = 0$$

with general solution

$$\phi(x) = A\cos\mu x + B\sin\mu x \quad \text{and} \quad \phi'(x) = -\mu A\sin\mu x + \mu B\cos\mu x$$

for $0 \leqslant x \leqslant 1$. From the first boundary condition,

$$\phi(0) - \phi'(0) = A - \mu B = 0$$

and $A = \mu B$. From the second boundary condition

$$\phi(1) + \phi'(1) = A\cos\mu + B\sin\mu - \mu A\sin\mu + \mu B\cos\mu = 0,$$

that is,

$$B\left[2\mu\cos\mu - (\mu^2 - 1)\sin\mu\right] = 0,$$

and the boundary value problem has a nontrivial solution if and only if

$$\tan\mu = \frac{2\mu}{\mu^2 - 1},$$

that is, if and only if λ satisfies the transcendental equation

$$\tan\sqrt{\lambda} = \frac{2\sqrt{\lambda}}{\lambda - 1}.$$

To determine the eigenvalues we sketch the graphs of the functions

$$f(\mu) = \tan\mu \qquad \text{and} \qquad g(\mu) = \frac{2\mu}{\mu^2 - 1}$$

for $\mu > 0$. Note that for $\mu > 0$, we have

$$g(\mu) = \frac{2\mu}{\mu^2 - 1} = \frac{1}{\mu + 1} + \frac{1}{\mu - 1},$$

so that

$$g'(\mu) = -\frac{1}{(\mu + 1)^2} - \frac{1}{(\mu - 1)^2} < 0$$

and g is decreasing on the interval $(0, 1)$ and on the interval $(1, \infty)$, and the line $\mu = 1$ is a vertical asymptote to the graph. The graphs of g and f are shown in Figure 12.3, from which it is clear that there are an infinite number of distinct solutions μ_n to the equation

$$\tan\mu = \frac{2\mu}{\mu^2 - 1},$$

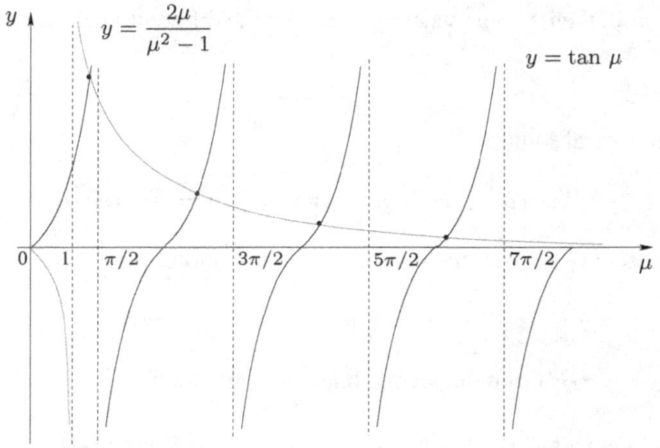

Figure 12.3 Intersection of graphs of $f(\mu) = \tan \mu$ and $g(\mu) = \dfrac{2\mu}{\mu^2 - 1}$

and the eigenvalues are $\lambda_n = \mu_n^2$ for $n \geqslant 1$. Since $\lim_{n\to\infty} \mu_n = +\infty$, then

$$\lim_{n\to\infty} \tan \mu_n = \lim_{n\to\infty} \frac{2\mu_n}{\mu_n^2 - 1} = 0,$$

and the roots of the equation $\tan \mu = 2\mu/(\mu^2 - 1)$ approach the roots of the equation $\tan \mu = 0$; that is, for large n,

$$\mu_n \approx n\pi,$$

and therefore

$$\lambda_n = \mu_n^2 \approx n^2\pi^2$$

for large n.

(d) We want to solve the boundary value–initial value problem

$$\frac{\partial u}{\partial t} = k\frac{\partial^2 u}{\partial x^2},$$

$$u(0,t) - \frac{\partial u}{\partial x}(0,t) = 0,$$

$$u(1,t) + \frac{\partial u}{\partial x}(1,t) = 0,$$

$$u(x,0) = f(x),$$

and since both the equation and the boundary conditions are linear and homogeneous, we can use separation of variables. Assuming a solution of the form $u(x,t) = \phi(x) \cdot h(t)$, and separating variables, we get the two ordinary differential equations

$$\begin{aligned} &\phi''(x) + \lambda\phi(x) = 0, \quad 0 < x < 1, \quad h'(t) + \lambda k h(t) = 0, \quad t > 0, \\ &\phi(0) - \phi'(0) = 0, \\ &\phi(1) + \phi'(1) = 0. \end{aligned}$$

From part (a) we know that we have a nontrivial solution if and only if $\lambda > 0$, in which case $\lambda = \mu^2$ where $\mu \neq 0$, and the differential equation becomes $\phi'' + \mu^2\phi = 0$ with general solution

$$\phi(x) = A\cos\mu x + B\sin\mu x \quad \text{with} \quad \phi'(x) = -\mu A\sin\mu x + \mu B\cos\mu x.$$

Applying the boundary conditions as in part (c), the only values of μ for which we have a nontrivial solution are those for which

$$\tan\mu = \frac{2\mu}{\mu^2 - 1},$$

and we have an infinite sequence of positive eigenvalues

$$0 < \lambda_1 < \lambda_2 < \cdots < \lambda_n < \cdots,$$

where $\lambda_n = \mu_n^2$. The corresponding eigenfunctions are

$$\phi_n(x) = \mu_n\cos\mu_n x + \sin\mu_n x$$

for $n \geqslant 1$. The corresponding solutions to the time equation are

$$h_n(t) = e^{-\lambda_n kt},$$

and from the superposition principle the sum

$$u(x,t) = \sum_{n=1}^{\infty} a_n\left(\mu_n\cos\mu_n x + \sin\mu_n x\right)e^{-\mu_n^2 kt}$$

satisfies the partial differential equation and the boundary conditions.

To satisfy the initial condition we use the orthogonality of the eigenfunctions from part (c), and write

$$f(x) = u(x,0) = \sum_{n=1}^{\infty} a_n\phi_n(x),$$

where

$$a_n = \frac{\displaystyle\int_0^1 f(x)\phi_n(x)\,dx}{\displaystyle\int_0^1 \phi_n(x)^2\,dx}$$

for $n \geqslant 1$. It is easily verified that the normalization constant is given by

$$\int_0^1 \phi_n(x)^2\,dx = \frac{\mu_n^2 + 3}{2},$$

so that

$$a_n = \frac{2}{\mu_n^2 + 3}\int_0^1 f(x)\phi_n(x)\,dx$$

for $n \geqslant 1$.

CHAPTER 13

HEAT EQUATION PROBLEMS

Exercise 13.1. XX

For each of the boundary value–initial value problems below, determine whether or not an equilibrium temperature distribution exists and find the values of β for which an equilibrium solution exists.

(a) $\dfrac{\partial u}{\partial t} = \dfrac{\partial^2 u}{\partial x^2} + 1, \qquad \dfrac{\partial u}{\partial x}(0,t) = 1, \qquad \dfrac{\partial u}{\partial x}(a,t) = \beta,$

(b) $\dfrac{\partial u}{\partial t} = \dfrac{\partial^2 u}{\partial x^2}, \qquad \dfrac{\partial u}{\partial x}(0,t) = 1, \qquad \dfrac{\partial u}{\partial x}(a,t) = \beta,$

(c) $\dfrac{\partial u}{\partial t} = \dfrac{\partial^2 u}{\partial x^2} + x - \beta, \qquad \dfrac{\partial u}{\partial x}(0,t) = 0, \qquad \dfrac{\partial u}{\partial x}(a,t) = 0.$

Solution. If the temperature has reached an equilibrium distribution, u no longer depends on the time t, so that $\partial u/\partial t = 0$ and $u = \phi(x)$ is a function of x alone.

(a) In this case the boundary value problem for the equilibrium temperature distribution is

$$\begin{aligned}
&\phi''(x) + 1 = 0, \quad 0 < x < a, \\
&\phi'(0) = 1, \\
&\phi'(a) = \beta.
\end{aligned}$$

The general solution is

$$\phi(x) = -\frac{x^2}{2} + Ax + B \qquad \text{with} \qquad \phi'(x) = -x + A$$

for $0 < x < a$. From the first boundary condition we have $\phi'(0) = A = 1$, while from the second boundary condition we have $\phi'(a) = \beta$; on the other hand, $\phi'(a) = -a + 1$. Therefore, there exists an equilibrium temperature distribution if and only if $\beta = 1 - a$. In this case we have

$$u(x) = \phi(x) = -\frac{x^2}{2} + x + B, \quad 0 < x < a,$$

where B is a constant.

(b) In this case the boundary value problem for the equilibrium temperature distribution is

$$\begin{aligned}
&\phi''(x) = 0, \quad 0 < x < a, \\
&\phi'(0) = 1, \\
&\phi'(a) = \beta.
\end{aligned}$$

The general solution is

$$\phi(x) = Ax + B \qquad \text{with} \qquad \phi'(x) = A$$

for $0 < x < a$. From the first boundary condition we have $\phi'(0) = A = 1$, while from the second boundary condition we have $\phi'(a) = \beta$, and therefore there exists an equilibrium temperature distribution if and only if $\beta = 1$. In this case the equilibrium temperature distribution is

$$u(x) = \phi(x) = x + B, \quad 0 < x < a,$$

where B is a constant.

(c) In this case the boundary value problem for the equilibrium temperature distribution is

$$\begin{aligned}
&\phi''(x) + x - \beta = 0, \quad 0 < x < a, \\
&\phi'(0) = 0, \\
&\phi'(a) = 0.
\end{aligned}$$

The general solution is

$$\phi(x) = -\frac{x^3}{6} + \frac{\beta x^2}{2} + Ax + B \qquad \text{with} \qquad \phi'(x) = -\frac{x^2}{2} + \beta x + A,$$

for $0 < x < a$. From the first boundary condition, we have $\phi'(0) = A = 0$, while from the second boundary condition

$$\phi'(a) = -\frac{a^2}{2} + \beta a = 0,$$

which implies that an equilibrium temperature distribution exists if and only if $\beta = a/2$. In this case the equilibrium temperature distribution is

$$u(x) = \phi(x) = -\frac{x^3}{6} + \frac{ax^2}{4} + B, \quad 0 < x < a,$$

where B is a constant.

Exercise 13.2. XX

Solve the homogeneous Dirichlet problem for the heat equation

$$\frac{\partial u}{\partial t} = k\frac{\partial^2 u}{\partial x^2}, \quad 0 < x < a, \quad t > 0,$$

subject to the boundary conditions

$$u(0,t) = 0, \qquad \text{and} \qquad u(a,t) = 0$$

for $t > 0$, with initial conditions

$$u(x,0) = \begin{cases} 1, & \text{for } 0 < x < \frac{a}{2}, \\ 2, & \text{for } \frac{a}{2} \leqslant x < a. \end{cases}$$

Solution. Since both the differential equation and the boundary conditions are linear and homogeneous, we can use separation of variables and write

$$u(x,t) = \phi(x) \cdot h(t),$$

where ϕ depends only on x and h depends only on t. Substituting this into the partial differential equation, we have

$$\phi \cdot h' = k\phi'' \cdot h,$$

and separating variables yields

$$\frac{\phi''}{\phi} = \frac{h'}{kh} = -\lambda$$

for a yet-unspecified constant λ. We obtain the two ordinary differential equations

$$\phi'' + \lambda\phi = 0 \qquad \text{and} \qquad h' + \lambda k\, h = 0.$$

Since

$$u(0,t) = \phi(0)\cdot h(t) \qquad \text{and} \qquad u(a,t) = \phi(a)\cdot h(t),$$

we can satisfy the boundary conditions by requiring that $\phi(0) = \phi(a) = 0$, so that $\phi(x)$ must satisfy the boundary value problem

$$\begin{aligned} &\phi'' + \lambda\phi = 0, \quad 0 < x < a, \\ &\phi(0) = 0, \\ &\phi(a) = 0. \end{aligned}$$

Therefore, this problem with homogeneous Dirichlet boundary conditions has a nontrivial solution only for the eigenvalues λ_n, where

$$\lambda_n = \mu_n^2 = \frac{n^2\pi^2}{a^2},$$

with corresponding eigenfunctions

$$\phi_n(x) = \sin\frac{n\pi x}{a}$$

for each integer $n \geqslant 1$. The corresponding solution to the time equation

$$h_n'(t) + \lambda_n k\, h_n(t) = 0$$

is

$$h_n(t) = e^{-\lambda_n kt} = e^{-\frac{n^2\pi^2}{a^2}kt}$$

for $n \geqslant 1$. For each $n \geqslant 1$, the function

$$u_n(x,t) = \phi_n(x)\cdot h_n(t) = e^{-\frac{n^2\pi^2}{a^2}kt}\sin\frac{n\pi x}{a}$$

is a solution to the partial differential equation satisfying both boundary conditions. Using the superposition principle, we write

$$u(x,t) = \sum_{n=1}^{\infty} b_n \sin\frac{n\pi x}{a}\, e^{-\frac{n^2\pi^2}{a^2}kt}$$

for $0 < x < a$, and $t > 0$.

We determine the constants b_n from the initial condition

$$u(x,0) = \begin{cases} 1, & \text{for } \ 0 < x < \dfrac{a}{2}, \\ 2, & \text{for } \ \dfrac{a}{2} \leqslant x < a. \end{cases}$$

Setting $t = 0$ in the expression for $u(x,t)$ above, we have

$$u(x,0) = \sum_{n=1}^{\infty} b_n \sin \frac{n\pi x}{a}.$$

Hence we need to find the Fourier sine series of $u(x,0)$ on the interval $[0,a]$, and the coeffcieints are

$$b_n = \frac{2}{a}\int_0^a u(x,0) \sin \frac{n\pi x}{a}\, dx$$

for $n \geqslant 1$. Now

$$\int_0^a u(x,0)\sin\frac{n\pi x}{a}\,dx = \int_0^{a/2} \sin\frac{n\pi x}{a}\,dx + \int_{a/2}^a 2\sin\frac{n\pi x}{a}\,dx,$$

so that

$$b_n = \frac{2}{n\pi} + \frac{2}{n\pi}\cos\frac{n\pi}{2} - \frac{4}{n\pi}(-1)^n$$

for all $n \geqslant 1$, and the solution to the initial value problem is

$$u(x,t) = \sum_{n=1}^{\infty} b_n \sin\frac{n\pi x}{a}\, e^{-\frac{n^2\pi^2}{a^2}kt}$$

for $0 < x < a$, and $t > 0$.

Exercise 13.3. XX

Solve the following boundary value–initial value problem for the heat equation

$$\frac{\partial u}{\partial t} = k\frac{\partial^2 u}{\partial x^2},$$

$$u(0,t) = 0, \qquad \text{and} \qquad u(a,t) = 0,$$

$$u(x,0) = 3\sin\frac{\pi x}{a} - \sin\frac{3\pi x}{a}$$

for $0 < x < a,\ t > 0$.

Solution. Assuming a solution of the form

$$u(x,t) = \phi(x)\cdot G(t)$$

and separating variables, we get

$$\frac{\phi''(x)}{\phi(x)} = \frac{G'(t)}{k\,G(t)} = -\lambda,$$

where λ is a constant. The partial differential equation is reduced to two ordinary differential equations:

- **Spatial equation**:

$$\begin{aligned} &\phi''(x) + \lambda\phi(x) = 0, \quad 0 < x < a,\\ &\phi(0) = 0,\\ &\phi(a) = 0. \end{aligned}$$

- **Time equation**:

$$G'(t) + k\lambda G(t) = 0, \quad t > 0.$$

Since it has a complete set of homogeneous Dirichlet boundary conditions, we solve the spatial equation first.

The eigenvalues and corresponding eigenfunctions for this problem are (see Exercise 13.2)

$$\lambda_n = \frac{n^2\pi^2}{a^2} \qquad \text{and} \qquad \phi_n(x) = \sin\frac{n\pi x}{a}$$

for $n = 1, 2, 3, \ldots$. The solutions to the time equation corresponding to these non-trivial solutions are

$$G_n(t) = e^{-\frac{kn^2\pi^2 t}{a^2}}$$

for $n = 1, 2, 3, \ldots$. For $n \geqslant 1$, the functions

$$u_n(x,t) = \phi_n(x) \cdot G_n(t) = \sin\frac{n\pi x}{a} \cdot e^{-\frac{kn^2\pi^2 t}{a^2}}$$

are also solutions to the partial differential equation satisfying the boundary conditions, and since the partial differential equation and the boundary conditions are linear and homogeneous, by the superposition principle, the function

$$u(x,t) = \sum_{n=1}^{\infty} b_n \sin\frac{n\pi x}{a} \cdot e^{-\frac{kn^2\pi^2 t}{a^2}}$$

is also a solution. We determine the constants b_n for $n \geqslant 1$ using the initial condition: namely,

$$u(x,0) = \sum_{n=1}^{\infty} b_n \sin\frac{n\pi x}{a} = 3\sin\frac{\pi x}{a} - \sin\frac{3\pi x}{a}.$$

From the orthogonality of the eigenfunctions on the interval $[0, a]$ we see immediately that the coefficients of the Fourier sine series are

$$b_1 = 3, \quad b_3 = -1 \quad \text{and} \quad b_n = 0 \quad \text{for} \quad n \neq 1, 3.$$

The solution is therefore

$$u(x,t) = 3\sin\frac{\pi x}{a} e^{-\frac{k\pi^2 t}{a^2}} - \sin\frac{3\pi x}{a} e^{-\frac{9k\pi^2 t}{a^2}}$$

for $0 < x < a, \ t > 0$.

Exercise 13.4. XX

Solve the boundary value problem for the one-dimensional heat equation

$$\begin{aligned}&\frac{\partial u}{\partial t} = \frac{\partial^2 u}{\partial x^2}, \quad 0 < x < \pi, \quad t > 0,\\ &u(0,t) = 0,\\ &u(\pi,t) = 0,\\ &u(x,0) = 30 \sin x,\end{aligned}$$

and give a brief physical explanation of the problem.

Solution. Using separation of variables, since we have homogeneous Dirichlet boundary conditions, we obtain the solution ($k = 1$ and $a = \pi$ here)

$$u(x,t) = \sum_{n=1}^{\infty} b_n e^{-n^2 t} \sin nx,$$

where $b_n = 2/\pi \int_0^{\pi} f(x) \sin nx \, dx$ for $n \geqslant 1$. Now

$$u(x,0) = f(x) = 30 \sin x$$

for $0 < x < \pi$; that is, $f(x)$ is its own Fourier sine series, so that $b_1 = 30$, and $b_n = 0$ for all $n \geqslant 2$. The solution is

$$u(x,t) = 30 e^{-t} \sin x,$$

and this gives the temperature in a bar whose sides are insulated and whose ends $x = 0$ and $x = \pi$ are kept at zero temperature, with an initial temperature distribution given by $u(x,0) = 30 \sin x, \ 0 < x < \pi$.

Exercise 13.5. XX

Solve the following homogeneous Dirichlet problem for the heat equation:

$$\begin{aligned}&\frac{\partial u}{\partial t} = \frac{\partial^2 u}{\partial x^2}, \quad 0 < x < 1, \quad t > 0,\\ &u(0,t) = 0,\\ &u(1,t) = 0,\\ &u(x,0) = e^{-x},\end{aligned}$$

and give a brief physical explanation of the problem.

Solution. After separating variables, applying the initial conditions, and using the superposition principle, we obtain the solution

$$u(x,t) = \sum_{n=1}^{\infty} b_n e^{-n^2 \pi^2 t} \sin n\pi x,$$

where

$$b_n = 2\int_0^1 e^{-x} \sin n\pi x \, dx$$

for $n \geqslant 1$. Integrating by parts, we get

$$\begin{aligned}\int_0^1 e^{-x} \sin n\pi x \, dx &= -\frac{e^{-x}}{1+n^2\pi^2}\left(\sin n\pi x + n\pi \cos n\pi x\right)\Big|_0^1 \\ &= \frac{n\pi}{1+n^2\pi^2}\left[1+(-1)^{n+1}e^{-1}\right],\end{aligned}$$

so that

$$u(x,t) = 2\pi \sum_{n=1}^{\infty} \frac{n}{1+n^2\pi^2}\left[1+(-1)^{n+1}e^{-1}\right] e^{-n^2\pi^2 t} \sin n\pi x,$$

and this gives the temperature in a bar whose sides are insulated and whose ends $x = 0$ and $x = 1$ are kept at zero temperature, with an initial temperature distribution given by $u(x,0) = e^{-x}$, $0 < x < 1$.

Exercise 13.6. XXX

Consider the Neumann problem for the heat equation

$$\begin{aligned}&\frac{\partial u}{\partial t} = k\frac{\partial^2 u}{\partial x^2} \quad 0 < x < a, \quad t > 0,\\ &\frac{\partial u}{\partial x}(0,t) = 0,\\ &\frac{\partial u}{\partial x}(a,t) = 0,\\ &u(x,0) = f(x).\end{aligned}$$

(a) Give a short physical interpretation of this problem.

(b) Use the method of separation of variables to solve this problem. First show that there are no separated solutions which grow exponentially in time.

Hint: The solution can be written as

$$u(x,t) = A_0 + \sum_{n=1}^{\infty} A_n e^{-\lambda_n kt} \cos \frac{n\pi x}{a}.$$

Find the λ_n.

(c) Show that the initial condition, $u(x,0) = f(x)$, leads to the Fourier cosine series:

$$f(x) = A_0 + \sum_{n=1}^{\infty} A_n \cos \frac{n\pi x}{a}.$$

(d) Solve for A_n by using

$$\int_0^a \cos \frac{n\pi x}{a} \cos \frac{m\pi x}{a}\, dx = \begin{cases} \frac{a}{2}, & \text{if} \quad n = m, \\ 0, & \text{if} \quad n \neq m \end{cases}$$

for n, $m \geqslant 1$.

(e) Consider the limit $\lim_{t\to\infty} u(x,t)$. Find the steady-state solution and show that $u(x,t)$ approaches this steady-state solution as $t \to \infty$.

Solution.

(a) The partial differential equation describes heat flow in a one-dimensional rod where

- all thermal properties are constant in any cross section,
- there is no heat flow through the lateral sides; that is, the lateral sides are perfectly insulated,
- there are no internal heat sources in the rod,
- the boundary conditions imply that there is no heat flow through the ends of the rod, and
- the initial temperature distribution in the rod is $u(x,0) = f(x)$ for $0 < x < a$.

(b) To show that there are no separated solutions which grow exponentially in time, we have to show that there are no negative eigenvalues. Assuming a solution of the form

$$u(x,t) = \phi(x) \cdot G(t)$$

and separating variables, we get

$$\frac{\phi''(x)}{\phi(x)} = \frac{G'(t)}{k\,G(t)} = -\lambda,$$

where λ is a constant. The partial differential equation is reduced to two ordinary differential equations:

- **Spatial equation**:

$$\begin{aligned}
&\phi''(x) + \lambda\phi(x) = 0, \quad 0 < x < a, \\
&\phi'(0) = 0, \\
&\phi'(a) = 0.
\end{aligned}$$

- **Time equation**:

$$G'(t) + k\lambda G(t) = 0, \quad t > 0.$$

We consider the spatial equation first. This boundary value problem with homogeneous Neumann conditions has eigenvalues

$$\lambda_n = \mu_n^2 = \frac{n^2\pi^2}{a^2},$$

with corresponding eigenfunctions

$$\phi_n(x) = \cos\frac{n\pi x}{a}$$

for $n = 0, 1, 2, 3, \ldots$. The solutions to the time equation corresponding to these nonnegative eigenvalues are

$$G_n(t) = e^{-\frac{kn^2\pi^2 t}{a^2}}$$

for $n = 0, 1, 2, 3, \ldots$.

For $n \geqslant 0$, the functions

$$u_n(x,t) = \phi_n(x) \cdot G_n(t) = \cos\frac{n\pi x}{a}\, e^{-\frac{kn^2\pi^2 t}{a^2}}$$

are also solutions to the partial differential equation satisfying the boundary conditions. The partial differential equation and the boundary conditions are homogeneous, and from the superposition principle, the solution is

$$u(x,t) = A_0 + \sum_{n=1}^{\infty} A_n \cos\frac{n\pi x}{a}\, e^{-\frac{kn^2\pi^2 t}{a^2}}.$$

Note that the constant A_0 corresponds to the $n = 0$ term.

(c) The initial condition is

$$u(x,0) = f(x)$$

for $0 < x < a$, and setting $t = 0$ in the solution above, we need

$$f(x) = A_0 + \sum_{n=1}^{\infty} A_n \cos\frac{n\pi x}{a}$$

for $0 < x < a$.

(d) To determine the coefficients A_n we use the fact that $\cos(n\pi x/a)$ and $\cos(m\pi x/a)$ are orthogonal on the interval $[0, a]$ in the sense that

$$\int_0^a \cos\frac{n\pi x}{a}\cos\frac{m\pi x}{a}\,dx = \begin{cases} 0, & \text{for} \quad n, m \geqslant 0,\ n \neq m, \\ \dfrac{a}{2}, & \text{for} \quad n = m \neq 0, \\ a, & \text{for} \quad n = m = 0. \end{cases}$$

From the initial condition

$$f(x) = A_0 + \sum_{n=1}^{\infty} A_n \cos\frac{n\pi x}{a} = \sum_{n=0}^{\infty} A_n \cos\frac{n\pi x}{a}$$

for $0 < x < a$, we multiply both sides of this equation by $\cos(m\pi x/a)$, and integrate over the interval $[0, a]$ to get

$$\int_0^a f(x)\cos\frac{m\pi x}{a}\,dx = \sum_{n=0}^{\infty}\int_0^a A_n \cos\frac{n\pi x}{a}\cos\frac{m\pi x}{a}\,dx,$$

and from the orthogonality of the eigenfunctions on the interval $[0, a]$, we have

$$\int_0^a f(x)\cos\frac{m\pi x}{a}\,dx = \begin{cases} \displaystyle\int_0^a A_0\,dx = aA_0, & \text{for} \quad m = 0, \\ \dfrac{aA_m}{2}, & \text{otherwise.} \end{cases}$$

Therefore,

$$A_0 = \frac{1}{a}\int_0^a f(x)\,dx \quad \text{and} \quad A_n = \frac{2}{a}\int_0^a f(x)\cos\frac{n\pi x}{a}\,dx, \quad n \geqslant 1.$$

(e) The solution is

$$u(x,t) = A_0 + \sum_{n=1}^{\infty} A_n \cos\frac{n\pi x}{a}\, e^{-\frac{kn^2\pi^2 t}{a^2}},$$

where

$$A_0 = \frac{1}{a}\int_0^a f(x)\,dx \quad \text{and} \quad A_n = \frac{2}{a}\int_0^a f(x)\cos\frac{n\pi x}{a}\,dx, \quad n \geqslant 1.$$

Taking the limit as $t \to \infty$, we have

$$\begin{aligned} \lim_{t\to\infty} u(x,t) &= \lim_{t\to\infty}\left(A_0 + \sum_{n=1}^{\infty} A_n \cos\frac{n\pi x}{a}\, e^{-\frac{kn^2\pi^2 t}{a^2}}\right) \\ &= A_0 + \lim_{t\to\infty}\left(\sum_{n=1}^{\infty} A_n \cos\frac{n\pi x}{a}\, e^{-\frac{kn^2\pi^2 t}{a^2}}\right) \\ &= A_0 \end{aligned}$$

and

$$\lim_{t \to \infty} u(x,t) = \frac{1}{a} \int_0^a f(x)\, dx.$$

Let v denote the steady-state solution; then v satisfies the boundary value problem

$$\frac{d^2v}{dx^2} = 0, \quad 0 < x < a,$$

$$\frac{dv}{dx}(0) = 0,$$

$$\frac{dv}{dx}(a) = 0.$$

The general solution is

$$v(x) = Ax + B$$

and

$$\frac{dv}{dx}(x) = A.$$

Applying the boundary conditions, we have

$$A = \frac{dv}{dx}(0) = \frac{dv}{dx}(a) = 0,$$

and the steady-state solution is

$$v(x) = B, \quad 0 < x < a.$$

To evaluate the constant B, we note that since the total heat energy in the bar is constant,

$$\int_0^a B\, dx = \int_0^a v(x)\, dx = \int_0^a u(x,0)\, dx = \int_0^a f(x)\, dx,$$

so that

$$B \cdot a = \int_0^a f(x)\, dx;$$

that is,

$$v(x) = B = \frac{1}{a} \int_0^a f(x)\, dx$$

for $0 < x < a$.

Exercise 13.7. ✗✗

Solve the problem of heat transfer in a bar of length 1 with initial heat distribution

$$f(x) = \cos \pi x, \quad 0 < x < 1,$$

and no heat loss at either end, where the thermal diffusivity is $k = 1$, that is, solve the boundary value–initial value problem

$$\frac{\partial u}{\partial t} = \frac{\partial^2 u}{\partial x^2}, \quad 0 < x < 1, \quad t > 0,$$

$$\frac{\partial u}{\partial x}(0, t) = \frac{\partial u}{\partial x}(1, t) = 0,$$

$$u(x, 0) = \cos \pi x.$$

Solution. From Exercise 13.6 with $a = 1$ and $k = 1$, the solution is

$$u(x, t) = a_0 + \sum_{n=1}^{\infty} a_n e^{-n^2\pi^2 t} \cos n\pi x$$

and all we need to do now is find the coefficients a_n for $n \geqslant 0$, so that the initial condition is also satisfied. Setting $t = 0$ in the series above, we have

$$\cos \pi x = u(x, 0) = a_0 + \sum_{n=1}^{\infty} a_n \cos n\pi x;$$

that is, the a_n's are just the coefficients in the Fourier cosine series for $\cos \pi x$ on the interval $[0, 1]$. Since $\cos \pi x$ is its own Fourier cosine series on the interval $[0, 1]$, then

$$a_n = \begin{cases} 0, & \text{for } \; n \neq 1, \\ 1, & \text{for } \; n = 1, \end{cases}$$

and the solution is

$$u(x, y) = e^{-\pi^2 t} \cos \pi x$$

for $0 < x < 1, \; t > 0$.

Exercise 13.8. XXX

Solve the problem of heat transfer in a bar of length $a = \pi$ and thermal diffusivity $k = 1$, with initial heat distribution $u(x, 0) = \sin x$, where one end of the bar is kept at a constant temperature $u(0, t) = 0$, while there is no heat loss at the other end of the bar, so that $\partial u(\pi, t)/\partial x = 0$, that is, solve the boundary value–initial value problem

$$\frac{\partial u}{\partial t} = \frac{\partial^2 u}{\partial x^2}, \quad 0 < x < \pi, \quad t > 0,$$

$$u(0, t) = 0,$$

$$\frac{\partial u}{\partial x}(\pi, t) = 0,$$

$$u(x, 0) = \sin x.$$

Solution. Assuming that $u(x, t) = X(x) \cdot T(t)$ and separating variables, we get the two ordinary differential equations $X'' + \lambda X = 0$ and $T' + \lambda T = 0$, and the boundary conditions lead to the following boundary value problem for X:

$$\begin{aligned} X'' + \lambda X &= 0, \quad 0 < x < \pi, \\ X(0) &= 0, \\ X'(\pi) &= 0. \end{aligned}$$

Arguing as in previous problems, this problem has homogeneous mixed boundary conditions with eigenvalues

$$\lambda_n = \mu_n^2 = \frac{(2n-1)^2}{4}$$

and corresponding eigenfunctions

$$X_n(x) = \sin \frac{(2n-1)x}{2}$$

for $n \geqslant 1$. The corresponding solutions to the T equation $T' + \mu_n^2 T = 0$ are

$$T_n(t) = e^{-\frac{(2n-1)^2 t}{4}}$$

for $n \geqslant 1$. For each $n \geqslant 1$, the function

$$u_n(x, t) = X_n(x) \cdot T_n(t) = e^{-\frac{(2n-1)^2 t}{4}} \sin \frac{(2n-1)x}{2}$$

satisfies the heat equation and the boundary conditions, and using the superposition principle, we write

$$u(x,t) = \sum_{n=1}^{\infty} b_n e^{-\frac{(2n-1)^2 t}{4}} \sin\frac{(2n-1)x}{2}$$

for $0 < x < \pi,\ t > 0$. Setting $t = 0$, to satisfy the initial condition we need

$$\sin x = u(x,0) = \sum_{n=1}^{\infty} b_n \sin\frac{(2n-1)x}{2} \tag{13.1}$$

for $0 < x < \pi$. Note that $\sin x$ is not of the form

$$\sin\frac{(2n-1)x}{2};$$

hence, we cannot just say that $\sin x$ is its own Fourier series. Here we have to compute the full generalized Fourier series of $\sin x$ in terms of $\sin[(2n-1)x/2]$. To determine the coefficients b_n, we use the fact that the functions

$$\left\{\sin\frac{(2n-1)x}{2}\right\}_{n\geqslant 1}$$

are orthogonal on the interval $[0,\pi]$. To see this, note that if $n \neq m$, then

$$\begin{aligned}
\int_0^\pi \sin\mu_m x \sin\mu_n x\, dx &= \frac{1}{2}\int_0^\pi [\cos(\mu_m-\mu_n)x - \cos(\mu_m+\mu_n)x]\; dx \\
&= \left.\frac{\sin(\mu_m-\mu_n)x}{2(\mu_m-\mu_n)}\right|_0^\pi - \left.\frac{\sin(\mu_m+\mu_n)x}{2(\mu_m+\mu_n)}\right|_0^\pi \\
&= \frac{\sin(m-n)\pi}{2(m-n)} - \frac{\sin(m+n-1)\pi}{2(m+n-1)} \\
&= 0.
\end{aligned}$$

Also, if $m = n$, then

$$\begin{aligned}
\int_0^\pi \sin^2\mu_m x\, dx &= \int_0^\pi \left(\frac{1}{2} - \frac{1}{2}\cos 2\mu_m x\right) dx \\
&= \frac{\pi}{2} - \left.\frac{\sin 2\mu_m x}{4\mu_m}\right|_0^\pi \\
&= \frac{\pi}{2} - \frac{\sin(2m-1)\pi}{2(2m-1)} \\
&= \frac{\pi}{2}.
\end{aligned}$$

Multiplying equation (13.1) by $\sin \mu_m x$, integrating from 0 to π, and using the orthogonality result just proven, we have

$$\int_0^{\pi} \sin x \sin \mu_m x \, dx = b_m \int_0^{\pi} \sin^2 \mu_m \, dx = \frac{\pi}{2} \cdot b_m;$$

that is,

$$b_m = \frac{2}{\pi} \int_0^{\pi} \sin x \sin \mu_m x \, dx = \frac{1}{\pi} \int_0^{\pi} \left[\cos(\mu_m - 1)x - \cos(\mu_m + 1)x\right] dx$$

$$= \left.\frac{\sin(\mu_m - 1)x}{\pi(\mu_m - 1)}\right|_0^{\pi} - \left.\frac{\sin(\mu_m + 1)x}{\pi(\mu_m + 1)}\right|_0^{\pi} = \frac{\sin(\mu_m - 1)\pi}{\pi(\mu_m - 1)} - \frac{\sin(\mu_m + 1)\pi}{\pi(\mu_m + 1)}$$

$$= \frac{2}{\pi}\left[\frac{\sin\dfrac{(2m-3)\pi}{2}}{2m-3} - \frac{\sin\dfrac{(2m+1)\pi}{2}}{2m+1}\right] = \frac{2}{\pi}\left[\frac{(-1)^m}{2m-3} - \frac{(-1)^m}{2m+1}\right];$$

so

$$b_m = \frac{8}{\pi} \frac{(-1)^m}{(2m-3)(2m+1)},$$

since $\sin[(2m+1)\pi/2] = (-1)^m$.
Therefore, the solution is

$$u(x,t) = \frac{8}{\pi} \sum_{n=1}^{\infty} \frac{(-1)^n}{(2n-3)(2n+1)} e^{-\frac{(2n-1)^2 t}{4}} \sin \frac{(2n-1)x}{2}$$

for $0 < x < \pi, \ t > 0$.

Exercise 13.9. XX

Consider the homogeneous Neumann problem for the heat equation:

$$\frac{\partial u}{\partial t} = k \frac{\partial^2 u}{\partial x^2}, \quad 0 < x < a, \quad t > 0,$$

$$\frac{\partial u}{\partial x}(0,t) = 0,$$

$$\frac{\partial u}{\partial x}(a,t) = 0,$$

$$u(x,0) = f(x).$$

Solve this problem by looking for a solution as a Fourier cosine series in terms of $\cos(n\pi x/a)$, $n \geqslant 0$. Assume that u and $\partial u/\partial x$ are continuous and $\partial^2 u/\partial x^2$ and $\partial u/\partial t$ are piecewise smooth.

Solution. We assume a solution of the form

$$u(x,t) = \sum_{n=0}^{\infty} a_n(t) \cos \frac{n\pi x}{a},$$

and assuming that all derivatives are continuous, we have

$$\frac{\partial^2 u}{\partial x^2} = -\sum_{n=0}^{\infty} a_n(t) \left(\frac{n\pi}{a}\right)^2 \cos\frac{n\pi x}{a},$$

and since $u(x,t)$ satisfies the heat equation,

$$\frac{\partial u}{\partial t} = k\frac{\partial^2 u}{\partial x^2},$$

we have

$$\sum_{n=0}^{\infty} a_n'(t) \cos\frac{n\pi x}{a} = -k\sum_{n=0}^{\infty} a_n(t) \left(\frac{n\pi}{a}\right)^2 \cos\frac{n\pi x}{a}.$$

Collecting terms that multiply $\cos(n\pi x/a)$ for $n \geqslant 0$, and using the fact that these trigonometric functions are linearly independent (they are orthogonal on the interval $[0,a]$), we get

$$a_n'(t) = -ka_n(t)\left(\frac{n\pi}{a}\right)^2,$$

and we can solve these first- order linear ordinary differential equations for $a_n(t)$ to get

$$a_n(t) = A_n e^{-\left(\frac{n\pi}{a}\right)^2 kt},$$

and the solution $u(x,t)$ becomes

$$u(x,t) = \sum_{n=0}^{\infty} A_n e^{-\left(\frac{n\pi}{a}\right)^2 kt} \cos\frac{n\pi x}{a}.$$

Note that if we started the solution using separation of variables, we would arrive at the same formula as above. Differentiating this with respect to x, we get

$$\frac{\partial u}{\partial x}(x,t) = -\sum_{n=0}^{\infty} A_n e^{-\left(\frac{n\pi}{a}\right)^2 kt}\left(\frac{n\pi}{a}\right)\sin\frac{n\pi x}{a},$$

and setting $x = 0$, we get

$$0 = \frac{\partial u}{\partial x}(0,t),$$

and the first boundary condition is satisfied.

The solution is now

$$u(x,t) = \sum_{n=0}^{\infty} A_n e^{-\left(\frac{n\pi}{a}\right)^2 kt} \cos\frac{n\pi x}{a},$$

and we note that the second boundary condition $\partial u(a,t)/\partial x = 0$ is also satisfied, so we only need to find the constants A_n to satisfy the initial condition

$$u(x,0) = f(x).$$

Setting $t = 0$ in the expression above for $u(x,t)$, we have

$$f(x) = u(x,0) = \sum_{n=0}^{\infty} A_n \cos\frac{n\pi x}{a},$$

and the A_n are the Fourier cosine series coefficients of $f(x)$, so that

$$A_n = \frac{2}{a}\int_0^a f(x)\cos\frac{n\pi x}{a}\,dx$$

for $n \geqslant 1$, and for $n = 0$,

$$A_0 = \frac{1}{a}\int_0^a f(x)\,dx.$$

Exercise 13.10. XXX

Consider the partial differential equation that describes the temperature u in the problem of heat transport with convection:

$$\frac{\partial u}{\partial t} = k\frac{\partial^2 u}{\partial x^2} - V_0\frac{\partial u}{\partial x},$$

where k denotes the specific heat and V_0 the convective velocity.

(a) Use separation of variables and show that the resulting spatial equation is **not** of Sturm-Liouville form. Find an appropriate multiplier to get it into Sturm-Liouville form.

(b) Solve the problem above with the boundary values and initial value given by

$$u(0,t) = 0, \quad t > 0,$$

$$u(a,t) = 0, \quad t > 0,$$

$$u(x,0) = f(x), \quad 0 < x < a.$$

Solution.

(a) Assuming a solution of the form $u(x,t) = \phi(x)\cdot h(t)$, the partial differential equation becomes

$$\phi\, h' = (k\phi'' - V_0\phi')h,$$

and separating variables yields

$$\frac{h'}{k\,h} = \frac{\phi''}{\phi} - \frac{V_0}{k}\frac{\phi'}{\phi} = -\lambda,$$

where λ is the separation constant. Thus, we have the following two ordinary differential equations:

$$\phi''(x) - \frac{V_0}{k}\,\phi'(x) + \lambda\,\phi(x) = 0, \quad 0 < x < a$$

and

$$h'(t) + \lambda\,k\,h(t) = 0, \quad t > 0.$$

The spatial equation is not of the form

$$\frac{d}{dx}\Big(p(x)\phi'(x)\Big) + [q(x) + \lambda\sigma(x)]\phi(x) = 0, \quad 0 < x < a,$$

where p, q, and σ satisfy the conditions for a Sturm-Liouville problem, since we would need

$$p(x) = 1 \qquad \text{and} \qquad p'(x) = -\frac{V_0}{k},$$

and that doesn't work. We can, however, multiply the spatial equation by $e^{-\frac{V_0 x}{k}}$, and obtain

$$\frac{d}{dx}\left(e^{-\frac{V_0 x}{k}}\frac{d\phi}{dx}\right) + \lambda e^{-\frac{V_0 x}{k}}\phi = 0,$$

which is of Sturm-Liouville form.

(b) The heat equation with convection satisfies the boundary value–initial value problem

$$\begin{aligned} &\frac{\partial u}{\partial t} = k\frac{\partial^2 u}{\partial x^2} - V_0\frac{\partial u}{\partial x}, \quad 0 < x < a, \quad t > 0,\\ &u(0,t) = 0,\\ &u(a,t) = 0,\\ &u(x,0) = f(x). \end{aligned}$$

Assuming a solution of the form $u(x,t) = \phi(x)\cdot h(t)$ and separating variables, we get the two problems

$$\begin{aligned} &\phi''(x) - \frac{V_0}{k}\,\phi'(x) + \lambda\phi(x) = 0, \qquad h'(t) + \lambda\,k\,h(t) = 0,\\ &\phi(0) = 0,\\ &\phi(a) = 0. \end{aligned}$$

Making the transformation

$$y = e^{-\frac{V_0 x}{2k}}\,\phi,$$

y satisfies the boundary value problem

$$y'' + \left(\lambda - \frac{V_0^2}{4k^2}\right) y = 0, \quad 0 < x < a,$$

$$y(0) = 0,$$

$$y(a) = 0,$$

which has nontrivial solutions if and only if

$$\lambda - \frac{V_0^2}{4k^2} > 0$$

and

$$\lambda - \frac{V_0^2}{4k^2} = \frac{n^2\pi^2}{a^2}$$

for some integer $n \geqslant 1$, and the corresponding nontrivial solutions are

$$y_n(x) = \sin \frac{n\pi x}{a}.$$

Therefore, for the Sturm-Liouville problem, the eigenvalues are

$$\lambda_n = \frac{V_0^2}{4k^2} + \frac{n^2\pi^2}{a^2}$$

with corresponding eigenfunctions

$$\phi_n(x) = e^{\frac{V_0 x}{2k}} y_n(x) = e^{\frac{V_0 x}{2k}} \sin \frac{n\pi x}{a}, \quad 0 < x < a$$

for $n \geqslant 1$.

The corresponding solutions to the time equation are

$$h_n(t) = e^{-\lambda_n k t}$$

for $n \geqslant 1$. For each $n \geqslant 1$, the products

$$u_n(x,t) = e^{\frac{V_0}{2k}x} e^{-\lambda_n k t} \sin \frac{n\pi x}{a}$$

satisfy the partial differential equation as well as the boundary conditions. From the superposition principle, we write

$$u(x,t) = \sum_{n=1}^{\infty} b_n e^{\frac{V_0}{2k}x} e^{-\lambda_n k t} \sin \frac{n\pi x}{a},$$

and we can satisfy the initial condition

$$f(x) = u(x,0) = \sum_{n=1}^{\infty} b_n e^{\frac{V_0}{2k}x} \sin \frac{n\pi x}{a}$$

using the orthogonality of the eigenfunctions on the interval $[0, a]$ with respect to the weight function

$$\sigma(x) = e^{-\frac{V_0}{k}x}.$$

We have

$$b_n = \frac{2}{a} \int_0^a f(x) e^{-\frac{V_0}{2k}x} \sin \frac{n\pi x}{a}\, dx$$

for $n \geqslant 1$. Therefore, the solution to the boundary value–initial value problem is

$$u(x,t) = \sum_{n=1}^{\infty} b_n e^{\frac{V_0}{2k}x} e^{-\lambda_n k t} \sin \frac{n\pi x}{a},$$

where

$$b_n = \frac{2}{a} \int_0^a f(x) e^{-\frac{V_0}{2k}x} \sin \frac{n\pi x}{a}\, dx$$

for $n \geqslant 1$.

Exercise 13.11. XXX

Consider the Neumann problem for heat transport in a nonhomogeneous rod of length a :

$$c\rho \frac{\partial u}{\partial t} = \frac{\partial}{\partial x}\left(K_0 \frac{\partial u}{\partial x}\right) + \alpha u, \quad 0 < x < a, \quad t > 0,$$

$$\frac{\partial u}{\partial x}(0,t) = 0,$$

$$\frac{\partial u}{\partial x}(a,t) = 0,$$

$$u(x,0) = f(x),$$

where $c(x)$, $\rho(x)$, $K_0(x)$, and $\alpha(x)$ are continuous functions on the interval $[0, a]$, and K_0, c, and ρ are nonnegative. Assume that the appropriate eigenfunctions of the spatial problem are known and denote them by $\phi_n(x)$. Show that the eigenvalues of the spatial problem are positive if $\alpha < 0$ and solve the boundary value–initial value problem, briefly discussing $\lim_{t\to\infty} u(x,t)$.

Solution. We use separation of variables. Assume a solution of the form

$$u(x,t) = \phi(x) \cdot h(t)$$

and substitute this into the partial differential equation to get

$$c\,\rho\,\phi\,h' = (K_0\,\phi')'\,h + \alpha\,\phi\,h.$$

Separating variables gives us

$$\frac{(K_0\,\phi')'}{c\,\rho\,\phi} + \frac{\alpha}{c\,\rho} = \frac{h'}{h} = -\lambda,$$

where λ is the separation constant. This leads to the two ordinary differential equations

$$\begin{aligned}&(K_0(x)\,\phi'(x))' + \alpha(x)\,\phi(x) + \lambda\,c(x)\,\rho(x)\,\phi(x) = 0, \quad 0 < x < a,\\ &\phi'(0) = 0,\\ &\phi'(a) = 0\end{aligned}$$

and

$$h'(t) + \lambda\,h(t) = 0, \quad t > 0.$$

The spatial equation is a regular Sturm-Liouville problem with

$$p(x) = K_0(x),\ q(x) = \alpha(x) \quad \text{and} \quad \sigma(x) = c(x)\,\rho(x),$$

all of which are assumed continuous on the closed interval $[0, a]$. In addition, on physical grounds we assume that K_0, c, and ρ are nonnegative and not identically zero on $[0, a]$. If λ is an eigenvalue with corresponding eigenvector $\phi(x)$, $0 < x < a$, from the boundary conditions we have

$$\left[-\,p(x)\,\phi(x)\,\phi'(x)\right]\Big|_0^a = 0,$$

and the Rayleigh quotient reduces to

$$\lambda = R(\phi) = \frac{\displaystyle\int_0^a \left[K_0(x)\,\phi'(x)^2 - \alpha(x)\,\phi(x)^2\right]\,dx}{\displaystyle\int_0^a \phi(x)^2 \rho(x)\,c(x)\,dx},$$

and if $\alpha(x) < 0$ for $0 \leqslant x \leqslant a$, then $\lambda \geqslant 0$. Note that in this case, $\lambda = 0$ is impossible, since that would imply that $\phi(x) = 0$ for all $0 < x < a$, which is a contradiction. Therefore, all the eigenvalues are strictly positive.

The boundary value problem for ϕ is a regular Sturm-Liouville problem and has an infinite sequence of eigenvalues and corresponding eigenfunctions $\{(\phi_n, \lambda_n)\}_{n\geqslant 1}$, where the ϕ_n's form a complete orthogonal set of functions in the linear space of piecewise continuous functions on $[0, a]$ with respect to the inner product with weight function $\sigma(x) = c(x)\,\rho(x)$. The corresponding solutions to the time equation are

$$h_n(t) = c_n e^{-\lambda_n t}, \quad t > 0$$

for $n \geqslant 1$. Using the superposition principle, we can write

$$u(x,t) = \sum_{n=1}^{\infty} c_n e^{-\lambda_n t} \phi_n(x)$$

for $0 < x < a,\ t > 0$, and this satisfies the partial differential equation and the boundary conditions.

To satisfy the initial condition, we use the orthogonality of the eigenfunctions to write the generalized Fourier series

$$f(x) = u(x,0) = \sum_{n=1}^{\infty} c_n \phi_n(x),$$

where

$$c_n = \frac{\displaystyle\int_0^a f(x)\,\phi_n(x)\,c(x)\,\rho(x)\,dx}{\displaystyle\int_0^a \phi_n(x)^2\,c(x)\,\rho(x)\,dx}.$$

Since $\lambda_n > 0$ for all $n \geqslant 1$, then for each term in the series,

$$e^{-\lambda_n t} \longrightarrow 0$$

as $t \to \infty$, and therefore

$$\lim_{t\to\infty} u(x,t) = 0$$

for each $x \in (0,a)$.

Exercise 13.12. ✗✗✗

Find the temperature distribution in a thin two-dimensional plate in the shape of a unit square, with thermal diffusivity $k = 1$ and with insulated faces and edges kept at zero temperature with an initial temperature distribution given by

$$f(x,y) = xy(1-x)(1-y)$$

for $0 < x, y < 1$; that is, solve the boundary value–initial value problem

$$\frac{\partial u}{\partial t} = \frac{\partial^2 u}{\partial x^2} + \frac{\partial^2 u}{\partial y^2}, \quad 0 < x < 1, \quad 0 < y < 1, \quad t > 0,$$

$$u(0,y,t) = u(1,y,t) = 0,$$

$$u(x,0,t) = u(x,1,t) = 0,$$

$$u(x,y,0) = xy(1-x)(1-y).$$

Solution. After separating variables, applying the boundary conditions, and using the superposition principle, we find that the solution has the form

$$u(x,y,t) = \sum_{n=1}^{\infty}\sum_{m=1}^{\infty} B_{n,m} \sin n\pi x \sin m\pi y \, e^{-(n^2+m^2)\pi^2 t}.$$

We evaluate $B_{n,m}$ using the initial condition

$$\begin{aligned} B_{n,m} &= 4\int_0^1\int_0^1 x(1-x)y(1-y)\sin n\pi x \sin m\pi y \, dy\, dx \\ &= \left(2\int_0^1 x(1-x)\sin n\pi x\, dx\right)\cdot\left(2\int_0^1 y(1-y)\sin m\pi y\, dy\right) \\ &= \frac{16\,[1-(-1)^n]\cdot[1-(-1)^m]}{n^3m^3\pi^6} \end{aligned}$$

for $n, m \geqslant 1$. Thus,

$$B_{n,m} = \begin{cases} \dfrac{64}{n^3m^3\pi^6}, & \text{if both } n, m \text{ are odd,} \\ 0, & \text{otherwise.} \end{cases}$$

The solution is therefore

$$\begin{aligned} u(x,y,t) = \sum_{n=1}^{\infty}\sum_{m=1}^{\infty} \Bigg\{ & \frac{64}{\pi^6(2n-1)^3(2m-1)^3}\sin(2n-1)\pi x \sin(2m-1)\pi y \\ & \times e^{-\left[(2n-1)^2+(2m-1)^2\right]\pi^2 t} \Bigg\} \end{aligned}$$

for $0 < x < 1,\ 0 < y < 1,\ t > 0$.

Exercise 13.13. XXX

Consider the heat equation with a steady source

$$\frac{\partial u}{\partial t} = k\frac{\partial^2 u}{\partial x^2} + Q(x),$$

subject to the boundary and initial conditions

$$\begin{aligned} u(0,t) &= 0, \\ u(a,t) &= 0, \\ u(x,0) &= f(x). \end{aligned}$$

Obtain the solution by the method of eigenfunction expansions. Show that the solution approaches a steady-state solution.

Solution. Since the problem already has homogeneous boundary conditions, we consider the corresponding homogeneous problem:

$$\begin{aligned}
&\frac{\partial u}{\partial t} = k\frac{\partial^2 u}{\partial x^2}, \quad 0 < x < a, \quad t > 0,\\
&u(0,t) = 0,\\
&u(a,t) = 0.
\end{aligned}$$

The eigenvalues and eigenfunctions for this problem are

$$\lambda_n = \frac{n^2\pi^2}{a^2} \qquad \text{and} \qquad \phi_n(x) = \sin\frac{n\pi x}{a}$$

for $n \geqslant 1$. We write the solution to the nonhomogeneous problem as an expansion in terms of these eigenfunctions:

$$u(x,t) = \sum_{n=1}^{\infty} a_n(t)\sin\frac{n\pi x}{a}$$

and determine the coefficients $a_n(t)$ which force this to be a solution to the nonhomogeneous problem. We will need the eigenfunction expansions for $Q(x)$ and $f(x)$:

$$Q(x) = \sum_{n=1}^{\infty} q_n \sin\frac{n\pi x}{a} \qquad \text{with} \qquad q_n = \frac{2}{a}\int_0^a Q(x)\sin\frac{n\pi x}{a}\,dx,$$

$$f(x) = \sum_{n=1}^{\infty} f_n \sin\frac{n\pi x}{a} \qquad \text{with} \qquad f_n = \frac{2}{a}\int_0^a f(x)\sin\frac{n\pi x}{a}\,dx.$$

Substituting these expansions into the nonhomogeneous equation

$$\frac{\partial u}{\partial t} = k\frac{\partial^2 u}{\partial x^2} + Q(x),$$

we obtain

$$\sum_{n=1}^{\infty}\frac{d\,a_n(t)}{dt}\sin\frac{n\pi x}{a} = -\sum_{n=1}^{\infty} k\frac{n^2\pi^2}{a^2}a_n(t)\sin\frac{n\pi x}{a} + \sum_{n=1}^{\infty} q_n\sin\frac{n\pi x}{a},$$

and using the orthogonality of the eigenfunctions on the interval $[0,a]$, the coefficients $a_n(t)$ satisfy the initial value problem

$$\begin{aligned}
&\frac{d\,a_n(t)}{dt} + \frac{n^2\pi^2}{a^2}k\,a_n(t) = q_n, \quad t > 0,\\
&a_n(0) = f_n
\end{aligned}$$

for $n \geqslant 1$.

The solution to this initial value problem is

$$a_n(t) = f_n e^{-\frac{n^2\pi^2}{a^2}kt} + q_n \int_0^t e^{-\frac{n^2\pi^2}{a^2}k(t-s)}\, ds;$$

that is,

$$a_n(t) = \frac{q_n}{k(n^2\pi^2/a^2)} + \left(f_n - \frac{q_n}{k(n^2\pi^2/a^2)}\right) e^{-\frac{n^2\pi^2}{a^2}kt}, \quad t > 0$$

for $n \geqslant 1$.

Note that since $k > 0$, we have $\lim\limits_{t\to\infty} a_n(t) = \dfrac{q_n}{k\frac{n^2\pi^2}{a^2}}$ for $n \geqslant 1$.

The solution to the heat equation with a steady source is therefore

$$u(x,t) = \sum_{n=1}^{\infty} \left[\frac{q_n}{k(n^2\pi^2/a^2)} + \left(f_n - \frac{q_n}{k(n^2\pi^2/a^2)}\right) e^{-\frac{n^2\pi^2}{a^2}kt}\right] \sin\frac{n\pi x}{a}$$

for $0 < x < a$ and $t > 0$. For large values of t, this solution approaches $r(x)$ where

$$r(x) = \lim_{t\to\infty} u(x,t) = \sum_{n=1}^{\infty} \frac{q_n}{k(n^2\pi^2/a^2)} \sin\frac{n\pi x}{a}$$

for $0 < x < a$, and where

$$q_n = \frac{2}{a}\int_0^a Q(x)\sin\frac{n\pi x}{a}\, dx$$

for $n \geqslant 1$. Differentiating this twice with respect to x, we see that

$$r''(x) = -\sum_{n=1}^{\infty} \frac{q_n}{k}\sin\frac{n\pi x}{a} = -\frac{1}{k}Q(x),$$

and since $r(0) = r(a) = 0$, the function $r(x)$ satisfies the boundary value problem

$$k\frac{d^2r}{dx^2} + Q = 0, \quad 0 < x < a,$$

$$r(0) = 0,$$
$$r(a) = 0,$$

which is exactly the boundary value problem for the steady-state solution; that is, $r(x)$ is the steady-state or equilibrium solution to the original heat flow problem.

Exercise 13.14. XXXX

Solve the two-dimensional heat equation inside a disk with circularly symmetric time-independent sources, boundary conditions, and initial conditions:

$$\frac{\partial u}{\partial t} = \frac{k}{r}\frac{\partial}{\partial r}\left(r\frac{\partial u}{\partial r}\right) + Q(r), \quad 0 < r < a, \quad t > 0,$$

with

$$u(r,0) = f(r), \qquad \text{and} \qquad u(a,t) = T.$$

Solution. As usual with problems involving polar coordinates, we seek solutions that are bounded as $r \to 0^+$, so that $|u(r,t)| \leqslant M$ as $r \to 0^+$. We first convert the problem into one with homogeneous boundary conditions and then use the method of eigenfunction expansions to solve the nonhomogeneous equation that results.

Step 1: To get a problem with homogeneous boundary conditions, we write

$$u(r,t) = v(r) + w(r,t),$$

where $v(r)$ satisfies

$$\nabla^2 v = \frac{1}{r}\frac{\partial}{\partial r}\left(r\frac{\partial v}{\partial r}\right) = 0, \quad 0 < r < a,$$

$$v(a) = T,$$

$$|v(r)| \leqslant M, \quad \text{as } r \to 0^+.$$

Then

$$\frac{\partial u}{\partial t} = \frac{\partial v}{\partial t} + \frac{\partial w}{\partial t} = \frac{\partial w}{\partial t}$$

and

$$\nabla^2 u = \nabla^2 v + \nabla^2 w = \nabla^2 w.$$

Therefore, $w(r,t)$ satisfies the boundary value–initial value problem

$$\frac{\partial w}{\partial t} = \frac{k}{r}\frac{\partial}{\partial r}\left(r\frac{\partial w}{\partial r}\right) + Q(r),$$

$$w(a,t) = 0,$$

$$w(r,0) = f(r) - v(r).$$

We solve the v equation

$$\frac{1}{r}\frac{\partial}{\partial r}\left(r\frac{\partial v}{\partial r}\right) = 0$$

to get

$$v(r) = c_1 \log r + c_2,$$

with two integration constants, c_1 and c_2. From the boundedness condition, we have $c_1 = 0$, while from the boundary condition $v(a) = c_2 = T$, so that

$$v(r) = T$$

for $0 < r < a$. Therefore, $u(r,t) = w(r,t) + T$, and w satisfies the nonhomogeneous equation with homogeneous boundary conditions

$$\begin{aligned} \frac{\partial w}{\partial t} &= \frac{k}{r}\frac{\partial}{\partial r}\left(r\frac{\partial w}{\partial r}\right) + Q(r), \qquad (13.2) \\ w(a,t) &= 0, \\ w(r,0) &= f(r) - T. \end{aligned}$$

Step 2: Next we find the eigenvalues and eigenfunctions for the corresponding homogeneous problem:

$$\begin{aligned} \frac{\partial w}{\partial t} &= \frac{k}{r}\frac{\partial}{\partial r}\left(r\frac{\partial w}{\partial r}\right), \qquad (13.3) \\ w(a,t) &= 0, \\ |w(r,t)| &\quad \text{bounded at} \quad r = 0. \end{aligned}$$

Using separation of variables,we assume that $w(r,t) = \phi(r) \cdot T(t)$, and separating variables we get

$$\begin{aligned} (r\phi')' + \lambda\, r\, \phi &= 0, \quad 0 < r < a, \\ T' + \lambda\, k\, T &= 0, \quad t > 0. \end{aligned}$$

The boundary condition $w(a,t) = 0$ for all $t > 0$ is satisfied if we require that

$$\phi(a) = 0.$$

Also, since $r = 0$ is a singular point of the differential equation for ϕ, we add the requirement

$$|\phi(r)| \quad \text{bounded at} \quad r = 0,$$

which is equivalent to requiring that $|w(r,t)|$ be bounded at $r = 0$. Thus, ϕ satisfies the boundary value problem

$$\begin{aligned} (r\phi')' + \lambda\, r\, \phi &= 0, \quad 0 < r < a, \qquad (13.4) \\ \phi(a) &= 0, \\ |\phi(r)| &\quad \text{bounded at} \quad r = 0. \end{aligned}$$

If (ϕ, λ) is an eigenpair of this equation, the Rayleigh quotient is

$$\lambda = \frac{-r\phi(r)\phi'(r)\Big|_0^a + \displaystyle\int_0^a r\phi'(r)^2\, dr}{\displaystyle\int_0^a \phi(r)^2 r\, dr}.$$

From the boundary condition and the boundedness condition, we have

$$-r\phi(r)\phi'(r)\Big|_0^a = -a\phi(a)\phi'(a) + \lim_{r\to 0^+} r\phi(r)\phi'(r) = 0,$$

and therefore all the eigenvalues are nonnegative, and, in fact, positive.

We multiply the equation by r and recognize the equation

$$r^2\phi'' + r\phi' + \lambda r^2\phi = 0$$

as Bessel's equation of order zero, for which the function

$$\phi(r) = J_0(\sqrt{\lambda}\, r)$$

is the solution bounded at $r = 0$. To satisfy the boundary condition $\phi(a) = 0$, we must have

$$J_0(\sqrt{\lambda}\, a) = 0$$

or

$$\sqrt{\lambda_n}\, a = z_n, \quad n = 1, 2, \ldots,$$

where z_n are the positive zeros of the Bessel function J_0. Therefore, the eigenvalues and eigenfunctions of the boundary value problem satisfied by $\phi(r)$ are

$$\lambda_n = \frac{z_n^2}{a^2} \qquad \text{and} \qquad \phi_n(r) = J_0(\sqrt{\lambda_n}\, r)$$

for $n \geqslant 1$.

Step 3: Now we use an eigenfunction expansion for $w(r, t)$ as

$$w(r, t) = \sum_{n=1}^{\infty} a_n(t) J_0(\sqrt{\lambda_n}\, r)$$

and determine the coefficients $a_n(t)$ so that $w(r, t)$ is a solution to the nonhomogeneous equation

$$\frac{\partial w}{\partial t} = \frac{k}{r}\frac{\partial}{\partial r}\left(r\frac{\partial w}{\partial r}\right) + Q(r), \tag{13.5}$$

$$w(a, t) = 0,$$

$$w(r, 0) = f(r) - T,$$

and this means that we will need the Fourier-Bessel series for $Q(r)$ and $f(r) - T$:

$$Q(r) = \sum_{n=1}^{\infty} q_n J_0(\sqrt{\lambda_n}\, r) \quad \text{with} \quad q_n = \frac{\displaystyle\int_0^a J_0(\sqrt{\lambda_n}\, r) Q(r) r\, dr}{\displaystyle\int_0^a J_0(\sqrt{\lambda_n}\, r)^2 r\, dr},$$

$$f(r) - T = \sum_{n=1}^{\infty} f_n J_0(\sqrt{\lambda_n}\, r) \quad \text{with} \quad f_n = \frac{\displaystyle\int_0^a J_0(\sqrt{\lambda_n}\, r)(f(r) - T) r\, dr}{\displaystyle\int_0^a J_0(\sqrt{\lambda_n}\, r)^2 r\, dr}.$$

Note that the Sturm-Liouville problem (13.4) has the weight function $\sigma(r) = r$, hence the factor r in the integrals. Substituting these expansions into (13.5), we have

$$\sum_{n=1}^{\infty} \frac{da_n(t)}{dt} J_0(\sqrt{\lambda_n}\, r) = \sum_{n=1}^{\infty} a_n(t)(-\lambda_n k) J_0(\sqrt{\lambda_n}\, r) + \sum_{n=1}^{\infty} q_n J_0(\sqrt{\lambda_n}\, r),$$

and using the orthogonality of the eigenfunctions, the coefficients $a_n(t)$ satisfy the linear differential equation

$$\frac{da_n(t)}{dt} + \lambda_n k\, a_n(t) = q_n, \quad t > 0$$

for $n \geqslant 1$. From the initial condition

$$w(r, 0) = \sum_{n=1}^{\infty} a_n(0) J_0(\sqrt{\lambda_n}\, r) = f(r) - T = \sum_{n=1}^{\infty} f_n J_0(\sqrt{\lambda_n}\, r),$$

using the orthogonality again, we have

$$a_n(0) = f_n$$

for $n \geqslant 1$. Therefore, $a_n(t)$ satisfies the initial value problem

$$\frac{da_n(t)}{dt} + \lambda_n k\, a_n(t) = q_n, \quad t > 0,$$

$$a_n(0) = f_n,$$

for $n \geqslant 1$. Multiplying by the integrating factor $e^{\lambda_n k t}$, the differential equation becomes

$$\frac{d}{dt}\left(a_n(t) e^{\lambda_n k t}\right) = q_n e^{\lambda_n k t},$$

and integrating gives us

$$a_n(t) e^{\lambda_n k t} - a_n(0) = \int_0^t q_n e^{\lambda_n k s}\, ds,$$

so that for $t > 0$,

$$a_n(t) = a_n(0)e^{-\lambda_n kt} + \int_0^t q_n e^{-\lambda_n k(t-s)}\, ds$$
$$= a_n(0)e^{-\lambda_n kt} + \frac{q_n}{\lambda_n k}\left(1 - e^{-\lambda_n kt}\right)$$

for $n \geqslant 1$.
Step 4: Putting everything together, the solution is

$$u(r,t) = v(r) + w(r,t) = T + \sum_{n=1}^{\infty} a_n(t) J_0(\sqrt{\lambda_n}\, r);$$

that is,

$$u(r,t) = T + \sum_{n=1}^{\infty} \left[\frac{q_n}{\lambda_n k} + \left(f_n - \frac{q_n}{\lambda_n k}\right) e^{-\lambda_n kt}\right] J_0(\sqrt{\lambda_n}\, r)$$

for $0 < r < a,\ t > 0$, where $\lambda_n = z_n^2/a^2$, and

$$q_n = \frac{\displaystyle\int_0^a J_0(\sqrt{\lambda_n}\, r) Q(r) r\, dr}{\displaystyle\int_0^a J_0(\sqrt{\lambda_n}\, r)^2 r\, dr}$$

and

$$f_n = \frac{\displaystyle\int_0^a J_0(\sqrt{\lambda_n}\, r)(f(r) - T) r\, dr}{\displaystyle\int_0^a J_0(\sqrt{\lambda_n}\, r)^2 r\, dr}$$

for $n \geqslant 1$.

Exercise 13.15. XXX

A thin homogeneous bar of length π has poorly insulated sides, so that heat radiates freely from the bar along its entire length. Assuming that the heat transfer coefficient A is constant and that the temperature T of the surrounding medium is also constant, the temperature $u(x,t)$ in the bar satisfies the following partial differential equation:

$$\frac{\partial u}{\partial t} = \frac{\partial^2 u}{\partial x^2} - A(u - T), \quad 0 < x < \pi, \quad t > 0.$$

The ends of the bar are kept at temperature T, and the initial temperature is

$$u(x,0) = x + T, \quad 0 < x < \pi.$$

(a) State the boundary value–initial value problem satisfied by $u(x,t)$.

(b) Transform this problem into a familiar one by setting

$$v(x,t) = e^{At}\left[u(x,t) - T\right]$$

and then finding the boundary value–initial value problem satisfied by $v(x,t)$.

(c) Use the method of separation of variables to solve the problem in part (b), and hence obtain the solution $u(x,t)$ to the original problem.

Solution.

(a) The problem satisfied by the temperature function is

$$\begin{aligned}
&\frac{\partial u}{\partial t} = \frac{\partial^2 u}{\partial x^2} - A(u - T), \quad 0 < x < \pi, \quad t > 0, \\
&u(0,t) = T, \\
&u(\pi,t) = T, \\
&u(x,0) = x + T.
\end{aligned}$$

(b) We let $v(x,t) = e^{At}\left[u(x,t) - T\right]$, so that

$$\begin{aligned}
u &= T + e^{-At}v, \\
\frac{\partial u}{\partial t} &= e^{-At}\frac{\partial v}{\partial t} - Ae^{-At}v, \\
\frac{\partial^2 u}{\partial x^2} &= e^{-At}\frac{\partial^2 v}{\partial x^2}.
\end{aligned}$$

If u is a solution to the partial differential equation in part (a),

$$e^{-At}\frac{\partial v}{\partial t} - Ae^{-At}v = e^{-At}\frac{\partial^2 v}{\partial x^2} - Ae^{-At}v,$$

and since e^{-At} is never zero, v satisfies the one-dimensional heat equation

$$\frac{\partial v}{\partial t} = \frac{\partial^2 v}{\partial x^2}.$$

As for the boundary conditions, we have

$$v(0,t) = e^{At}\left[u(0,t) - T\right] = e^{At}\left[T - T\right] = 0$$

and

$$v(\pi,t) = e^{At}\left[u(\pi,t) - T\right] = e^{At}\left[T - T\right] = 0,$$

while for the initial condition, we have

$$v(x,0) = u(x,0) - T = x + T - T = x.$$

Therefore, $v(x,t) = e^{At}\left[u(x,t) - T\right]$ satisfies the boundary value–initial value problem

$$\begin{aligned}
&\frac{\partial v}{\partial t} = \frac{\partial^2 v}{\partial x^2}, \quad 0 < x < \pi, \quad t > 0,\\
&v(0,t) = 0,\\
&v(\pi,t) = 0,\\
&v(x,t) = x.
\end{aligned}$$

(c) Using separation of variables, the solution to the Dirichlet problem in part (b) is

$$v(x,t) = \sum_{n=1}^{\infty} b_n \sin nx\, e^{-n^2 t},$$

where

$$b_n = \frac{2}{\pi}\int_0^{\pi} x \sin nx\, dx = \frac{2(-1)^{n+1}}{n}$$

for $n \geqslant 1$. Therefore, the solution to the original heat transfer problem is

$$u(x,t) = T + e^{-At}\sum_{n=1}^{\infty} \frac{2(-1)^{n+1}}{n} \sin nx\, e^{-n^2 t}$$

for $0 < x < \pi,\ t > 0$.

Exercise 13.16. XXXX

Solve the homogeneous Robin problem for heat flow in a homogeneous rod of length a where we have convection at the ends into a medium at zero temperature and where the initial temperature is $f(x)$:

$$\frac{\partial u}{\partial t} = k\frac{\partial^2 u}{\partial x^2}, \quad 0 < x < a, \; t > 0,$$

$$\frac{\partial u}{\partial x}(0,t) - hu(0,t) = 0,$$

$$\frac{\partial u}{\partial x}(a,t) + hu(a,t) = 0,$$

$$u(x,0) = f(x),$$

where $0 < h < \pi/2a$.

Solution. We use separation of variables. Assume a solution of the form

$$u(x,t) = \phi(x) \cdot T(t)$$

and substitute this into the partial differential equation to get

$$\phi T' = k\phi'' T.$$

Separating variables yields

$$\frac{\phi''}{\phi} = \frac{T'}{kT} = -\lambda,$$

where λ is the separation constant. This leads to the two ordinary differential equations

$$\begin{aligned} &\phi''(x) + \lambda\,\phi(x) = 0, \quad 0 < x < a, \qquad && T'(t) + \lambda k T(t) = 0, \quad t > 0, \\ &\phi'(0) - h\phi(0) = 0, && \\ &\phi'(a) + h\phi(a) = 0. && \end{aligned}$$

The spatial equation is a regular Sturm-Liouville problem with

$$p(x) = 1, \quad q(x) = 0, \quad \text{and} \quad \sigma(x) = 1,$$

all of which are continuous on the closed interval $[0,a]$. If λ is an eigenvalue with corresponding eigenvector $\phi(x)$, $0 < x < a$, from the boundary conditions we have

$$\big[-p(x)\,\phi(x)\,\phi'(x)\big]\Big|_0^a = h\big[\phi(a)^2 + \phi(0)^2\big],$$

and the Rayleigh quotient reduces to

$$\lambda = R(\phi) = \frac{h\big[\phi(a)^2 + \phi(0)^2\big] + \displaystyle\int_0^a \phi'(x)^2\,dx}{\displaystyle\int_0^a \phi(x)^2\,dx},$$

and $\lambda \geqslant 0$. Note that $\lambda = 0$ is impossible, since that would imply that $\phi(x) = 0$ for all $0 < x < a$, which is a contradiction. Therefore, all the eigenvalues are strictly positive.

The boundary value problem for ϕ is a regular Sturm-Liouville problem and has an infinite sequence of eigenvalues and corresponding eigenfunctions $\{(\phi_n, \lambda_n)\}_{n\geqslant 1}$, where the ϕ_n's form a complete orthogonal set of functions in the linear space of piecewise continuous functions on $[0, a]$ with respect to the inner product with weight function $\sigma(x) = 1$.

In fact, if we write $\lambda = \mu^2$, where $\mu > 0$, the boundary value problem becomes

$$\phi'' + \mu^2\phi = 0,$$

$$\phi'(0) - h\phi(0) = 0,$$

$$\phi'(a) + h\phi(a) = 0,$$

with general solution

$$\phi(x) = A\cos\mu x + B\sin\mu x \quad \text{and} \quad \phi'(x) = -\mu A\sin\mu x + \mu B\cos\mu x.$$

Applying the first boundary condition, we have

$$\frac{A}{B} = \frac{\mu}{h},$$

and applying the second boundary condition, we have

$$\frac{h^2 - \mu^2}{h}\sin\mu a + 2\mu\cos\mu a = 0,$$

and the boundary value problem has a nontrivial solution if and only if

$$\tan\mu a = \frac{2\mu h}{\mu^2 - h^2}.$$

We determine the eigenvalues from the graphs of the functions

$$f(\mu) = \tan\mu a \qquad \text{and} \qquad g(\mu) = \frac{2\mu h}{\mu^2 - h^2}$$

for $\mu > 0$.
Note that for $\mu > 0$, we have

$$g(\mu) = \frac{2\mu h}{\mu^2 - h^2} = \frac{h}{\mu + h} + \frac{h}{\mu - h},$$

so that

$$g'(\mu) = -\frac{h}{(\mu + h)^2} - \frac{h}{(\mu - h)^2} < 0$$

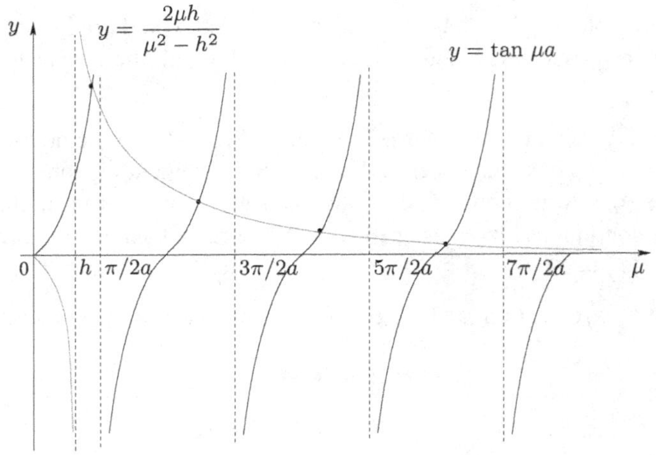

Figure 13.1 Intersection of graphs of $f(\mu) = \tan \mu a$ and $g(\mu) = \dfrac{2\mu h}{\mu^2 - h^2}$

and g is decreasing on the interval $(0, h)$ and on the interval (h, ∞), and the line $\mu = h$ is a vertical asymptote to the graph. The graphs of g and f are shown in Figure 13.1, from which it is clear that there are an infinite number of distinct positive solutions $\mu = \mu_n$ to the transcendental equation

$$\tan \mu a = \frac{2\mu h}{\mu^2 - h^2}, \tag{13.6}$$

and the eigenvalues are $\lambda_n = \mu_n^2$ for $n \geqslant 1$. Since $\lim_{n\to\infty} \mu_n = +\infty$, then

$$\lim_{n\to\infty} \tan \mu_n a = \lim_{n\to\infty} \frac{2\mu_n h}{\mu_n^2 - h^2} = 0,$$

and the roots of the equation $\tan \mu a = 2\mu h/(\mu^2 - h^2)$ approach the roots of the equation $\tan \mu a = 0$; that is, for large n,

$$\mu_n a \approx n\pi,$$

and therefore

$$\lambda_n = \mu_n^2 \approx \frac{n^2\pi^2}{a^2}$$

for large n. The corresponding eigenfunctions are

$$\phi_n(x) = \cos \mu_n x + \frac{h}{\mu_n} \sin \mu_n x \tag{13.7}$$

for $n \geqslant 1$. The corresponding solutions to the time equation are

$$T_n(t) = e^{-k\mu_n^2 t}, \quad t > 0,$$

for $n \geqslant 1$. Using the superposition principle, we can write

$$u(x,t) = \sum_{n=1}^{\infty} a_n \phi_n(x)\, e^{-k\mu_n^2 t}$$

for $0 < x < a,\ t > 0$, and this satisfies the partial differential equation and the boundary conditions.

To satisfy the initial condition, we use the orthogonality of the eigenfunctions to write the generalized Fourier series

$$f(x) = u(x,0) = \sum_{n=1}^{\infty} a_n \phi_n(x),$$

where

$$a_n = \frac{\displaystyle\int_0^a f(x)\,\phi_n(x)\,dx}{\displaystyle\int_0^a \phi_n(x)^2\,dx}.$$

To determine the normalization constant $\displaystyle\int_0^a \phi_n(x)^2\,dx$, we note that

$$\phi_n'' + \mu_n^2 \phi_n = 0,$$

so that

$$\mu_n^2 \int_0^a \phi_n^2\,dx = -\int_0^a \phi_n \phi_n''\,dx = -\phi_n\phi_n'\Big|_0^a + \int_0^a \left(\phi_n'\right)^2 dx.$$

However,

$$\mu_n \phi_n(x) = \mu_n \cos \mu_n x + h \sin \mu_n x$$

and

$$\phi_n'(x) = -\mu_n \sin \mu_n x + h \cos \mu_n x,$$

so that

$$\mu_n^2 \phi_n(x)^2 + \left(\phi_n'(x)\right)^2 = \mu_n^2 + h^2$$

for $0 \leqslant x \leqslant a$, and integrating, we have

$$\mu_n^2 \int_0^a \phi_n(x)^2\,dx + \int_0^a \left(\phi_n'(x)\right)^2 dx = \left(\mu_n^2 + h^2\right) a.$$

Also,

$$\mu_n^2 \int_0^a \phi_n(x)^2\,dx - \int_0^a \left(\phi_n'(x)\right)^2 dx = -\phi_n\phi_n'\Big|_0^a,$$

and adding, we obtain

$$
\begin{aligned}
2\mu_n^2 \int_0^a \phi_n(x)^2\,dx &= \left(\mu_n^2 + h^2\right) a - \phi_n \phi_n' \Big|_0^a \\
&= \left(\mu_n^2 + h^2\right) a + h\left(\phi_n(a)^2 + \phi_n(0)^2\right). \qquad (13.8)
\end{aligned}
$$

Since

$$
\mu_n^2 \phi_n(x)^2 + \left(\phi_n'(x)\right)^2 = \mu_n^2 + h^2
$$

for $0 \leqslant x \leqslant a$, from the first boundary condition we have

$$
\mu_n^2 \phi_n(0)^2 + h^2 \phi_n(0)^2 = \left(\mu_n^2 + h^2\right) \phi_n(0)^2 = \mu_n^2 + h^2,
$$

so that $\phi_n(0)^2 = 1$. Similarly, from the second boundary condition we have

$$
\mu_n^2 \phi_n(a)^2 + h^2 \phi_n(a)^2 = \left(\mu_n^2 + h^2\right) \phi_n(a)^2 = \mu_n^2 + h^2,
$$

so that $\phi_n(a)^2 = 1$, and from (13.8) we have

$$
\int_0^a \phi_n(x)^2\,dx = \frac{\left(\mu_n^2 + h^2\right) a + 2h}{2\mu_n^2}
$$

for $n \geqslant 1$. The solution to the homogeneous Robin problem is

$$
u(x,t) = \sum_{n=1}^{\infty} a_n \left(\cos \mu_n x + \frac{h}{\mu_n} \sin \mu_n x\right) e^{-k\mu_n^2 t}
$$

for $0 < x < a,\ t > 0$, where the μ_n's are the positive roots of the transcendental equation

$$
\tan \mu a = \frac{2\mu h}{\mu^2 - h^2}
$$

and

$$
a_n = \frac{2\mu_n^2}{\left(\mu_n^2 + h^2\right) a + 2h} \int_0^a f(x) \left(\cos \mu_n x + \frac{h}{\mu_n} \sin \mu_n x\right) dx
$$

for $n \geqslant 1$.

Exercise 13.17. XXX

Solve the following boundary value problem for the steady-state temperature $u(x, y)$ in a thin plate in the shape of a semi-infinite strip when heat transfer to the surroundings at temperature zero takes place at the faces of the plate,

$$\frac{\partial^2 u}{\partial x^2} + \frac{\partial^2 u}{\partial y^2} - b\,u = 0, \quad 0 < x < \infty, \quad 0 < y < 1,$$

$$\frac{\partial u}{\partial x}(0, y) = 0,$$

$$u(x, 0) = 0,$$

$$u(x, 1) = f(x),$$

where b is a positive constant and $f(x) = \begin{cases} 1, & 0 < x < a, \\ 0, & x > a. \end{cases}$

Solution. Using separation of variables, we write

$$u(x, y) = X(x) \cdot Y(y),$$

and the partial differential equation becomes

$$X'' \cdot Y + X \cdot Y'' - bX \cdot Y = 0;$$

that is,

$$\frac{X''}{X} = -\frac{Y''}{Y} + b = p,$$

where p is the separation constant. Since we want bounded solutions, we obtain the two ordinary differential equations

$$X'' - pX = 0, \quad 0 < x < \infty, \qquad Y'' + (p - b)\,Y = 0, \quad 0 < y < 1,$$

$$X'(0) = 0, \qquad Y(0) = 0,$$

$$|X(x)| \quad \text{bounded as} \quad x \to \infty.$$

Case (i): $p = 0$. The general solution to the equation $X'' = 0$ is

$$X(x) = c_1\, x + c_2,$$

and the condition $X'(0) = 0$ implies that $c_1 = 0$. The solution is therefore $X(x) = 1$.

Case (ii): $p > 0$, say $p = \lambda^2$. The general solution to the equation $X'' - \lambda^2 X = 0$ is

$$X(x) = c_1 \cosh \lambda x + c_2 \sinh \lambda x,$$

and the condition $X'(0) = 0$ implies that $c_2 = 0$, while the condition $|X(x)|$ bounded as $x \to \infty$ implies that $c_1 = 0$. There are no nontrivial solutions in this case.

Case (iii): $p < 0$, say $p = -\lambda^2$. The general solution to the equation $X'' + \lambda^2 X = 0$ is

$$X(x) = c_1 \cos \lambda x + c_2 \sin \lambda x,$$

the condition $X'(0) = 0$ implies that $c_2 = 0$, and the solution is $X(x) = c_1 \cos \lambda x$, which is bounded as $x \to \infty$. Therefore, for **any** $\lambda \geqslant 0$, the function $X_\lambda(x) = \cos \lambda x$ satisfies the differential equation, the boundary condition, and the boundedness condition. In this case we no longer have a discrete spectrum, that is, a discrete set of eigenvalues, and every $\lambda \geqslant 0$ is an eigenvalue.

The corresponding equation for Y is given by

$$Y'' - (\lambda^2 + b)\, Y = 0,$$

$$Y(0) = 0$$

and has general solution

$$Y(y) = c_1 \sinh\left((1-y)\sqrt{\lambda^2 + b}\right) + c_2 \sinh\left(y\sqrt{\lambda^2 + b}\right).$$

The condition $Y(0) = 0$ implies that $c_1 = 0$, and the solutions are

$$Y_\lambda(y) = \sinh\left(y\sqrt{\lambda^2 + b}\right).$$

Using the superposition principle, we write

$$u(x, y) = \int_0^\infty A(\lambda) \cos \lambda x \, \sinh\left(y\sqrt{\lambda^2 + b}\right) d\lambda$$

and $u(x, 1) = f(x)$ implies that

$$\begin{aligned} A(\lambda) \sinh \sqrt{\lambda^2 + b} &= \frac{2}{\pi} \int_0^\infty f(x) \cos \lambda x \, dx \\ &= \frac{2}{\pi} \int_0^a \cos \lambda x \, dx \\ &= \frac{2}{\pi \lambda} \sin \lambda a. \end{aligned}$$

Therefore,

$$u(x, y) = \frac{2}{\pi} \int_0^\infty \frac{\sin \lambda a \, \cos \lambda x \, \sinh(y\sqrt{\lambda^2 + b})}{\lambda \sinh \sqrt{\lambda^2 + b}} \, d\lambda$$

for $0 < x < \infty, \quad 0 < y < 1.$

Note that this problem can also be solved using the Fourier cosine transform with x as the transform variable. To obtain the inverse Fourier cosine transform, the method of images has to be used (see Section 9.4.2).

Exercise 13.18. XXXX

Solve the problem of heat flow in a circular disk if the material properties are constant and there is no internal heat generation. The problem governing the temperature of the disk is

$$\frac{\partial u}{\partial t} = k\nabla^2 u, \qquad 0 < r < a, \ -\pi < \theta < \pi, \ t > 0,$$

$$u(a, \theta, t) = f(\theta),$$

$$u(r, \theta, 0) = g(r, \theta),$$

$$u(r, \pi, t) = u(r, -\pi, t),$$

$$\frac{\partial u}{\partial \theta}(r, \pi, t) = \frac{\partial u}{\partial \theta}(r, -\pi, t),$$

$$|u(r, \theta, t)| \quad \text{bounded as} \quad r \to 0^+.$$

Solution. Since the boundary conditions are not homogeneous, we look for a solution of the form

$$u(r, \theta, t) = v(r, \theta) + w(r, \theta, t),$$

where v will satisfy a homogeneous equation with time-independent boundary and periodicity conditions, and w will satisfy the heat equation with periodicity conditions and homogeneous boundary conditions. This is analogous to the method we employed earlier for nonhomogeneous problems. We substitute this expression for u in the problem above to get

$$\frac{\partial w}{\partial t} = k(\nabla^2 v + \nabla^2 w),$$

$$v(a, \theta) + w(a, \theta, t) = f(\theta),$$

$$v(r, \theta) + w(r, \theta, 0) = g(r, \theta),$$

and split this into two problems in the obvious way:

$$\nabla^2 v = 0, \qquad\qquad \frac{\partial w}{\partial t} = k\nabla^2 w,$$

$$v(a,\theta) = f(\theta), \qquad\qquad w(a,\theta,t) = 0,$$

$$v(r,\pi) = v(r,-\pi), \qquad\qquad w(r,\pi,t) = w(r,-\pi,t),$$

$$\frac{\partial v}{\partial \theta}(r,\pi) = \frac{\partial v}{\partial \theta}(r,-\pi), \qquad\qquad \frac{\partial w}{\partial \theta}(r,\pi,t) = \frac{\partial w}{\partial \theta}(r,-\pi,t),$$

$$|v(r,\theta)| \quad \text{bounded as} \quad r \to 0^+, \qquad\qquad |w(r,\theta,t)| \quad \text{bounded as} \quad r \to 0^+,$$

$$w(r,\theta,0) = g(r,\theta) - v(r,\theta) = h(r,\theta).$$

The problem for v is Laplace's equation in a disk, and the solution to this problem was obtained earlier, namely, Poisson's integral formula:

$$v(r,\theta) = \frac{a^2 - r^2}{2\pi} \int_{-\pi}^{\pi} \frac{f(\varphi)}{a^2 - 2ar\cos(\theta - \varphi) + r^2}\, d\varphi$$

for $0 < r < a, \; -\pi < \theta < \pi$.

In the problem for w, writing

$$w(r,\theta,t) = \Psi(r,\theta) \cdot T(t) \tag{13.9}$$

and separating variables leads to

$$\frac{T'}{kT} = \frac{\nabla^2 \Psi}{\Psi} = -\lambda,$$

where λ is the separation constant. We get two differential equations, an ordinary differential equation for T and a partial differential equation for Ψ. From (13.9), we can satisfy the boundary and periodicity conditions by requiring that

$$\Psi(a,\theta) = 0,$$

$$\Psi(r,\pi) = \Psi(r,-\pi),$$

$$\frac{\partial \Psi}{\partial \theta}(r,\pi) = \frac{\partial \Psi}{\partial \theta}(r,-\pi).$$

Thus, T and Ψ satisfy the following problems:

$$T' + \lambda k T = 0, \qquad\qquad
\begin{aligned}
&\nabla^2 \Psi = -\lambda \Psi,\\
&\Psi(a,\theta) = 0,\\
&\Psi(r,\pi) = \Psi(r,-\pi),\\
&\frac{\partial \Psi}{\partial \theta}(r,\pi) = \frac{\partial \Psi}{\partial \theta}(r,-\pi),\\
&|\Psi(r,\theta)| \quad \text{bounded as} \quad r \to 0.
\end{aligned}$$

Clearly, the solution to the equation for T is

$$T(t) = e^{-\lambda k t}$$

for $t > 0$. The equation for Ψ is

$$\frac{1}{r}\frac{\partial}{\partial r}\left(r\frac{\partial \Psi}{\partial r}\right) + \frac{1}{r^2}\frac{\partial^2 \Psi}{\partial \theta^2} = -\lambda \Psi.$$

Writing

$$\Psi(r,\theta) = R(r)\cdot S(\theta), \tag{13.10}$$

and separating variables again leads to

$$\frac{r(rR')'}{R} + \lambda r^2 = -\frac{S''}{S} = \nu,$$

where ν is another separation constant. We get two ordinary differential equations, one for R and one for S:

$$\begin{aligned} &r^2R'' + rR' + (\lambda r^2 - \nu)R = 0, &\qquad &S'' + \nu S = 0,\\ &R(a) = 0, & &S(\pi) = S(-\pi),\\ &|R(r)| < \infty, & &S'(\pi) = S'(-\pi). \end{aligned}$$

We have solved the equation for S before, nontrivial solutions exist only for nonnegative ν, and the eigenvalues are

$$\nu_n = \omega_n^2 = n^2,$$

where n is an integer. The eigenfunctions are the **circular harmonics**:

$$S_n(\theta) = a_n \cos n\theta + b_n \sin n\theta$$

for $n \geqslant 0$, where a_n and b_n are arbitrary constants.

We need to solve the R equation only for those values of ν for which we have a nontrivial solution to the S equation, and for these values, the problem for R becomes

$$\begin{aligned} &r^2R'' + rR' + (\lambda r^2 - n^2)R = 0, \quad 0 < r < a,\\ &R(a) = 0,\\ &|R(r)| < \infty \quad \text{as} \quad r \to 0^+. \end{aligned}$$

If (R,λ) is an eigenpair of the radial equation, the Rayleigh quotient is

$$\lambda = \frac{-rR(r)R'(r)\Big|_0^a + \displaystyle\int_0^a \left[rR'(r)^2 + \frac{n^2}{r}R(r)^2\right]dr}{\displaystyle\int_0^a R(r)^2 r\,dr}.$$

From the boundary condition and the boundedness condition, we have

$$-rR(r)R'(r)\Big|_0^a = -aR(a)R'(a) + \lim_{r\to 0^+} rR(r)R'(r) = 0,$$

and therefore all the eigenvalues are nonnegative, and, in fact, positive. We write $\lambda = \mu^2$, where $\mu \neq 0$. Thus, the problem for R is now

$$r^2R'' + rR' + (\mu^2 r^2 - n^2)R = 0, \quad 0 < r < a,$$

$$R(a) = 0,$$

$$|R(r)| < \infty \quad \text{as} \quad r \to 0^+.$$

The equation for R resembles, but is not quite, Bessel's equation. If we make the change of independent variable $x = \mu r$, then $x(d/dx) = r(d/dr)$ and the equation becomes

$$x^2\frac{d^2R}{dx^2} + x\frac{dR}{dx} + (x^2 - n^2)R = 0,$$

which is Bessel's equation of order n. Thus, the problem for R has general solution

$$R_n(r) = A_nJ_n(\mu r) + B_nY_n(\mu r)$$

for $n \geqslant 0$, where J_n and Y_n are Bessel functions of the first and second kind, respectively.

The boundedness condition $|R_n(r)| < \infty$ as $r \to 0^+$ implies that $B_n = 0$, since Y_n is singular at $r = 0$, and therefore,

$$R_n(r) = J_n(\mu r)$$

for $n \geqslant 0$. The boundary condition $R_n(a) = 0$ yields

$$J_n(\mu a) = 0,$$

so that

$$\mu = \mu_{nk} = \frac{z_{nk}}{a},$$

for $k \geqslant 1$, where z_{nk} is the kth zero of J_n.

Thus we have the nontrivial solutions to the R and S equations:

$$R_{nk}(r) = J_n(\mu_{nk}r) \qquad \text{and} \qquad S_n(\theta) = a_n\cos n\theta + b_n\sin n\theta,$$

and for these values of $\lambda = \mu_{nk}^2$, the solution to the T equation is

$$T_{nk}(t) = e^{-k\mu_{nk}^2 t}$$

for $n \geqslant 0$, $k \geqslant 1$. We write the product solutions as

$$w_{nk}(r,\theta,t) = J_n(\mu_{nk}r)\left[a_{nk}\cos n\theta + b_{nk}\sin n\theta\right] e^{-k\mu_{nk}^2 t},$$

for $n \geqslant 0,\ k \geqslant 1$, and note that they satisfy all the conditions except the initial condition. To satisfy the initial condition, we use the superposition principle to write the solution to the w problem as

$$w(r,\theta,t) = \sum_{n=0}^{\infty}\sum_{k=1}^{\infty} J_n(\mu_{nk}r)\left[a_{nk}\cos n\theta + b_{nk}\sin n\theta\right] e^{-k\mu_{nk}^2 t} \tag{13.11}$$

for $0 < r < a,\ -\pi < \theta < \pi$, and $t > 0$.

Finally, to determine the constants a_{nk} and b_{nk}, we apply the initial condition

$$w(r,\theta,0) = g(r,\theta) - v(r,\theta) = h(r,\theta).$$

Setting $t = 0$ in the expression for $w(r,\theta,t)$, we have

$$\sum_{n=0}^{\infty}\sum_{k=1}^{\infty} J_n(\mu_{nk}r)\left[a_{nk}\cos n\theta + b_{nk}\sin n\theta\right] = h(r,\theta);$$

that is,

$$\sum_{n=0}^{\infty}\left(\sum_{k=1}^{\infty} J_n(\mu_{nk}r)a_{nk}\right)\cos n\theta + \sum_{n=0}^{\infty}\left(\sum_{k=1}^{\infty} J_n(\mu_{nk}r)b_{nk}\right)\sin n\theta = h(r,\theta),$$

and this is a Fourier series for $h(r,\theta)$ on the interval $[-\pi,\pi]$ holding r fixed. Therefore,

$$\sum_{k=1}^{\infty} a_{0k}J_0\left(\sqrt{\lambda_{0k}}\,r\right) = \frac{1}{2\pi}\int_{-\pi}^{\pi} h(r,\theta)\,d\theta, \quad \text{for } n = 0,$$

$$\sum_{k=1}^{\infty} a_{nk}J_n\left(\sqrt{\lambda_{nk}}\,r\right) = \frac{1}{\pi}\int_{-\pi}^{\pi} h(r,\theta)\cos n\theta\,d\theta, \quad \text{for } n \geqslant 1,$$

$$\sum_{k=1}^{\infty} b_{nk}J_n\left(\sqrt{\lambda_{nk}}\,r\right) = \frac{1}{\pi}\int_{-\pi}^{\pi} h(r,\theta)\sin n\theta\,d\theta, \quad \text{for } n \geqslant 1.$$

Note that these Fourier series coefficients are actually Fourier-Bessel series, so that

$$a_{0k} = \frac{\dfrac{1}{2\pi}\displaystyle\int_{-\pi}^{\pi}\int_{0}^{a} h(r,\theta) J_0\big(\sqrt{\lambda_{0k}}\, r\big)\, r\, dr\, d\theta}{\displaystyle\int_0^a J_0\big(\sqrt{\lambda_{0k}}\, r\big)^2 r\, dr}, \qquad \text{for } n = 0,\ k \geqslant 1,$$

$$a_{nk} = \frac{\dfrac{1}{\pi}\displaystyle\int_{-\pi}^{\pi}\int_{0}^{a} h(r,\theta) J_n\big(\sqrt{\lambda_{nk}}\, r\big) \cos n\theta\, r\, dr\, d\theta}{\displaystyle\int_0^a J_n\big(\sqrt{\lambda_{nk}}\, r\big)^2 r\, dr}, \qquad \text{for } n \geqslant 1,\ k \geqslant 1,$$

$$b_{nk} = \frac{\dfrac{1}{\pi}\displaystyle\int_{-\pi}^{\pi}\int_{0}^{a} h(r,\theta) J_n\big(\sqrt{\lambda_{nk}}\, r\big) \sin n\theta\, r\, dr\, d\theta}{\displaystyle\int_0^a J_n\big(\sqrt{\lambda_{nk}}\, r\big)^2 r\, dr}, \qquad \text{for } n \geqslant 1,\ k \geqslant 1,$$

where the normalization constants are given by

$$\int_0^a J_n\big(\sqrt{\lambda_{nk}}\, r\big)^2 r\, dr = \int_0^a J_n\left(\frac{z_{nk}}{a} r\right)^2 r\, dr = \frac{a^2}{2} J_{n+1}(z_{nk})$$

for $n \geqslant 0$ and $k \geqslant 1$, by Theorem 6.9.

Suppose that we consider the special case with no angular dependence, in other words, the case with

$$f(\theta) = 0 \qquad \text{and} \qquad g(r,\theta) = g(r);$$

then

$$v(r,\theta) = 0 \qquad \text{and} \qquad h(r,\theta) = g(r),$$

and for $k \geqslant 1$ we have

$$\begin{aligned} a_{0k} &= \frac{1}{\pi a^2 J_1(z_{0k})} \int_{-\pi}^{\pi}\int_0^a g(r) J_0\left(\frac{z_{0k}}{a} r\right) r\, dr\, d\theta \\ &= \frac{2}{a^2 J_1(z_{0k})} \int_0^a g(r) J_0\left(\frac{z_{0k}}{a} r\right) r\, dr, \end{aligned}$$

and since

$$\int_{-\pi}^{\pi} \cos n\theta\, d\theta = \int_{-\pi}^{\pi} \sin n\theta\, d\theta = 0$$

for $n \geqslant 1$, then $a_{nk} = b_{nk} = 0$ for $n \geqslant 1$. In this case the final solution to the problem reduces to

$$u(r,\theta,t) = \sum_{k=1}^{\infty} a_{0k} J_0\left(\frac{z_{0k}}{a} r\right) e^{-k\left(\frac{z_{0k}}{a}\right)^2 t}, \tag{13.12}$$

where

$$a_{0k} = \frac{2}{a^2\, J_1(z_{0k})} \int_0^a g(r) J_0\left(\frac{z_{0k}}{a} r\right) r\, dr$$

for $k \geqslant 1$.

Exercise 13.19. Heat Flow on a Spherical Shell XXXX

Consider the flow of heat on a thin conducting spherical shell

$$S = \{\, (r, \theta, \phi) \,|\, r = 1,\ 0 \leqslant \theta \leqslant \pi,\ -\pi \leqslant \phi \leqslant \pi \,\}.$$

We want to find the temperature distribution $u(\theta, t)$ on the shell if we are given the initial temperature distribution $u(\theta, 0) = f(\theta)$.

Solution. We want to solve the heat equation on the spherical shell; here the appropriate coordinate system consists of spherical coordinates r, θ, and ϕ, where

$$x = \sin\theta \cos\phi, \qquad y = \sin\theta \sin\phi, \qquad z = \cos\theta.$$

Note that since we are on the surface of the unit sphere, then $r = 1$, and since the initial condition is independent of the polar angle ϕ, we assume a solution of the form $u = u(\theta, t)$, that is, independent of ϕ. Also, note that there are no boundary conditions, since there is no physical boundary. However, there are nonphysical boundaries at $\theta = 0$ and $\theta = \pi$, since $\sin\theta$ in the denominator of the Laplacian vanishes there.

Thus, to ensure the continuity and boundedness of the solution, we need the following additional assumptions:

$$|u(0, t)| < \infty \qquad \text{and} \qquad |u(\pi, t)| < \infty.$$

Thus, the heat flow problem is given by

$$\begin{aligned}
&\frac{1}{k}\frac{\partial u}{\partial t} = \frac{1}{\sin\theta}\frac{\partial}{\partial\theta}\left(\sin\theta \frac{\partial u}{\partial\theta}\right), \qquad 0 < \theta < \pi, \quad t > 0,\\
&u(\theta, 0) = f(\theta),\\
&|u(0, t)| < \infty,\\
&|u(\pi, t)| < \infty.
\end{aligned}$$

We apply the method of separation of variables and assume a product solution of the form

$$u(\theta, t) = \Theta(\theta) \cdot T(t),$$

and substituting this expression into the PDE, we have

$$\frac{1}{k}\,\Theta \cdot T' = T \cdot \frac{(\Theta' \sin\theta)'}{\sin\theta}.$$

Separating variables, we have

$$\frac{T'}{kT} = \frac{(\Theta' \sin\theta)'}{\Theta \sin\theta} = -\lambda,$$

where λ is the separation constant. This gives rise to the following ODEs:

$$(\Theta' \sin\theta)' + \lambda \sin\theta\, \Theta = 0, \quad 0 < \theta < \pi, \qquad T' + \lambda k T = 0, \quad t > 0,$$

$$|\Theta(0)| < \infty,$$

$$|\Theta(\pi)| < \infty.$$

If we let $x = \cos\theta$ and $v(x) = \Theta(\theta)$, the equation for Θ becomes

$$(1 - x^2)v'' - 2xv' + \lambda v = 0, \quad -1 < x < 1,$$

which is Legendre's equation, and Legendre's equation has bounded solutions if and only if $\lambda = n(n+1)$, where n is an integer. Therefore, v satisfies

$$(1 - x^2)v'' - 2xv' + n(n+1)v = 0$$

with general solution

$$v(x) = a_n P_n(x) + b_n Q_n(x),$$

where $n = 0, 1, 2, \ldots$. Since we want $|v(x)|$ to be bounded, we must have $b_n = 0$ for all $n \geqslant 0$, and hence

$$v(x) = P_n(x)$$

for $n \geqslant 0$.

The eigenvalues for the singular Sturm-Liouville problem for Θ are

$$\lambda_n = n(n+1),$$

with corresponding eigenfunctions

$$\Theta_n(\theta) = P_n(\cos\theta)$$

for $n \geqslant 0$. The corresponding equation for T becomes

$$T' + n(n+1)\, kT = 0,$$

with solution

$$T_n(t) = e^{-n(n+1)kt}$$

for $n \geqslant 0$.

Therefore, each of the following products is a bounded solution to the heat equation:

$$u_n(\theta, t) = \Theta_n(\theta) \cdot T_n(t) = P_n(\cos\theta)\, e^{-n(n+1)kt}.$$

Applying the superposition principle, the solution is

$$u(\theta, t) = \sum_{n=0}^{\infty} a_n P_n(\cos\theta) e^{-n(n+1)kt}$$

for $0 < \theta < \pi$, and $t > 0$.

We determine the constants a_n from the initial condition $u(\theta, 0) = f(\theta)$, and we want

$$f(\theta) = \sum_{n=0}^{\infty} a_n P_n(\cos\theta)$$

for $0 < \theta < \pi$. The coefficients in this Fourier-Legendre expansion are given by

$$a_n = \frac{\langle f, P_n \rangle}{\|P_n\|^2} = \frac{\langle f, P_n \rangle}{\langle P_n, P_n \rangle},$$

where the inner product is given by

$$\langle f, g \rangle = \int_0^{\pi} f(\theta) g(\theta) \sin\theta \, d\theta.$$

Thus, we have

$$u(\theta, t) = \sum_{n=0}^{\infty} a_n P_n(\cos\theta) e^{-n(n+1)kt}$$

for $0 < \theta < \pi$, and $t > 0$, where

$$a_n = \frac{2n+1}{2} \int_0^{\pi} f(\theta) P_n(\cos\theta) \sin\theta \, d\theta$$

for $n \geqslant 0$.

Exercise 13.20. Heat Flow in a Hemisphere ✗✗✗✗

A solid isotropic hemisphere of constant diffusivity k and radius a has its flat base kept at temperature $u = 0$ and its hemispherical surface kept at temperature $u = u_0$. Determine the evolution of temperature in the interior if the temperature is initially zero throughout the hemisphere.

Solution. The problem is a Dirichlet problem for the heat equation in a hemispherical solid, and the solid hemisphere can be described in spherical coordinates as

$$H(a) = \{ (r, \theta, \phi) \mid 0 \leq r \leq a, \ -\pi \leq \phi \leq \pi, \ 0 \leq \theta \leq \pi/2 \}.$$

Given the symmetry of the problem about the z-axis, it is appropriate to look for a solution that exhibits this same symmetry; hence, we look for a solution independent of the polar angle ϕ, so that $u = u(r, \theta, t)$. Also, since the Laplacian operator in spherical coordinates is singular at $r = 0$, and also along the z-axis, that is, at $\theta = 0$, and $\theta = \pi$, to ensure the boundedness and continuity of the solution we require the following boundedness conditions:

$$|u(0, \theta, t)| < \infty \quad \text{and} \quad |u(r, 0, t)| < \infty$$

for all $(r, \theta, \phi) \in H(a)$ and all $t \geqslant 0$. Thus, the boundary value–initial value problem is

$$\begin{aligned}
\frac{\partial u}{\partial t} = k\left(\frac{1}{r^2}\frac{\partial}{\partial r}\left(r^2\frac{\partial u}{\partial r}\right) + \frac{1}{r^2 \sin\theta}\frac{\partial}{\partial \theta}\left(\sin\theta\frac{\partial u}{\partial \theta}\right)\right)&, \quad (r, \theta, \phi) \in H(a),\ t > 0,\\
u(a, \theta, t) &= u_0,\\
u(r, \pi/2, t) &= 0,\\
u(r, \theta, 0) &= 0,\\
|u(0, \theta, t)| &< \infty,\\
|u(r, 0, t)| &< \infty.
\end{aligned}$$

In addition, since this problem has nonhomogeneous boundary conditions, we split the solution into a steady-state part $v(r, \theta)$ and a transient part $w(r, \theta, t)$:

$$u(r, \theta, t) = v(r, \theta) + w(r, \theta, t),$$

where v satisfies

$$\begin{aligned}
\frac{1}{r^2}\frac{\partial}{\partial r}\left(r^2\frac{\partial v}{\partial r}\right) + \frac{1}{r^2 \sin\theta}\frac{\partial}{\partial \theta}\left(\sin\theta\frac{\partial v}{\partial \theta}\right) = 0&, \quad (r, \theta, \phi) \in H(a),\\
v(a, \theta) &= u_0,\\
v(r, \pi/2) &= 0,\\
|v(r, \theta)| &< \infty,
\end{aligned}$$

and w satisfies

$$\begin{aligned}
\frac{\partial w}{\partial t} = k\left(\frac{1}{r^2}\frac{\partial}{\partial r}\left(r^2\frac{\partial w}{\partial r}\right) + \frac{1}{r^2 \sin\theta}\frac{\partial}{\partial \theta}\left(\sin\theta\frac{\partial w}{\partial \theta}\right)\right)&, \quad (r, \theta, \phi) \in H(a),\ t > 0,\\
w(a, \theta, t) &= 0,\\
w(r, \pi/2, t) &= 0,\\
w(r, \theta, 0) &= -v(r, \theta),\\
|w(r, \theta, t)| &< \infty.
\end{aligned}$$

The problem for the steady-state temperature has been solved in Exercise 15.13, and the solution given there is

$$v(r, \theta) = u_0 \sum_{n=0}^{\infty} \frac{(-1)^n}{2^{2n+1}} \left(\frac{4n+3}{n+1}\right) \binom{2n}{n} \left(\frac{r}{a}\right)^{2n+1} P_{2n+1}(\cos\theta).$$

We now proceed to solve the problem for the transient temperature w. Applying separation of variables, we look for a solution of the form

$$w(r,\theta,t) = W(r,\theta)\cdot T(t),$$

and substituting this expression into the problem for w yields

$$\frac{1}{k}\frac{T'}{T} = \frac{1}{W}\left(\frac{1}{r^2}\frac{\partial}{\partial r}\left(r^2\frac{\partial W}{\partial r}\right) + \frac{1}{r^2\sin\theta}\frac{\partial}{\partial\theta}\left(\sin\theta\frac{\partial W}{\partial\theta}\right)\right) = -\lambda,$$

where λ is the first separation constant. The homogeneous boundary conditions can easily be separated, leading to two problems:

- **Spatial problem**:

$$\nabla^2 W = \frac{1}{r^2}\frac{\partial}{\partial r}\left(r^2\frac{\partial W}{\partial r}\right) + \frac{1}{r^2\sin\theta}\frac{\partial}{\partial\theta}\left(\sin\theta\frac{\partial W}{\partial\theta}\right) = -\lambda W,$$

$$W(a,\theta) = 0,$$
$$W(r,\pi/2) = 0,$$
$$|W(r,\theta)| < \infty.$$

- **Time problem**:

$$T' + \lambda kT = 0.$$

From Theorem 5.7 we know that the spatial problem, that is, the Helmholtz eigenvalue problem for the surface Laplacian with Dirichlet boundary conditions, has positive eigenvalues. Therefore, we write $\lambda = \nu^2$, and the equation for W becomes

$$\frac{1}{r^2}\frac{\partial}{\partial r}\left(r^2\frac{\partial W}{\partial r}\right) + \frac{1}{r^2\sin\theta}\frac{\partial}{\partial\theta}\left(\sin\theta\frac{\partial W}{\partial\theta}\right) = -\nu^2 W, \quad r>0,\ 0<\theta<\frac{\pi}{2},$$

$$W(a,\theta) = 0,$$
$$W(r,\pi/2) = 0,$$
$$|W(0,\theta)| < \infty,$$
$$|W(r,0)| < \infty.$$

As above, the r and $\sin\theta$ that appear in the denominator of the Laplacian mean that we require the boundedness conditions. Applying separation of variables, we assume a solution of the form

$$W(r,\theta) = R(r)\cdot S(\theta)$$

and obtain

$$\frac{(r^2R')'\cdot S}{r^2} + \frac{R\cdot(\sin\theta S')'}{r^2\sin\theta} = -\nu^2 R\cdot S,$$

and separating variables, we get

$$\frac{(r^2R')'}{R} + \nu^2 r^2 = -\frac{(\sin\theta S')'}{\sin\theta S} = \mu,$$

where μ is the second separation constant.

Again, the homogeneous boundary conditions can easily be separated, and also applying the boundedness assumptions yields the following boundary value problems:

- **Radial problem**:

$$\begin{aligned}&\left(r^2R'\right)' + \left(\nu^2r^2 - \mu\right)R = 0, \quad 0 < r < a,\\ &|R(0)| < \infty,\\ &R(a) = 0.\end{aligned}$$

- **Azimuthal angular problem**:

$$\begin{aligned}&\left(S'\sin\theta\right)' + \mu\sin\theta\, S = 0, \quad 0 < \theta < \frac{\pi}{2},\\ &|S(0)| < \infty,\\ &S(\pi/2) = 0.\end{aligned}$$

Now we solve these ODEs sequentially, from the bottom up.

(a) Solution to the azimuthal problem:

Letting $x = \cos\theta$ and $S(\theta) = \varphi(x(\theta))$, the azimuthal problem becomes

$$\begin{aligned}&(1 - x^2)\varphi'' - 2x\varphi' + \mu\varphi = 0, \quad 0 < x < 1,\\ &\varphi(0) = 0,\\ &|\varphi(1)| < \infty,\end{aligned}$$

which is Legendre's equation, and this equation has bounded solutions if and only if $\mu = n(n+1)$, where n is an integer. Thus, the eigenvalues are

$$\mu_n = n(n+1)$$

with corresponding eigenfunctions

$$\varphi_n(x) = P_n(x)$$

for $n = 0,\, 1,\, 2,\, \ldots$, where P_n is the Legendre polynomial of degree n. The boundary condition $\varphi(0) = 0$ gives

$$P_n(0) = 0$$

and therefore $n = 2m + 1$; that is, n is an odd integer.

Therefore, the solution to the azimuthal problem is

$$S_m(\theta) = P_{2m+1}(\theta)$$

for $m = 0,\, 1,\, 2,\, \ldots$.

(b) Solution to the radial problem:

Letting $\mu = n(n+1) = (2m+1)(2m+2)$ in the radial equation, making the change of variables $x = \nu r$ and $\varphi(x) = R(r)$, the radial problem becomes

$$\begin{aligned}
&\left(x^2\varphi'\right)' + \left(x^2 - (2m+1)(2m+2)\right)\varphi = 0, \\
&\varphi(\nu a) = 0, \\
&|\varphi(0)| < \infty,
\end{aligned}$$

which is the spherical Bessel's equation of order $2m+1$. The general solution to this equation is

$$\varphi(x) = A\, j_{2m+1}(x) + B\, y_{2m+1}(x),$$

where j_{2m+1} and y_{2m+1} are the spherical Bessel functions of the first and second kind, respectively. From the boundedness condition we have $B = 0$, and the solution reduces to

$$\varphi(x) = A\, j_{2m+1}(x);$$

that is,

$$R(r) = A\, j_{2m+1}(\nu r).$$

The boundary condition $R(a) = 0$ implies that $j_{2m+1}(\nu a) = 0$, which in turn implies that νa is a positive root of the spherical Bessel function j_{2m+1}. If we denote the kth positive root of j_{2m+1} by α_{mk}, the eigenvalues are

$$\nu_{mk}\, a = \alpha_{mk}$$

and the corresponding eigenfunctions are

$$R_{mk}(r) = j_{2m+1}\left(\frac{\alpha_{mk} r}{a}\right).$$

Therefore, the solution to the radial problem is

$$R_{mk}(r) = j_{2m+1}\left(\frac{\alpha_{mk} r}{a}\right)$$

for $m = 0,\ 1,\ 2,\ \ldots$ and $k = 1,\ 2,\ \ldots$.

(c) Solution to the time problem:

Since $\lambda = \nu^2$, letting $\lambda_{mk} = \nu_{mk}^2$ in the time equation, the time problem becomes

$$T' + \nu_{mk}^2 kT = 0,$$

and the solution to this equation is

$$T_{mk}(t) = e^{-\nu_{mk}^2 kt}$$

for $m = 0, 1, 2, \ldots$ and $k = 1, 2, \ldots$.

(d) Solution to the spatial problem:

Combining the results from the radial and azimuthal problems, we have

$$W_{mk}(r,\theta) = R_{mk}(r)S_m(\theta) = j_{2m+1}\left(\frac{\alpha_{mk}r}{a}\right)P_{2m+1}(\cos\theta)$$

for $m = 0, 1, 2, \ldots$ and $k = 1, 2, \ldots$.

Gathering the solutions to the various ODEs and applying the superposition principle, we have

$$w(r,\theta,t) = \sum_{m=0}^{\infty}\sum_{k=1}^{\infty} a_{mk}W_{mk}(r,\theta)T_{mk}(t);$$

that is,

$$w(r,\theta,t) = \sum_{m=0}^{\infty}\sum_{k=1}^{\infty} a_{mk}\, j_{2m+1}\left(\frac{\alpha_{mk}r}{a}\right)P_{2m+1}(\cos\theta)\, e^{-\nu_{mk}^2 kt}.$$

To determine the coefficients a_{mk}, we apply the initial condition

$$w(r,\theta,0) = -v(r,\theta),$$

and obtain

$$-v(r,\theta) = \sum_{m=0}^{\infty}\sum_{k=1}^{\infty} a_{mk}W_{mk}(r,\theta).$$

The coefficients a_{mk} are given by

$$a_{mk} = -\frac{\langle v, W_{mk}\rangle}{||W_{mk}||^2},$$

where the inner product is

$$\langle f,g\rangle = \int_0^a\int_0^{\pi/2} f(r,\theta)g(r,\theta)\, r\sin\theta\, dr\, d\theta.$$

Note that if the integrands are separable, in the sense that

$$f(r,\theta) = f_1(r)f_2(\theta) \qquad \text{and} \qquad g(r,\theta) = g_1(r)g_2(\theta),$$

then

$$\begin{aligned}\langle f, g\rangle &= \int_0^a \int_0^{\pi/2} f_1(r) f_2(\theta) g_1(r) g_2(\theta) r \sin\theta \, dr \, d\theta \\ &= \left(\int_0^a f_1(r) g_1(r) r \, dr\right)\left(\int_0^{\pi/2} f_2(\theta) g_2(\theta) \sin\theta \, d\theta\right) \\ &= \langle f_1, g_1\rangle_R \langle f_2, g_2\rangle_A,\end{aligned}$$

where

$$\langle f, g\rangle_R = \int_0^a f(r)\, g(r)\, r \, dr \qquad \text{and} \qquad \langle f, g\rangle_A = \int_0^{\pi/2} f(\theta)\, g(\theta) \sin\theta \, d\theta$$

are the one-dimensional inner products associated with the radial and azimuthal problems, respectively.

Now we have

$$\begin{aligned}||W_{mk}||^2 &= \langle W_{mk}, W_{mk}\rangle = \langle R_{mk} S_m, R_{mk} S_m\rangle \\ &= \langle R_{mk}, R_{mk}\rangle_R \langle S_m, S_m\rangle_A = ||R_{mk}||_R^2 \, ||S_m||_A^2,\end{aligned}$$

and if we define

$$A_n = \frac{(-1)^n}{2^{2n+1} a^{2n+1}} \frac{4n+3}{n+1} \binom{2n}{n},$$

then

$$\begin{aligned}\langle v, W_{mk}\rangle &= \langle \sum_{n=0}^{\infty} A_n r^{2n+1} P_{2n+1}(\cos\theta), W_{mk}\rangle \\ &= \sum_{n=0}^{\infty} A_n \langle r^{2n+1} S_n, R_{mk} S_m\rangle \\ &= \sum_{n=0}^{\infty} A_n \langle r^{2n+1}, R_{mk}\rangle_R \langle S_n, S_m\rangle_A \\ &= A_m \langle r^{2m+1}, R_{mk}\rangle_R ||S_m||_A^2.\end{aligned}$$

Therefore, the final solution is

$$u(r, \theta, t) = v(r, \theta) + w(r, \theta, t),$$

where

$$v(r, \theta) = \sum_{n=0}^{\infty} A_n r^{2n+1} P_{2n+1}(\cos\theta)$$

and

$$w(r, \theta, t) = \sum_{m=0}^{\infty} \sum_{k=1}^{\infty} a_{mk} R_{mk}(r) S_m(\theta) e^{-\nu_{mk}^2 kt},$$

and where

$$R_{mk}(r) = j_{2m+1}(\nu_{mk} r),$$

$$S_m(\theta) = P_{2m+1}(\cos\theta),$$

$$a_{mk} = -A_m \frac{\langle r^{2m+1}, R_{mk} \rangle_R}{||R_{mk}||_R^2}$$

and

$$A_m = \frac{(-1)^m}{2^{2m+1} a^{2m+1}} \frac{4m+3}{m+1} \binom{2m}{m}.$$

CHAPTER 14

WAVE EQUATION PROBLEMS

Exercise 14.1. *XX*

Use the method of separation of variables to solve the one-dimensional wave equation with homogeneous Dirichlet boundary conditions

$$\begin{aligned}
&\frac{\partial^2 u}{\partial t^2} = \frac{1}{\pi^2}\frac{\partial^2 u}{\partial x^2}, \quad 0 < x < 1, \quad t > 0,\\
&u(0,t) = 0,\\
&u(1,t) = 0,\\
&u(x,0) = \sin \pi x \cos \pi x,\\
&\frac{\partial u}{\partial t}(x,0) = 0.
\end{aligned}$$

Solution. We assume a solution of the form $u(x,t) = X(x) \cdot T(t)$ to obtain two ordinary differential equations:

$$X'' + \lambda\pi^2 X = 0 \qquad \text{and} \qquad T'' + \lambda T = 0,$$

Partial Differential Equations: Theory and Completely Solved Problems, First Edition. By T. Hillen, I. E. Leonard, H. van Roessel

with separation constant λ. We can satisfy the two boundary conditions by requiring that

$$X(0) = 0 \quad \text{and} \quad X(1) = 0,$$

so that X satisfies the boundary value problem

$$\begin{aligned} &X'' + \lambda\pi^2 X = 0, \quad 0 < x < 1, \\ &X(0) = 0, \\ &X(1) = 0. \end{aligned}$$

The cases $\lambda = 0$ and $\lambda < 0$ both result in the trivial solution $X(x) = 0$ for all $x \in [0,1]$, and the only nontrivial solution arises when $\lambda > 0$, say $\lambda = \mu^2$, where $\mu \neq 0$. In this case the boundary value problem

$$\begin{aligned} &X'' + \mu^2\pi^2 X = 0, \quad 0 < x < 1, \\ &X(0) = 0, \\ &X(1) = 0 \end{aligned}$$

has general solution

$$X(x) = A\cos \mu\pi x + B\sin \mu\pi x,$$

and applying the first boundary condition, we have $X(0) = A = 0$. Applying the second boundary condition, we have $X(1) = B\sin \mu\pi = 0$, and to get a nontrivial solution we must have $\sin \mu\pi = 0$, but this can happen only if $\mu\pi = n\pi$, where n is an integer. The eigenvalues and corresponding eigenfunctions are

$$\lambda_n = n^2 \qquad \text{and} \qquad X_n(x) = \sin n\pi x$$

for $n \geqslant 1$. The corresponding T-equation is

$$T'' + n^2 T = 0,$$

with solutions

$$T_n(t) = a_n \cos nt + b_n \sin nt$$

for $n \geqslant 1$. For each integer $n \geqslant 1$, the function

$$u_n(x,t) = X_n(x) \cdot T_n(t) = \sin n\pi x\,(a_n \cos nt + b_n \sin nt)$$

satisfies the wave equation and the two homogeneous boundary conditions. Using the superposition principle, we write

$$u(x,t) = \sum_{n=1}^{\infty} \sin n\pi x\,(a_n \cos nt + b_n \sin nt)\,.$$

To satisfy the initial conditions, we need

$$u(x,0) = \sum_{n=1}^{\infty} a_n \sin n\pi x \tag{14.1}$$

and

$$\frac{\partial u}{\partial t}(x,0) = \sum_{n=1}^{\infty} nb_n \sin n\pi x, \tag{14.2}$$

that is, the Fourier sine series of $u(x,0)$ and $\partial u(x,0)/\partial t$.
Therefore, from (14.1) we have

$$a_n = 2\int_0^1 u(x,0)\sin n\pi x\,dx,$$

and from (14.2) we have

$$nb_n = 2\int_0^1 \frac{\partial u}{\partial t}(x,0)\sin n\pi x\,dx$$

for $n \geqslant 1$.

Note that $b_n = 0$ for all $n \geqslant 1$, since $\partial u(x,0)/\partial t = 0$ for $0 < x < 1$. Also, we have

$$u(x,0) = \sin \pi x \cos \pi x = \frac{1}{2}\sin 2\pi x,$$

so that $u(x,0)$ is its own Fourier sine series, and

$$a_n = \begin{cases} \dfrac{1}{2}, & \text{if } n = 2,\\ 0, & \text{if } n \neq 2. \end{cases}$$

Therefore, the solution is

$$u(x,t) = \frac{1}{2}\sin 2\pi x \cos 2t$$

for $0 < x < 1,\ t > 0$.

Exercise 14.2. ✗✗
Solve the following boundary value–initial value problem for the wave equation:

$$\begin{aligned} \frac{\partial^2 u}{\partial t^2} = \frac{\partial^2 u}{\partial x^2}, \quad 0 < x < 1, \quad t > 0,\\ u(0,t) = 0,\\ u(1,t) = 0,\\ u(x,0) = \sin \pi x + \tfrac{1}{2}\sin 3\pi x + 3\sin 7\pi x,\\ \frac{\partial u}{\partial t}(x,0) = \sin 2\pi x. \end{aligned}$$

Solution. Similar to Exercise 14.1 we use separation of variables and the superposition principle to get the general solution

$$u(x,t)=\sum_{n=1}^{\infty}\sin n\pi x\,(a_n\cos n\pi t+b_n\sin n\pi t)\,,$$

where the coefficients are to be determined using the initial conditions. Differentiating, we have

$$\frac{\partial u}{\partial t}(x,t)=\sum_{n=1}^{\infty}\sin n\pi x\,(-n\pi a_n\sin n\pi t+n\pi b_n\cos n\pi t)\,,$$

and setting $t=0$, we get

$$u(x,0)=\sum_{n=1}^{\infty}a_n\sin n\pi x \qquad \text{and} \qquad \frac{\partial u}{\partial t}(x,0)=\sum_{n=1}^{\infty}n\pi b_n\sin n\pi x,$$

and again these are just the Fourier sine series of $f(x)$ and $g(x)$, the initial displacement and initial velocity.

From the first initial condition,

$$u(x,0)=\sin\pi x+\tfrac{1}{2}\sin 3\pi x+3\sin 7\pi x,$$

we see that

$$a_1=1,\quad a_3=\frac{1}{2},\quad a_7=3,$$

and $a_n=0$ for all other values of n. From the second initial condition

$$\frac{\partial u}{\partial t}(x,0)=\sin 2\pi x,$$

so that

$$b_n=\begin{cases}\dfrac{1}{2\pi}, & \text{if } n=2,\\[2mm] 0, & \text{if } n\neq 2.\end{cases}$$

Therefore, the solution is

$$\begin{aligned}u(x,t)&=\sin\pi x\cos\pi t+\frac{1}{2\pi}\sin 2\pi x\sin 2\pi t\\&\quad+\frac{1}{2}\sin 3\pi x\cos 3\pi t+3\sin 7\pi x\cos 7\pi t\end{aligned}$$

for $0<x<1,\ t>0$.

Exercise 14.3. XX
Show that the solution to the wave equation

$$\begin{aligned}
&\frac{\partial^2 u}{\partial t^2} = c^2 \frac{\partial^2 u}{\partial x^2}, \quad 0 < x < a, \quad t > 0,\\
&u(0,t) = 0,\\
&u(a,t) = 0,\\
&u(x,0) = f(x),\\
&\frac{\partial u}{\partial t}(x,0) = g(x)
\end{aligned}$$

is given by

$$u(x,t) = \frac{1}{2}\sum_{n=1}^{\infty} a_n \left[\sin\frac{n\pi(x-ct)}{a} + \sin\frac{n\pi(x+ct)}{a}\right],$$

where $a_n = (2/a)\int_0^a f(x)\sin(n\pi x/a)\,dx$, $n = 1, 2, \ldots$, if $g(x) = 0$ for $0 < x < a$.

Solution. Using separation of variables, the solution to this problem is given by

$$u(x,t) = \sum_{n=1}^{\infty} \sin\frac{n\pi x}{a}\left(a_n \cos\frac{n\pi ct}{a} + b_n \sin\frac{n\pi ct}{a}\right)$$

where

$$a_n = \frac{2}{a}\int_0^a f(x)\sin\frac{n\pi x}{a}\,dx \qquad \text{and} \qquad b_n = \frac{2}{n\pi c}\int_0^a g(x)\sin\frac{n\pi x}{a}\,dx$$

for $n \geqslant 1$. Since $g(x) = 0$ for $0 < x < a$, then $b_n = 0$ for all $n \geqslant 1$, and the solution is given by

$$u(x,t) = \sum_{n=1}^{\infty} a_n \sin\frac{n\pi x}{a}\cos\frac{n\pi ct}{a},$$

and since

$$\sin A\cos B = \frac{1}{2}\left[\sin(A-B) + \sin(A+B)\right],$$

then

$$u(x,t) = \frac{1}{2}\sum_{n=1}^{\infty} a_n \left[\sin\frac{n\pi(x-ct)}{a} + \sin\frac{n\pi(x+ct)}{a}\right],$$

where $a_n = (2/a)\int_0^a f(x)\sin(n\pi x/a)\,dx$.

Exercise 14.4. XXX

Consider the homogeneous Dirichlet problem for the wave equation

$$\frac{\partial^2 u}{\partial t^2} = c^2 \frac{\partial^2 u}{\partial x^2},$$

$$u(0,t) = 0, \quad u(a,t) = 0, \quad \text{for} \quad t > 0,$$

$$u(x,0) = f(x), \quad \frac{\partial u}{\partial t}(x,0) = 0, \quad \text{for} \quad 0 < x < a.$$

Show that the solution can be written as

$$u(x,t) = \frac{1}{2}\Big[F(x-ct) + F(x+ct)\Big],$$

where $F(x)$ is the odd periodic extension of $f(x)$.

Hint: Use separation of variables and

$$\sin A \cos B = \frac{1}{2}\big[\sin(A+B) + \sin(A-B)\big].$$

Solution. We assume a solution of the form

$$u(x,t) = X(t) \cdot T(t)$$

and separate variables to obtain the two ordinary differential equations

$$X''(x) + \lambda X(x) = 0, \quad 0 < x < a, \qquad T''(t) + \lambda c^2 T(t) = 0, \quad t > 0,$$

$$X(0) = 0, \qquad T'(0) = 0.$$

$$X(a) = 0,$$

The eigenvalues and eigenfunctions for the X-equation are

$$\lambda_n = \frac{n^2\pi^2}{a^2} \qquad \text{and} \qquad X_n(x) = \sin \frac{n\pi x}{a}$$

for $n \geqslant 1$, and the corresponding solutions of the T-equation are

$$T_n(t) = a_n \cos \frac{n\pi ct}{a} + b_n \sin \frac{n\pi ct}{a}$$

for $n \geqslant 1$. Using the superposition principle, we write the solution as

$$u(x,t) = \sum_{n=1}^{\infty} \left(a_n \cos \frac{n\pi ct}{a} + b_n \sin \frac{n\pi ct}{a} \right) \sin \frac{n\pi x}{a}$$

and

$$\frac{\partial u}{\partial t}(x,t) = \sum_{n=1}^{\infty} \left(-a_n \frac{n\pi c}{a} \sin \frac{n\pi ct}{a} + b_n \frac{n\pi c}{a} \cos \frac{n\pi ct}{a} \right) \sin \frac{n\pi x}{a}.$$

We determine the coefficients using the initial conditions and the orthogonality of the eigenfunctions on the interval $[0, a]$.

From the first initial condition we have

$$f(x) = u(x,0) = \sum_{n=1}^{\infty} a_n \sin \frac{n\pi x}{a},$$

so that

$$a_n = \frac{2}{a} \int_0^a f(x) \sin \frac{n\pi x}{a}\, dx$$

for $n \geqslant 1$. From the second initial condition we have

$$0 = \frac{\partial u}{\partial t}(x,0) = \sum_{n=1}^{\infty} b_n \frac{n\pi c}{a} \sin \frac{n\pi x}{a},$$

so that $b_n = 0$ for $n \geqslant 1$. The solution is

$$u(x,t) = \sum_{n=1}^{\infty} a_n \left\{ \frac{1}{2} \sin \frac{n\pi(x-ct)}{a} + \frac{1}{2} \sin \frac{n\pi(x+ct)}{a} \right\}$$

for $0 < x < a$ and $t > 0$. Note that if $f \in PWS[0, a]$, that is, f is piecewise smooth on the interval $[0, a]$, the series

$$\sum_{n=1}^{\infty} a_n \sin \frac{n\pi x}{a}\, dx$$

is the Fourier sine series for f and converges for all real numbers x, and, except for at most countably many values of x, it converges to the odd periodic extension F of f.

Therefore, assuming that the odd periodic extension F is continuous, the solution is

$$u(x,t) = \frac{1}{2} \sum_{n=1}^{\infty} a_n \sin \frac{n\pi(x-ct)}{a} + \frac{1}{2} \sum_{n=1}^{\infty} a_n \sin \frac{n\pi(x+ct)}{a};$$

that is,

$$u(x,t) = \frac{1}{2}F(x-ct) + \frac{1}{2}F(x+ct)$$

for $0 < x < a$ and $t > 0$.

Exercise 14.5. XXX

Consider the homogeneous Dirichlet problem for the wave equation with given initial velocity

$$\frac{\partial^2 u}{\partial t^2} = c^2 \frac{\partial^2 u}{\partial x^2},$$

$$u(0,t) = 0, \quad u(a,t) = 0, \quad \text{for} \quad t > 0,$$

$$u(x,0) = 0, \quad \frac{\partial u}{\partial t}(x,0) = g(x), \quad \text{for} \quad 0 < x < a.$$

Show that

$$u(x,t) = \frac{1}{2c} \int_{x-ct}^{x+ct} G(s)\, ds,$$

where $G(x)$ is the odd periodic extension of $g(x)$.

Hint: Use separation of variables and

$$\sin A \sin B = \frac{1}{2}\big[\cos(A-B) - \cos(A+B)\big].$$

Solution. As in Exercise 14.4, we assume a solution of the form

$$u(x,t) = X(t) \cdot T(t)$$

and separate variables to obtain two ordinary differential equations:

$$\begin{aligned} &X''(x) + \lambda X(x) = 0, \quad 0 < x < a, && T''(t) + \lambda c^2 T(t) = 0, \quad t > 0, \\ &X(0) = 0, && T(0) = 0. \\ &X(a) = 0, && \end{aligned}$$

As before, the eigenvalues and eigenfunctions for the X-equation are

$$\lambda_n = \frac{n^2\pi^2}{a^2} \qquad \text{and} \qquad X_n(x) = \sin\frac{n\pi x}{a}$$

for $n \geqslant 1$, and the corresponding solutions of the T-equation are

$$T_n(t) = a_n \cos\frac{n\pi ct}{a} + b_n \sin\frac{n\pi ct}{a}$$

for $n \geqslant 1$. Using the superposition principle, we write the solution as

$$u(x,t) = \sum_{n=1}^{\infty} \left(a_n \cos\frac{n\pi ct}{a} + b_n \sin\frac{n\pi ct}{a} \right) \sin\frac{n\pi x}{a}$$

and

$$\frac{\partial u}{\partial t}(x,t) = \sum_{n=1}^{\infty} \left(-a_n \frac{n\pi c}{a} \sin \frac{n\pi ct}{a} + b_n \frac{n\pi c}{a} \cos \frac{n\pi ct}{a} \right) \sin \frac{n\pi x}{a}.$$

We determine the coefficients using the initial conditions and the orthogonality of the eigenfunctions on the interval $[0, a]$.

From the first initial condition we have

$$0 = u(x,0) = \sum_{n=1}^{\infty} a_n \sin \frac{n\pi x}{a},$$

so that $a_n = 0$ for $n \geqslant 1$. From the second initial condition we have

$$g(x) = \frac{\partial u}{\partial t}(x,0) = \sum_{n=1}^{\infty} b_n \frac{n\pi c}{a} \sin \frac{n\pi x}{a},$$

so that

$$b_n = \frac{2}{n\pi c} \int_0^a g(x) \sin \frac{n\pi x}{a}\, dx$$

for $n \geqslant 1$, and the solution is

$$u(x,t) = \sum_{n=1}^{\infty} b_n \left\{ \frac{1}{2} \cos \frac{n\pi(x-ct)}{a} - \frac{1}{2} \cos \frac{n\pi(x+ct)}{a} \right\}$$

for $0 < x < a$ and $t > 0$. Now

$$\frac{\partial u}{\partial t}(x,t) = \frac{1}{2} \sum_{n=1}^{\infty} b_n \frac{n\pi c}{a} \sin \frac{n\pi(x-ct)}{a} + \frac{1}{2} \sum_{n=1}^{\infty} b_n \frac{n\pi c}{a} \sin \frac{n\pi(x+ct)}{a}$$

and if $g \in PWS[0,a]$, that is, g is piecewise smooth on the interval $[0, a]$, then the series

$$\sum_{n=1}^{\infty} b_n \frac{n\pi c}{a} \sin \frac{n\pi x}{a}$$

is the Fourier sine series for g and converges for <u>all</u> real numbers x, and, except for at most countably many values of x, it converges to the odd periodic extension G of g; that is,

$$\frac{\partial u}{\partial t}(x,t) = \frac{1}{2} G(x-ct) + \frac{1}{2} G(x+ct).$$

Integrating this from 0 to t, we have

$$\int_0^t \frac{\partial u}{\partial \tau}(x,\tau)\, d\tau = u(x,t) - u(x,0) = u(x,t)$$

since $u(x, 0) = 0$. Therefore,

$$u(x,t) = \frac{1}{2}\int_0^t G(x - c\tau)\,d\tau + \frac{1}{2}\int_0^t G(x + c\tau)\,d\tau$$
$$= -\frac{1}{2c}\int_x^{x-ct} G(s)\,ds + \frac{1}{2c}\int_x^{x+ct} G(s)\,ds,$$

where we made the substitution $s = x - c\tau$ in the first integral and $s = x + c\tau$ in the second integral. Assuming that the odd periodic extension of g is continuous, the solution is

$$u(x,t) = \frac{1}{2c}\int_{x-ct}^{x+ct} G(s)\,ds$$

for $0 < x < a$ and $t > 0$.

Exercise 14.6. XXX
Derive d'Alembert's solution to the wave equation

$$\frac{\partial^2 u}{\partial t^2} = c^2\frac{\partial^2 u}{\partial x^2}, \qquad -\infty < x < \infty, \quad t > 0,$$

and use it and the superposition principle to solve the wave equation with initial data

$$u(x,0) = e^{-x^2}, \qquad \frac{\partial u}{\partial t}(x,0) = \frac{x}{(1+x^2)^2}$$

for $-\infty < x < \infty$.

Solution. In Part I we have seen at least three different methods of deriving d'Alembert's solution to the one-dimensional wave equation: separation of variables, Fourier transforms, and the method of characteristics. Here we give a fourth method, based on the change of variables

$$\alpha = x + ct \qquad \text{and} \qquad \beta = x - ct.$$

From the chain rule we have

$$\frac{\partial u}{\partial x} = \frac{\partial u}{\partial \alpha}\frac{\partial \alpha}{\partial x} + \frac{\partial u}{\partial \beta}\frac{\partial \beta}{\partial x} = \frac{\partial u}{\partial \alpha} + \frac{\partial u}{\partial \beta},$$

and replacing u by $\partial u/\partial x$, we get

$$\frac{\partial^2 u}{\partial x^2} = \frac{\partial}{\partial x}\left(\frac{\partial u}{\partial \alpha} + \frac{\partial u}{\partial \beta}\right) = \frac{\partial}{\partial \alpha}\left(\frac{\partial u}{\partial \alpha} + \frac{\partial u}{\partial \beta}\right) + \frac{\partial}{\partial \beta}\left(\frac{\partial u}{\partial \alpha} + \frac{\partial u}{\partial \beta}\right);$$

that is,

$$\frac{\partial^2 u}{\partial x^2} = \frac{\partial^2 u}{\partial \alpha^2} + 2\frac{\partial^2 u}{\partial \alpha \partial \beta} + \frac{\partial^2 u}{\partial \beta^2}.$$

Again, from the chain rule, we have

$$\frac{\partial u}{\partial t} = \frac{\partial u}{\partial \alpha}\frac{\partial \alpha}{\partial t} + \frac{\partial u}{\partial \beta}\frac{\partial \beta}{\partial t} = c\frac{\partial u}{\partial \alpha} - c\frac{\partial u}{\partial \beta},$$

and replacing u by $\partial u/\partial t$, we get

$$\frac{\partial^2 u}{\partial t^2} = \frac{\partial}{\partial t}\left(c\frac{\partial u}{\partial \alpha} - c\frac{\partial u}{\partial \beta}\right) = c\frac{\partial}{\partial \alpha}\left(c\frac{\partial u}{\partial \alpha} - c\frac{\partial u}{\partial \beta}\right) - c\frac{\partial}{\partial \beta}\left(c\frac{\partial u}{\partial \alpha} - c\frac{\partial u}{\partial \beta}\right),$$

that is,

$$\frac{\partial^2 u}{\partial t^2} = c^2\frac{\partial^2 u}{\partial \alpha^2} - 2c^2\frac{\partial^2 u}{\partial \alpha \partial \beta} + c^2\frac{\partial^2 u}{\partial \beta^2},$$

and substituting these expressions into the wave equation, we obtain

$$\frac{\partial^2 u}{\partial \alpha \partial \beta} = 0.$$

This equation says that $\partial u/\partial \beta$ does not depend on α, and therefore

$$\frac{\partial u}{\partial \beta} = g(\beta),$$

where g is an arbitrary differentiable function. Now, integrating this equation with respect to β, holding α fixed, we get

$$u = \int \frac{\partial u}{\partial \beta}\, d\beta + F(\alpha) = \int g(\beta)\, d\beta + F(\alpha) = F(\alpha) + G(\beta),$$

where F is an arbitrary differentiable function and G is an antiderivative of g. Finally, using the fact that $\alpha = x + ct$ and $\beta = x - ct$, we get **d'Alembert's general solution** to the one-dimensional wave equation:

$$u(x,t) = F(x+ct) + G(x-ct),$$

where F and G are arbitrary differentiable functions.

Now, to solve the original question, we solve the following boundary value–initial value problems and use the superposition principle to combine them to get a solution to the original problem:

$$\begin{aligned} &\frac{\partial^2 v}{\partial t^2} = c^2\frac{\partial^2 v}{\partial x^2}, \qquad -\infty < x < \infty, \quad t > 0, \\ &v(x,0) = e^{-x^2}, \\ &\frac{\partial v}{\partial t}(x,0) = 0 \end{aligned} \tag{14.3}$$

and

$$\begin{aligned}
&\frac{\partial^2 w}{\partial t^2} = c^2 \frac{\partial^2 w}{\partial x^2}, \qquad -\infty < x < \infty, \quad t > 0, \\
&w(x,0) = 0, \\
&\frac{\partial w}{\partial t}(x,0) = \frac{x}{(1+x^2)^2},
\end{aligned} \tag{14.4}$$

and from the linearity of the PDE and side conditions, the solution to the original problem is $u = v + w$.

For problem (14.3), we use the initial conditions to write

$$v(x,0) = e^{-x^2} = F(x) + G(x),$$

so that $F(x) + G(x) = e^{-x^2}$, and

$$\frac{\partial v}{\partial t} = 0 = cF'(x) - cG'(x),$$

so that

$$F(x) - G(x) = C,$$

where C is an arbitrary constant. Therefore,

$$2F(x) = e^{-x^2} + C \qquad \text{and} \qquad 2G(x) = e^{-x^2} - C,$$

and the solution to the first problem is

$$v(x,t) = F(x+ct) + G(x-ct) = \frac{1}{2}\left[e^{-(x+ct)^2} + e^{-(x-ct)^2}\right]. \tag{14.5}$$

For problem (14.4), we use the initial conditions to write

$$w(x,0) = 0 = F(x) + G(x),$$

so that $G(x) = -F(x)$, and

$$\frac{\partial w}{\partial t}(x,0) = \frac{x}{(1+x^2)^2} = cF'(x) - cG'(x),$$

so that $cF'(x) - cG'(x) = 2cF'(x) = x/(1+x^2)^2$, and integrating, we have

$$2cF(x) = \frac{1}{2} \cdot \frac{-1}{1+x^2} + 2cC,$$

where C is an arbitrary constant. Therefore,

$$F(x) = \frac{-1}{4c(1+x^2)} + C \qquad \text{and} \qquad G(x) = \frac{1}{4c(1+x^2)} - C$$

and the solution to the second problem is

$$w(x,t) = \frac{1}{4c}\left[\frac{-1}{1+(x+ct)^2} + \frac{1}{1+(x-ct)^2}\right]. \tag{14.6}$$

The solution to the original boundary value–initial value problem is then

$$u(x,t) = v(x,t) + w(x,t);$$

that is,

$$u(x,t) = \frac{1}{2}\left[e^{-(x+ct)^2} + e^{-(x-ct)^2}\right] + \frac{1}{4c}\left[\frac{-1}{1+(x+ct)^2} + \frac{1}{1+(x-ct)^2}\right].$$

Exercise 14.7. XX

Use d'Alembert's solution of the wave equation to solve the boundary value–initial value problem

$$\begin{aligned} &\frac{\partial^2 u}{\partial x^2} = \frac{1}{c^2}\frac{\partial^2 u}{\partial t^2}, \qquad -\infty < x < \infty, \quad t > 0, \\ &u(x,0) = f(x), \\ &\frac{\partial u}{\partial t}(x,0) = g(x), \end{aligned}$$

with $f(x) = 0$ and $g(x) = 1/(1+x^2)$.

Solution. The boundary value–initial value problem for the displacement of an infinite vibrating string is

$$\begin{aligned} &\frac{\partial^2 u}{\partial x^2} = \frac{1}{c^2}\frac{\partial^2 u}{\partial t^2}, \qquad -\infty < x < \infty, \quad t > 0, \\ &u(x,0) = f(x), \\ &\frac{\partial u}{\partial t}(x,0) = g(x), \end{aligned}$$

and the general solution, that is, d'Alembert's solution to the wave equation, is

$$u(x,t) = \frac{1}{2}\left[f(x-ct) + f(x+ct)\right] + \frac{1}{2c}\int_{x-ct}^{x+ct} g(s)\, ds$$

for $-\infty < x < \infty$ and $t > 0$, and since $f(x) = 0$ for $-\infty < x < \infty$, then

$$u(x,t) = \frac{1}{2c}\int_{x-ct}^{x+ct} g(s)\, ds = \frac{1}{2c}\int_{x-ct}^{x+ct} \frac{1}{1+s^2}\, ds.$$

The solution is therefore

$$u(x,t) = \frac{1}{2c}\left[\tan^{-1}(x+ct) - \tan^{-1}(x-ct)\right]$$

for $-\infty < x < \infty$ and $t > 0$.

Exercise 14.8. XX

Use d'Alembert's solution to solve the boundary value–initial value problem for the wave equation:

$$\begin{aligned} &\frac{\partial^2 u}{\partial t^2} = \frac{\partial^2 u}{\partial x^2}, \quad 0 < x < 1, \quad t > 0, \\ &u(0,t) = 0, \\ &u(1,t) = 0, \\ &u(x,0) = 0, \\ &\frac{\partial u}{\partial t}(x,0) = 1. \end{aligned}$$

Solution. d'Alembert's solution to the wave equation is

$$u(x,t) = \frac{1}{2}\left[f_{\text{odd}}(x-ct) + f_{\text{odd}}(x+ct)\right] + \frac{1}{2c}\int_{x-ct}^{x+ct} g_{\text{odd}}(s)\,ds,$$

where f_{odd} and g_{odd} are the the odd 2-periodic extensions of f and g. For this problem we have $c = 1$ and $f(x) = 0$ for $0 < x < 1$, so that $f_{\text{odd}}(x) = 0$ for all $x \in \mathbb{R}$. Also, we have $g(x) = 1$ for $0 < x < 1$, so that

$$g_{\text{odd}}(x) = \begin{cases} 1 & \text{for} \quad 0 < x < 1, \\ -1 & \text{for} \quad -1 < x < 0, \end{cases} \quad \text{and} \quad g_{\text{odd}}(x+2) = g_{\text{odd}}(x) \quad \text{otherwise.}$$

We take an antiderivative of $g_{\text{odd}}(x)$ on the interval $[-1,1]$ as

$$G(x) = \begin{cases} x & \text{for} \quad 0 < x < 1, \\ -x & \text{for} \quad -1 < x < 0, \end{cases}$$

and $G(x+2) = G(x)$ otherwise. Therefore, the solution is

$$u(x,t) = \frac{1}{2}\left[G(x+t) - G(x-t)\right],$$

where G is as above.

Exercise 14.9. ✗✗

Use d'Alembert's solution to solve the boundary value-initial value problem for the wave equation:

$$\begin{aligned}
&\frac{\partial^2 u}{\partial t^2} = \frac{\partial^2 u}{\partial x^2}, \quad 0 < x < 1, \quad t > 0,\\
&u(0,t) = 0,\\
&u(1,t) = 0,\\
&u(x,0) = 0,\\
&\frac{\partial u}{\partial t}(x,0) = \sin \pi x.
\end{aligned}$$

Solution. As in previous problems, d'Alembert's solution to the wave equation is

$$u(x,t) = \frac{1}{2}\left[f_{\text{odd}}(x-ct) + f_{\text{odd}}(x+ct)\right] + \frac{1}{2c}\int_{x-ct}^{x+ct} g_{\text{odd}}(s)\,ds,$$

where f_{odd} and g_{odd} are the the odd 2-periodic extensions of f and g. For this problem, we have $c = 1$, and $f(x) = 0$ for $0 < x < 1$, so that $f_{\text{odd}}(x) = 0$ for all $x \in \mathbb{R}$. Also, we have $g(x) = \sin \pi x$ for $0 < x < 1$, so that

$$g_{\text{odd}}(x) = \sin \pi x$$

for $x \in \mathbb{R}$. An antiderivative of $g_{\text{odd}}(x)$ is given by

$$G(x) = -\frac{1}{\pi}\cos \pi x$$

for $x \in \mathbb{R}$. Therefore, the solution is

$$u(x,t) = \frac{1}{2\pi}\left[\cos \pi(x-t) - \cos \pi(x+t)\right] = \frac{1}{\pi}\sin \pi x \sin \pi t$$

for $0 < x < 1,\ t > 0$.

Exercise 14.10. ✗✗

Use d'Alembert's solution to solve the boundary value–initial value problem for the wave equation:

$$\begin{aligned}
&\frac{\partial^2 u}{\partial t^2} = 25\frac{\partial^2 u}{\partial x^2}, \quad -\infty < x < \infty, \quad t > 0,\\
&u(x,0) = x^2,\\
&\frac{\partial u}{\partial t}(x,0) = 3.
\end{aligned}$$

Solution. d'Alembert's solution to the wave equation is

$$u(x,t) = \frac{1}{2}\left[f(x+ct) + f(x-ct)\right] + \frac{1}{2c}\int_{x-ct}^{x+ct} g(s)\,ds,$$

where $c = 5$, $f(x) = x^2$, and $g(x) = 3$, so that

$$\begin{aligned} u(x,t) &= \frac{1}{2}\left[(x+5t)^2 + (x-5t)^2\right] + \frac{1}{10}\int_{x-5t}^{x+5t} 3\,ds \\ &= \frac{1}{2}\left(x^2 + 10xt + 25t^2 + x^2 - 10xt + 25t^2\right) + \frac{3}{10}\left(x + 5t - x + 5t\right) \\ &= x^2 + 25t^2 + 3t, \end{aligned}$$

and the solution is

$$u(x,t) = x^2 + 25t^2 + 3t$$

for $-\infty < x < \infty$ and $t > 0$.

Exercise 14.11. XX

Using the one-dimensional wave equation governing the small vertical displacements of a uniform vibrating string,

$$\frac{\partial^2 u}{\partial t^2} = c^2\frac{\partial^2 u}{\partial x^2}, \quad 0 < x < L, \quad t > 0,$$

derive the conservation of energy for a vibrating string,

$$\frac{dE}{dt} = \rho c^2 \left.\frac{\partial u}{\partial x}\frac{\partial u}{\partial t}\right|_0^L,$$

where the total energy E is the sum of the kinetic energy and the potential energy, and ρ is the linear density, that is, the mass per unit length of the string (assumed constant),

$$E(t) = \frac{\rho}{2}\int_0^L \left(\frac{\partial u}{\partial t}\right)^2 dx + \frac{\rho c^2}{2}\int_0^L \left(\frac{\partial u}{\partial x}\right)^2 dx.$$

Solution. The total energy (potential energy plus kinetic energy) of the string at time t is given by

$$E(t) = \frac{1}{2}\int_0^L \left[T\left(\frac{\partial u}{\partial x}\right)^2 + \rho\left(\frac{\partial u}{\partial t}\right)^2\right] dx,$$

that is,

$$E(t) = \frac{\rho}{2}\int_0^L \left[c^2\left(\frac{\partial u}{\partial x}\right)^2 + \left(\frac{\partial u}{\partial t}\right)^2\right] dx.$$

Using Leibniz's rule, we have

$$
\begin{aligned}
E'(t) &= \frac{d}{dt}\left(\frac{\rho}{2}\int_0^L \left[c^2\left(\frac{\partial u}{\partial x}\right)^2 + \left(\frac{\partial u}{\partial t}\right)^2\right] dx\right) \\
&= \frac{\rho}{2}\int_0^L \frac{\partial}{\partial t}\left[c^2\left(\frac{\partial u}{\partial x}\right)^2 + \left(\frac{\partial u}{\partial t}\right)^2\right] dx \\
&= \rho\int_0^L \left[c^2\frac{\partial u}{\partial x}\cdot\frac{\partial^2 u}{\partial t\partial x} + \frac{\partial u}{\partial t}\cdot\frac{\partial^2 u}{\partial t^2}\right] dx \\
&= \rho\int_0^L \left[c^2\frac{\partial u}{\partial x}\cdot\frac{\partial^2 u}{\partial x\partial t} + \frac{\partial u}{\partial t}\cdot c^2\frac{\partial^2 u}{\partial x^2}\right] dx \\
&= \rho c^2\int_0^L \frac{\partial}{\partial x}\left(\frac{\partial u}{\partial x}\cdot\frac{\partial u}{\partial t}\right) dx \\
&= \rho c^2\frac{\partial u}{\partial x}\cdot\frac{\partial u}{\partial t}\bigg|_0^L,
\end{aligned}
$$

that is,

$$
E'(t) = \rho c^2\frac{\partial u}{\partial x}\cdot\frac{\partial u}{\partial t}\bigg|_0^L.
$$

Note that if the string is fixed at both ends, so that

$$
\frac{\partial u}{\partial t}(0,t) = \frac{\partial u}{\partial t}(L,t) = 0
$$

for all $t > 0$, then $E'(t) = 0$ for all $t > 0$; that is, the total energy of the string is conserved.

Exercise 14.12. XXXX

Consider the boundary value–initial value problem (with $h > 0$) given by

$$\frac{\partial^2 u}{\partial t^2} = c^2 \frac{\partial^2 u}{\partial x^2}, \quad 0 < x < a, \quad t > 0,$$

$$\frac{\partial u}{\partial x}(0, t) - hu(0, t) = 0,$$

$$\frac{\partial u}{\partial x}(a, t) = 0,$$

$$u(x, 0) = f(x),$$

$$\frac{\partial u}{\partial t}(x, 0) = g(x).$$

Use separation of variables to do the following:

(a) Show that there are an infinite number of different frequencies of oscillation.

(b) Estimate the large frequencies of oscillation.

(c) Solve the boundary value–initial value problem.

Solution. Since the partial differential equation is linear and homogeneous and the boundary conditions are linear and homogeneous, we can use separation of variables. Assuming a solution of the form

$$u(x, t) = \phi(x) \cdot G(t), \quad 0 < x < a, \quad t > 0,$$

and separating variables, we have two ordinary differential equations:

$$\begin{aligned} &\phi''(x) + \lambda\phi(x) = 0, \quad 0 < x < a, \qquad && G''(t) + \lambda c^2 G(t) = 0, \quad t > 0. \\ &\phi'(0) - h\phi(0) = 0, && \\ &\phi'(a) = 0, && \end{aligned}$$

(a) We use the Rayleigh quotient to show that $\lambda > 0$ for all eigenvalues λ. Let λ be an eigenvalue of the Sturm-Liouville problem, and let $\phi(x)$ be the corresponding eigenfunction; then

$$-p(x)\phi(x)\phi'(x)\Big|_0^a = -\phi(a)\phi'(a) + \phi(0)\phi'(0) = h\phi(0)^2 \geqslant 0,$$

and since $q(x) = 0 \leqslant 0$ for all $0 \leqslant x \leqslant a$, then

$$\lambda = \frac{h\phi(0)^2 + \displaystyle\int_0^a \phi'(x)^2\,dx}{\displaystyle\int_0^a \phi(x)^2\,dx} \geqslant 0$$

since $p(x) = \sigma(x) = 1$ for $0 \leqslant x \leqslant a$. Note that if $\lambda = 0$, then

$$h\phi(0)^2 + \int_0^a \phi'(x)^2\,dx = 0$$

implies that

$$h\phi(0)^2 = 0 \qquad \text{and} \qquad \int_0^a \phi'(x)^2\,dx = 0.$$

Since $h > 0$, this implies that $\phi(0) = 0$, and since ϕ' is continuous on $[0, a]$, that $\phi'(x) = 0$ for $0 \leqslant x \leqslant a$. Therefore, $\phi(x)$ is constant on $[0, a]$, so that $\phi(x) = \phi(0) = 0$ for $0 < x < a$, and $\lambda = 0$ is not an eigenvalue. Hence, all of the eigenvalues λ of this Sturm-Liouville problem satisfy $\lambda > 0$.

If $\lambda > 0$, then $\lambda = \mu^2$, where $\mu \neq 0$, and the differential equation becomes $\phi'' + \mu^2\phi = 0$ with general solution

$$\phi(x) = A\cos\mu x + B\sin\mu x \quad \text{with} \quad \phi'(x) = -\mu A\sin\mu x + \mu B\cos\mu x$$

for $0 < x < a$.

Applying the first boundary condition gives us

$$\phi'(0) - h\phi(0) = -hA + B\mu = 0,$$

so that $B = (h/\mu)A$. Applying the second boundary condition yields

$$\phi'(a) = -\mu A\sin\mu a + \mu B\cos\mu a = A\left(-\mu\sin\mu a + h\cos\mu a\right) = 0,$$

and the boundary value problem has a nontrivial solution if and only if

$$\tan\mu a = \frac{h}{\mu}.$$

From Figure 14.1 it is clear that there are an infinite number of distinct eigenvalues $\lambda_n = \mu_n^2$, and that $\lim_{n\to\infty}\lambda_n = +\infty$.

(b) Note that since $\lim_{n\to\infty}\mu_n = +\infty$, then

$$\lim_{n\to\infty}\tan\mu_n a = \lim_{n\to\infty}\frac{h}{\mu_n} = 0,$$

and the roots of the equation

$$\tan\mu_n a = \frac{h}{\mu_n}$$

are approaching the roots of the equation $\tan\mu_n a = 0$; that is, for large n, we have

$$\mu_n \approx \frac{n\pi}{a},$$

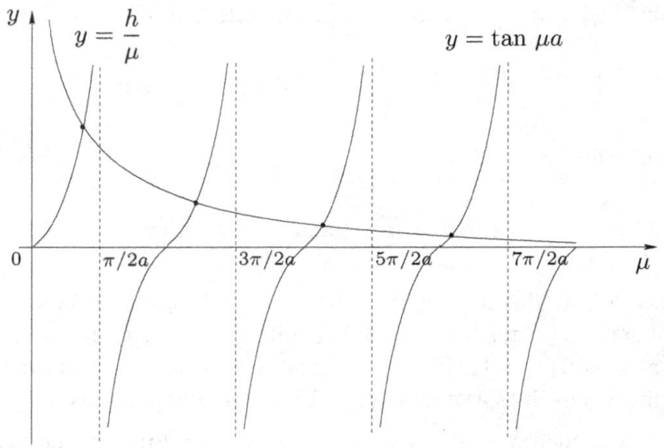

Figure 14.1 Intersection of graphs of $f(\mu) = \tan \mu a$ and $g(\mu) = h/\mu$.

and therefore for large n the eigenvalues are

$$\lambda_n = \mu_n^2 \approx \frac{n^2\pi^2}{a^2}.$$

The frequency of oscillation refers to the frequency arising from the solution of the corresponding time equation

$$G_n(t) = a_n \cos \mu_n ct + b_n \sin \mu_n ct,$$

and one period of oscillation corresponds to $\mu_n cT = 2\pi$; that is,

$$T = \frac{2\pi}{\mu_n c}.$$

Thus, the frequency of oscillation is

$$\nu = \frac{1}{T} = \frac{\mu_n c}{2\pi},$$

and for large n,

$$\nu \approx \frac{nc}{2a}.$$

(c) The solutions to the spatial problem are

$$\phi_n(x) = \cos \mu_n x + \frac{h}{\mu_n} \sin \mu_n x$$

for $0 < x < a$, and the corresponding solutions to the time equation are

$$G_n(t) = a_n \cos \mu_n ct + b_n \sin \mu_n ct$$

for $t > 0$.

Using the superposition principle, the function

$$u(x,t)=\sum_{n=1}^{\infty}\phi_n(x)\cdot G_n(t),$$

that is,

$$u(x,t)=\sum_{n=1}^{\infty}\left(\cos\mu_n x+\frac{h}{\mu_n}\sin\mu_n x\right)(a_n\cos\mu_n ct+b_n\sin\mu_n ct),$$

satisfies the partial differential equation and the boundary conditions.

Since the spatial problem is a regular Sturm-Liouville problem, the eigenfunctions are orthogonal on the interval $[0,a]$, and we use this fact to satisfy the initial conditions

$$f(x)=\sum_{n=1}^{\infty}a_n\phi_n(x)\qquad\text{and}\qquad g(x)=\sum_{n=1}^{\infty}b_n\mu_n c\phi_n(x),$$

where the generalized Fourier coefficients are given by

$$a_n=\frac{\displaystyle\int_0^a f(x)\phi_n(x)\,dx}{\displaystyle\int_0^a \phi_n(x)^2\,dx}\qquad\text{and}\qquad b_n=\frac{\displaystyle\int_0^a g(x)\phi_n(x)\,dx}{\displaystyle\mu_n c\int_0^a \phi_n(x)^2\,dx}$$

for $n\geqslant 1$.

Exercise 14.13. XXX

Solve the problem for a vibrating square membrane with side length 1, where the vibrations are governed by the following two-dimensional wave equation:

$$\frac{\partial^2 u}{\partial t^2}=\frac{1}{\pi^2}\left(\frac{\partial^2 u}{\partial x^2}+\frac{\partial^2 u}{\partial y^2}\right),\quad 0<x<1,\quad 0<y<1,\quad t>0,$$

$$u(0,y,t)=u(1,y,t)=0,$$

$$u(x,0,t)=u(x,1,t)=0,$$

$$u(x,y,0)=\sin\pi x\sin\pi y,$$

$$\frac{\partial u}{\partial t}(x,y,0)=\sin\pi x.$$

Solution. Separating variables, we write $u(x,y,t)=\phi(x,y)\cdot T(t)$, and substituting this into the wave equation, we get

$$\pi^2\frac{T''}{T}=\frac{1}{\phi}\left(\frac{\partial^2\phi}{\partial x^2}+\frac{\partial^2\phi}{\partial y^2}\right)=-\lambda$$

and obtain the two equations

$$T'' + \frac{\lambda}{\pi^2} T = 0 \qquad \text{and} \qquad \frac{\partial^2 \phi}{\partial x^2} + \frac{\partial^2 \phi}{\partial y^2} = -\lambda\phi.$$

Separating variables again in the second equation, we write $\phi(x, y) = X(x) \cdot Y(y)$, and substituting this into the equation, we obtain

$$\frac{X''}{X} + \frac{Y''}{Y} = -\lambda;$$

that is,

$$\frac{X''}{X} = -\frac{Y''}{Y} - \lambda = -\kappa,$$

where κ is a second separation constant. The boundary conditions give rise to the two boundary value problems

$$\begin{aligned} &X'' + \kappa X = 0, &\qquad &Y'' + (\lambda - \kappa)Y = 0,\\ &X(0) = 0, & &Y(0) = 0,\\ &X(1) = 0, & &Y(1) = 0. \end{aligned}$$

We find nontrivial solutions to the X equation first, since it involves only one separation constant.

As in previous problems, there are nontrivial solutions if and only if $\kappa_n = n^2\pi^2$ and the eigenfunctions are

$$X_n(x) = \sin n\pi x$$

for $n \geqslant 1$. For each $n \geqslant 1$, the function Y satisfies the boundary value problem

$$\begin{aligned} &Y'' + (\lambda - n^2\pi^2)Y = 0,\\ &Y(0) = 0,\\ &Y(1) = 0, \end{aligned}$$

and as in previous problems, this has nontrivial solutions if and only if

$$\lambda - n^2\pi^2 = m^2\pi^2,$$

that is, $\lambda = (n^2 + m^2)\pi^2$, and the eigenfunctions are

$$Y_m(y) = \sin m\pi y$$

for $m \geqslant 1$. For each $n, m \geqslant 1$, the function

$$\phi_{n,m}(x, y) = \sin n\pi x \cdot \sin m\pi y$$

satisfies the equation for ϕ, as well as the four boundary conditions.

The solutions of the equation

$$T'' + \frac{\lambda}{\pi^2} T = 0$$

corresponding to the separation constant $\lambda = (n^2 + m^2)\pi^2$ are

$$T_{n,m} = A_{n,m} \cos \sqrt{n^2 + m^2}\, t + B_{n,m} \sin \sqrt{n^2 + m^2}\, t$$

and for each $n, m \geqslant 1$, the function

$$u_{n,m}(x, y, t) = \phi_{n,m}(x, y) \cdot T_{n,m}(t),$$

that is,

$$u_{n,m} = \sin n\pi x \sin m\pi y \left(A_{n,m} \cos \sqrt{n^2 + m^2}\, t + B_{n,m} \sin \sqrt{n^2 + m^2}\, t \right),$$

satisfies the wave equation and all four boundary conditions. Using the superposition principle, we write the solution as

$$\begin{aligned} u(x, y, t) = \sum_{n=1}^{\infty} \sum_{m=1}^{\infty} & \sin n\pi x \sin m\pi y \\ & \times \left(A_{n,m} \cos \sqrt{n^2 + m^2}\, t + B_{n,m} \sin \sqrt{n^2 + m^2}\, t \right). \end{aligned}$$

We evaluate the constants $A_{n,m}$ and $B_{n,m}$ using the initial conditions. Setting $t = 0$ in the expression above for $u(x, y, t)$, we see that

$$\sin \pi x \sin \pi y = u(x, y, 0) = \sum_{n=1}^{\infty} \sum_{m=1}^{\infty} A_{n,m} \sin n\pi x \sin m\pi y,$$

so that

$$A_{n,m} = \begin{cases} 1, & \text{for } \quad n = m = 1, \\ 0, & \text{otherwise.} \end{cases}$$

Differentiating the expression for $u(x, y, t)$ with respect to t, and setting $t = 0$, we see that

$$\sin \pi x = \frac{\partial u}{\partial t}(x, y, 0) = \sum_{n=1}^{\infty} \sum_{m=1}^{\infty} \sqrt{n^2 + m^2}\, B_{n,m} \sin n\pi x \sin m\pi y,$$

that is,

$$\sin \pi x = \sum_{n=1}^{\infty} \sin n\pi x \left(\sum_{m=1}^{\infty} \sqrt{n^2 + m^2}\, B_{n,m} \sin m\pi y \right),$$

and for $n = 1$ we need

$$\sum_{m=1}^{\infty} \sqrt{1 + m^2}\, B_{1,m} \sin m\pi y = 1,$$

while for $n \neq 1$ we need

$$\sum_{m=1}^{\infty} \sqrt{n^2+m^2}\, B_{n,m} \sin m\pi y = 0.$$

Therefore, we may take $B_{n,m} = 0$ for all $n \neq 1$, while for $n = 1$ we want

$$\sqrt{1+m^2}\, B_{1,m}$$

to be the coefficients in the Fourier sine series of the function

$$f(x) = 1,\ 0 \leqslant x \leqslant 1;$$

that is,

$$B_{1,m} = \frac{2}{\sqrt{1+m^2}} \int_0^1 \sin m\pi y\, dy = \frac{2}{m\pi\sqrt{1+m^2}} [1-(-1)^m]$$

for $m \geqslant 1$. Therefore, the solution is

$$u(x,y,t) = \sin \pi x \sin \pi y \cos \sqrt{2}\pi t + \sum_{m=1}^{\infty} \frac{2[1-(-1)^m]}{m\pi\sqrt{1+m^2}} \sin \pi x\, \sin m\pi y\, \sin \sqrt{1+m^2}\, t$$

for $0 < x, y < 1,\ t > 0$.

Exercise 14.14. XXX

Solve the wave equation for a vibrating radially symmetric circular membrane

$$\frac{\partial^2 u}{\partial t^2} = \frac{4}{r}\frac{\partial}{\partial r}\left(r\frac{\partial u}{\partial r}\right), \quad 0 < r < 1, \quad t > 0,$$

$$u(1,t) = 0,$$

$$u(r,0) = 5J_0(z_3 r),$$

$$\frac{\partial u}{\partial t}(r,0) = 0,$$

where $J_0(z)$ denotes the Bessel function of the first kind of order zero, and z_n denotes the nth zero of $J_0(z)$.

Solution. We use separation of variables; assuming that $u(r,t) = \phi(r) \cdot T(t)$, the wave equation above becomes

$$\frac{4}{r}(r\phi')' \cdot T = \phi \cdot T'',$$

and dividing by $4\,\phi \cdot T$, the variables are separated, and we get

$$\frac{(r\phi'(r))'}{r\phi(r)} = \frac{T''(t)}{4\,T(t)}.$$

The two sides of this equation must be a constant, say $-\lambda$, which yields two ordinary differential equations

$$(r\phi')' + \lambda\, r\, \phi = 0, \quad 0 < r < 1,$$

$$T'' + 4\lambda\, T = 0, \quad t > 0.$$

The boundary condition $u(1, t) = 0$ for all $t > 0$ is satisfied if we require that

$$\phi(1) = 0.$$

Also, since $r = 0$ is a singular point of the differential equation for ϕ, we add the requirement

$$|\phi(r)| \quad \text{bounded at} \quad r = 0,$$

which is equivalent to requiring that $|u(r, t)|$ be bounded at $r = 0$. Thus, ϕ satisfies the boundary value problem

$$(r\phi')' + \lambda\, r\, \phi = 0, \quad 0 < r < 1,$$

$$\phi(1) = 0,$$

$$|\phi(r)| \quad \text{bounded at} \quad r = 0,$$

and it is easy to see using the Rayleigh quotient that $\lambda > 0$. We multiply the equation by r and recognize the equation

$$r^2\phi'' + r\phi' + \lambda\, r^2\phi = 0$$

as Bessel's equation of order zero, for which the function

$$\phi(r) = J_0(\sqrt{\lambda}\, r)$$

is the solution bounded at $r = 0$. To satisfy the boundary condition $\phi(1) = 0$, we must have

$$J_0(\sqrt{\lambda}) = 0$$

or

$$\sqrt{\lambda_n} = z_n, \quad n = 1, 2, \ldots ,$$

where z_n are the positive zeros of the function J_0. Therefore, the eigenvalues and eigenfunctions of the boundary value problem satisfied by $\phi(r)$ are

$$\lambda_n = z_n^2 \qquad \text{and} \qquad \phi_n(r) = J_0(\sqrt{\lambda_n}\, r)$$

for $n \geqslant 1$. For these values of λ which give a nontrivial solution to the boundary value problem for ϕ, the differential equation for T is

$$T''(t) + 4\lambda_n T(t) = 0$$

with general solution

$$T_n(t) = a_n \cos 2\sqrt{\lambda_n}\, t + b_n \sin 2\sqrt{\lambda_n}\, t$$

for $n \geqslant 1$. For each $n \geqslant 1$, the product solution

$$u_n(r,t) = \phi_n(r) \cdot T_n(t)$$

to the original partial differential equation satisfies the boundary condition $u(1,t) = 0$ and the boundedness condition $|u(r,t)|$ bounded at $r = 0$, for all $t > 0$.
Using the superposition principle, we write the solution as

$$u(r,t) = \sum_{n=1}^{\infty} J_0(\sqrt{\lambda_n}\, r) \left(a_n \cos 2\sqrt{\lambda_n}\, t + b_n \sin 2\sqrt{\lambda_n}\, t\right).$$

The initial conditions are satisfied if

$$u(r,0) = \sum_{n=1}^{\infty} a_n J_0(\sqrt{\lambda_n}\, r) = 5J_0\left(z_3\, r\right),$$

$$\frac{\partial u}{\partial t}(r,0) = \sum_{n=1}^{\infty} 2b_n \sqrt{\lambda_n} J_0(\sqrt{\lambda_n}\, r) = 0$$

for $0 < r < 1$. Using the fact that the eigenfunctions $\{J_0(\sqrt{\lambda_n}\, r)\}_{n \geqslant 1}$ are orthogonal on the interval $[0,1]$ with respect to the weight function $\sigma(r) = r$, we see that $a_n = 0$ for all $n \neq 3$, and $a_3 = 5$, while $b_n = 0$ for all $n \geqslant 1$. Therefore, the solution is

$$u(r,t) = 5J_0\left(z_3\, r\right) \cos 2z_3 t$$

for $0 < r < 1,\ t > 0$, where z_3 is the third positive zero of $J_0(z)$.

Exercise 14.15. XXXX

Find the solution for the vibrating circular membrane in polar coordinates:

$$\frac{\partial^2 u}{\partial t^2} = 100\left(\frac{\partial^2 u}{\partial r^2} + \frac{1}{r}\frac{\partial u}{\partial r}\right), \quad 0 < r < 1, \quad t > 0,$$

$$u(1,t) = 0,$$

$$u(r,0) = 1 - r^2,$$

$$\frac{\partial u}{\partial t}(r,0) = 1.$$

You may use the formula

$$\int_0^a x^{p+1} J_p\left(\frac{\alpha}{a}x\right)\,dx = \frac{a^{p+2}}{\alpha} J_{p+1}(\alpha),$$

where α is a positive zero of $J_p(z)$, the Bessel function of the first kind of order p.

Solution. Since $f(r) = 1 - r^2$ and $g(r) = 1$ are radially symmetric, we may assume that the solution does not depend on θ (we can show this by separating variables and applying periodicity conditions in θ). Also, we expect periodic functions in t, and to separate variables we write $u(r,t) = R(r) \cdot T(t)$, and as usual this leads to positive eigenvalues, so we obtain the radial problem

$$rR'' + R' + \lambda^2 rR = 0, \quad 0 < r < 1,$$
$$R(1) = 0,$$
$$|R(r)| < M, \text{ for } r \to 0^+,$$

where the last condition is added to exclude singular solutions (physically, a singular solution would correspond to a ripping membrane). The time problem gives

$$T'' + 100\lambda^2 T = 0, \quad t > 0.$$

The differential equation in the radial problem is Bessel's equation of order 0; hence, the solutions to the first problem are

$$R(r) = J_0(\lambda r), \quad r > 0,$$

where J_0 is the Bessel function of order 0 of the first kind. The boundary condition $u(1,t) = 0$ for all $t > 0$ can be satisfied by requiring that $R(1) = 0$, that is, $J_0(\lambda) = 0$, so that λ must be a root of the Bessel function J_0. Now, J_0 has infinitely many positive zeros, and we write them as

$$z_1 < z_2 < z_3 < \cdots < z_n < \cdots,$$

and therefore we have nontrivial solutions to the boundary value problem if and only if

$$\lambda_n^2 = z_n^2,$$

$n = 1, 2, 3, \ldots$. These are the eigenvalues of the boundary value problem, and the corresponding eigenfunctions are

$$R_n(r) = J_0\left(z_n r\right)$$

for $n = 1,\ 2,\ 3,\ \ldots$.

The solution to the differential equation for T corresponding to $\lambda_n = z_n$ is given by

$$T_n(t) = A_n \cos 10 z_n t + B_n \sin 10 z_n t,$$

and the functions

$$u_n(r, t) = \left(A_n \cos 10 z_n t + B_n \sin 10 z_n t\right) J_0(z_n r)$$

satisfy the wave equation and the boundary condition for each $n = 1, 2, \ldots$. Using the superposition principle, we write the solution as a Fourier-Bessel series

$$u(r, t) = \sum_{n=1}^{\infty} \left(A_n \cos 10 z_n t + B_n \sin 10 z_n t\right) J_0(z_n r), \tag{14.7}$$

and evaluate the coefficients A_n and B_n from the initial conditions. To do this, we need the orthogonality conditions

$$\int_0^1 r J_0(z_n r) J_0(z_m r)\, dr = 0$$

for $n \neq m$. To see this we recall that R_n and R_m satisfy the equations

$$\begin{aligned}(rR_n')' + z_n^2 r R_n &= 0,\\ (rR_m')' + z_m^2 r R_m &= 0,\end{aligned}$$

and multiplying the first equation by R_m and the second equation by R_n and subtracting, we get

$$(rR_n')' R_m - (rR_m')' R_n = (z_m^2 - z_n^2) r R_n R_m,$$

that is,

$$\left(r(R_m R_n' - R_n R_m')\right)' = (z_m^2 - z_n^2) r R_n R_m,$$

and integrating this last equation from 0 to 1, and using the fact that

$$R_m(1) = R_n(1) = 0,$$

we have

$$(z_n^2 - z_m^2)\int_0^1 rR_n(r)R_m(r)\,dr = 0$$

for $n \neq m$, and since $z_n \neq z_m$, we have

$$\int_0^1 rJ_0(z_n r)\,J_0(z_m r)\,dr = 0 \tag{14.8}$$

for $n \neq m$, and the eigenfunctions are orthogonal with respect to the weight function r on the interval $[0, 1]$.

To determine the coefficient A_n from the initial condition, we also need to know the value of

$$\int_0^1 rR_n(r)^2\,dr,$$

and we can determine this by considering the differential equation satisfied by R_n, namely,

$$(rR_n')' + z_n^2 rR_n = 0,$$

and multiplying this by $2rR_n'$ to get

$$\frac{d}{dr}\left[(rR_n')^2\right] + 2z_n^2 r^2 R_n R_n' = 0,$$

and integrating both terms we get

$$\left(rR_n'(r)\right)^2\Big|_0^1 + z_n^2\left[r^2R_n(r)^2\Big|_0^1 - \int_0^1 2rR_n(r)^2\,dr\right] = 0,$$

where we integrated by parts in the second integral. Since $R_n(1) = 0$, we get

$$R_n'(1)^2 - z_n^2\int_0^1 2rR_n(r)^2\,dr = 0,$$

that is,

$$\int_0^1 rR_n(r)^2\,dr = \frac{1}{2z_n^2}R_n'(1)^2 = \frac{1}{2}J_0'(z_n)^2 = \frac{1}{2}J_1(z_n)^2 \tag{14.9}$$

for $n = 1, 2, 3, \ldots$, where we have used the identity $J_0'(r) = -J_1(r)$.

Now we can use the initial conditions to determine the coefficients in the solution (14.7). Setting $t = 0$, multiplying by $rR_m(r)$, and integrating from 0 to 1, we get

$$\int_0^1 rf(r)R_m(r)\,dr = A_m\int_0^1 rR_m(r)^2\,dr = A_m\frac{J_1(z_m)^2}{2},$$

and since $f(r) = 1 - r^2$, we have

$$A_m = \frac{2}{J_1(z_m)^2}\int_0^1 r(1-r^2)R_m(r)\,dr = \frac{2}{J_1(z_m)^2}\int_0^1 r(1-r^2)J_0(z_m r)\,dr$$

for $m = 1, 2, 3, \ldots$. If we make the substitution $s = z_m r$ in the last integral, we get

$$\int_0^1 r(1-r^2)J_0(z_m r)\,dr = \frac{1}{z_m^4}\int_0^{z_m} s(z_m^2 - s^2)J_0(s)\,ds,$$

and integrating by parts with $u = z_m^2 - s^2$ and $dv = J_0(s)s\,ds$ so that

$$v = \int sJ_0(s)\,ds = sJ_1(s),$$

we get

$$\begin{aligned}\int_0^1 r(1-r^2)J_0(z_m r)\,dr &= \frac{2}{z_m^4}\int_0^{z_m} J_1(s)s^2\,ds \\ &= \frac{2}{z_m^4}s^2 J_2(s)\Big|_0^{z_m} = \frac{2}{z_m^2}J_2(z_m)\end{aligned}$$

for $m = 1, 2, 3, \ldots$, where we used the identity

$$\int x^{p+1}J_p(x)\,dx = x^{p+1}J_{p+1}(x) + C.$$

Therefore,

$$A_m = \frac{2}{J_1(z_m)^2}\int_0^1 r(1-r^2)J_0(z_m r)\,dr = \frac{4J_2(z_m)}{z_m^2 J_1(z_m)^2},$$

and finally, since z_m is a zero of J_0, from the identity

$$J_0(x) + J_2(x) = \frac{2}{x}J_1(x),$$

we have

$$A_m = \frac{8}{z_m^3 J_1(z_m)}$$

for $m = 1, 2, 3, \ldots$, and

$$1 - r^2 = f(r) = \sum_{n=1}^{\infty} \frac{8}{z_n^3 J_1(z_n)} J_0(z_n r), \quad 0 < r < 1$$

is the Fourier-Bessel expansion for the initial displacement.

To compute the B_n's, we differentiate (14.7) with respect to t and then set $t = 0$ to get

$$1 = g(r) = \frac{\partial u}{\partial t}(r, 0) = \sum_{n=1}^{\infty} 10 z_n B_n J_0(z_n r),$$

and a similar argument to that above shows that

$$B_m = \frac{1}{5z_m^2 J_1(z_m)}$$

for $m = 1, 2, 3, \ldots$. Therefore, the solution is

$$u(r,t) = \sum_{n=1}^{\infty} \frac{J_0(z_n r)}{5z_n^3 J_1(z_n)} \left[40\cos(10z_n t) + z_n \sin(10z_n t)\right]$$

for $0 < r < 1$, and $t > 0$.

Exercise 14.16. XXX

Find a solution for the vibrating circular membrane

$$\frac{\partial^2 u}{\partial t^2} = \frac{\partial^2 u}{\partial r^2} + \frac{1}{r}\frac{\partial u}{\partial r}, \quad 0 < r < 1, \quad t > 0,$$

$$u(1,t) = 0,$$

$$u(r,0) = 0,$$

$$\frac{\partial u}{\partial t}(r,0) = J_0(z_3 r),$$

where z_3 denotes the third root of the Bessel function J_0. You may use the formula

$$\int_0^a J_p^2\left(\frac{\alpha}{a}x\right) x\,dx = \frac{a^2}{2} J_{p+1}^2(\alpha),$$

where α is a positive zero of $J_p(z)$.

Solution. As in Exercise 14.15, the solution is

$$u(r,t) = \sum_{n=1}^{\infty} \left(A_n \cos z_n t + B_n \sin z_n t\right) J_0(z_n r),$$

where z_n is the nth positive root of the Bessel function J_0. In this case, however, $u(r,0) = f(r) = 0$ for $0 < r < 1$, so that $A_n = 0$ for all $n \geqslant 1$. We use the initial condition

$$\frac{\partial u}{\partial t}(r,0) = J_0(z_3 r), \quad 0 < r < 1,$$

and the orthogonality to determine the B_n's, as in Exercise 14.15, we have

$$B_n = \frac{2}{z_n J_1(z_n)^2} \int_0^1 r J_0(z_3 r) J_0(z_n r)\,dr = 0$$

for all $n \neq 3$, while for $n = 3$, we have

$$B_3 = \frac{2}{z_3 J_1(z_3)^2} \int_0^1 r J_0(z_3 r)^2 \, dr = \frac{2}{z_3 J_1(z_3)^2} \cdot \frac{1}{2} J_1(z_3)^2 = \frac{1}{z_3}$$

and the solution is

$$u(r,t) = \frac{1}{z_3} J_0(z_3 r) \sin z_3 t$$

for $0 < r < 1$, and $t > 0$.

Exercise 14.17. XXX
Solve the wave equation on a disk of radius $a > 0$:

$$\frac{\partial^2 u}{\partial t^2} = c^2 \nabla^2 u \qquad \text{subject to} \qquad \frac{\partial u}{\partial r}(a, \theta, t) = 0,$$

with initial conditions

$$u(r, \theta, 0) = 0, \qquad \frac{\partial u}{\partial t}(r, \theta, 0) = \beta(r).$$

Solution. Since neither the boundary conditions nor the initial conditions depend on the variable θ, we look for a solution that is also independent of θ, say $u = u(r,t)$. In this case the problem becomes

$$\frac{\partial^2 u}{\partial t^2} = \frac{c^2}{r} \frac{\partial}{\partial r}\left(r \frac{\partial u}{\partial r}\right) \qquad \text{subject to} \qquad \frac{\partial u}{\partial r}(a, t) = 0,$$

with initial conditions

$$u(r, 0) = 0, \qquad \frac{\partial u}{\partial t}(r, 0) = \beta(r).$$

Separating variables, we write $u(r,t) = R(r) \cdot T(t)$; then R and T satisfy the following boundary value and initial value problems, respectively:

$$\begin{aligned} &(rR')' + \lambda r R = 0, \quad 0 < r < a, \qquad && T'' + \lambda c^2 T = 0, \quad t > 0, \\ &R'(a) = 0, && T(0) = 0. \\ &|R(0)| < \infty, && \end{aligned}$$

We solve the singular Sturm-Liouville problem for R first. The Rayleigh quotient is

$$\lambda = \frac{-rRR'\Big|_0^a + \displaystyle\int_0^a r(R')^2 \, dr}{\displaystyle\int_0^a r\, R^2 \, dr},$$

and from the boundedness and boundary conditions,

$$-rRR'\Big|_0^a = -aR(a)R'(a) = 0, \tag{14.10}$$

so that $\lambda \geqslant 0$; that is, there are no negative eigenvalues. Also, we note that using the boundary and boundedness conditions in (14.10), the proof of the orthogonality of the eigenfunctions for this singular Sturm-Liouville problem follows exactly as in the proof of Theorem 4.5.

Case (i): If $\lambda = 0$, the differential equation is

$$(rR')' = 0,$$

with general solution

$$R(r) = A\log r + B.$$

Applying the boundedness condition, we have $A = 0$, and the eigenfunction is

$$R_0(r) = 1$$

for $0 < r < a$.

Case (ii): If $\lambda > 0$, the differential equation is

$$(rR')' + \lambda rR = 0,$$

which is Bessel's equation of order 0, with general solution

$$R(r) = AJ_0(\sqrt{\lambda}r) + BY_0(\sqrt{\lambda}r).$$

Applying the boundedness condition, we have $B = 0$, and the solution is

$$R(r) = AJ_0(\sqrt{\lambda}r),$$

and applying the boundary condition, we have

$$R'(a) = A\sqrt{\lambda}J_0'(\sqrt{\lambda}a) = 0,$$

so that $\sqrt{\lambda}a = \omega_n$, the nth positive root of $J_0'(z)$.

The eigenvalues and corresponding eigenfunctions in this case are

$$\lambda_n = \left(\frac{\omega_n}{a}\right)^2 \qquad \text{and} \qquad R_n(r) = J_0\left(\sqrt{\lambda_n}r\right)$$

for $n = 1, 2, 3, \ldots$. The corresponding time equation is

$$\begin{aligned} &T''(t) + \lambda c^2 T = 0, \quad t > 0, \\ &T(0) = 0. \end{aligned}$$

If $\lambda = 0$, the equation is $T''(t) = 0$, with general solution

$$T(t) = At + B.$$

From the initial condition we have $T(0) = B = 0$, and we may take $A = 1$, so that

$$T_0(t) = t$$

for $t > 0$.
If $\lambda > 0$, the equation is $T'' + \lambda_n c^2 T = 0$ with general solution

$$T(t) = A\cos\sqrt{\lambda_n}ct + B\sin\sqrt{\lambda_n}ct,$$

and from the initial condition we have $T(0) = A = 0$, and we may take

$$T_n(t) = \sin\sqrt{\lambda_n}ct$$

for $t > 0$. Using the superposition principle, we write

$$u(r,t) = c_0 t + \sum_{n=1}^{\infty} c_n J_0(\sqrt{\lambda_n}r) \cdot \sin\sqrt{\lambda_n}ct$$

for $0 < r < a$, $t > 0$. Finally, applying the nonhomogeneous initial condition, we have

$$\beta(r) = \frac{\partial u}{\partial t}(r,0) = c_0 + \sum_{n=1}^{\infty} c_n J_0(\sqrt{\lambda_n}\,r)\sqrt{\lambda_n}\,c$$

for $0 < r < a$, and using the orthogonality of the eigenfunctions (they are the eigenfunctions for a singular Sturm-Liouville problem)

$$c_0 = \frac{\displaystyle\int_0^a \beta(r)\,r\,dr}{\displaystyle\int_0^a r\,dr} = \frac{2}{a^2}\int_0^a \beta(r)\,r\,dr$$

and

$$c_n = \frac{\displaystyle\int_0^a \beta(r)J_0(\sqrt{\lambda_n}\,r)\,r\,dr}{\displaystyle\sqrt{\lambda_n}\,c\int_0^a J_0(\sqrt{\lambda_n}\,r)^2 r\,dr}$$

for $n = 1, 2, 3, \ldots$.

Exercise 14.18. XXXX

Solve the wave equation for a "pie-shaped" membrane of radius a and angle $\pi/3$ $(= 60^\circ)$:

$$\frac{\partial^2 u}{\partial t^2} = c^2 \nabla^2 u.$$

Show that the eigenvalues are all positive. Determine the natural frequencies of oscillation if the boundary conditions are

$$u(r, 0, t) = 0, \qquad u\left(r, \frac{\pi}{3}, t\right) = 0, \qquad \frac{\partial u}{\partial r}(a, \theta, t) = 0.$$

Solution. The wave equation in polar coordinates is

$$\frac{\partial^2 u}{\partial t^2} = c^2 \left(\frac{1}{r}\frac{\partial}{\partial r}\left(r\frac{\partial u}{\partial r}\right) + \frac{1}{r^2}\frac{\partial^2 u}{\partial \theta^2}\right),$$

and assuming a solution of the form $u(r, \theta, t) = \Psi(r, \theta) \cdot T(t)$, we have

$$\frac{T''}{c^2 T} = \frac{1}{\Psi}\left(\frac{1}{r}\frac{\partial}{\partial r}\left(r\frac{\partial \Psi}{\partial r}\right) + \frac{1}{r^2}\frac{\partial^2 \Psi}{\partial \theta^2}\right) = -\lambda,$$

where λ is the separation constant. This gives two equations, one of which is the Helmholtz equation,

$$\nabla^2 \Psi = -\lambda \Psi,$$

and we know from Theorem 5.7 that the eigenvalues are nonnegative; that is, $\lambda \geqslant 0$. Now, assuming that $\Psi(r, \theta) = R(r) \cdot \Theta(\theta)$, and separating variables, we have

$$\frac{T''}{c^2 T} = \frac{1}{rR}\left(rR'\right)' + \frac{1}{r^2 \Theta}\Theta'' = -\lambda,$$

and therefore,

$$\frac{\Theta''}{\Theta} = -\lambda r^2 - \frac{r}{R}\left(rR'\right)' = -\tau,$$

where τ is another separation constant.

We can satisfy the boundary conditions in θ,

$$u(r, 0, t) = 0 \qquad \text{and} \qquad u\left(r, \frac{\pi}{3}, t\right) = 0$$

for all $0 < r < a$ and $t > 0$, by requiring that

$$\Theta(0) = 0 \qquad \text{and} \qquad \Theta\left(\frac{\pi}{3}\right) = 0.$$

Also, if the solution is to be bounded we need to require the boundedness condition

$$|u(0, \theta, t)| < \infty,$$

for all $0 \leqslant \theta \leqslant \pi/3$, and $t > 0$. We can satisfy the boundedness condition as well as the boundary condition that

$$\frac{\partial u}{\partial r}(a, \theta, t) = 0$$

for $0 \leqslant \theta \leqslant \pi/3$ and $t > 0$, by requiring that

$$R'(a) = 0 \qquad \text{and} \qquad |R(0)| < \infty.$$

Therefore, R and Θ satisfy the boundary value problems

$$\begin{aligned} &r(rR')' + (\lambda r^2 - \tau)R = 0, \quad 0 < r < a, && \Theta'' + \tau\Theta = 0, \quad 0 \leqslant \theta \leqslant \frac{\pi}{3}, \\ &R'(a) = 0, && \Theta(0) = 0, \\ &|R(0)| < \infty, && \Theta\left(\frac{\pi}{3}\right) = 0, \end{aligned}$$

while T satisfies the differential equation

$$T'' + \lambda c^2 T = 0, \quad t > 0.$$

Since the problem for Θ has a complete set of homogeneous boundary conditions, we solve this first. The eigenvalues and corresponding eigenfunctions are

$$\tau_m = (3m)^2 \qquad \text{and} \qquad \Theta_m(\theta) = \sin 3m\theta$$

for $m = 1, 2, \ldots$. The boundary value problem for the corresponding functions $R(r)$ is

$$\begin{aligned} &r(rR')' + (\lambda r^2 - 9m^2)R = 0, \quad 0 < r < a, \\ &R'(a) = 0, \\ &|R(0)| < \infty. \end{aligned}$$

If $\lambda = 0$, the differential equation is

$$r(rR')' - 9m^2 R = 0,$$

which is a Cauchy-Euler equation with solution

$$R(r) = Ar^{3m} + \frac{B}{r^{3m}}.$$

Applying the boundedness condition, we have $B = 0$, and the solution is

$$R(r) = Ar^{3m} \qquad \text{with} \qquad R'(r) = 3mAr^{3m-1},$$

and since $m \geqslant 1$, $R'(a) = 0$ implies that $A = 0$. Thus, $\lambda = 0$ is not an eigenvalue.

If $\lambda > 0$, making the substitution $z = \sqrt{\lambda}\, r$, the differential equation becomes

$$z^2 \frac{d^2 R}{dz^2} + z\frac{dR}{dz} + (z^2 - 9m^2)R = 0,$$

which is Bessel's equation of order $3m$ for $m = 1, 2, \ldots$, and the general solution to the R-equation is

$$R(r) = c_1 J_{3m}(\sqrt{\lambda}\, r) + c_2 Y_{3m}(\sqrt{\lambda}\, r),$$

where $J_{3m}(\sqrt{\lambda}\, r)$ is the Bessel function of the first kind of order $3m$, and $Y_{3m}(\sqrt{\lambda}\, r)$ is the Bessel function of the second kind of order $3m$. Applying the boundedness condition $|R(0)| < \infty$, $c_2 = 0$, and the solutions are

$$R(r) = c_1 J_{3m}(\sqrt{\lambda}\, r)$$

for $0 \leqslant r \leqslant a$. Applying the boundary condition $R'(a) = 0$, we have

$$R'(a) = c_1 \sqrt{\lambda}\, J'_{3m}(\sqrt{\lambda}\, a) = 0,$$

and we have a nontrivial solution if and only if $J'_{3m}(\sqrt{\lambda}\, a) = 0$; that is,

$$\sqrt{\lambda}\, a = \omega_{3m,n}$$

for $n = 1, 2, 3, \ldots$, where $\omega_{3m,n}$ denotes the value of z for which $J_{3m}(z)$ has minima and maxima in increasing order, that is, the positive roots of $J'_{3m}(z)$. Therefore, the eigenvalues are

$$\lambda_{3m,n} = \left(\frac{\omega_{3m,n}}{a}\right)^2$$

with corresponding eigenfunctions

$$R_{3m,n} = J_{3m}\left(\sqrt{\lambda_{3m,n}}\, r\right)$$

for $m \geqslant 1$, $n \geqslant 1$. We note that $\lambda_{3m,n} > 0$ and for a fixed m, the eigenfunctions

$$R_{3m,n} = J_{3m}\left(\frac{\omega_{3m,n}}{a} r\right), \quad n = 1, 2, 3, \ldots$$

are orthogonal on the interval $0 \leqslant r \leqslant a$ with weight function $\sigma(r) = r$, so that

$$\int_0^a J_{3m}\left(\frac{\omega_{3m,k}}{a} r\right) J_{3m}\left(\frac{\omega_{3m,\ell}}{a} r\right) r\, dr = 0$$

for $k \neq \ell$. The corresponding solutions to the time equation

$$T'' + \lambda_{3m,n} c^2 T = 0, \quad t \geqslant 0,$$

are given by

$$T_{3m,n}(t) = A_{3m,n} \cos\left(\sqrt{\lambda_{3m,n}}\, ct\right) + B_{3m,n} \sin\left(\sqrt{\lambda_{3m,n}}\, ct\right), \quad t \geqslant 0.$$

From the superposition principle, the solution can be written as

$$u(r,\theta,t) = \sum_{m=1}^{\infty}\sum_{n=1}^{\infty} J_{3m}\left(\sqrt{\lambda_{3m,n}}\, r\right) \sin 3m\theta$$

$$\times \left[A_{3m,n}\cos\left(\sqrt{\lambda_{3m,n}}\, ct\right) + B_{3m,n}\sin\left(\sqrt{\lambda_{3m,n}}\, ct\right)\right],$$

where the coefficients $A_{3m,n}$ and $B_{3m,n}$ can be determined from any initial conditions. Thus, the frequencies of oscillation $\nu_{m,n}$ satisfy

$$\frac{\sqrt{\lambda_{3m,n}}\, c}{\nu_{m,n}} = 2\pi;$$

that is,

$$\nu_{m,n} = \frac{\sqrt{\lambda_{3m,n}}\, c}{2\pi} = \frac{\omega_{3m,n}\, c}{2\pi a}$$

for $m \geqslant 1$ and $n \geqslant 1$.

Exercise 14.19. XXXX

(a) Show that small displacements $u(x,t)$ of a hanging chain of length a are governed by the boundary value–initial value problem

$$\frac{\partial^2 u}{\partial t^2} = g\left(x\frac{\partial^2 u}{\partial x^2} + \frac{\partial u}{\partial x}\right), \quad 0 < x < a, \quad t > 0,$$

$$u(a,t) = 0,$$

$$u(x,t) \text{ bounded, as } x \to 0^+,$$

$$u(x,0) = f(x),$$

$$\frac{\partial u}{\partial t}(x,0) = v(x),$$

where g is the acceleration due to gravity (assumed constant).

(b) Solve the boundary value–initial value problem above.

Solution.

(a) First we give a more detailed description of the **hanging chain problem**. A flexible chain of length a and density (mass per unit length) ρ is fixed at the upper end ($x = a$) and allowed to make small vibrations in a vertical plane. We let $u(x,t)$ be the deflection from the vertical in this plane. In the equilibrium

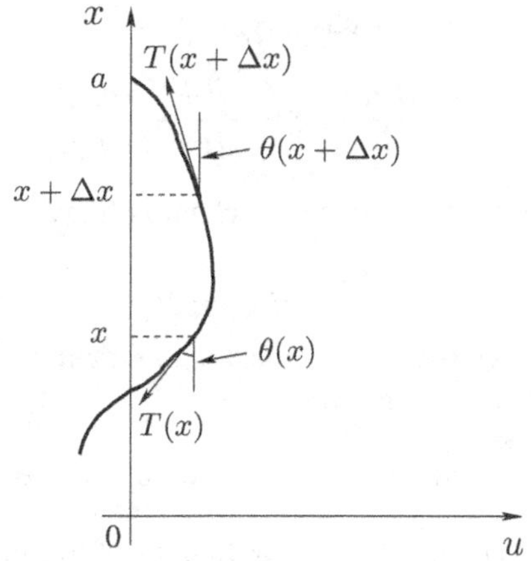

Figure 14.2 The hanging chain.

position, the weight of the chain below a point x is equal to the tension in the chain:

$$T_0(x) = \rho g x,$$

where g is the gravitational acceleration (assumed constant and acting vertically downward).

At time $t > 0$, the horizontal displacement of the chain at the point x is given by $u = u(x, t)$, and applying Newton's second law to the portion of the chain between x and $x + \Delta x$, we obtain

$$\rho \Delta x \frac{\partial^2 u}{\partial t^2} \doteq T(x + \Delta x)\sin(\theta(x + \Delta x)) - T(x)\sin(\theta(x)), \tag{14.11}$$

where $\theta(x)$ is the angle between the vertical and the tension $T(x)$, as in Figure 14.2. For small displacements, we have

$$\sin(\theta(x)) \doteq \frac{\partial u}{\partial x}(x, t) \qquad \text{and} \qquad \sin(\theta(x + \Delta x)) \doteq \frac{\partial u}{\partial x}(x + \Delta x, t),$$

and dividing by Δx in (14.11), we have

$$\rho \frac{\partial^2 u}{\partial t^2} \doteq \frac{1}{\Delta x}\left[T(x + \Delta x)\frac{\partial u}{\partial x}(x + \Delta x, t) - T(x)\frac{\partial u}{\partial x}(x, t)\right]. \tag{14.12}$$

Letting $\Delta x \to 0$ in (14.12), in the limit we get the equation

$$\rho \frac{\partial^2 u}{\partial t^2} = \frac{\partial}{\partial x}\left[T(x)\frac{\partial u}{\partial x}\right].$$

For small displacements, the tension $T(x)$ is approximately the equilibrium tension $T_0(x) = \rho g x$, and the partial differential equation governing *small* displacements of the hanging chain is

$$\frac{\partial^2 u}{\partial t^2} = g\left(x\frac{\partial^2 u}{\partial x^2} + \frac{\partial u}{\partial x}\right), \tag{14.13}$$

for $0 < x < a$ and $t > 0$. Since the chain is fixed at $x = a$, one boundary condition is

$$u(a,t) = 0$$

for $t > 0$. Also, since the displacement is to remain bounded as $x \to 0^+$, we require the boundedness condition

$$|u(x,t)| < \infty, \quad \text{for all} \quad t > 0$$

as $x \to 0^+$. Finally, to obtain a unique solution to this problem, we need to use the initial conditions,

$$u(x,0) = f(x),$$

the initial shape of the chain, and

$$\frac{\partial u}{\partial t}(x,0) = v(x),$$

the initial velocity of the chain. The displacement of the hanging chain satisfies the boundary value–initial value problem:

$$\begin{aligned}
&\frac{\partial^2 u}{\partial t^2} = g\left(x\frac{\partial^2 u}{\partial x^2} + \frac{\partial u}{\partial x}\right), \quad 0 < x < a, \quad t > 0,\\
&u(a,t) = 0,\\
&\lim_{x\to 0^+} |u(x,t)| < \infty,\\
&u(x,0) = f(x),\\
&\frac{\partial u}{\partial t}(x,0) = v(x).
\end{aligned}$$

(b) Since the partial differential equation and the boundary condition are both linear and homogeneous, we can use separation of variables.

Step 1: Assuming a solution of the form $u(x,t) = X(x)\cdot G(t)$ and substituting this into the partial differential equation, we get

$$X \cdot G'' = g(xX'' \cdot G + X' \cdot G),$$

and separating variables gives us

$$\frac{xX'' + X'}{X} = \frac{G''}{gG} = -\lambda,$$

where λ is the separation constant. We obtain the two problems

$$xX''(x) + X'(x) + \lambda X(x) = 0, 0 < x < a, \quad G''(t) + \lambda g\, G(t) = 0, \; t > 0,$$
$$X(a) = 0,$$
$$|X(0)| < \infty.$$

The spatial problem can be written in Sturm-Liouville form,

$$\frac{d}{dx}\left(x\frac{dX}{dx}\right) + \lambda X = 0,$$

where $p(x) = x$, $q(x) = 0$, and $\sigma(x) = 1$, for $0 < x < a$. Note that this is a singular Sturm-Liouville problem since $p(0) = 0$, and the boundary conditions are not of Sturm-Liouville type.

However, if λ is an eigenvalue of this problem with corresponding eigenfunction X, the relationship between λ and X still holds; that is, λ is given by the Rayleigh quotient

$$\lambda = R(X) = \frac{-xX(x)X'(x)\Big|_0^a + \displaystyle\int_0^a xX'(x)^2\,dx}{\displaystyle\int_0^a X(x)^2\,dx},$$

and from the boundary condition and the boundedness condition we have

$$\lambda = \frac{\displaystyle\int_0^a xX'(x)^2\,dx}{\displaystyle\int_0^a X(x)^2\,dx} \geqslant 0,$$

so that all of the eigenvalues of this singular Sturm-Liouville problem are nonnegative. We leave it as an exercise to show that $\lambda = 0$ is not an eigenvalue, so that all of the eigenvalues are positive.

Therefore, we can write $\lambda = \mu^2$ where $\mu \neq 0$, and the boundary value problem for X becomes

$$xX'' + X' + \mu^2 X = 0, \quad 0 < x < a,$$
$$X(a) = 0,$$
$$|X(0)| < \infty.$$

We can transform this into Bessel's equation of order zero, by letting

$$s = 2\sqrt{x} \qquad \text{and} \qquad \varphi(s) = X(x(s));$$

then from the chain rule, we have

$$\frac{dX}{dx} = \frac{d\varphi}{ds} \cdot \frac{ds}{dx} = \frac{1}{\sqrt{x}} \cdot \frac{d\varphi}{ds}$$

and

$$\frac{d^2X}{dx^2} = \frac{d}{dx}\left(\frac{1}{\sqrt{x}} \cdot \frac{d\varphi}{ds}\right) = \frac{1}{x} \cdot \frac{d^2\varphi}{ds^2} - \frac{1}{2x^{\frac{3}{2}}} \cdot \frac{d\varphi}{ds}.$$

Therefore,

$$\frac{dX}{dx} = \frac{2}{s} \cdot \frac{d\varphi}{ds},$$

$$\frac{d^2X}{dx^2} = \frac{4}{s^2} \cdot \frac{d^2\varphi}{ds^2} - \frac{4}{s^3} \cdot \frac{d\varphi}{ds},$$

and

$$xX'' + X' + \mu^2 X = 0$$

implies that

$$\frac{s^2}{4}\left(\frac{4}{s^2} \cdot \frac{d^2\varphi}{ds^2} - \frac{4}{s^3} \cdot \frac{d\varphi}{ds}\right) + \frac{2}{s} \cdot \frac{d\varphi}{ds} + \mu^2\varphi = 0,$$

that is,

$$\frac{d^2\varphi}{ds^2} - \frac{1}{s} \cdot \frac{d\varphi}{ds} + \frac{2}{s} \cdot \frac{d\varphi}{ds} + \mu^2\varphi = 0,$$

so that

$$s^2\frac{d^2\varphi}{ds^2} + s\frac{d\varphi}{ds} + \mu^2 s^2\varphi = 0,$$

which is Bessel's equation of order zero. The general solution is

$$\varphi(s) = AJ_0(\mu s) + BY_0(\mu s),$$

and from the boundedness condition at $x = 0$, we must have $B = 0$, so that

$$\varphi(s) = AJ_0(\mu s)$$

for $0 < s < 2\sqrt{a}$.

Applying the boundary condition $\varphi(2\sqrt{a}) = 0$, we have a nontrivial solution if and only if

$$\varphi(2\sqrt{a}) = AJ_0(\mu 2\sqrt{a}) = 0$$

and $A \neq 0$, that is, if and only if $\mu_n 2\sqrt{a} = z_n$, the nth positive zero of J_0. The eigenvalues are

$$\lambda_n = \mu_n^2 = \left(\frac{z_n}{2\sqrt{a}}\right)^2,$$

and the corresponding eigenfunctions are

$$\varphi_n(s) = J_0\left(\frac{z_n s}{2\sqrt{a}}\right)$$

for $n \geqslant 1$. Note that the functions

$$\left\{J_0\left(\frac{z_n s}{2\sqrt{a}}\right)\right\}_{n\geqslant 1}$$

are orthogonal on the interval $0 \leqslant s \leqslant 2\sqrt{a}$ with respect to the weight function $w(s) = s$, so that

$$\int_0^{2\sqrt{a}} J_0\left(\frac{z_n s}{2\sqrt{a}}\right) J_0\left(\frac{z_m s}{2\sqrt{a}}\right) s\,ds = 0$$

if $n \neq m$.

Since $x = s^2/4$, then $dx = \frac{1}{2}s\,ds$, and we have

$$\int_0^a J_0\left(z_n\sqrt{\frac{x}{a}}\right) J_0\left(z_m\sqrt{\frac{x}{a}}\right) dx = 0$$

if $n \neq m$. Therefore, the eigenfunctions

$$X_n(x) = J_0\left(z_n\sqrt{\frac{x}{a}}\right)$$

for $n \geqslant 1$ are orthogonal on the interval $0 \leqslant x \leqslant a$ with respect to the weight function $\sigma(x) = 1$.

Note that after the substitution $x = s^2/4$, the normalization

$$\int_0^{2\sqrt{a}} J_0^2\left(\frac{z_n s}{2\sqrt{a}}\right) s\,ds = 2aJ_1^2(z_n)$$

becomes

$$\int_0^a J_0^2\left(z_n\sqrt{\frac{x}{a}}\right) dx = aJ_1^2(z_n)$$

for $n \geqslant 1$. The corresponding time equation is

$$G'' + g\left(\frac{z_n}{2\sqrt{a}}\right)^2 G = 0,$$

with general solution

$$G_n(t) = A_n \cos\left(\sqrt{\frac{g}{a}}\frac{z_n}{2}t\right) + B_n \sin\left(\sqrt{\frac{g}{a}}\frac{z_n}{2}t\right)$$

for $n \geqslant 1$. Using the superposition principle we can write the solution as

$$u(x,t) = \sum_{n=1}^{\infty} X_n(x) \cdot G_n(t),$$

that is,

$$u(x,t) = \sum_{n=1}^{\infty} J_0\left(z_n\sqrt{\frac{x}{a}}\right)\left[A_n \cos\left(\sqrt{\frac{g}{a}}\frac{z_n}{2}t\right) + B_n \sin\left(\sqrt{\frac{g}{a}}\frac{z_n}{2}t\right)\right]$$

for $0 < x < a,\ t > 0$.

Finally, we use the initial conditions and the orthogonality of the eigenfunctions to determine the constants A_n and B_n.

Step 2: For $t = 0$, we have

$$f(x) = u(x,0) = \sum_{n=1}^{\infty} A_n J_0\left(z_n\sqrt{\frac{x}{a}}\right)$$

with

$$A_n = \frac{\displaystyle\int_0^a f(x) J_0\left(z_n\sqrt{\frac{x}{a}}\right) dx}{\displaystyle\int_0^a J_0^2\left(z_n\sqrt{\frac{x}{a}}\right) dx} = \frac{\displaystyle\int_0^a f(x) J_0\left(z_n\sqrt{\frac{x}{a}}\right) dx}{a J_1^2(z_n)}$$

for $n \geqslant 1$, and

$$v(x) = \frac{\partial u}{\partial t}(x,0) = \sum_{n=1}^{\infty} B_n\sqrt{\frac{g}{a}}\frac{z_n}{2} J_0\left(z_n\sqrt{\frac{x}{a}}\right)$$

with

$$B_n = \frac{2}{z_n J_1^2(z_n)\sqrt{ga}} \int_0^a v(x) J_0\left(z_n\sqrt{\frac{x}{a}}\right) dx$$

for $n \geqslant 1$.

Exercise 14.20. ✗

Show that the equation

$$\sin\theta \frac{d}{d\theta}\left(\sin\theta\frac{dS}{d\theta}\right) + (\nu \sin^2\theta - \mu)S = 0$$

for $0 \leq \theta \leq \pi$, where ν and μ are constant, can be transformed into the associated Legendre equation

$$(1-x^2)\frac{d^2v}{dx^2} - 2x\frac{dv}{dx} + \left(\nu - \frac{\mu}{1-x^2}\right)v = 0$$

by means of the transformation $x = \cos\theta$, where $v(x) = S(\theta)$.

Solution. Set $x = \cos\theta$ and write $S(\theta) = v\big(x(\theta)\big)$; then from the chain rule,

$$\frac{dS}{d\theta} = \frac{dv}{d\theta} = \frac{dv}{dx}\frac{dx}{d\theta} = -\sin\theta\frac{dv}{dx} \tag{14.14}$$

and

$$\frac{d^2S}{d\theta^2} = \frac{d}{d\theta}\left(\frac{dS}{d\theta}\right) = -\cos\theta\frac{dv}{dx} - \sin\theta\frac{d^2v}{dx^2}\frac{dx}{d\theta} = -\cos\theta\frac{dv}{dx} + \sin^2\theta\frac{d^2v}{dx^2}.$$

Therefore,

$$\begin{aligned}\sin\theta\frac{d}{d\theta}\left(\sin\theta\frac{dS}{d\theta}\right) &= \sin^2\theta\frac{d^2S}{d\theta^2} + \sin\theta\cos\theta\frac{dS}{d\theta}\\ &= -2\sin^2\theta\cos\theta\frac{dv}{dx} + \sin^4\theta\frac{d^2v}{dx^2}\end{aligned}$$

and

$$\sin\theta\frac{d}{d\theta}\left(\sin\theta\frac{dS}{d\theta}\right) + (\nu\sin^2\theta - \mu)S = 0$$

becomes

$$\sin^4\theta\frac{d^2v}{dx^2} - 2\sin^2\theta\cos\theta\frac{dv}{dx} + (\nu\sin^2\theta - \mu)v = 0.$$

Dividing by $\sin^2\theta$, we obtain

$$\sin^2\theta\frac{d^2v}{dx^2} - 2\cos\theta\frac{dv}{dx} + \left(\nu - \frac{\mu}{\sin^2\theta}\right)v = 0,$$

and since $x = \cos\theta$ and $\sin^2\theta = 1 - \cos^2\theta = 1 - x^2$, we have

$$(1 - x^2)\frac{d^2v}{dx^2} - 2x\frac{dv}{dx} + \left(\nu - \frac{\mu}{1 - x^2}\right)v = 0$$

which is the associated Legendre equation.

Alternatively, starting with (14.14), we have

$$\sin\theta\frac{dS}{d\theta} = -\sin^2\theta\frac{dv}{dx} = -(1 - x^2)\frac{dv}{dx}$$

and

$$\frac{d}{d\theta}\left(\sin\theta\frac{dS}{d\theta}\right) = \sin\theta\frac{d}{dx}\left[(1 - x^2)\frac{dv}{dx}\right],$$

so that

$$\sin\theta \frac{d}{d\theta}\left(\sin\theta \frac{dS}{d\theta}\right) = \sin^2\theta \frac{d}{dx}\left[(1-x^2)\frac{dv}{dx}\right]$$

and

$$\sin\theta \frac{d}{d\theta}\left(\sin\theta \frac{dS}{d\theta}\right) = (1-x^2)\left\{(1-x^2)\frac{d^2v}{dx^2} - 2x\frac{dv}{dx}\right\}.$$

Therefore,

$$(1-x^2)\frac{d^2v}{dx^2} - 2x\frac{dv}{dx} + \left(\nu - \frac{\mu}{1-x^2}\right)v = 0.$$

Exercise 14.21. ✗✗✗✗

Consider the boundary value–initial value problem for the solution $u(r,\theta,\phi,t)$ of the homogeneous wave equation on a hemispherical solid H:

$$\begin{aligned}
&\frac{\partial^2 u}{\partial t^2} = \nabla^2 u, \quad (r,\theta,\phi) \in H, \quad t > 0,\\
&u(1,\theta,\phi,t) = 0,\\
&u(r,\theta,-\frac{\pi}{2},t) = u(r,\theta,\frac{\pi}{2},t) = 0,\\
&u(r,\theta,\phi,0) = f(r,\theta,\phi),\\
&\frac{\partial u}{\partial t}(r,\theta,\phi,0) = 0,
\end{aligned}$$

where the hemisphere is parametrized using spherical coordinates as

$$H = \left\{ (r,\theta,\phi) \,\middle|\, 0 \leqslant r \leqslant 1, \quad -\frac{\pi}{2} \leqslant \phi \leqslant \frac{\pi}{2}, \ 0 \leqslant \theta \leqslant \pi \right\}.$$

(a) Supplement the problem above with an appropriate number of additional side conditions which will guarantee that there is a unique bounded solution.

(b) Assume a separated solution of the form

$$u(r,\theta,\phi,t) = R(r)\cdot S(\theta)\cdot \Phi(\phi)\cdot T(t),$$

and separate the differential equation and side conditions to derive a radial problem for R, an azimuthal problem for S, a polar problem for Φ, and a time problem for T.

(c) The radial, azimuthal, and polar problems are complete, that is, have the right number of side conditions, while the temporal problem is incomplete. The polar problem depends on one separation constant, and this problem should be solved first.

(d) Solve the azimuthal problem and identify the second separation constant.

(e) Solve the radial problem and identify the third separation constant.

(f) Solve the time problem and use the superposition principle to write the general solution.

(g) Solve the boundary value–initial value problem with the following initial condition:

$$f(r,\theta,\phi) = 4j_2(\alpha_{23}r)P_2^1(\cos\theta)\cos\phi,$$

where j_2 is the spherical Bessel function of the first kind of order 2, α_{23} is the third positive zero of this function, and $P_2^1(\cos\theta)$ is the associated Legendre function of degree 2 and order 1.

Solution.

(a) The wave equation above is second order in all variables r, θ, ϕ, and t, and according to our rule of thumb for side conditions, we should expect eight side conditions. Five are given in the problem. The missing three are boundedness conditions:

$$|u(0,\theta,\phi,t)|, \quad |u(r,0,\phi,t)|, \quad |u(r,\pi,\phi,t)|$$

which are all bounded.

(b) The wave equation in spherical coordinates can be written as

$$\frac{1}{c^2}\frac{\partial^2 u}{\partial t^2} = \frac{1}{r^2}\frac{\partial}{\partial r}\left(r^2\frac{\partial u}{\partial r}\right) + \frac{1}{r^2\sin\theta}\frac{\partial}{\partial\theta}\left(\sin\theta\frac{\partial u}{\partial\theta}\right) + \frac{1}{r^2\sin^2\theta}\frac{\partial^2 u}{\partial\phi^2} \tag{14.15}$$

for $(r,\theta,\phi) \in H$ and $t > 0$. We could look for a separated solution of the form

$$u(r,\theta,\phi,t) = W(r,\theta,\phi)\cdot T(t)$$

as we did in Exercise 13.20, and then substitute this into the PDE to get

$$\frac{1}{c^2}\frac{T''}{T} = \frac{1}{W}\nabla^2 W = -\lambda,$$

where λ is a separation constant. Then from Theorem 7.1, we know that the eigenvalues λ of the Helmholtz equation

$$\nabla^2 W = -\lambda W$$

are nonnegative, and in fact, are positive.

However, instead, we look for a separated solution of the form

$$u(r, \theta, \phi, t) = R(r) \cdot S(\theta) \cdot \Phi(\phi) \cdot T(t),$$

and keep in mind that the first separation constant λ is positive. Substituting the expression above into the wave equation (14.15), and separating variables, we have

$$\frac{1}{c^2} R \cdot S \cdot \Phi \cdot T'' = \frac{1}{r^2} \left(r^2 R'\right)' \cdot S \cdot \Phi \cdot T + \frac{R \cdot \Phi \cdot T}{r^2 \sin\theta} \left(\sin\theta\, S'\right)' + \frac{R \cdot S \cdot T}{r^2 \sin^2\theta} \cdot \Phi'';$$

that is,

$$\frac{1}{c^2} \frac{T''}{T} = \frac{1}{r^2} \frac{\left(r^2 R'\right)'}{R} + \frac{1}{r^2 \sin\theta} \frac{\left(\sin\theta S'\right)'}{S} + \frac{1}{r^2 \sin^2\theta} \frac{\Phi''}{\Phi} = -\lambda,$$

where $\lambda > 0$ is the first separation constant. We multiply by r^2 to obtain

$$\frac{(rR')'}{R} + \frac{1}{\sin\theta} \frac{\left(\sin\theta S'\right)'}{S} + \frac{1}{\sin^2\theta} \frac{\Phi''}{\Phi} = -\lambda r^2,$$

and separating variables again, we have

$$\frac{(r^2 R')'}{R} + \lambda r^2 = -\frac{1}{\sin\theta} \frac{\left(\sin\theta S'\right)'}{S} - \frac{1}{\sin^2\theta} \frac{\Phi''}{\Phi} = \nu,$$

where ν is the second separation constant. Multiplying the equation

$$-\frac{1}{\sin\theta} \frac{(\sin\theta S')'}{S} = \frac{1}{\sin^2\theta} \frac{\Phi''}{\Phi} + \nu$$

by $\sin^2\theta$, we have

$$\frac{\sin\theta\left(\sin\theta S'\right)'}{S} + \nu \sin^2\theta = -\frac{\Phi''}{\Phi} = \mu,$$

where μ is the third separation constant.

The homogeneous side conditions can easily be separated, and we obtain the following four problems:

- **Radial problem:**

$$(r^2 R')' + (\lambda r^2 - \nu) R = 0,$$

$$R(1) = 0,$$

$$|R(0)| \quad \text{bounded.}$$

- **Azimuthal angular problem:**

$$\sin\theta\big(\sin\theta S'\big)' + (\nu\sin^2\theta - \mu)S = 0,$$

$$|S(0)| \text{ and } |S(\pi)| \text{ bounded.}$$

- **Polar angular problem:**

$$\Phi'' + \mu\Phi = 0,$$

$$\Phi\Big(\frac{\pi}{2}\Big) = 0,$$

$$\Phi\Big(-\frac{\pi}{2}\Big) = 0.$$

- **Time problem:**

$$T'' + \lambda c^2 T = 0,$$

$$T'(0) = 0.$$

(c) The polar angular problem is easily solved, and the eigenvalues and corresponding eigenfunctions are

$$\mu_n = n^2 \qquad \text{and} \qquad \Phi_n(\phi) = \begin{cases} \cos n\phi, & \text{for odd } n, \\ \sin n\phi, & \text{for even } n \end{cases}$$

for $n = 1, 2, \ldots$.

(d) The azimuthal problem can be solved by letting $x = \cos\theta$ and $S(\theta) = v\big(x(\theta)\big)$; then, according to Exercise 14.20, the azimuthal problem becomes the associated Legendre equation

$$(1-x^2)v'' - 2xv' + \Big(\nu - \frac{n^2}{1-x^2}\Big)v = 0,$$

where we replaced μ by $\mu_n = n^2$. It has bounded solutions if and only if the eigenvalues are given by

$$\nu_m = m(m+1)$$

for $m = 1, 2, \ldots$. Thus, the azimuthal problem becomes

$$(1-x^2)v'' - 2xv' + \left(m(m+1) - \frac{n^2}{1-x^2}\right)v = 0,$$

$$|v(1)| \text{ and } |v(-1)| \text{ bounded,}$$

with general solution

$$v(x) = A_{nm}P_m^n(x) + B_{nm}Q_m^n(x),$$

where P_m^n and Q_m^n are the associated Legendre functions of the first and second kind. Since $|v(\pm 1)|$ must be bounded, we have $B_{nm} = 0$, and therefore

$$S_{nm}(\theta) = P_m^n(\cos\theta)$$

for $n = 1, 2, \ldots$ and $m = 1, 2, \ldots$.

(e) The radial problem is now

$$(r^2R')' + \big(\lambda r^2 - m(m+1)\big)R = 0,$$

$$R(1) = 0,$$

$$|R(0)| \text{ bounded.}$$

Letting $x = \sqrt{\lambda}\, r$ (recall that $\lambda > 0$), this becomes the spherical Bessel equation of order m, and the bounded solutions are the spherical Bessel functions of first kind of order m:

$$R(r) = j_m\big(\sqrt{\lambda}\, r\big).$$

The side condition $R(1) = 0$ implies that

$$j_m\big(\sqrt{\lambda}\big) = 0,$$

and therefore $\sqrt{\lambda}$ has to be a positive zero of the spherical Bessel function j_m. If α_{mk} denotes the kth positive zero of $j_m(\alpha)$, the eigenvalues are

$$\lambda_{mk} = \alpha_{mk}^2,$$

and the corresponding eigenfunctions are

$$R_{mk}(r) = j_m\big(\alpha_{mk} r\big)$$

for $m = 1, 2, \ldots$ and $k = 1, 2, \ldots$.

(f) The time problem is now

$$T'' + \lambda_{mk}c^2T = 0$$

and has the general solution

$$T_{mk}(t) = c_1 \sin(\alpha_{mk}ct) + c_2 \cos(\alpha_{mk}ct),$$

and since $T'(0) = 0$, we must have $c_1 = 0$, and hence

$$T_{mk}(t) = \cos(\alpha_{mk}ct)$$

for $m = 1, 2, \ldots$ and $k = 1, 2, \ldots$. Using the superposition principle, we can write the general solution of the wave equation above as

$$u(r,\theta,\phi,t) = \sum_{n=1}^{\infty}\sum_{m=1}^{\infty}\sum_{k=1}^{\infty} a_{nmk}\cos(\alpha_{mk}ct)j_m\big(z_{mk}r\big) \times P_m^n(\cos\theta)\,\Phi_n(\phi).$$

(g) With the initial condition

$$u(r,\theta,\phi,0) = 4j_2\big(\alpha_{23}r\big)P_2^1(\cos\theta)\cos\phi,$$

we find that the only nonzero coefficient arises for $n = 1, m = 2, k = 3$; hence,

$$a_{123} = 4 \qquad \text{and} \qquad a_{nmk} = 0 \quad \text{otherwise.}$$

The final solution is

$$u(r,\theta,\phi,t) = 4\cos(\alpha_{23}ct)j_2\big(\alpha_{23}r\big)P_2^1(\cos\theta)\cos\phi.$$

Exercise 14.22. Spherical Waves XXX

Find the solution to the wave equation in spherical coordinates if $u = u(r,t)$ depends only on the distance r from the origin in spherical coordinates and the time t,

$$c^2\frac{\partial^2 u}{\partial t^2} = \nabla^2 u, \quad r > 0, \quad t > 0,$$

where the initial displacement is $u(r,0) = f(r)$ and the initial velocity is $\partial u(r,0)/\partial t = g(r)$ for $r > 0$.

Solution. If u is independent of θ and ϕ, the Laplacian in spherical coordinates becomes

$$\nabla^2 u = \frac{1}{r^2}\frac{\partial}{\partial r}\left(r^2\frac{\partial u}{\partial r}\right),$$

and since

$$\frac{\partial}{\partial r}\left(r^2\frac{\partial u}{\partial r}\right) = 2r\frac{\partial u}{\partial r} + r^2\frac{\partial^2 u}{\partial r^2},$$

then

$$\frac{1}{r^2}\frac{\partial}{\partial r}\left(r^2\frac{\partial u}{\partial r}\right) = \frac{2}{r}\frac{\partial u}{\partial r} + \frac{\partial^2 u}{\partial r^2}. \tag{14.16}$$

However,

$$\frac{1}{r}\frac{\partial^2}{\partial r^2}(ru) = \frac{1}{r}\frac{\partial}{\partial r}\left(u + r\frac{\partial u}{\partial r}\right) = \frac{2}{r}\frac{\partial u}{\partial r} + \frac{\partial^2 u}{\partial r^2},$$

and the wave equation becomes

$$\frac{1}{r}\frac{\partial^2}{\partial r^2}(ru) = c^2\frac{\partial^2 u}{\partial t^2}$$

or

$$\frac{\partial^2(ru)}{\partial r^2} = rc^2\frac{\partial^2 u}{\partial t^2} = c^2\frac{\partial^2(ru)}{\partial t^2}. \tag{14.17}$$

This is just the one-dimensional wave equation, and from d'Alembert's solution we have

$$ru(r,t) = \frac{1}{2}\Big[f(r+ct) + f(r-ct)\Big] + \frac{1}{2c}\int_{r-ct}^{r+ct} g(s)\,ds;$$

that is,

$$u(r,t) = \frac{1}{2r}\Big[f(r+ct) + f(r-ct)\Big] + \frac{1}{2rc}\int_{r-ct}^{r+ct} g(s)\,ds$$

for $r > 0$ and $t > 0$.

CHAPTER 15

LAPLACE EQUATION PROBLEMS

Exercise 15.1. ✗
Show that the function

$$u = \frac{1}{\sqrt{x^2 + y^2 + z^2}}$$

is harmonic; that is, it is a solution to the three-dimensional Laplace equation $\Delta u = 0$.

Solution. By symmetry, we need only calculate the derivatives with respect to one of the variables, say x, and obtain the other derivatives by permuting the variables. For example,

$$\frac{\partial u}{\partial x} = \frac{\partial}{\partial x}\left(\frac{1}{\sqrt{x^2 + y^2 + z^2}}\right) = \frac{-x}{(x^2 + y^2 + z^2)^{3/2}},$$

so that

$$\frac{\partial u}{\partial y} = \frac{-y}{(x^2 + y^2 + z^2)^{3/2}} \qquad \text{and} \qquad \frac{\partial u}{\partial z} = \frac{-z}{(x^2 + y^2 + z^2)^{3/2}}.$$

Partial Differential Equations: Theory and Completely Solved Problems, First Edition. By T. Hillen, I. E. Leonard, H. van Roessel

Similarly,

$$\frac{\partial^2 u}{\partial x^2} = \frac{\partial}{\partial x}\left(\frac{-x}{(x^2+y^2+z^2)^{3/2}}\right) = \frac{2x^2-y^2-z^2}{(x^2+y^2+z^2)^{5/2}},$$

so that

$$\frac{\partial^2 u}{\partial y^2} = \frac{2y^2-x^2-z^2}{(x^2+y^2+z^2)^{5/2}} \quad \text{and} \quad \frac{\partial^2 u}{\partial z^2} = \frac{2z^2-x^2-y^2}{(x^2+y^2+z^2)^{5/2}}.$$

Therefore,

$$\frac{\partial^2 u}{\partial x^2}+\frac{\partial^2 u}{\partial y^2}+\frac{\partial^2 u}{\partial z^2} = \frac{(2x^2-y^2-z^2)+(2y^2-x^2-z^2)+(2z^2-x^2-y^2)}{(x^2+y^2+z^2)^{5/2}} = 0;$$

that is, u satisfies Laplace's equation $\Delta u = 0$.

Exercise 15.2. ✗✗✗

In this exercise we derive the form of the Laplacian in polar and cylindrical coordinate systems and obtain the following theorem.

Theorem 15.1. The Laplacian in the Cartesian coordinate system is defined as

$$\nabla^2 u = \frac{\partial^2 u}{\partial x^2}+\frac{\partial^2 u}{\partial y^2}+\frac{\partial^2 u}{\partial z^2}.$$

(a) If $x = r\cos\theta$, $y = r\sin\theta$, and $z = z$, the Laplacian can be written in the cylindrical coordinate system as

$$\nabla^2 u = \frac{1}{r}\frac{\partial}{\partial r}\left(r\frac{\partial u}{\partial r}\right)+\frac{1}{r^2}\frac{\partial^2 u}{\partial \theta^2}+\frac{\partial^2 u}{\partial z^2}.$$

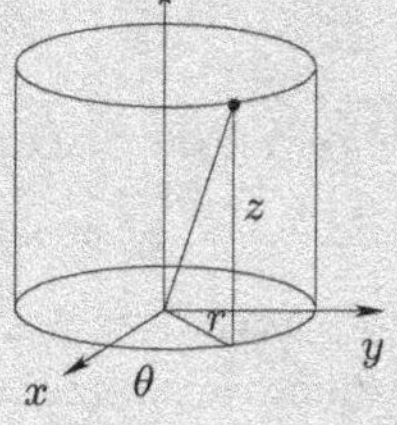

(b) If $x = r\cos\theta$ and $y = r\sin\theta$, the Laplacian can be written in the planar polar coordinate system as

$$\nabla^2 u = \frac{1}{r}\frac{\partial}{\partial r}\left(r\frac{\partial u}{\partial r}\right)+\frac{1}{r^2}\frac{\partial^2 u}{\partial \theta^2}$$

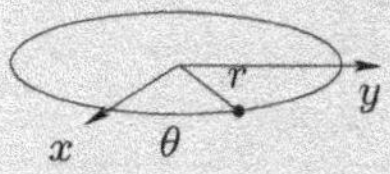

Solution:

(a) For cylindrical coordinates

$$x = r\cos\theta, \quad y = r\sin\theta, \quad z = z,$$

we let $\hat{u}(r,\theta,z) = u(r\cos\theta, r\sin\theta, z)$; then from the chain rule we have

$$\frac{\partial \hat{u}}{\partial r} = \frac{\partial u}{\partial x}\frac{\partial x}{\partial r} + \frac{\partial u}{\partial y}\frac{\partial y}{\partial r} + \frac{\partial u}{\partial z}\frac{\partial z}{\partial r} = \cos\theta\frac{\partial u}{\partial x} + \sin\theta\frac{\partial u}{\partial y}$$

$$\frac{\partial \hat{u}}{\partial \theta} = \frac{\partial u}{\partial x}\frac{\partial x}{\partial \theta} + \frac{\partial u}{\partial y}\frac{\partial y}{\partial \theta} + \frac{\partial u}{\partial z}\frac{\partial z}{\partial \theta} = -r\sin\theta\frac{\partial u}{\partial x} + r\cos\theta\frac{\partial u}{\partial y}$$

$$\frac{\partial \hat{u}}{\partial z} = \frac{\partial u}{\partial x}\frac{\partial x}{\partial z} + \frac{\partial u}{\partial y}\frac{\partial y}{\partial z} + \frac{\partial u}{\partial z}\frac{\partial z}{\partial z} = \frac{\partial u}{\partial z},$$

and in matrix form:

$$\begin{pmatrix} \dfrac{\partial \hat{u}}{\partial r} \\ \dfrac{1}{r}\dfrac{\partial \hat{u}}{\partial \theta} \\ \dfrac{\partial \hat{u}}{\partial z} \end{pmatrix} = \begin{pmatrix} \cos\theta & \sin\theta & 0 \\ -\sin\theta & \cos\theta & 0 \\ 0 & 0 & 1 \end{pmatrix} \begin{pmatrix} \dfrac{\partial u}{\partial x} \\ \dfrac{\partial u}{\partial y} \\ \dfrac{\partial u}{\partial z} \end{pmatrix}.$$

Now we note that the coefficient matrix is orthogonal, so that its inverse is the same as its transpose, and we get

$$\begin{pmatrix} \dfrac{\partial u}{\partial x} \\ \dfrac{\partial u}{\partial y} \\ \dfrac{\partial u}{\partial z} \end{pmatrix} = \begin{pmatrix} \cos\theta & -\sin\theta & 0 \\ \sin\theta & \cos\theta & 0 \\ 0 & 0 & 1 \end{pmatrix} \begin{pmatrix} \dfrac{\partial \hat{u}}{\partial r} \\ \dfrac{1}{r}\dfrac{\partial \hat{u}}{\partial \theta} \\ \dfrac{\partial \hat{u}}{\partial z} \end{pmatrix},$$

so that

$$\frac{\partial u}{\partial x} = \cos\theta\frac{\partial \hat{u}}{\partial r} - \frac{\sin\theta}{r}\frac{\partial \hat{u}}{\partial \theta}$$

$$\frac{\partial u}{\partial y} = \sin\theta\frac{\partial \hat{u}}{\partial r} + \frac{\cos\theta}{r}\frac{\partial \hat{u}}{\partial \theta}$$

$$\frac{\partial u}{\partial z} = \frac{\partial \hat{u}}{\partial z}.$$

We can use these last equations as operator equations as follows:

$$\begin{aligned}\frac{\partial^2 u}{\partial x^2} = \frac{\partial}{\partial x}\left(\frac{\partial u}{\partial x}\right) &= \cos\theta\frac{\partial}{\partial r}\left(\frac{\partial u}{\partial x}\right) - \frac{\sin\theta}{r}\frac{\partial}{\partial\theta}\left(\frac{\partial u}{\partial x}\right)\\
&= \cos\theta\frac{\partial}{\partial r}\left(\cos\theta\frac{\partial\hat{u}}{\partial r} - \frac{\sin\theta}{r}\frac{\partial\hat{u}}{\partial\theta}\right)\\
&\qquad - \frac{\sin\theta}{r}\frac{\partial}{\partial\theta}\left(\cos\theta\frac{\partial\hat{u}}{\partial r} - \frac{\sin\theta}{r}\frac{\partial\hat{u}}{\partial\theta}\right)\\
&= \cos^2\theta\frac{\partial^2\hat{u}}{\partial r^2} + \frac{\sin^2\theta}{r}\frac{\partial\hat{u}}{\partial r} + \frac{2\sin\theta\cos\theta}{r^2}\frac{\partial\hat{u}}{\partial\theta}\\
&\qquad - \frac{2\sin\theta\cos\theta}{r}\frac{\partial^2\hat{u}}{\partial r\partial\theta} + \frac{\sin^2\theta}{r^2}\frac{\partial^2\hat{u}}{\partial\theta^2}.\end{aligned}$$

Similarly,

$$\begin{aligned}\frac{\partial^2 u}{\partial y^2} = \frac{\partial}{\partial y}\left(\frac{\partial u}{\partial y}\right) &= \sin\theta\frac{\partial}{\partial r}\left(\frac{\partial u}{\partial y}\right) + \frac{\cos\theta}{r}\frac{\partial}{\partial\theta}\left(\frac{\partial u}{\partial y}\right)\\
&= \sin\theta\frac{\partial}{\partial r}\left(\sin\theta\frac{\partial\hat{u}}{\partial r} + \frac{\cos\theta}{r}\frac{\partial\hat{u}}{\partial\theta}\right)\\
&\qquad + \frac{\cos\theta}{r}\frac{\partial}{\partial\theta}\left(\sin\theta\frac{\partial\hat{u}}{\partial r} + \frac{\cos\theta}{r}\frac{\partial\hat{u}}{\partial\theta}\right)\\
&= \sin^2\theta\frac{\partial^2\hat{u}}{\partial r^2} + \frac{\cos^2\theta}{r}\frac{\partial\hat{u}}{\partial r} - \frac{2\sin\theta\cos\theta}{r^2}\frac{\partial\hat{u}}{\partial\theta}\\
&\qquad + \frac{2\sin\theta\cos\theta}{r}\frac{\partial^2\hat{u}}{\partial r\partial\theta} + \frac{\cos^2\theta}{r^2}\frac{\partial^2\hat{u}}{\partial\theta^2}.\end{aligned}$$

Adding these two expressions, we have

$$\frac{\partial^2 u}{\partial x^2} + \frac{\partial^2 u}{\partial y^2} + \frac{\partial^2 u}{\partial z^2} = \frac{\partial^2\hat{u}}{\partial r^2} + \frac{1}{r}\frac{\partial\hat{u}}{\partial r} + \frac{1}{r^2}\frac{\partial^2\hat{u}}{\partial\theta^2} + \frac{\partial^2\hat{u}}{\partial z^2},$$

or equivalently,

$$\frac{\partial^2 u}{\partial x^2} + \frac{\partial^2 u}{\partial y^2} + \frac{\partial^2 u}{\partial z^2} = \frac{1}{r}\frac{\partial}{\partial r}\left(r\frac{\partial\hat{u}}{\partial r}\right) + \frac{1}{r^2}\frac{\partial^2\hat{u}}{\partial\theta^2} + \frac{\partial^2\hat{u}}{\partial z^2}.$$

Note that if we introduce the differential operator $\mathbf{D}$, defined by

$$\mathbf{D} = \begin{pmatrix}\cos\theta & -\sin\theta & 0\\ \sin\theta & \cos\theta & 0\\ 0 & 0 & 1\end{pmatrix}\begin{pmatrix}\dfrac{\partial}{\partial r}\\ \dfrac{1}{r}\dfrac{\partial}{\partial\theta}\\ \dfrac{\partial}{\partial z}\end{pmatrix},$$

then the gradient of u is just $\mathbf{D}u$, while the Laplacian of u is $\mathbf{D}\cdot\mathbf{D}u$.

(b) The Laplacian in plane polar coordinates can be obtained from the previous result by eliminating the z-dependence, and therefore the Laplacian of u is given by

$$\nabla^2 u = \frac{1}{r}\frac{\partial}{\partial r}\left(r\frac{\partial u}{\partial r}\right) + \frac{1}{r^2}\frac{\partial^2 u}{\partial \theta^2}.$$

Exercise 15.3. ✗
Compute the Laplacian of the function

$$u(x, y) = \tan^{-1}\left(\frac{y}{x}\right)$$

in polar coordinates. Decide if the given function satisfies Laplace's equation $\Delta u = 0$.

Solution. Note that in polar coordinates $\theta = \tan^{-1}(y/x)$, so that

$$u(r, \theta) = \theta,$$

and since

$$\frac{\partial u}{\partial r} = \frac{\partial^2 u}{\partial r^2} = 0,$$

Laplace's equation becomes

$$\frac{\partial^2 u}{\partial r^2} + \frac{1}{r}\frac{\partial u}{\partial r} + \frac{\partial^2 u}{\partial \theta^2} = \frac{\partial^2 u}{\partial \theta^2} = 0,$$

and $u(x, y) = \tan^{-1}(y/x)$ does satisfy Laplace's equation.

Exercise 15.4. ✗
Compute the Laplacian of the function

$$u(x, y) = \log(x^2 + y^2)$$

in an appropriate coordinate system and decide if the given function satisfies Laplace's equation $\nabla^2 u = 0$.

Solution. Note that in polar coordinates, $r^2 = x^2 + y^2$, so that

$$u(r, \theta) = \log r^2 = 2\log r$$

and

$$\frac{1}{r}\frac{\partial u}{\partial r} = \frac{2}{r^2} \qquad \text{and} \qquad \frac{\partial^2 u}{\partial r^2} = -\frac{2}{r^2},$$

and since $\partial^2 u/\partial\theta^2 = 0$, then

$$\frac{\partial^2 u}{\partial r^2} + \frac{1}{r}\frac{\partial u}{\partial r} + \frac{\partial^2 u}{\partial \theta^2} = -\frac{2}{r^2} + \frac{2}{r^2} = 0$$

and $u(x,y) = \log(x^2 + y^2)$ does satisfy Laplace's equation.

Exercise 15.5. XX

Solve Laplace's equation inside a rectangle:

$$\nabla^2 u = \frac{\partial^2 u}{\partial x^2} + \frac{\partial^2 u}{\partial y^2}, \qquad 0 < x < a,\ 0 < y < b$$

subject to the boundary conditions

$$u(0,y) = g(y), \quad u(x,0) = 0,$$
$$u(a,y) = 0, \quad u(x,b) = 0$$

for $0 < x < a,\ 0 < y < b$.

Solution. Since the boundary conditions at $y = 0$ and $y = b$ are homogeneous, we can find a solution using the method of separation of variables. Writing $u(x,y) = X(x)\,Y(y)$, we obtain

$$\frac{X''}{X} = -\frac{Y''}{Y} = \lambda,$$

where λ is the separation constant. Hence, we obtain two ordinary differential equations:

$$X'' - \lambda X = 0, \quad 0 < x < a, \qquad Y'' + \lambda Y = 0, \quad 0 < y < b,$$
$$X(a) = 0, \qquad Y(0) = 0,$$
$$Y(b) = 0.$$

Solving the boundary value problem for Y, and writing $\lambda_n = \mu_n^2$, the eigenvalues are

$$\lambda_n = \mu_n^2 = \frac{n^2\pi^2}{b^2}$$

and the corresponding eigenfunctions are

$$Y_n(y) = \sin \mu_n y.$$

For convenience, the corresponding solutions to the equation $X'' - \mu_n^2 X = 0$ are taken to be

$$X_n(x) = a_n \cosh \mu_n(a - x) + b_n \sinh \mu_n(a - x)$$

for $n \geqslant 1$.

In this form, the boundary condition $X(a) = 0$, implies that $a_n = 0$ for all $n \geqslant 1$. Using the superposition principle, we write

$$u(x,y) = \sum_{n=1}^{\infty} b_n \sinh \mu_n(a-x) \sin \mu_n y.$$

From the boundary condition $u(0,y) = g(y)$, we have

$$g(y) = \sum_{n=1}^{\infty} b_n \sinh \frac{n\pi a}{b} \sin \mu_n y,$$

so that

$$b_n \sinh \frac{n\pi a}{b} = \frac{2}{b}\int_0^b g(y) \sin \mu_n y \, dy$$

and

$$b_n = \frac{2}{b \sinh(n\pi a/b)} \int_0^b g(y) \sin \mu_n y \, dy$$

for $n \geqslant 1$.

Exercise 15.6. ✗✗✗

Solve Laplace's equation inside a rectangle $0 < x < a$, $0 < y < b$, with the following boundary conditions:

(a)

$$\frac{\partial u}{\partial x}(0,y) = g(y), \qquad \frac{\partial u}{\partial x}(a,y) = 0,$$
$$u(x,0) = 0, \qquad u(x,b) = 0.$$

(b)

$$\frac{\partial u}{\partial x}(0,y) = 0, \qquad \frac{\partial u}{\partial x}(a,y) = 0,$$
$$u(x,0) = \begin{cases} 1 & \text{for } 0 < x < a/2, \\ 0 & \text{for } a/2 < x < a, \end{cases} \qquad \frac{\partial u}{\partial y}(x,b) = 0.$$

Solution.

(a) We assume a solution of the form $u(x,y) = X(x) \cdot Y(y)$, and substitute this into Laplace's equation to obtain

$$X''(x) \cdot Y(y) + X(x) \cdot Y''(y) = 0,$$

so that

$$\frac{X''(x)}{X(x)} = -\frac{Y''(y)}{Y(y)} = \lambda,$$

where λ is the separation constant, and we have two ordinary differential equations,

$$X''(x) - \lambda X(x) = 0 \qquad \text{and} \qquad Y''(y) + \lambda Y(y) = 0.$$

We can satisfy the (homogeneous) boundary conditions by requiring that

$$Y(0) = 0, \quad Y(b) = 0 \qquad \text{and} \qquad X'(a) = 0.$$

Therefore, X and Y satisfy the boundary value problems

$$\begin{aligned} &X''(x) - \lambda X(x) = 0, \; 0 < x < a, && Y''(y) + \lambda Y(y) = 0, \; 0 < y < b, \\ &X'(a) = 0, && Y(0) = 0, \\ & && Y(b) = 0. \end{aligned}$$

We solve the complete Dirichlet boundary value problem for Y first. The eigenvalues are

$$\lambda_n = \left(\frac{n\pi}{b}\right)^2$$

with corresponding eigenfunctions

$$Y_n(y) = \sin\frac{n\pi y}{b}$$

for $n \geqslant 1$. The corresponding functions $X(x)$ satisfy the boundary value problem

$$\begin{aligned} &X_n'' - \lambda_n X_n = 0, \quad 0 < x < a \\ &X_n'(a) = 0, \end{aligned}$$

and since the boundary condition at $x = a$ is homogeneous, we choose the following representation for the general solution:

$$X_n(x) = A\cosh\frac{n\pi(a-x)}{b} + B\sinh\frac{n\pi(a-x)}{b}.$$

The condition $X_n'(a) = 0$ implies that $B = 0$, and therefore the solution to the boundary value problem for X is

$$X_n(x) = \cosh\frac{n\pi(a-x)}{b}, \quad 0 < x < a$$

for $n \geqslant 1$.

From the superposition principle, the function

$$u(x,y)=\sum_{n=1}^{\infty}B_n\sin\frac{n\pi y}{b}\cosh\frac{n\pi(a-x)}{b} \tag{15.1}$$

satisfies Laplace's equation in the region $0<x<a,\ 0<y<b$, and satisfies all of the boundary conditions except $\partial u(0,y)/\partial x=g(y)$.

To satisfy this last condition, we use the orthogonality of the eigenfunctions on the interval $0\leqslant y\leqslant b$. Differentiating (15.1) with respect to x, and setting $x=0$ we get

$$g(y)=\frac{\partial u}{\partial x}(0,y)=-\sum_{n=1}^{\infty}\frac{n\pi}{b}B_n\sin\frac{n\pi y}{b}\sinh\frac{n\pi a}{b}.$$

Multiplying both sides of this equation by $\sin(m\pi y/b)$, and integrating over the interval $0\leqslant y\leqslant b$, we obtain

$$\int_0^b g(y)\sin\frac{m\pi y}{b}\,dy=-\sum_{n=1}^{\infty}\frac{n\pi}{b}\sinh\frac{n\pi a}{b}B_n\int_0^b\sin\frac{m\pi y}{b}\sin\frac{n\pi y}{b}\,dy$$

and using the orthogonality of the eigenfunctions, we have

$$B_m=\frac{-2}{m\pi\sinh(m\pi a/b)}\int_0^b g(y)\sin\frac{m\pi y}{b}\,dy \tag{15.2}$$

for $m\geqslant 1$. The solution to Laplace's equation satisfying the given boundary conditions is given by (15.1), where the coefficients $B_m,\ m\geqslant 1$, are given by (15.2).

(b) As above, we assume a solution of the form $u(x,y)=X(x)\cdot Y(y)$ and separate variables. We get the boundary value problems

$$X''(x)+\lambda X(x)=0,\ 0<x<a,\quad Y''(y)-\lambda Y(y)=0,\ 0<y<b,$$
$$X'(0)=0,\qquad Y'(b)=0.$$
$$X'(a)=0,$$

We solve the complete Neumann boundary value problem for X first. The eigenvalues are

$$\lambda_n=\left(\frac{n\pi}{a}\right)^2$$

with corresponding eigenfunctions

$$X_n(x)=\cos\frac{n\pi x}{a}$$

for $n\geqslant 0$.

The corresponding functions $Y_n(y)$ satisfy the boundary value problem

$$\begin{aligned}&Y_n'' - \lambda_n Y_n = 0, \quad 0 < y < b,\\ &Y_n'(b) = 0,\end{aligned}$$

and since the boundary condition at $y = b$ is homogeneous, we choose to represent the general solution as follows:

$$Y_n(y) = A\cosh\frac{n\pi(b-y)}{a} + B\sinh\frac{n\pi(b-y)}{a}.$$

The condition $Y_n'(b) = 0$ implies that $B = 0$, and therefore the solution to the boundary value problem for Y is

$$Y_n(y) = \cosh\frac{n\pi(b-y)}{a}, \quad 0 < y < b$$

for $n \geqslant 0$.

From the superposition principle, the function

$$u(x,y) = \sum_{n=0}^{\infty} A_n \cos\frac{n\pi x}{a}\cosh\frac{n\pi(b-y)}{a} \tag{15.3}$$

satisfies Laplace's equation in the region $0 < x < a,\ 0 < y < b$, and satisfies all of the boundary conditions except

$$u(x,0) = f(x) = \begin{cases} 1, & \text{for } \ 0 < x < a/2,\\ 0, & \text{for } \ a/2 < x < a.\end{cases}$$

To satisfy this condition, we use the orthogonality of the eigenfunctions on the interval $0 \leqslant x \leqslant a$, and setting $y = 0$ we get

$$f(x) = u(x,0) = \sum_{n=0}^{\infty} A_n \cos\frac{n\pi x}{a}\cosh\frac{n\pi b}{a}.$$

Multiplying both sides of this equation by $\cos(m\pi x/a)$, and integrating over the interval $0 \leqslant x \leqslant a$, we obtain

$$\int_0^a f(x)\cos\frac{m\pi x}{a}\,dx = \sum_{n=0}^{\infty}\cosh\frac{n\pi b}{a}A_n\int_0^a \cos\frac{m\pi x}{a}\cos\frac{n\pi x}{a}\,dx,$$

so that

$$A_0 = \frac{1}{2} \quad \text{and} \quad A_m = \frac{2\sin(m\pi/2)}{m\pi\cosh(m\pi b/a)} \tag{15.4}$$

for $m \geqslant 1$.

From (15.3) and (15.4) the solution to Laplace's equation satisfying the given boundary conditions is given by

$$u(x, y) = \frac{1}{2} + \sum_{n=1}^{\infty} \frac{2\sin(n\pi/2)}{n\pi\cosh(n\pi b/a)} \cos\frac{n\pi x}{a} \cosh\frac{n\pi(b-y)}{a}$$

for $0 < x < a$ and $0 < y < b$.

Exercise 15.7. XX

Solve Laplace's equation for the square $[0, 1] \times [0, 1]$ with the boundary conditions

$$\frac{\partial^2 u}{\partial x^2} + \frac{\partial^2 u}{\partial y^2} = 0, \quad 0 < x < 1, \quad 0 < y < 1,$$

$$u(x, 0) = 0,$$

$$u(x, 1) = 100,$$

$$u(0, y) = 0,$$

$$u(1, y) = 100.$$

Solution. We split the original problem into two problems:

$$\frac{\partial^2 v}{\partial x^2} + \frac{\partial^2 v}{\partial y^2} = 0, \quad 0 < x < 1, \quad 0 < y < 1,$$

$$v(x, 0) = 0,$$

$$v(x, 1) = 100,$$

$$v(0, y) = 0,$$

$$v(1, y) = 0$$

and

$$\frac{\partial^2 w}{\partial x^2} + \frac{\partial^2 w}{\partial y^2} = 0, \quad 0 < x < 1, \quad 0 < y < 1,$$

$$w(x, 0) = 0,$$

$$w(x, 1) = 0,$$

$$w(0, y) = 0,$$

$$w(1, y) = 100,$$

each with one pair of homogeneous boundary conditions (so we can use separation of variables) and the soluton to the original problem is then $u(x,y) = v(x,y) + w(x,y)$.

Now note that we only have to solve one of these problems, say the first, for $v(x,y)$, since we can get the solution to the second problem by interchanging x and y in the solution to the first problem, that is, $w(x,y) = v(y,x)$, so that the solution to the original problem is $u(x,y) = v(x,y) + v(y,x)$.
Writing $v(x,y) = X(x) \cdot Y(y)$, after substituting this into Laplace's equation and separating variables, we have

$$\frac{X''}{X} = -\frac{Y''}{Y} = -\lambda,$$

with separation constant λ. We get the following boundary value problems for X and Y :

$$\begin{aligned} &X'' + \lambda X = 0, && Y'' - \lambda Y = 0, \\ &X(0) = 0, && Y(0) = 0. \\ &X(1) = 0, && \end{aligned}$$

We solve the complete X-problem first, since it has two boundary conditions. The eigenvalues are $\mu_n^2 = n^2\pi^2$, and the eigenfunctions are

$$X_n(x) = \sin n\pi x$$

for $n \geqslant 1$. For each $n \geqslant 1$, the corresponding equation for Y is $Y'' - \mu_n^2 Y = 0$, with general solution

$$Y(y) = A \cosh n\pi y + B \sinh n\pi y,$$

and applying the boundary condition $Y(0) = 0$, we get $A = 0$, so the corresponding solutions are

$$Y_n(y) = \sinh n\pi y, \quad n \geqslant 1.$$

For each $n \geqslant 1$, the function

$$v_n(x,y) = X_n(x) \cdot Y_n(y) = \sin n\pi x \sinh n\pi y$$

satisfies Laplace's equation and all of the boundary conditions except $v(x,1) = 100$.

Now we use the superposition principle to write

$$v(x,y) = \sum_{n=1}^{\infty} b_n \sin n\pi x \sinh n\pi y$$

and determine the constants b_n using this last boundary condition, that is,

$$100 = v(x,1) = \sum_{n=1}^{\infty} b_n \sin n\pi x \sinh n\pi = \sum_{n=1}^{\infty} (b_n \sinh n\pi) \sin n\pi x,$$

and we recognize the constant $b_n \sinh n\pi$ as the Fourier sine series coefficient of the constant function 100 on the interval $[0, 1]$; therefore,

$$b_n \sinh n\pi = 2\int_0^1 100 \sin n\pi x \, dx = \frac{200}{n\pi}\left[1 - (-1)^n\right] = \begin{cases} \dfrac{400}{n\pi}, & \text{if } n \text{ is odd,} \\ 0, & \text{if } n \text{ is even.} \end{cases}$$

The solution to the first problem is therefore

$$v(x, y) = \frac{400}{\pi}\sum_{n=1}^{\infty} \frac{1}{(2n-1)\sinh(2n-1)\pi} \sin(2n-1)\pi x \sinh(2n-1)\pi y$$

for $0 \leqslant x, y \leqslant 1$. Interchanging x and y in this solution, we get the solution to the second problem,

$$w(x, y) = \frac{400}{\pi}\sum_{n=1}^{\infty} \frac{1}{(2n-1)\sinh(2n-1)\pi} \sin(2n-1)\pi y \sinh(2n-1)\pi x,$$

and the solution to the original problem is therefore

$$u(x, y) = \frac{400}{\pi}\sum_{n=1}^{\infty} \frac{1}{(2n-1)\sinh(2n-1)\pi} \Big[\sin(2n-1)\pi x \sinh(2n-1)\pi y + \sin(2n-1)\pi y \sinh(2n-1)\pi x \Big]$$

for $0 < x, y < 1$.

Exercise 15.8. XXX

Solve Laplace's equation in the unit square with boundary conditions

$$\frac{\partial^2 u}{\partial x^2} + \frac{\partial^2 u}{\partial y^2} = 0, \quad 0 < x < 1, \quad 0 < y < 1,$$

$$u(x, 0) = 1 - x,$$

$$u(x, 1) = x,$$

$$u(0, y) = 0,$$

$$u(1, y) = 0.$$

Solution. As in Exercise 15.7, we divide the problem into two problems:

$$\frac{\partial^2 v}{\partial x^2} + \frac{\partial^2 v}{\partial y^2} = 0, \quad 0 < x, y < 1, \qquad \frac{\partial^2 w}{\partial x^2} + \frac{\partial^2 w}{\partial y^2} = 0, \quad 0 < x, y < 1,$$

$$v(x,0) = 0, \qquad w(x,0) = 1 - x,$$

$$v(x,1) = x, \qquad w(x,1) = 0,$$

$$v(0,y) = 0, \qquad w(0,y) = 0,$$

$$v(1,y) = 0, \qquad w(1,y) = 0,$$

each with one pair of homogeneous boundary conditions (so we can use separation of variables) and the solution to the original problem is then $u(x,y) = v(x,y) + w(x,y)$. Note that if we find the solution $v(x,y)$ to the first problem, the solution to the second problem is

$$w(x,y) = v(1-x, 1-y).$$

We leave it to you to check that $w(x,y)$ satisfies Laplace's equation, and for the boundary conditions, note that

$$\begin{aligned} w(x,0) &= v(1-x,1) = 1-x, \\ w(x,1) &= v(1-x,0) = 0, \\ w(0,y) &= v(1,1-y) = 0, \\ w(1,y) &= v(0,1-y) = 0, \end{aligned}$$

so that $w(x,y)$ is a solution to the second problem.

We can use separation of variables as in Exercise 15.7 to find the solution $v(x,y)$, and the result is

$$v(x,y) = \sum_{n=1}^{\infty} b_n \sin n\pi x \sinh n\pi y.$$

The constants b_n are determined from the second boundary condition

$$x = v(x,1) = \sum_{n=1}^{\infty} b_n \sinh n\pi \sin n\pi x,$$

so that

$$b_n \sinh n\pi = 2\int_0^1 x \sin n\pi x \, dx = \frac{2(-1)^{n+1}}{n\pi}$$

and

$$b_n = \frac{2(-1)^{n+1}}{n\pi \sinh n\pi}$$

for $n \geqslant 1$.

The solution to the first problem is

$$v(x, y) = \frac{2}{\pi} \sum_{n=1}^{\infty} \frac{(-1)^{n+1}}{n \sinh n\pi} \sin n\pi x \sinh n\pi y,$$

and the solution to the second problem is

$$w(x, y) = \frac{2}{\pi} \sum_{n=1}^{\infty} \frac{(-1)^{n+1}}{n \sinh n\pi} \sin n\pi(1 - x) \sinh n\pi(1 - y).$$

The solution to the original problem is given by

$$u(x, y) = \frac{2}{\pi} \sum_{n=1}^{\infty} \frac{(-1)^{n+1}}{n \sinh n\pi} \big[\sin n\pi x \sinh n\pi y + \sin n\pi(1 - x) \sinh n\pi(1 - y) \big]$$

for $0 < x < 1,\ 0 < y < 1$.

Exercise 15.9. XX
Approximate the temperature at the center of the plate from Exercise 15.8.

Solution. Note that at the center of the plate $x = y = \frac{1}{2}$, and from Exercise 15.8,

$$u(\tfrac{1}{2}, \tfrac{1}{2}) = \frac{4}{\pi} \sum_{n=1}^{\infty} \frac{(-1)^{n+1}}{n \sinh n\pi} \sin \frac{n\pi}{2} \sinh \frac{n\pi}{2}.$$

Now,

$$\sinh n\pi = 2 \sinh \frac{n\pi}{2} \cosh \frac{n\pi}{2}$$

and

$$\sin \frac{n\pi}{2} = 0$$

if n is even, while

$$\sin \frac{(2k+1)\pi}{2} = (-1)^k$$

if $n = 2k + 1$ is odd. Therefore,

$$u(\tfrac{1}{2}, \tfrac{1}{2}) = \frac{2}{\pi} \sum_{k=0}^{\infty} \frac{(-1)^k}{(2k+1) \cosh \dfrac{(2k+1)\pi}{2}}.$$

A geometric symmetry argument as given below shows that this series converges to $\frac{1}{4}$; that is,

$$u(\tfrac{1}{2}, \tfrac{1}{2}) = \tfrac{1}{4}.$$

To see this, note that we can decompose the problem when the solution is identically 1 on the boundary of the square into four separate problems, as shown in Figure 15.1.

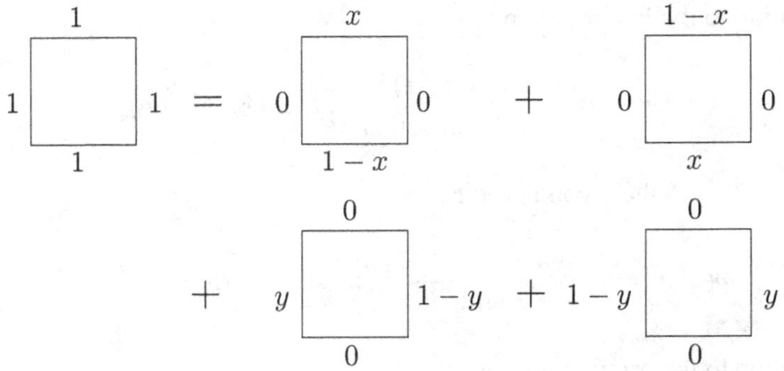

Figure 15.1 Decomposition of the problem.

By symmetry, each of the four problems has exactly the same value of the solution at the center $(\frac{1}{2}, \frac{1}{2})$, and since the solution to the original problem is identically 1 on the square, then

$$4u(\tfrac{1}{2}, \tfrac{1}{2}) = 1;$$

that is, $u(\frac{1}{2}, \frac{1}{2}) = \frac{1}{4}$.

Exercise 15.10. ✗✗✗✗

Solve Laplace's equation inside a circular annulus $(0 < a < r < b)$

$$\Delta u = \frac{1}{r}\frac{\partial}{\partial r}\left(r\frac{\partial u}{\partial r}\right) + \frac{1}{r^2}\frac{\partial^2 u}{\partial \theta^2} = 0, \quad a < r < b, \quad -\pi < \theta < \pi,$$

subject to the boundary conditions

$$\frac{\partial u}{\partial r}(a, \theta) = f(\theta), \qquad \frac{\partial u}{\partial r}(b, \theta) = g(\theta)$$

for $-\pi < \theta < \pi$.

Solution. Note that we need to include two periodicity conditions to ensure that the solution is single valued and continuous (and to get the right number of boundary conditions):

$$u(r, -\pi) = u(r, \pi) \qquad \text{and} \qquad \frac{\partial u}{\partial \theta}(r, -\pi) = \frac{\partial u}{\partial \theta}(r, \pi)$$

for $a \leqslant r \leqslant b$. We assume a solution of the form $u(r, \theta) = \phi(\theta) \cdot G(r)$, and substitute this into Laplace's equation to get

$$\frac{r}{G}\frac{d}{dr}\left(r\frac{dG}{dr}\right) = -\frac{1}{\phi}\frac{d^2\phi}{d\theta^2} = \lambda,$$

and we have two boundary value problems:

$$r\frac{d}{dr}\left(r\frac{dG}{dr}\right) - \lambda G = 0, \quad a < r < b, \qquad \phi''(\theta) + \lambda\phi(\theta) = 0, \quad -\pi < \theta < \pi,$$

$$\phi(-\pi) = \phi(\pi),$$

$$\phi'(-\pi) = \phi'(\pi),$$

where we satisfy the periodicity conditions by requiring that

$$\phi(-\pi) = \phi(\pi) \qquad \text{and} \qquad \phi'(-\pi) = \phi'(\pi).$$

We solve the complete (two periodicity conditions) boundary value problem for ϕ first. The eigenvalues are $\lambda_n = n^2$, with corresponding eigenfunctions

$$\phi_n(\theta) = \cos n\theta \qquad \text{and} \qquad \psi_n(\theta) = \sin n\theta$$

for $n \geqslant 1$. We assume a solution to the corresponding G-equation

$$r\frac{d}{dr}\left(r\frac{dG}{dr}\right) - n^2 G = 0$$

of the form $G(r) = r^\alpha$ for $n \geqslant 1$. Substituting this into the equation, we get

$$r\frac{d}{dr}\left(\alpha r^\alpha\right) - n^2 r^\alpha = 0;$$

that is,

$$\alpha^2 r^\alpha - n^2 r^\alpha = 0,$$

so that $\alpha = \pm n$. Hence, we get two linearly independent solutions,

$$G_{1n}(r) = r^n \qquad \text{and} \qquad G_{2n}(r) = \frac{1}{r^n},$$

and the general solution is

$$G_n(r) = Ar^n + \frac{B}{r^n}$$

for $n \geqslant 1$. For $n = 0$, the corresponding differential equation for $G(r)$ is

$$r\frac{d}{dr}\left(r\frac{dG}{dr}\right) = 0,$$

and we get two linearly independent solutions,

$$G_{10}(r) = 1 \qquad \text{and} \qquad G_{20}(r) = \log r,$$

and the general solution is

$$G_0(r) = A + B\log r,$$

From the superposition principle, the function

$$u(r,\theta) = A_0 + B_0 \log r + \sum_{n=1}^{\infty} \left[r^n (A_n \cos n\theta + B_n \sin n\theta) + \frac{1}{r^n}(C_n \cos n\theta + D_n \sin n\theta) \right] \tag{15.5}$$

with

$$\frac{\partial u}{\partial r}(r,\theta) = \frac{B_0}{r} + \sum_{n=1}^{\infty} \left[nr^{n-1}(A_n \cos n\theta + B_n \sin n\theta) - \frac{n}{r^{n+1}}(C_n \cos n\theta + D_n \sin n\theta) \right]$$

satisfies the periodicity conditions and Laplace's equation in the region $a < r < b$, and $-\pi < \theta < \pi$.

We can satisfy the boundary conditions

$$\frac{\partial u}{\partial r}(a,\theta) = f(\theta) \qquad \text{and} \qquad \frac{\partial u}{\partial r}(b,\theta) = g(\theta)$$

for $-\pi < \theta < \pi$, by requiring that

$$f(\theta) = \frac{B_0}{a} + \sum_{n=1}^{\infty} \left[na^{n-1}(A_n \cos n\theta + B_n \sin n\theta) - \frac{n}{a^{n+1}}(C_n \cos n\theta + D_n \sin n\theta) \right] \tag{15.6}$$

and

$$g(\theta) = \frac{B_0}{b} + \sum_{n=1}^{\infty} \left[nb^{n-1}(A_n \cos n\theta + B_n \sin n\theta) - \frac{n}{b^{n+1}}(C_n \cos n\theta + D_n \sin n\theta) \right] \tag{15.7}$$

where the coefficients are determined using the orthogonality of the eigenfunctions

$$\{1,\ \cos\theta,\ \sin\theta,\ \cos 2\theta,\ \sin 2\theta,\ \cos 3\theta,\ \sin 3\theta,\ \ldots\}$$

on the interval $-\pi \leqslant \theta \leqslant \pi$.

Multiplying equations (15.6) and (15.7) above by the eigenfunction 1 and integrating over the interval $[-\pi, \pi]$, we obtain

$$B_0 = \frac{a}{2\pi} \int_{-\pi}^{\pi} f(\theta)\, d\theta \qquad \text{and} \qquad B_0 = \frac{b}{2\pi} \int_{-\pi}^{\pi} g(\theta)\, d\theta. \tag{15.8}$$

Thus, there is **no solution** unless (15.8) holds; that is,

$$\int_{-\pi}^{\pi} f(\theta)\, a\, d\theta = \int_{-\pi}^{\pi} g(\theta)\, b\, d\theta.$$

If this is the case, then multiplying equations (15.6) and (15.7) by the appropriate eigenfunctions and integrating over the interval $[-\pi, \pi]$, we get

$$\int_{-\pi}^{\pi} f(\theta) \cos n\theta\, d\theta = n\pi \left(a^{n-1} A_n - \frac{1}{a^{n+1}} C_n \right),$$

$$\int_{-\pi}^{\pi} g(\theta) \cos n\theta\, d\theta = n\pi \left(b^{n-1} A_n - \frac{1}{b^{n+1}} C_n \right),$$

$$\int_{-\pi}^{\pi} f(\theta) \sin n\theta\, d\theta = n\pi \left(a^{n-1} B_n - \frac{1}{a^{n+1}} D_n \right),$$

$$\int_{-\pi}^{\pi} g(\theta) \sin n\theta\, d\theta = n\pi \left(b^{n-1} B_n - \frac{1}{b^{n+1}} D_n \right),$$

and solving for A_n, B_n, C_n, and D_n, we have

$$A_n = \frac{1}{n\pi(b^{2n} - a^{2n})} \left[b^n \int_{-\pi}^{\pi} g(\theta) \cos n\theta\, b\, d\theta - a^n \int_{-\pi}^{\pi} f(\theta) \cos n\theta\, a\, d\theta \right],$$

$$B_n = \frac{1}{n\pi(b^{2n} - a^{2n})} \left[b^n \int_{-\pi}^{\pi} g(\theta) \sin n\theta\, b\, d\theta - a^n \int_{-\pi}^{\pi} f(\theta) \sin n\theta\, a\, d\theta \right],$$

$$C_n = \frac{a^n b^n}{n\pi(b^{2n} - a^{2n})} \left[a^n \int_{-\pi}^{\pi} g(\theta) \cos n\theta\, b\, d\theta - b^n \int_{-\pi}^{\pi} f(\theta) \cos n\theta\, a\, d\theta \right],$$

$$D_n = \frac{a^n b^n}{n\pi(b^{2n} - a^{2n})} \left[a^n \int_{-\pi}^{\pi} g(\theta) \sin n\theta\, b\, d\theta - b^n \int_{-\pi}^{\pi} f(\theta) \sin n\theta\, a\, d\theta \right]$$

for $n \geqslant 1$. Thus, the solution to the Neumann problem for Laplace's equation in the annulus $a < r < b$ is given by (15.5), where the coefficients A_n, B_n, C_n, and D_n for $n \geqslant 1$ are given above, while

$$B_0 = \frac{a}{2\pi} \int_{-\pi}^{\pi} f(\theta)\, d\theta = \frac{b}{2\pi} \int_{-\pi}^{\pi} g(\theta)\, d\theta,$$

and A_0 is an arbitrary constant.

We reiterate that there is **no solution** to the Neumman problem for Laplace's equation in the annulus D:

$$\Delta u = \frac{1}{r}\frac{\partial}{\partial r}\left(r\frac{\partial u}{\partial r}\right) + \frac{1}{r^2}\frac{\partial^2 u}{\partial \theta^2} = 0, \quad a < r < b, \ -\pi < \theta < \pi$$

subject to the boundary conditions

$$\frac{\partial u}{\partial r}(a,\theta) = f(\theta), \qquad \frac{\partial u}{\partial r}(b,\theta) = g(\theta)$$

for $-\pi < \theta < \pi$, **unless the compatibility condition**

$$\int_{\partial D} \text{grad u} \cdot \mathbf{n}\, ds = 0$$

holds, where $\mathbf{n}$ is the outward unit normal to the boundary of D. This follows from the divergence theorem, since

$$\begin{aligned}
\int_{-\pi}^{\pi} g(\theta)\, b\, d\theta - \int_{-\pi}^{\pi} f(\theta)\, a\, d\theta &= \int_{-\pi}^{\pi} \frac{\partial u}{\partial r}(b,\theta)\, b\, d\theta - \int_{-\pi}^{\pi} \frac{\partial u}{\partial r}(a,\theta)\, a\, d\theta \\
&= \int_{\partial D} \text{grad u} \cdot \mathbf{n}\, ds \\
&= \iint_D \Delta u\, r dr d\theta = 0,
\end{aligned}$$

since $\Delta u = 0$ in the closed annular region D between the circles $r = a$ and $r = b$, and where $\mathbf{n}$ is the outward unit normal to the boundary of D.

Exercise 15.11. ✗✗✗✗

Find the solution of the *exterior Dirichlet problem for a disk*, that is, find a bounded solution to the problem

$$\begin{aligned}
&\frac{1}{r}\frac{\partial}{\partial r}\left(r\frac{\partial u}{\partial r}\right) + \frac{1}{r^2}\frac{\partial^2 u}{\partial \theta^2} = 0, \quad a < r < \infty, \quad -\pi < \theta < \pi, \\
&u(r,\pi) = u(r,-\pi), \\
&\frac{\partial u}{\partial \theta}(r,\pi) = \frac{\partial u}{\partial \theta}(r,-\pi), \\
&u(a,\theta) = f(\theta).
\end{aligned}$$

Solution. As in Exercise 15.10, using separation of variables, a solution to Laplace's equation in polar coordinates which satisfies the periodicity conditions is given by

$$u(r,\theta) = A_0 + B_0 \log r + \sum_{n=1}^{\infty} \Big\{ r^n \big(A_n \cos n\theta + B_n \sin n\theta\big) + \frac{1}{r^n}\big(C_n \cos n\theta + D_n \sin n\theta\big) \Big\},$$

and to satisfy the boundedness condition

$$\lim_{r\to\infty} |u(r,\theta)| < \infty,$$

we need $B_0 = A_n = B_n = 0$, for $n = 1, 2, 3, \ldots$. Therefore,

$$u(r,\theta) = A_0 + \sum_{n=1}^{\infty} \frac{1}{r^n}\big(C_n \cos n\theta + D_n \sin n\theta\big).$$

When $r = a$ we have

$$f(\theta) = u(a,\theta) = A_0 + \sum_{n=1}^{\infty} \frac{1}{a^n}\big(C_n \cos n\theta + D_n \sin n\theta\big),$$

where

$$A_0 = \frac{1}{2\pi}\int_{-\pi}^{\pi} f(\phi)\, d\phi,$$

$$C_n = \frac{a^n}{\pi}\int_{-\pi}^{\pi} f(\phi) \cos n\phi \, d\phi,$$

$$D_n = \frac{a^n}{\pi}\int_{-\pi}^{\pi} f(\phi) \sin n\phi \, d\phi$$

for $n = 1,\ 2,\ 3,\ \ldots$. Therefore,

$$u(r,\theta) = \frac{1}{2\pi}\int_{-\pi}^{\pi} f(\phi)\, d\phi + \frac{1}{\pi}\sum_{n=1}^{\infty}\left(\frac{a}{r}\right)^n \int_{-\pi}^{\pi} f(\phi)\Big\{ \cos n\phi \cos n\theta + \sin n\phi \sin n\theta \Big\}\, d\phi;$$

that is,

$$u(r,\theta) = \frac{1}{2\pi}\int_{-\pi}^{\pi} f(\phi)\Big\{1 + 2\sum_{n=1}^{\infty}\left(\frac{a}{r}\right)^n \cos n(\theta-\phi)\Big\}\, d\phi.$$

To evaluate the sum we could use the same method as we used to derive the Poisson integral formula for the interior Dirichlet problem for the disk in Section 6.1.1. However, here we give an alternative method, and set

$$z = \frac{a}{r}\, e^{i(\theta-\phi)},$$

so that

$$z^n = \left(\frac{a}{r}\right)^n e^{in(\theta-\phi)} = \left(\frac{a}{r}\right)^n \Big[\cos n(\theta-\phi) + i\,\sin n(\theta-\phi)\Big]$$

and

$$1 + 2\sum_{n=1}^{\infty}\left(\frac{a}{r}\right)^n \cos n(\theta-\phi) = \text{Re}\left(1 + 2\sum_{n=1}^{\infty} z^n\right).$$

And since $|z| = a/r < 1$, then

$$1 + 2\sum_{n=1}^{\infty}\left(\frac{a}{r}\right)^n \cos n(\theta-\phi) = \text{Re}\left(1 + \frac{2z}{1-z}\right) = \text{Re}\left(\frac{1+z}{1-z}\right)$$

$$= \frac{r^2 - a^2}{a^2 - 2ar\cos(\theta-\phi) + r^2}.$$

The solution to the exterior Dirichlet problem for the disk is therefore

$$u(r,\theta) = \frac{r^2 - a^2}{2\pi}\int_{-\pi}^{\pi} \frac{f(\phi)}{a^2 - 2ar\cos(\theta-\phi) + r^2}\,d\phi$$

for $a < r < \infty$, $-\pi < \theta < \pi$. Notice the similarity to the solution for the interior Dirichlet problem for the disk obtained in Chapter 6.

Exercise 15.12. ✗✗✗

Solve the problem of steady-state heat conduction in a quarter-plane; that is, find the function $u(x,y)$ that is bounded and harmonic in the first quadrant of the plane, that is, $x > 0,\ y > 0$, and satisfies the problem

$$\nabla^2 u = \frac{\partial^2 u}{\partial x^2} + \frac{\partial^2 u}{\partial y^2} = 0, \quad 0 < x < \infty, \quad 0 < y < \infty,$$

$$u(0,y) = g(y),$$

$$\frac{\partial u}{\partial y}(x,0) = f(x),$$

$|u(x,y)|$ bounded as $x, y \to \infty$.

Solution. We write $u(x, y) = v(x, y) + w(x, y)$, where v and w are solutions to the two problems

$$\frac{\partial^2 v}{\partial x^2} + \frac{\partial^2 v}{\partial y^2} = 0, \quad 0 < x < \infty, \quad 0 < y < \infty,$$

$$v(0, y) = 0, \tag{I}$$

$$\frac{\partial v}{\partial y}(x, 0) = f(x),$$

$$|v(x, y)| \text{ bounded as } x, y \to \infty$$

and

$$\frac{\partial^2 w}{\partial x^2} + \frac{\partial^2 w}{\partial y^2} = 0, \quad 0 < x < \infty, \quad 0 < y < \infty,$$

$$w(0, y) = g(y), \tag{II}$$

$$\frac{\partial w}{\partial y}(x, 0) = 0,$$

$$|w(x, y)| \text{ bounded as } x, y \to \infty,$$

and use the superposition principle to get the solution to the original problem.

• Problem I: We consider first the problem

$$\begin{aligned} &\frac{\partial^2 v}{\partial x^2} + \frac{\partial^2 v}{\partial y^2} = 0, \quad 0 < x < \infty, \quad 0 < y < \infty, \\ &v(0, y) = 0, \\ &\frac{\partial v}{\partial y}(x, 0) = f(x), \\ &|v(x, y)| \text{ bounded as } x, y \to \infty. \end{aligned}$$

Using separation of variables, we assume that $v(x, y) = X(x) \cdot Y(y)$, so that

$$\frac{X''}{X} = -\frac{Y''}{Y} = \mu^2,$$

where we have chosen the separation constant to be nonnegative so that the two differential equations

$$\begin{aligned} &X'' + \lambda^2 X = 0, && Y'' - \lambda^2 Y = 0, \\ &X(0) = 0, && |Y(y)| \text{ bounded as } y \to \infty, \\ &|X(x)| \text{ bounded as } x \to \infty, \end{aligned}$$

have bounded solutions

$$X_\mu(x) = \sin \mu x, \quad x > 0$$

and

$$Y_\mu(y) = e^{-\mu y}, \quad y > 0$$

for $\mu \geqslant 0$. Using the superposition principle, we write

$$v(x, y) = \int_0^\infty B(\mu) \sin \mu x \, e^{-\mu y} \, d\mu$$

for $x > 0, y > 0$. From the boundary condition at $y = 0$, we have

$$f(x) = \frac{\partial v}{\partial y}(x, 0) = -\int_0^\infty \mu B(\mu) \sin \mu x \, d\mu,$$

so that $-\mu B(\mu)$ is the Fourier sine integral of $f(x)$; that is,

$$-\mu B(\mu) = \frac{2}{\pi} \int_0^\infty f(x) \sin \mu x \, dx$$

and

$$B(\mu) = -\frac{2}{\mu\pi} \int_0^\infty f(x) \sin \mu x \, dx$$

for $\mu > 0$. Therefore, the solution to Problem I is

$$v(x, y) = \int_0^\infty B(\mu) \sin \mu x \, e^{-\mu y} \, d\mu, \quad x > 0, \; y > 0, \tag{15.9}$$

where

$$B(\mu) = -\frac{2}{\mu\pi} \int_0^\infty f(x) \sin \mu x \, dx \tag{15.10}$$

for $\mu > 0$.

- Problem II: Next we consider the problem

$$\begin{aligned}
&\frac{\partial^2 w}{\partial x^2} + \frac{\partial^2 w}{\partial y^2} = 0, \quad 0 < x < \infty, \quad 0 < y < \infty, \\
&w(0, y) = g(y), \\
&\frac{\partial w}{\partial y}(x, 0) = 0, \\
&|w(x, y)| \text{ bounded as } x, y \to \infty.
\end{aligned}$$

Using separation of variables, we assume that $w(x, y) = X(x) \cdot Y(y)$, so that

$$\frac{X''}{X} = -\frac{Y''}{Y} = \mu^2,$$

where again we have chosen the separation constant so that the two differential equations

$$X'' - \mu^2 X = 0, \qquad\qquad Y'' + \mu^2 Y = 0,$$

$$|X(x)| \text{ bounded as } x \to \infty, \qquad\qquad Y'(0) = 0,$$

$$|Y(y)| \text{ bounded as } y \to \infty,$$

have bounded solutions

$$X_\mu(x) = e^{-\mu x}, \quad x > 0$$

and

$$Y_\mu(y) = \cos \mu y, \quad y > 0.$$

Using the superposition principle, we write

$$w(x, y) = \int_0^\infty A(\mu) \cos \mu y \, e^{-\mu x} \, d\mu$$

for $x > 0, y > 0$. From the boundary condition at $x = 0$, we have

$$g(y) = w(0, y) = \int_0^\infty A(\mu) \cos \mu x \, d\mu,$$

so that $A(\mu)$ is the Fourier cosine integral of $g(y)$; that is,

$$A(\mu) = \frac{2}{\pi} \int_0^\infty g(y) \cos \mu y \, dy,$$

for $\mu \geqslant 0$. Therefore, the solution to Problem II is

$$w(x, y) = \int_0^\infty A(\mu) \cos \mu y \, e^{-\mu x} \, d\mu, \quad x > 0, \ y > 0, \tag{15.11}$$

where

$$A(\mu) = \frac{2}{\pi} \int_0^\infty g(y) \cos \mu y \, dy \tag{15.12}$$

for $\mu > 0$. The solution to Laplace's equation in the quarter-plane is therefore

$$u(x, y) = \int_0^\infty \left(A(\mu) \cos \mu y \, e^{-\mu x} + B(\mu) \sin \mu x \, e^{-\mu y} \right) d\mu \tag{15.13}$$

for $x > 0, \ y > 0$, where

$$A(\mu) = \frac{2}{\pi} \int_0^\infty g(y) \cos \mu y \, dy \quad \text{and} \quad B(\mu) = -\frac{2}{\mu\pi} \int_0^\infty f(x) \sin \mu x \, dx \tag{15.14}$$

for $\mu > 0$.

Exercise 15.13. Temperatures in a Hemisphere XXXX

A solid hemisphere of radius a has its flat base kept at temperature $u = 0$ and its hemispherical surface kept at temperature $u = u_0$. Find the steady-state temperature distribution in the interior of the hemisphere.

Hint: Use the method of images and solve a corresponding problem for the entire sphere, where the boundary temperature is $-u_0$ on the lower half of the sphere. To find the coefficients of the resulting Fourier-Legendre series, you might want to use the identity for Legendre functions of odd order:

$$\int_0^1 P_{2n+1}(x)\,dx = \frac{(-1)^n}{(n+1)2^{2n+1}}\binom{2n}{n}$$

for $n \geqslant 0$.

Solution. We want to solve the Dirichlet problem for Laplace's equation in the hemispherical solid

$$H(a) = \{\,(x,y,z)\,|\,x^2+y^2+z^2 \leqslant a^2,\ z \geqslant 0\,\}.$$

Here the appropriate coordinate system consists of spherical coordinates r, θ, and ϕ, where $x = r\sin\theta\cos\phi$, $y = r\sin\theta\sin\phi$, and $z = r\cos\theta$, and where the temperature $u = u(r,\theta)$ is independent of the polar angle ϕ. The boundary value problem is

$$\begin{aligned}
&\frac{1}{r^2}\frac{\partial}{\partial r}\left(r^2\frac{\partial u}{\partial r}\right) + \frac{1}{r^2\sin\theta}\frac{\partial}{\partial\theta}\left(\sin\theta\frac{\partial u}{\partial\theta}\right) = 0, \quad 0<r<a,\ 0<\theta<\frac{\pi}{2},\\
&u\left(r,\frac{\pi}{2}\right) = 0, \quad 0<r<a, \qquad\qquad (15.15)\\
&u(a,\theta) = u_0, \quad 0<\theta<\frac{\pi}{2}.
\end{aligned}$$

Following the hint, we will use the method of images. We reflect the hemisphere in the xy-plane and use the odd extension of the boundary data; that is, we will assume that $v = v(r,\theta)$ satisfies the boundary value problem

$$\begin{aligned}
&\frac{1}{r^2}\frac{\partial}{\partial r}\left(r^2\frac{\partial v}{\partial r}\right) + \frac{1}{r^2\sin\theta}\frac{\partial}{\partial\theta}\left(\sin\theta\frac{\partial v}{\partial\theta}\right) = 0, \quad 0<r<a, \quad 0<\theta<\pi,\\
&v(a,\theta) = \begin{cases} u_0, & \text{for } 0<\theta<\dfrac{\pi}{2},\\[2mm] -u_0, & \text{for } \dfrac{\pi}{2}<\theta<\pi. \end{cases} \qquad\qquad (15.16)
\end{aligned}$$

Once we have found the solution to this problem, we will show that

$$\lim_{\theta\to\frac{\pi}{2}^-} v(r,\theta) = 0 \qquad\qquad (15.17)$$

for $0 < r < a$. Therefore, the solution to the steady-state temperature distibution problem in the hemisphere will be given by

$$u(r, \theta) = v(r, \theta) \tag{15.18}$$

for $0 < r < a$ and $0 < \theta < \pi/2$.

For the problem in the sphere, we use separation of variables, and assume a solution of the form

$$v(r, \theta) = R(r) \cdot \Theta(\theta).$$

After substituting this into Laplace's equation and separating variables, we have the ODEs

$$r^2 R'' + 2r\, R' - \lambda\, R = 0, \quad 0 < r < a,$$

$$\sin\theta \frac{d}{d\theta}\big(\sin\theta\, \Theta' \big) + \lambda \sin^2\theta\, \Theta = 0, \quad 0 < \theta < \pi,$$

where λ is the separation constant. If we let $x = \cos\theta$ in the θ equation, we get

$$r^2 R'' + 2r\, R' - \lambda\, R = 0, \quad 0 < r < a, \tag{15.19}$$

$$\frac{d}{dx}\left[(1 - x^2) \frac{d\,\Theta}{dx} \right] + \lambda\, \Theta = 0, \quad -1 < x < 1. \tag{15.20}$$

The angular equation (15.20) is Legendre's equation, and the only bounded solutions occur when λ has the values

$$\lambda_n = n(n+1) \tag{15.21}$$

for $n = 0, 1, 2, \ldots$, and these are the eigenvalues. In this case, the corresponding eigenfunctions are the Legendre polynomials

$$\Theta_n(\theta) = P_n(x) = P_n(\cos\theta) \tag{15.22}$$

for $n = 0, 1, 2, \ldots$. For these values of λ_n, the radial equation (15.19) is a Cauchy-Euler equation,

$$r^2 R'' + 2r\, R' - n(n+1)\, R = 0, \tag{15.23}$$

with general solution

$$R_n(r) = a_n r^n + \frac{b_n}{r^{n+1}}$$

for $n = 0, 1, 2, \ldots$. Since the solution must be bounded, we must have $b_n = 0$ for $n \geqslant 0$, and we have

$$R_n(r) = a_n r^n \tag{15.24}$$

for $n \geqslant 0$. From the superposition principle we have

$$v(r,\theta) = \sum_{n=0}^{\infty} a_n r^n P_n(\cos\theta), \tag{15.25}$$

and this is a bounded solution to Laplace's equation in the sphere $0 < r < a$.

Now we determine the coefficients a_n from the boundary condition. We want

$$\sum_{n=0}^{\infty} a_n a^n P_n(\cos\theta) = v(a,\theta) = \begin{cases} u_0, & \text{for } 0 < \theta < \dfrac{\pi}{2}, \\ -u_0, & \text{for } \dfrac{\pi}{2} < \theta < \pi. \end{cases} \tag{15.26}$$

Therefore,

$$\begin{aligned} a_n a^n &= \frac{2n+1}{2} \int_0^{\pi} v(a,\theta) P_n(\cos\theta) \sin\theta \, d\theta \\ &= u_0 \frac{2n+1}{2} \left[\int_0^{\pi/2} P_n(\cos\theta) \sin\theta \, d\theta - \int_{\pi/2}^{\pi} P_n(\cos\theta) \sin\theta \, d\theta \right] \\ &= u_0 \frac{2n+1}{2} \left[\int_0^1 P_n(x)\, dx - \int_{-1}^0 P_n(x)\, dx \right]. \end{aligned}$$

Replacing x by $-x$ in the last integral, we have

$$a_n a^n = u_0 \frac{2n+1}{2} \left[\int_0^1 P_n(x)\, dx - \int_0^1 P_n(-x)\, dx \right],$$

and since $P_n(x)$ is an even function if n is even and an odd function if n is odd, then

$$a_{2n} = 0$$

while

$$a^{2n+1} a_{2n+1} = u_0 \,(4n+3) \int_0^1 P_{2n+1}(x)\, dx \tag{15.27}$$

for $n \geqslant 0$. Thus, the solution is

$$v(r,\theta) = \sum_{n=0}^{\infty} a^{2n+1} a_{2n+1} \left(\frac{r}{a} \right)^{2n+1} P_{2n+1}(\cos\theta) \tag{15.28}$$

for $0 < r < a$ and $0 < \theta < \pi$.

Note that since

$$P_{2n+1}(0) = 0$$

for all $n \geqslant 0$, the Fourier-Legendre series in (15.28) converges to 0 as $\theta \to \pi/2^-$, as promised; therefore, the steady-state temperature distribution in the hemisphere $H(a)$ is given by

$$u(r,\theta) = v(r,\theta) = \sum_{n=0}^{\infty} a^{2n+1} a_{2n+1} \left(\frac{r}{a}\right)^{2n+1} P_{2n+1}(\cos\theta) \tag{15.29}$$

for $0 < r < a$ and $0 < \theta < \pi/2$.

Note that since $u(a,\theta) = u_0$ for $0 < \theta < \pi/2$, then

$$u_0 = \sum_{n=0}^{\infty} a_{2n+1} a^{2n+1} P_{2n+1}(\cos\theta) = \sum_{n=0}^{\infty} b_n P_{2n+1}(\cos\theta),$$

where $b_n = a_{2n+1} a^{2n+1}$ for $n \geqslant 0$ and

$$b_n = u_0\,(4n+3) \int_0^{\pi/2} P_{2n+1}(\cos\theta)\sin\theta\, d\theta = u_0\,(4n+3) \int_0^1 P_{2n+1}(x)\, dx \tag{15.30}$$

for $n \geqslant 0$. We can evaluate this integral as follows: From Theorem 7.5, part (i), we have

$$P'_{n+1}(x) - P'_{n-1}(x) = (2n+1)P_n(x),$$

and for $0 < \epsilon < 1$, integrating this, we have

$$\begin{aligned}(2n+1)\int_\epsilon^1 P_n(x)\, dx &= \Big[P_{n+1}(1) - P_{n-1}(1)\Big] - \Big[P_{n+1}(\epsilon) - P_{n-1}(\epsilon)\Big] \\ &= P_{n-1}(\epsilon) - P_{n+1}(\epsilon)\end{aligned}$$

since $P_n(1) = 1$ for all $n \geqslant 0$. Letting $\epsilon \to 0^+$, we have

$$(2n+1)\int_0^1 P_n(x)\, dx = P_{n-1}(0) - P_{n+1}(0)$$

for $n \geqslant 1$. Therefore,

$$(4n+3)\int_0^1 P_{2n+1}(x)\, dx = P_{2n}(0) - P_{2n+2}(0) \tag{15.31}$$

for all $n \geqslant 0$. However, for each $n \geqslant 0$, from the Maclaurin series for $P_{2n}(x)$, or from Rodriques' formula, we have

$$P_{2n}(0) = (-1)^n \frac{(2n)!}{2^{2n}(n!)^2},$$

so that

$$P_{2n}(0) - P_{2n+2}(0) = \frac{4n+3}{2n+2}(-1)^n \frac{(2n)!}{2^{2n}(n!)^2}$$

and

$$(4n+3)\int_0^1 P_{2n+1}(x)\,dx = \frac{4n+3}{2n+2}(-1)^n \frac{(2n)!}{2^{2n}(n!)^2}$$

for $n \geqslant 0$; that is,

$$\int_0^1 P_{2n+1}(x)\,dx = \frac{(-1)^n}{(n+1)2^{2n+1}}\binom{2n}{n}$$

for $n \geqslant 0$. Finally, the solution can be written as

$$u(r,\theta) = u_0 \sum_{n=0}^{\infty} \frac{(-1)^n}{2^{2n+1}} \frac{4n+3}{n+1}\binom{2n}{n}\left(\frac{r}{a}\right)^{2n+1} P_{2n+1}(\cos\theta) \tag{15.32}$$

for $0 < r < a$ and $0 < \theta < \pi/2$.

Exercise 15.14. ✗
If $u(r,\theta,\phi)$ satisfies Laplace's equation $\nabla^2 u = 0$, show that

$$v(r,\theta,\phi) = \frac{a^2}{r}\,u\left(\frac{a^2}{r},\theta,\phi\right)$$

satisfies Laplace's equation; that is, $\nabla^2 v = 0$ also. This inversion transformation is often called **Kelvin's transformation**.

Solution. Laplace's equation in spherical coordinates is

$$\frac{1}{r^2}\frac{\partial}{\partial r}\left(r^2\frac{\partial v}{\partial r}\right) + \frac{1}{r^2\sin\theta}\frac{\partial}{\partial\theta}\left(\sin\theta\frac{\partial v}{\partial\theta}\right) + \frac{1}{r^2\sin^2\theta}\frac{\partial^2 v}{\partial\phi^2} = 0.$$

Set $s = a^2/r$ and write

$$v(r,\theta,\phi) = \frac{a^2}{r}\,u\left(\frac{a^2}{r},\theta,\phi\right) = s\,u(s,\theta,\phi).$$

Note first that

$$\frac{\partial v}{\partial\theta} = s\,\frac{\partial u}{\partial\theta} \quad\text{and}\quad \frac{\partial^2 v}{\partial\theta^2} = s\,\frac{\partial^2 u}{\partial\theta^2}$$

and

$$\frac{\partial v}{\partial\phi} = s\,\frac{\partial u}{\partial\phi} \quad\text{and}\quad \frac{\partial^2 v}{\partial\phi^2} = s\,\frac{\partial^2 u}{\partial\phi^2};$$

thus, we only have to consider the r-term in the Laplacian:

$$\frac{1}{r^2}\frac{\partial}{\partial r}\left(r^2\frac{\partial v}{\partial r}\right).$$

From the chain rule,

$$\frac{\partial v}{\partial r} = \frac{\partial (s\,u)}{\partial s} \cdot \frac{ds}{dr} = -\frac{s^2}{a^2} \cdot \frac{\partial (s\,u)}{\partial s} = -\frac{s^3}{a^2} \cdot \frac{\partial u}{\partial s} - \frac{s^2}{a^2} \cdot u$$

and

$$r^2 \frac{\partial v}{\partial r} = -a^2 s \cdot \frac{\partial u}{\partial s} - a^2 \cdot u. \tag{15.33}$$

From the chain rule and (15.33), we have

$$\begin{aligned}
\frac{\partial}{\partial r}\left(r^2 \frac{\partial v}{\partial r} \right) &= \frac{\partial}{\partial s}\left(-a^2 s \cdot \frac{\partial u}{\partial s} - a^2 \cdot u \right) \cdot \frac{ds}{dr} \\
&= -\frac{s^2}{a^2} \cdot \frac{\partial}{\partial s}\left(-a^2 s \cdot \frac{\partial u}{\partial s} - a^2 \cdot u \right) \\
&= s^3 \cdot \frac{\partial^2 u}{\partial s^2} + 2s^2 \cdot \frac{\partial u}{\partial s}
\end{aligned}$$

and

$$\frac{1}{r^2}\frac{\partial}{\partial r}\left(r^2 \frac{\partial v}{\partial r} \right) = \frac{2s^4}{a^4} \cdot \frac{\partial u}{\partial s} + \frac{s^5}{a^4}\frac{\partial^2 u}{\partial s^2};$$

that is,

$$\frac{1}{r^2}\frac{\partial}{\partial r}\left(r^2 \frac{\partial v}{\partial r} \right) = \frac{s^3}{a^4} \cdot \frac{\partial}{\partial s}\left(s^2 \frac{\partial u}{\partial s} \right). \tag{15.34}$$

Therefore,

$$\begin{aligned}
&\frac{1}{r^2}\frac{\partial}{\partial r}\left(r^2 \frac{\partial v}{\partial r} \right) + \frac{1}{r^2 \sin\theta}\frac{\partial}{\partial \theta}\left(\sin\theta \frac{\partial v}{\partial \theta} \right) + \frac{1}{r^2 \sin^2\theta}\frac{\partial^2 v}{\partial \phi^2} \\
&\qquad = \frac{s^3}{a^4}\left[\frac{\partial}{\partial s}\left(s^2 \frac{\partial u}{\partial s} \right) + \frac{1}{\sin\theta}\frac{\partial}{\partial \theta}\left(\sin\theta \frac{\partial u}{\partial \theta} \right) + \frac{1}{\sin^2\theta}\frac{\partial^2 u}{\partial \phi^2} \right]
\end{aligned}$$

and

$$\nabla^2 v = \frac{s^5}{a^4}\left[\frac{1}{s^2}\frac{\partial}{\partial s}\left(s^2 \frac{\partial u}{\partial s} \right) + \frac{1}{s^2 \sin\theta}\frac{\partial}{\partial \theta}\left(\sin\theta \frac{\partial u}{\partial s} \right) + \frac{1}{s^2 \sin^2\theta}\frac{\partial^2 u}{\partial \phi^2} \right],$$

and since

$$\frac{1}{s^2}\frac{\partial}{\partial s}\left(s^2 \frac{\partial u}{\partial s} \right) + \frac{1}{s^2 \sin\theta}\frac{\partial}{\partial \theta}\left(\sin\theta \frac{\partial u}{\partial s} \right) + \frac{1}{s^2 \sin^2\theta}\frac{\partial^2 u}{\partial \phi^2} = 0,$$

then

$$\nabla^2 v = 0,$$

and v is harmonic.

Exercise 15.15. XXX

Solve the Neumann problem of steady-state heat conduction in a half-plane; that is, find a function $u(x, y)$ that is bounded and harmonic in the upper half-plane $-\infty < x < \infty,\ y > 0$ and that satisfies the boundary and boundedness conditions

$$\nabla^2 u = \frac{\partial^2 u}{\partial x^2} + \frac{\partial^2 u}{\partial y^2} = 0, \quad -\infty < x < \infty, \quad y > 0,$$

$$\frac{\partial u}{\partial y}(x, 0) = f(x),$$

$$|u(x, y)| \leqslant M \text{ and } \left|\frac{\partial u}{\partial y}(x, y)\right| \leqslant M,$$

where $M > 0$ is a constant.

Hint: Use Poisson's integral formula, (9.27).

Solution. If $u(x, y)$ is a solution to the boundary value problem above, and we let

$$v(x, y) = \frac{\partial u}{\partial y}(x, y),$$

then

$$\begin{aligned}
\frac{\partial^2 v}{\partial x^2} + \frac{\partial^2 v}{\partial y^2} &= \frac{\partial}{\partial y}\left(\frac{\partial^2 u}{\partial x^2}\right) + \frac{\partial}{\partial y}\left(\frac{\partial^2 u}{\partial y^2}\right) \\
&= \frac{\partial}{\partial y}\left(\frac{\partial^2 u}{\partial x^2} + \frac{\partial^2 u}{\partial y^2}\right) \\
&= \frac{\partial}{\partial y}\left(0\right) \\
&= 0;
\end{aligned}$$

that is, $v(x, y)$ is harmonic in the half-plane $-\infty < x < \infty,\ y > 0$. Therefore, $v(x, y)$ satisfies the following boundary value problem with Dirichlet boundary conditions:

$$\nabla^2 v = \frac{\partial^2 v}{\partial x^2} + \frac{\partial^2 v}{\partial y^2} = 0, \quad -\infty < x < \infty, \quad y > 0,$$

$$v(x, 0) = f(x),$$

$$|v(x, y)| \leqslant M,$$

where $M > 0$ is a constant.

The solution to this boundary value problem is given by Poisson's integral formula for the half-plane, equation (9.27); that is,

$$v(x,y)=\frac{1}{2\pi}\int_{-\infty}^{\infty} f(s)\frac{2y}{(x-s)^2+y^2}\,ds$$

for $-\infty < x < \infty,\ y > 0$. Now we let

$$u(x,y)=\frac{1}{2\pi}\int_{-\infty}^{\infty} f(s)\log\left[(x-s)^2+y^2\right]ds+K,$$

where K is an arbitrary constant, and assume that we can differentiate under the integral sign,

$$\frac{\partial^2 u}{\partial x^2}+\frac{\partial^2 u}{\partial y^2}=\frac{1}{2\pi}\int_{-\infty}^{\infty} f(s)\left[\frac{\partial^2}{\partial x^2}\left(\log\left[(x-s)^2+y^2\right]\right)\right.$$
$$\left.+\frac{\partial^2}{\partial y^2}\left(\log\left[(x-s)^2+y^2\right]\right)\right]ds$$

for $-\infty < x < \infty,\ y > 0$. From Exercise 15.4, we see that

$$\frac{\partial^2}{\partial x^2}\left(\log\left[(x-s)^2+y^2\right]\right)+\frac{\partial^2}{\partial y^2}\left(\log\left[(x-s)^2+y^2\right]\right)=0$$

for $-\infty < x < \infty,\ y > 0$.
Thus, $u(x,y)$ is harmonic in the half-plane and

$$\frac{\partial u}{\partial y}(x,y)=\frac{1}{2\pi}\int_{-\infty}^{\infty} f(s)\frac{2y}{(x-s)^2+y^2}\,ds=v(x,y),$$

so that

$$\frac{\partial u}{\partial y}(x,0)=v(x,0)=f(x)$$

for $-\infty < x < \infty$; that is, the solution to the Neumann problem in the half-plane is given by

$$u(x,y)=\frac{1}{2\pi}\int_{-\infty}^{\infty} f(s)\log\left[(x-s)^2+y^2\right]ds+K, \qquad (15.35)$$

where K is an arbitrary constant.

Note: The conditions needed to justify differentiating under the integral sign can be found in Bromwich's *Theory of Infinite Series* ([10], p. 484).

CHAPTER 16

FOURIER TRANSFORM PROBLEMS

Recall: The **Fourier transform** is defined as

$$\mathcal{F}\big[\, f \,\big](\omega) = \widehat{f}(\omega) = \frac{1}{2\pi} \int_{-\infty}^{\infty} f(t) e^{i\omega t} \, dt,$$

and the **inverse Fourier transform** is defined as

$$\mathcal{F}^{-1}\big[\, \widehat{f} \,\big](t) = \int_{-\infty}^{\infty} \widehat{f}(\omega) e^{-i\omega t} \, d\omega.$$

The **convolution** of two functions $f(x)$ and $g(x)$ is defined to be

$$f * g\,(x) = \int_{-\infty}^{\infty} f(x-t) g(t) \, dt$$

and the **convolution theorem** says that if $F(\omega)$ and $G(\omega)$ are the Fourier transforms of f and g, respectively, then

$$\mathcal{F}^{-1}\big[\, F(\omega) G(\omega) \,\big](x) = \frac{1}{2\pi} f * g\,(x) = \frac{1}{2\pi} \int_{-\infty}^{\infty} f(x-t) g(t) \, dt.$$

Partial Differential Equations: Theory and Completely Solved Problems, First Edition. By T. Hillen, I. E. Leonard, H. van Roessel

Exercise 16.1. ✗✗

Evaluate the Fourier integral formula for the function

$$f(x) = \begin{cases} 1 - \cos x, & \text{if } -\dfrac{\pi}{2} < x < \dfrac{\pi}{2}, \\ 0, & \text{otherwise.} \end{cases}$$

Solution. The Fourier integral representation of $f(x)$ is given by

$$f(x) \sim \int_0^\infty (A(\omega)\cos\omega x + B(\omega)\sin\omega x)\,d\omega,$$

where

$$A(\omega) = \frac{1}{\pi}\int_{-\infty}^{\infty} f(t)\cos\omega t\,dt \qquad \text{and} \qquad B(\omega) = \frac{1}{\pi}\int_{-\infty}^{\infty} f(t)\sin\omega t\,dt.$$

Since $f(x)$ is an even function, then $B(\omega) = 0$ for all ω.

Also, since $f(x)$ is even and $f(x) = 0$ for $|x| \geqslant \pi/2$, then for all $\omega \neq 0$ and $\omega \neq \pm 1$, we have

$$\begin{aligned}
A(\omega) &= \frac{2}{\pi}\int_0^{\pi/2} (1 - \cos t)\cos\omega t\,dt \\
&= \frac{2}{\pi}\int_0^{\pi/2} \cos\omega t\,dt - \frac{2}{\pi}\int_0^{\pi/2} \cos t\cos\omega t\,dt \\
&= \frac{2}{\pi}\frac{\sin(\omega\pi/2)}{\omega} - \frac{1}{\pi}\int_0^{\pi/2} [\cos(1-\omega)t + \cos(1+\omega)t]\,dt \\
&= \frac{2}{\pi}\frac{\sin(\omega\pi/2)}{\omega} - \frac{1}{\pi}\frac{\sin(1-\omega)t}{1-\omega}\bigg|_0^{\pi/2} - \frac{1}{\pi}\frac{\sin(1+\omega)t}{1+\omega}\bigg|_0^{\pi/2} \\
&= \frac{2}{\pi}\frac{\sin(\omega\pi/2)}{\omega} - \frac{1}{\pi}\frac{\sin((1-\omega)\pi/2)}{1-\omega} - \frac{1}{\pi}\frac{\sin((1+\omega)\pi/2)}{1+\omega} \\
&= \frac{2}{\pi}\frac{\sin(\omega\pi/2)}{\omega} - \frac{\cos(\omega\pi/2)}{\pi}\left[\frac{1}{1-\omega} + \frac{1}{1+\omega}\right] \\
&= \frac{2}{\pi}\left[\frac{\sin(\omega\pi/2)}{\omega} - \frac{\cos(\omega\pi/2)}{1-\omega^2}\right],
\end{aligned}$$

so that

$$A(\omega) = \frac{2}{\pi}\left[\frac{\sin(\omega\pi/2)}{\omega} - \frac{\cos(\omega\pi/2)}{1-\omega^2}\right]$$

for $\omega \neq 0, \pm 1$.

If $\omega = 0$, then

$$A(0) = \frac{2}{\pi}\int_0^{\pi/2}(1-\cos t)\,dt = \frac{2}{\pi}\left[\frac{\pi}{2} - \sin(\pi/2)\right] = 1 - \frac{2}{\pi}.$$

If $\omega = \pm 1$, then

$$\begin{aligned} A(\pm 1) &= \frac{2}{\pi}\frac{\sin(\pm\pi/2)}{\pm 1} - \frac{2}{\pi}\int_0^{\pi/2}\cos^2 t\,dt \\ &= \frac{2}{\pi} - \frac{2}{\pi}\int_0^{\pi/2}\left(\frac{1+\cos 2t}{2}\right)dt \\ &= \frac{2}{\pi} - \frac{1}{2}. \end{aligned}$$

Note that $A(w)$ is continuous for all ω.

From Dirichlet's theorem, the integral

$$\frac{2}{\pi}\int_0^{\infty}\left[\frac{\sin(\omega\pi/2)}{\omega} - \frac{\cos(\omega\pi/2)}{1-\omega^2}\right]\cos\omega x\,d\omega$$

converges to $1-\cos x$ for all $|x| < \pi/2$, converges to 0 for all $|x| > \pi/2$, and converges to $\frac{1}{2}$ for $x = \pm\pi/2$. Thus, if we redefine $f(\pm\pi/2) = \frac{1}{2}$, the Fourier integral representation of $f(x)$ is given by

$$\frac{2}{\pi}\int_0^{\infty}\left[\frac{\sin(\omega\pi/2)}{\omega} - \frac{\cos(\omega\pi/2)}{1-\omega^2}\right]\cos\omega x\,d\omega = \begin{cases} 1-\cos x, & \text{for } |x| < \dfrac{\pi}{2}, \\ 0, & \text{for } |x| > \dfrac{\pi}{2}, \\ \frac{1}{2}, & \text{for } x = \pm\dfrac{\pi}{2}. \end{cases}$$

Exercise 16.2. ✗✗

Find the Fourier integral representation of the function

$$f(x) = \begin{cases} x, & \text{if } -1 < x < 1, \\ 2-x, & \text{if } 1 < x < 2, \\ -2-x, & \text{if } -2 < x < -1, \\ 0, & \text{otherwise.} \end{cases}$$

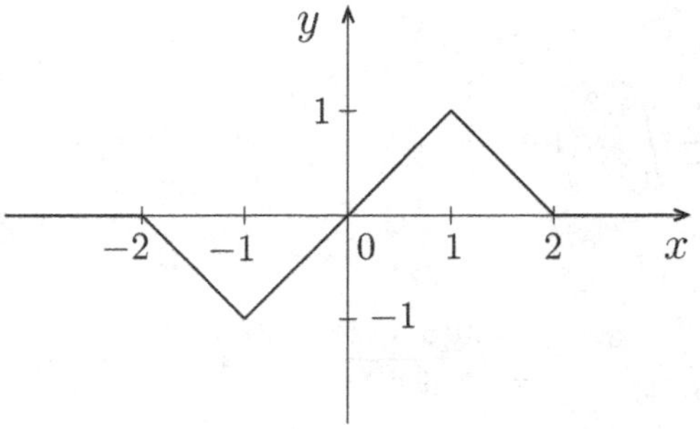

Figure 16.1

Solution. The graph of $f(x)$ is shown in Figure 16.1, and it is easy to see that the function $f(x)$ is an odd function. Therefore, $A(\omega) = 0$ for all ω, and

$$B(\omega) = \frac{2}{\pi}\int_0^2 f(t)\sin \omega t\, dt = \frac{2}{\pi}\int_0^1 t \sin \omega t\, dt + \frac{2}{\pi}\int_1^2 (2-t)\sin \omega t\, dt.$$

Integrating by parts, we have

$$\begin{aligned}
B(\omega) &= \frac{2}{\pi}\left[\frac{-t}{\omega}\cos \omega t\Big|_0^1 + \int_0^1 \frac{\cos \omega t}{\omega}\, dt\right] + \frac{2}{\pi}\left[\frac{-2+t}{\omega}\cos \omega t\Big|_1^2 - \int_1^2 \frac{\cos \omega t}{\omega}\, dt\right] \\
&= \frac{2}{\pi}\left[-\frac{\cos \omega}{\omega} + \frac{\sin \omega t}{\omega^2}\Big|_0^1\right] + \frac{2}{\pi}\left[\frac{\cos \omega}{\omega} - \frac{\sin \omega t}{\omega^2}\Big|_1^2\right] \\
&= \frac{2}{\pi}\left[\frac{2\sin \omega}{\omega^2} - \frac{\sin 2\omega}{\omega^2}\right] \\
&= \frac{2}{\pi}\left(\frac{2\sin \omega - \sin 2\omega}{\omega^2}\right);
\end{aligned}$$

that is,

$$B(\omega) = \frac{2}{\pi}\left(\frac{2\sin \omega - \sin 2\omega}{\omega^2}\right)$$

for all $\omega \neq 0$.

If $\omega = 0$, then

$$B(0) = \frac{2}{\pi}\int_0^2 f(t)\sin(0\cdot t)\, dt = 0.$$

Since $f(x)$ is continuous everywhere, from Dirichlet's theorem, the Fourier sine integral converges to $f(x)$ for all x, and therefore

$$\frac{2}{\pi}\int_0^\infty \left(\frac{2\sin \omega - \sin 2\omega}{\omega^2}\right)\sin \omega x\, d\omega = f(x)$$

for all $x \in \mathbb{R}$.

Exercise 16.3. ✗✗

Let

$$f(x) = \begin{cases} x, & \text{if } |x| < 1, \\ 0, & \text{otherwise.} \end{cases}$$

(a) Plot the function $f(x)$ and find its Fourier transform.

(b) If $\widehat{f}$ is real valued, plot it; otherwise, plot $|\widehat{f}|$.

Solution.

(a) The graph of the function $f(x)$ is plotted in Figure 16.2. The Fourier transform of $f(x)$ is computed as

$$\begin{aligned}
\widehat{f}(\omega) &= \frac{1}{2\pi}\int_{-\infty}^{\infty} f(t)e^{i\omega t}\,dt = \frac{1}{2\pi}\int_{-1}^{1} te^{i\omega t}\,dt \\
&= \frac{1}{2\pi}\left[\frac{t}{i\omega}e^{i\omega t}\bigg|_{-1}^{1} - \frac{1}{i\omega}\int_{-1}^{1} e^{i\omega t}\,dt\right] \\
&= \frac{1}{2\pi}\left[\frac{1}{i\omega}\left(e^{i\omega} + e^{-i\omega}\right) - \frac{1}{(i\omega)^2}e^{i\omega t}\bigg|_{-1}^{1}\right] \\
&= \frac{2}{2\pi i}\left[\left(\frac{e^{i\omega} + e^{-i\omega}}{2\omega}\right) - \left(\frac{e^{i\omega} - e^{-i\omega}}{2i\omega^2}\right)\right] \\
&= \frac{1}{\pi i}\left(\frac{\omega\cos\omega - \sin\omega}{\omega^2}\right),
\end{aligned}$$

so that

$$\widehat{f}(\omega) = \frac{1}{\pi i}\left(\frac{\omega\cos\omega - \sin\omega}{\omega^2}\right)$$

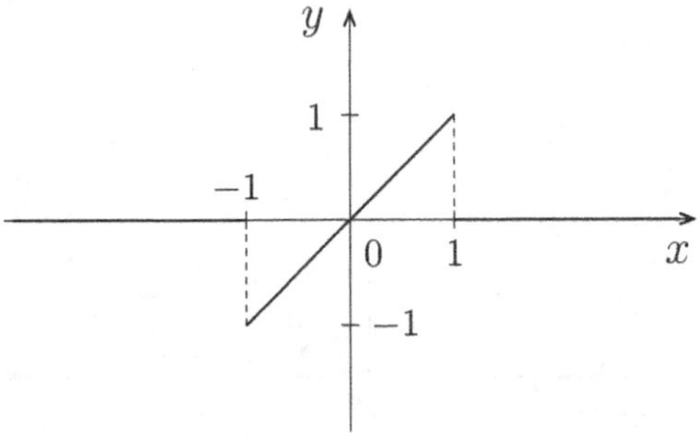

Figure 16.2

for all $\omega \neq 0$.

If $\omega = 0$, then

$$\widehat{f}(0) = \frac{1}{2\pi}\int_{-1}^{1} t\,dt = \frac{1}{2\pi}\frac{t^2}{2}\bigg|_{-1}^{1} = 0,$$

and from L'Hospital's rule, we see that $\lim_{\omega\to 0}\widehat{f}(\omega) = 0$ also, so that $\widehat{f}(\omega)$ is continuous at each ω.

(b) Since

$$\widehat{f}(\omega) = \frac{1}{\pi i}\left(\frac{\omega\cos\omega - \sin\omega}{w^2}\right),$$

then

$$\left|\widehat{f}(\omega)\right| = \frac{1}{\pi}\left|\frac{\sin\omega - \omega\cos\omega}{\omega^2}\right|$$

for all ω.

Note that the zeros of the function $g(\omega) = \sin\omega - \omega\cos\omega$ are precisely the roots of the equation $\tan\omega = \omega$, so the graph of $\left|\widehat{f}(\omega)\right|$ looks something like that shown in Figure 16.3.

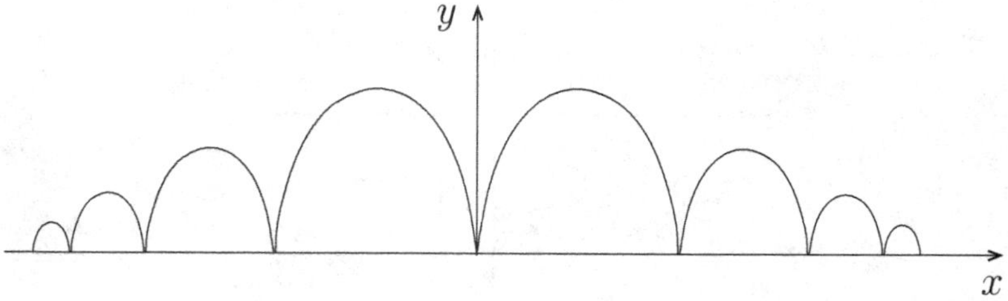

Figure 16.3

Exercise 16.4. Reciprocity Relation for the Fourier Transform XX

(a) From the definition of transforms, explain why

$$\mathcal{F}[\,f\,](x) = \frac{1}{2\pi}\mathcal{F}^{-1}[\,f\,](-x).$$

(b) Use (a) to derive the **reciprocity relation**

$$\mathcal{F}^2[\,f\,](x) = \frac{1}{2\pi}f(-x),$$

where $\mathcal{F}^2[\,f\,] = \mathcal{F}[\mathcal{F}[f]]$.

(c) Conclude the following: f is even if and only if $\mathcal{F}^2[\,f\,](x) = (1/2\pi)f(x)$; and f is odd if and only if $\mathcal{F}^2[\,f\,](x) = -(1/2\pi)f(x)$.

(d) Show that for any f, $\mathcal{F}^4[\,f\,] = (1/4\pi^2)f$.

Solution.

(a) Note that the Fourier transform of f is

$$\mathcal{F}[\,f\,](\omega) = \frac{1}{2\pi}\int_{-\infty}^{\infty} f(t)e^{i\omega t}\,dt,$$

and evaluating this transform at $\omega = x$ and making a change of variables, we get

$$\begin{aligned}\mathcal{F}[\,f\,](x) &= \frac{1}{2\pi}\int_{-\infty}^{\infty} f(t)e^{ixt}\,dt\\ &= \frac{1}{2\pi}\int_{-\infty}^{\infty} f(\omega)e^{-i\omega(-x)}\,d\omega\\ &= \frac{1}{2\pi}\mathcal{F}^{-1}[\,f\,](-x);\end{aligned}$$

that is,

$$\mathcal{F}[\,f\,](x) = \frac{1}{2\pi}\mathcal{F}^{-1}[\,f\,](-x)$$

for all $x \in \mathbb{R}$.

(b) Let $\widehat{f}$ be the Fourier transform of f; from part (a) we have

$$\mathcal{F}[\,\widehat{f}\,](x) = \frac{1}{2\pi}\mathcal{F}^{-1}[\,\widehat{f}\,](-x) = \frac{1}{2\pi}f(-x),$$

and therefore

$$\mathcal{F}^2[\,f\,](x) = \frac{1}{2\pi}f(-x)$$

for all $x \in \mathbb{R}$.

(c) The function f is even if and only if $f(-x) = f(x)$ for all $x \in \mathbb{R}$, but from part (b), we have that f is even if and only if

$$\mathcal{F}^2[\,f\,](x) = \mathcal{F}[\,\widehat{f}\,](x) = \frac{1}{2\pi} f(-x) = \frac{1}{2\pi} f(x)$$

for all $x \in \mathbb{R}$. Similarly, f is odd if and only if $f(-x) = -f(x)$ for all $x \in \mathbb{R}$, but again from part (b), we have that f is odd if and only if

$$\mathcal{F}^2[\,f\,](x) = \mathcal{F}[\,\widehat{f}\,](x) = \frac{1}{2\pi} f(-x) = -\frac{1}{2\pi} f(x)$$

for all $x \in \mathbb{R}$.

(d) For any integrable f, we have

$$\begin{aligned} \mathcal{F}^4[\,f\,](x) &= \mathcal{F}^2\left[\mathcal{F}^2[\,f\,]\right](x) = \frac{1}{2\pi}\mathcal{F}^2[\,f\,](-x) \\ &= \frac{1}{4\pi^2} f\big(-(-x)\big) = \frac{1}{4\pi^2} f(x) \end{aligned}$$

for all $x \in \mathbb{R}$.

Exercise 16.5. Basic Properties of Convolutions. ✗✗

Establish the following properties of convolutions. (These properties can be derived directly from the definitions or by using the operational properties of the Fourier transform.)

(a) $f * g = g * f$ (commutativity).

(b) $f * (g * h) = (f * g) * h$ (associativity).

(c) Let a be a real number and let f_a denote the translate of f by a; that is,

$$f_a(x) = f(x - a).$$

Show that

$$(f_a) * g = f * (g_a) = (f * g)_a.$$

This important property says that convolutions commute with translations.

Solution. The most convenient way to prove that these properties are true is to use the uniqueness of the Fourier transform; that is, if f and g are integrable and if $\widehat{f} = \widehat{g}$, then $f = g$. However, we will prove them directly from the definition of the convolution.

(a) Given absolutely integrable functions f and g, we make a simple substitution in the definition of the convolution to get

$$f * g\,(x) = \int_{-\infty}^{\infty} f(x-t)g(t)\,dt = \int_{-\infty}^{\infty} f(s)g(x-s)ds = g * f\,(x)$$

for all $x \in \mathbb{R}$, and therefore $f * g = g * f$.

(b) Let $f,\ g,$ and h be absolutely integrable; then for each $x \in \mathbb{R}$ we have

$$\begin{aligned}
(f * (g * h))\,(x) &= \int_{-\infty}^{\infty} f(x-t)(g*h)(t)\,dt \\
&= \int_{-\infty}^{\infty} f(x-t)\left(\int_{-\infty}^{\infty} g(t-s)h(s)\,ds\right)\,dt \\
&= \int_{-\infty}^{\infty}\left(\int_{-\infty}^{\infty} f(x-t)g(t-s)h(s)\,ds\right)\,dt \\
&= \int_{-\infty}^{\infty}\left(\int_{-\infty}^{\infty} h(s)f(x-t)g(t-s)\,dt\right)\,ds
\end{aligned}$$

(letting $v = x - s$)

$$= \int_{-\infty}^{\infty}\left(\int_{-\infty}^{\infty} h(x-v)f(x-t)g(t-(x-v))\,dt\right)\,dv$$

(letting $u = x - t$)

$$\begin{aligned}
&= \int_{-\infty}^{\infty}\left(\int_{-\infty}^{\infty} h(x-v)f(u)g(v-u)\,du\right)\,dv \\
&= \int_{-\infty}^{\infty} h(x-v)\left(\int_{-\infty}^{\infty} f(u)g(v-u)\,du\right)\,dv \\
&= \int_{-\infty}^{\infty} h(x-v)(g*f)\,(v)\,dv \\
&= \int_{-\infty}^{\infty} h(x-v)(f*g)\,(v)\,dv \\
&= (h * (f * g))\,(x) = ((f * g) * h)\,(x).
\end{aligned}$$

(c) We use the shift theorem,

$$\begin{aligned}
\mathcal{F}[\,f_a\,](\omega) &= \mathcal{F}[\,f(x-a)\,](\omega) \\
&= \frac{1}{2\pi}\int_{-\infty}^{\infty} f(t-a)e^{i\omega t}\,dt \\
&= \frac{1}{2\pi}\int_{-\infty}^{\infty} f(s)e^{i\omega(s+a)}\,ds \\
&= \frac{e^{i\omega a}}{2\pi}\int_{-\infty}^{\infty} f(s)e^{i\omega s}\,ds \\
&= e^{i\omega a}\mathcal{F}[\,f\,](\omega),
\end{aligned}$$

for all ω, so that

$$\mathcal{F}[\,f_a\,] = e^{i\omega a}\mathcal{F}[\,f\,].$$

We have

$$\begin{aligned}
\mathcal{F}[\,(f*g)_a(x)\,] &= \mathcal{F}[\,(f*g)(x-a)\,] \\
&= e^{i\omega a}\mathcal{F}[\,(f*g)(x)\,] \\
&= 2\pi\, e^{i\omega a}\mathcal{F}[\,f(x)\,]\mathcal{F}[\,g(x)\,] \\
&= 2\pi\, \mathcal{F}[\,f_a(x)\,]\mathcal{F}[\,g(x)\,] \\
&= \mathcal{F}[\,((f_a)*g)\,(x)\,],
\end{aligned}$$

and $\mathcal{F}[\,(f*g)_a\,] = \mathcal{F}[\,(f_a)*g\,]$. Since the Fourier transform is unique, $(f*g)_a = (f_a)*g$.

We can also prove this directly, as follows;

$$\begin{aligned}
(f*g)_a(x) &= (f*g)(x-a) \\
&= \int_{-\infty}^{\infty} f(x-a-t)g(t)\,dt \\
&= \int_{-\infty}^{\infty} f_a(x-t)g(t)\,dt \\
&= ((f_a)*g)(x)
\end{aligned}$$

for all $x \in \mathbb{R}$, so that $(f*g)_a = (f_a)*g$. Also, since $f*g = g*f$, we have

$$(f*g)_a = (g*f)_a = (g_a)*f = f*(g_a).$$

Exercise 16.6. XX

Determine the solution of the following boundary value–initial value problem for the heat equation

$$\frac{\partial u}{\partial t} = \frac{1}{4}\frac{\partial^2 u}{\partial x^2}, \qquad -\infty < x < \infty, \quad t > 0,$$

$$u(x,0) = e^{-x^2}.$$

Give your answer in the form of an inverse Fourier transform.

Solution. We hold t fixed and take the Fourier transform of the partial differential equation and the initial condition with respect to the space variable to get the initial value problem for $\widehat{u}(\omega, t) = \mathcal{F}\big[\, u(x,t)\,\big](\omega)$:

$$\frac{d\widehat{u}}{dt}(\omega, t) = -\frac{\omega^2}{4}\widehat{u}(\omega, t),$$

$$\widehat{u}(\omega, 0) = \mathcal{F}\big[\, e^{-x^2}\,\big](\omega) = \frac{1}{\sqrt{4\pi}}\, e^{-\frac{\omega^2}{4}}.$$

The general solution to this first-order linear equation is

$$\widehat{u}(\omega, t) = A(\omega)\, e^{-\frac{\omega^2}{4}t},$$

and we can determine the "constant" of integration $A(\omega)$ from the initial condition. Setting $t = 0$, we get

$$\widehat{u}(\omega, 0) = A(\omega) = \frac{1}{\sqrt{4\pi}}\, e^{-\frac{\omega^2}{4}},$$

so that

$$\widehat{u}(\omega, t) = \frac{1}{\sqrt{4\pi}}\, e^{-\frac{\omega^2}{4}}\, e^{-\frac{\omega^2}{4}t} = \frac{1}{\sqrt{4\pi}}\, e^{-\frac{\omega^2}{4}(1+t)}.$$

Taking the inverse transform, the solution is

$$u(x,t) = \frac{1}{2\sqrt{\pi}}\int_{-\infty}^{\infty} e^{-\frac{\omega^2}{4}(1+t)}\, e^{-i\omega x}\, d\omega$$

for $-\infty < x < \infty$, $t > 0$.

Exercise 16.7. X

Use the Fourier transform to solve the following boundary value–initial value problem:

$$\frac{\partial u}{\partial x} = \frac{\partial u}{\partial t}, \qquad -\infty < x < \infty, \quad t > 0,$$

$$u(x,0) = f(x).$$

Assume that the function f has a Fourier transform.

Solution. Taking the Fourier transform of the partial differential equation and the initial condition with respect to x, we have

$$\frac{d\widehat{u}}{dt}(\omega, t) + i\omega\widehat{u}(\omega, t) = 0,$$

$$\widehat{u}(\omega, 0) = \widehat{f}(\omega).$$

The general solution to this first-order linear equation is

$$\widehat{u}(\omega, t) = A(\omega)\, e^{-i\omega t},$$

and we can determine the "constant" of integration $A(\omega)$ from the transformed initial condition

$$\widehat{u}(\omega, 0) = A(\omega) = \widehat{f}(\omega).$$

Therefore,

$$\widehat{u}(\omega, t) = \widehat{f}(\omega)\, e^{-i\omega t},$$

and taking the inverse Fourier transform, we have

$$\begin{aligned} u(x,t) &= \int_{-\infty}^{\infty} \widehat{f}(\omega)\, e^{-i\omega t}\, e^{-i\omega x}\, d\omega \\ &= \int_{-\infty}^{\infty} \widehat{f}(\omega)\, e^{-i\omega(x+t)}\, d\omega \\ &= f(x+t), \end{aligned}$$

and the solution is

$$u(x,t) = f(x+t)$$

for $-\infty < x < \infty,\ t > 0$.

Exercise 16.8. ✗✗
Find the Fourier cosine transform of

$$f(x) = \begin{cases} 1 - x, & \text{if } \ 0 < x < 1, \\ 0, & \text{if } \ x \geqslant 1 \end{cases}$$

and write $f(x)$ as an inverse cosine transform. Use a known Fourier transform and the fact that if $f(x),\ x \geqslant 0$, is the restriction of an *even* function f_e, then

$$\mathcal{C}[\,f\,](\omega) = 2\mathcal{F}[\,f_e\,](\omega)$$

for all $\omega \geqslant 0$.

Solution. The Fourier cosine transform of the function f is given by

$$\widehat{f}_c(\omega) = \mathcal{C}[\,f\,](\omega) = \frac{2}{\pi}\int_0^{\infty} f(t)\cos\omega t\, dt = \frac{2}{\pi}\int_0^1 (1-t)\cos\omega t\, dt,$$

and this is the same as the Fourier transform of the *even* extension f_e of f to the whole real line $\mathbb{R}$. In this case, however, we can evaluate the last integral directly by integration by parts:

$$\begin{aligned}\int_0^1 (1-t)\cos\omega t\,dt &= \int_0^1 \cos\omega t\,dt - \int_0^1 t\cos\omega t\,dt\\ &= \left.\frac{\sin\omega t}{\omega}\right|_0^1 - \left[t\cdot\left.\frac{\sin\omega t}{\omega}\right|_0^1 - \frac{1}{\omega}\int_0^1 \sin\omega t\,dt\right]\\ &= \frac{\sin\omega}{\omega} - \frac{\sin\omega}{\omega} + \frac{1}{\omega}\left[-\frac{1}{\omega}\left.\cos\omega t\right|_0^1\right]\\ &= \frac{1-\cos\omega}{\omega^2},\end{aligned}$$

and therefore

$$\widehat{f_c}(\omega) = \frac{2}{\pi}\frac{1-\cos\omega}{\omega^2}$$

for $\omega > 0$.

Knowing that f_e is absolutely integrable implies that $\widehat{f_c}$ is continuous at $\omega = 0$, and we have

$$\widehat{f_c}(0) = \lim_{\omega\to 0+}\frac{2}{\pi}\frac{1-\cos\omega}{\omega^2} = \frac{2}{\pi}\lim_{\omega\to 0+}\frac{\sin\omega}{2\omega} = \frac{1}{\pi}$$

by L'Hospital's rule. Therefore, we have

$$\widehat{f_c}(\omega) = \begin{cases}\dfrac{2}{\pi}\dfrac{1-\cos\omega}{\omega^2}, & \text{for } \omega > 0\\[2ex] \dfrac{1}{\pi}, & \text{for } \omega = 0.\end{cases}$$

Since f_e is continuous for all $x \in \mathbb{R}$, from Dirichlet's theorem the inverse Fourier cosine transform of $\widehat{f_c}$ is given by

$$\frac{2}{\pi}\int_0^\infty \frac{1-\cos\omega}{\omega^2}\cos\omega x\,d\omega = \begin{cases}1-x, & \text{for } 0 \leqslant x < 1,\\[1ex] 0, & \text{for } x \geqslant 1.\end{cases}$$

Exercise 16.9. XX
Find the Fourier sine transform of

$$f(x)=\frac{x}{1+x^2},\qquad x>0,$$

and write $f(x)$ as an inverse sine transform. Use a known Fourier transform and the fact that if $f(x),\ x\geqslant 0$, is the restriction of an *odd* function f_o, then

$$\mathcal{S}[\,f\,](\omega)=-2i\mathcal{F}[\,f_o\,](\omega)$$

for all $\omega\geqslant 0$.
Hint: Consider the Fourier sine transform of $g(x)=e^{-x}$.

Solution. We can find the Fourier sine transform of the given function using the method suggested, or we can find it directly. To do this, we consider the function

$$g(x)=e^{-x},\qquad x>0$$

with Fourier sine transform given by

$$\widehat{g}_s(\omega)=\mathcal{S}[\,f\,](\omega)=\frac{2}{\pi}\int_0^{\infty}e^{-t}\sin\omega t\,dt$$

and we can evaluate this integral by integrating by parts:

$$\begin{aligned}\int_0^{\infty}e^{-t}\sin\omega t\,dt &= -\frac{e^{-t}\cos\omega t}{\omega}\bigg|_0^{\infty}-\frac{1}{\omega}\int_0^{\infty}e^{-t}\cos\omega t\,dt\\ &=\frac{1}{\omega}-\frac{1}{\omega}\left[e^{-t}\frac{\sin\omega t}{\omega}\bigg|_0^{\infty}+\frac{1}{\omega}\int_0^{\infty}e^{-t}\sin\omega t\,dt\right]\\ &=\frac{1}{\omega}-\frac{1}{\omega^2}\int_0^{\infty}e^{-t}\sin\omega t\,dt,\end{aligned}$$

so that

$$\left(1+\frac{1}{\omega^2}\right)\int_0^{\infty}e^{-t}\sin\omega t\,dt=\frac{1}{\omega}.$$

Therefore,

$$\int_0^{\infty}e^{-t}\sin\omega t\,dt=\frac{\omega}{1+\omega^2}$$

for $\omega\geqslant 0$, so that

$$\widehat{g}_s(\omega)=\frac{2}{\pi}\frac{\omega}{1+\omega^2}$$

for $\omega\geqslant 0$.
Taking the inverse Fourier sine transform of this, we have

$$g(x)=\int_0^{\infty}\widehat{g}_s(\omega)\sin\omega x\,d\omega=\frac{2}{\pi}\int_0^{\infty}\frac{\omega}{1+\omega^2}\sin\omega x\,d\omega.$$

Now interchanging x and ω, we have

$$e^{-\omega} = g(\omega) = \frac{2}{\pi}\int_0^\infty \frac{x}{1+x^2}\sin \omega x\, dx$$

and

$$\widehat{f}_s(\omega) = \frac{2}{\pi}\int_0^\infty \frac{x}{1+x^2}\sin \omega x\, dx = g(\omega) = e^{-\omega}$$

for $\omega \geqslant 0$. From the above, we can write $f(x)$ as an inverse Fourier sine transform:

$$f(x) = \frac{x}{1+x^2} = \int_0^\infty e^{-\omega}\sin \omega x\, d\omega$$

for $x > 0$.

Exercise 16.10. ✗✗
Show that

$$\int_0^\infty \frac{\cos \lambda x}{a^2+\lambda^2}\, d\lambda = \frac{\pi}{2a}e^{-ax}$$

for $a > 0, \quad x > 0$.

Solution. Let $f(x) = e^{-ax}$ for $x > 0$ and let f_e be the even extension of f to all of $(-\infty, \infty)$. Since f_e is piecewise smooth and agrees with f for $x > 0$, we can write (again from Dirichlet's theorem)

$$e^{-ax} = \int_0^\infty A(\lambda)\cos \lambda x\, d\lambda, \quad x > 0,$$

where

$$A(\lambda) = \frac{2}{\pi}\int_0^\infty e^{-ax}\cos \lambda x\, dx.$$

To evaluate $A(\lambda)$, integrating by parts twice, since $a > 0$ we have

$$\begin{aligned}
\int_0^\infty e^{-ax}\cos \lambda x\, dx &= \left.\frac{e^{-ax}\sin \lambda x}{\lambda}\right|_0^\infty - \int_0^\infty \frac{(-a)e^{-ax}\sin \lambda x}{\lambda}\, dx \\
&= \frac{a}{\lambda}\left[-\left.\frac{e^{-ax}\cos \lambda x}{\lambda}\right|_0^\infty + \int_0^\infty \frac{(-a)e^{-ax}\cos \lambda x}{\lambda}\, dx\right] \\
&= \frac{a}{\lambda}\left[\frac{1}{\lambda} - \frac{a}{\lambda}\int_0^\infty e^{-ax}\cos \lambda x\, dx\right],
\end{aligned}$$

so that

$$\left(1 + \frac{a^2}{\lambda^2}\right)\int_0^\infty e^{-ax}\cos \lambda x\, dx = \frac{a}{\lambda^2}.$$

Therefore,

$$\int_0^\infty e^{-ax} \cos \lambda x \, dx = \frac{a}{\lambda^2 + a^2}$$

for $\lambda > 0$ (you should check that this holds for $\lambda = 0$ also). Therefore,

$$A(\lambda) = \frac{2a}{\pi(a^2 + \lambda^2)}$$

for $\lambda \geqslant 0$, and

$$e^{-ax} = \frac{2}{\pi} \int_0^\infty \frac{a \cos \lambda x}{a^2 + \lambda^2} \, d\lambda$$

for $0 < x < \infty$. Thus,

$$\int_0^\infty \frac{\cos \lambda x}{a^2 + \lambda^2} \, d\lambda = \frac{\pi}{2a} e^{-ax}$$

for $x > 0$.

Exercise 16.11. XXX
Given the function

$$f(x) = \begin{cases} 0, & \text{for } x < 0, \\ 1, & \text{for } 0 < x < 1, \\ 2, & \text{for } 1 < x < 2, \\ 0, & \text{for } x > 2. \end{cases}$$

(a) Find the Fourier integral formula for $f(x)$.

(b) Find the Fourier sine integral formula for $f(x)$.

(c) Find the Fourier cosine integral formula for $f(x)$.

(d) Find the Fourier transform of $f(x)$.

(e) Find the Fourier sine transform of $f(x)$.

(f) Find the Fourier cosine transform of $f(x)$.

Solution.

(a) The Fourier integral representation of f is given by

$$\frac{f(x^+) + f(x^-)}{2} = \int_0^\infty (A(\omega) \cos \omega x + B(\omega) \sin \omega x) \, d\omega,$$

where

$$A(\omega) = \frac{1}{\pi} \int_{-\infty}^\infty f(x) \cos \omega x \, dx \quad \text{and} \quad B(\omega) = \frac{1}{\pi} \int_{-\infty}^\infty f(x) \sin \omega x \, dx$$

for $0 < \omega < \infty$.

We have

$$A(\omega) = \frac{1}{\pi}\int_0^2 f(x)\cos\omega x\,dx$$

$$= \frac{1}{\pi}\int_0^1 \cos\omega x\,dx + \frac{1}{\pi}\int_1^2 2\cos\omega x\,dx$$

$$= \frac{\sin\omega x}{\pi\omega}\bigg|_0^1 + \frac{2\sin\omega x}{\pi\omega}\bigg|_1^2$$

$$= \frac{1}{\pi}\left(-\frac{\sin\omega}{\omega} + \frac{2\sin 2\omega}{\omega}\right)$$

and

$$B(\omega) = \frac{1}{\pi}\int_0^2 f(x)\sin\omega x\,dx$$

$$= \frac{1}{\pi}\int_0^1 \sin\omega x\,dx + \frac{1}{\pi}\int_1^2 2\sin\omega x\,dx$$

$$= \frac{1}{\pi\omega}(1 - \cos\omega + 2\cos\omega - 2\cos 2\omega)$$

$$= \frac{1}{\pi\omega}(1 + \cos\omega - 2\cos 2\omega)\,,$$

so that

$$\frac{f(x^+) + f(x^-)}{2} = \int_0^\infty \left(\frac{\sin\omega}{\pi\omega}(4\cos\omega - 1)\cos\omega x \right.$$
$$\left. + \frac{1}{\pi\omega}(3 + \cos\omega - 4\cos^2\omega)\sin\omega x\right) d\omega$$

for $-\infty < x < \infty$.

(a′) For those of you who prefer the complex Fourier integral representation of f, we have

$$\frac{f(x^+) + f(x^-)}{2} = \int_{-\infty}^\infty F(\omega)e^{-i\omega x}\,d\omega,$$

where

$$F(\omega) = \frac{1}{2\pi}\int_{-\infty}^\infty f(\xi)e^{i\omega\xi}\,d\xi$$

and

$$F(\omega) = \frac{1}{2\pi}\int_0^1 e^{i\omega\xi}\,d\xi + \frac{1}{2\pi}\int_1^2 2e^{i\omega\xi}\,d\xi$$

$$= \frac{1}{2\pi i\omega}\left(2e^{2i\omega} - e^{i\omega} - 1\right),$$

so that

$$\frac{f(x^+) + f(x^-)}{2} = \frac{1}{2\pi i} \int_{-\infty}^{\infty} \left(\frac{2e^{2i\omega} - e^{i\omega} - 1}{\omega} e^{-i\omega x} \right) d\omega$$

for $-\infty < x < \infty$.

(b) The Fourier sine integral formula for $f(x)$ is

$$f(x) = \int_0^{\infty} B(\omega) \sin \omega x \, d\omega,$$

where

$$B(\omega) = \frac{2}{\pi} \int_0^{\infty} f(x) \sin \omega x \, dx.$$

For the function f above, we have

$$\begin{aligned} B(\omega) &= \frac{2}{\pi} \int_0^{\infty} f(x) \sin \omega x \, dx \\ &= \frac{2}{\pi} \int_0^{2} f(x) \sin \omega x \, dx \\ &= \frac{2}{\pi} \int_0^{1} \sin \omega x \, dx + \frac{2}{\pi} \int_1^{2} 2 \sin \omega x \, dx \\ &= -\frac{2}{\pi\omega} \cos \omega x \Big|_0^1 - \frac{4}{\pi\omega} \cos \omega x \Big|_1^2 \\ &= -\frac{2}{\pi\omega} \cos \omega + \frac{2}{\pi\omega} - \frac{4}{\pi\omega} (\cos 2\omega - \cos \omega), \end{aligned}$$

so that

$$B(\omega) = \frac{2}{\pi\omega} (1 + \cos \omega - 2 \cos 2\omega)$$

for $0 < \omega < \infty$. The Fourier sine integral formula for $f(x)$ is

$$f(x) = \frac{2}{\pi} \int_0^{\infty} \frac{1 + \cos \omega - 2 \cos 2\omega}{\omega} \sin \omega x \, d\omega$$

for $0 < x < \infty$.

(c) The Fourier cosine integral formula for $f(x)$ is

$$f(x) = \int_0^{\infty} A(\omega) \cos \omega x \, d\omega,$$

where

$$A(\omega) = \frac{2}{\pi} \int_0^{\infty} f(x) \cos \omega x \, dx.$$

For the function f above, we have

$$\begin{aligned}A(\omega) &= \frac{2}{\pi}\int_0^{\infty} f(x)\cos\omega x\,dx \\ &= \frac{2}{\pi}\int_0^{2} f(x)\cos\omega x\,dx \\ &= \frac{2}{\pi}\int_0^{1} \cos\omega x\,dx + \frac{2}{\pi}\int_1^{2} 2\cos\omega x\,dx \\ &= \frac{2}{\pi\omega}\sin\omega x\Big|_0^1 + \frac{4}{\pi\omega}\sin\omega x\Big|_1^2 \\ &= \frac{2}{\pi\omega}\sin\omega + \frac{4}{\pi\omega}(\sin 2\omega - \sin\omega),\end{aligned}$$

so that

$$A(\omega) = \frac{2}{\pi\omega}(2\sin 2\omega - \sin\omega)$$

for $0 < \omega < \infty$. The Fourier cosine integral formula for $f(x)$ is

$$f(x) = \frac{2}{\pi}\int_0^{\infty} \frac{2\sin 2\omega - \sin\omega}{\omega}\cos\omega x\,d\omega$$

for $0 < x < \infty$.

(d) The Fourier transform of $f(x)$ is

$$F(\omega) = \frac{1}{2\pi}\int_{-\infty}^{\infty} f(\xi)e^{i\omega\xi}\,d\xi$$

for $-\infty < x < \infty$, and for the function given we have

$$F(\omega) = \frac{1}{2\pi}\int_0^{1} e^{i\omega\xi}\,d\xi + \frac{1}{2\pi}\int_1^{2} 2e^{i\omega\xi}\,d\xi;$$

that is,

$$F(\omega) = \frac{1}{2\pi i\omega}\left(2e^{2i\omega} - e^{i\omega} - 1\right)$$

for $-\infty < x < \infty$.

(e) The Fourier sine transform of $f(x)$ is

$$\mathcal{S}[\,f(x)\,](\omega) = \frac{2}{\pi}\int_0^{\infty} f(x)\sin\omega x\,dx = \frac{2}{\pi}\left(\frac{1+\cos\omega - 2\cos 2\omega}{\omega}\right)$$

for $0 < \omega < \infty$.

(f) The Fourier cosine transform of $f(x)$ is

$$\mathcal{C}[\,f(x)\,](\omega) = \frac{2}{\pi}\int_0^\infty f(x)\cos\omega x\,dx = \frac{2}{\pi}\left(\frac{2\sin 2\omega - \sin\omega}{\omega}\right)$$

for $0 < \omega < \infty$.

Exercise 16.12. XX

Use Fourier transforms to find the solution to

$$\frac{\partial u}{\partial t} = \frac{\partial^2 u}{\partial x^2}, \qquad -\infty < x < \infty, \quad t > 0,$$

$$u(x,0) = \begin{cases} 100, & |x| < 1, \\ 0, & |x| > 1, \end{cases}$$

in terms of the error function.

Solution. Let $\widehat{u}(\omega,t) = \mathcal{F}\,[\,u(x,t)\,]\,(\omega)$; taking the transform of both sides of the partial differential, equation we have

$$\frac{d\widehat{u}(\omega,t)}{dt} = (i\omega)^2\widehat{u}(\omega,t) = -\omega^2\widehat{u}(\omega,t),$$

and if $u(x,0) = f(x)$, the initial condition gives

$$\widehat{u}(\omega,0) = \widehat{f}(\omega) = \frac{1}{2\pi}\int_{-1}^{1} 100\,e^{i\omega x}\,dx.$$

The solution to this ordinary differential equation is

$$\widehat{u}(\omega,t) = \widehat{u}(\omega,0)\,e^{-\omega^2 t}$$

and from the convolution theorem we have

$$u(x,t) = \frac{1}{2\pi}\mathcal{F}^{-1}\left[e^{-\omega^2 t}\right](x) * f(x) = \frac{1}{\sqrt{4\pi t}}\int_{-1}^{1} 100\,e^{-\frac{(x-s)^2}{4t}}\,ds.$$

The solution

$$u(x,t) = \frac{1}{\sqrt{4\pi t}}\int_{-1}^{1} 100\,e^{-\frac{(x-s)^2}{4t}}\,ds$$

for $-\infty < x < \infty,\ t > 0$, can be written in terms of the error function by making the substitution

$$z = \frac{x-s}{\sqrt{4t}} \qquad \text{and} \qquad dz = -\frac{ds}{\sqrt{4t}}.$$

When $s = -1$, $z = (x+1)/\sqrt{4t}$, and when $s = 1$, $z = (x-1)/\sqrt{4t}$, so that

$$\begin{aligned} u(x,t) &= \frac{100}{\sqrt{4\pi t}}(-\sqrt{4t}) \int_{(x+1)/\sqrt{4t}}^{(x-1)/\sqrt{4t}} e^{-z^2}\, dz \\ &= \frac{100}{\sqrt{\pi}} \left(\int_0^{(x+1)/\sqrt{4t}} e^{-t^2}\, dt - \int_0^{(x-1)/\sqrt{4t}} e^{-t^2}\, dt \right) \\ &= \frac{100}{2} \left[\operatorname{erf}\left(\frac{x+1}{\sqrt{4t}}\right) - \operatorname{erf}\left(\frac{x-1}{\sqrt{4t}}\right) \right]; \end{aligned}$$

therefore,

$$u(x,t) = 50 \left[\operatorname{erf}\left(\frac{x+1}{\sqrt{4t}}\right) - \operatorname{erf}\left(\frac{x-1}{\sqrt{4t}}\right) \right]$$

for $-\infty < x < \infty$, $t > 0$, where

$$\operatorname{erf}(x) = \frac{2}{\sqrt{\pi}} \int_0^x e^{-z^2}\, dz$$

is the error function.

Exercise 16.13. XX

(a) Show that the Fourier transform is a linear operator; that is, show that

$$\mathcal{F}\left[c_1 f(x) + c_2 g(x)\right](\omega) = c_1 F(\omega) + c_2 G(\omega).$$

(b) Show that $\mathcal{F}\left[f(x)g(x)\right](\omega) \neq F(\omega)G(\omega)$.

Solution.

(a) If the Fourier transforms of f and g both exist, and c_1 and c_2 are constants, then

$$\begin{aligned} \mathcal{F}\left[c_1 f(x) + c_2 g(x)\right](\omega) &= \frac{1}{2\pi} \int_{-\infty}^{\infty} \left(c_1 f(x) + c_2 g(x)\right) e^{i\omega x}\, dx \\ &= \frac{c_1}{2\pi} \int_{-\infty}^{\infty} f(x) e^{i\omega x}\, dx + \frac{c_2}{2\pi} \int_{-\infty}^{\infty} g(x) e^{i\omega x}\, dx \\ &= c_1 \mathcal{F}\left[\, f(x)\, \right](\omega) + c_2 \mathcal{F}\left[\, g(x)\, \right](\omega) \end{aligned}$$

for all ω; that is,

$$\mathcal{F}\left[c_1 f(x) + c_2 g(x)\right] = c_1 \mathcal{F}\left[\, f(x)\, \right] + c_2 \mathcal{F}\left[\, g(x)\, \right]$$

and the Fourier transform is a linear operator.

(b) Let f and g be functions such that $\mathcal{F}[f](\omega) = F(\omega)$, $\mathcal{F}[g](\omega) = G(\omega)$, both exist; for example,

$$f(x) = g(x) = \begin{cases} 1, & \text{for } |x| < a, \\ 0, & \text{for } |x| > a. \end{cases}$$

Then

$$F(\omega) = G(\omega) = \frac{1}{2\pi}\int_{-a}^{a} e^{i\omega x}\,dx = \frac{\sin \omega a}{\pi\omega}.$$

Now let $h(x) = f(x) \cdot g(x)$ for $-\infty < x < \infty$. Clearly, $h(x) = f(x) = g(x)$ for all x, and

$$H(\omega) = \frac{\sin \omega a}{\pi\omega} \neq \frac{\sin^2 \omega a}{\pi^2\omega^2} = F(\omega) \cdot G(\omega).$$

Exercise 16.14. ✗

If $F(\omega)$ is the Fourier transform of $f(x)$, show that the inverse Fourier transform of $e^{i\omega\beta}F(\omega)$ is $f(x-\beta)$. This result is known as the **shift theorem** for Fourier transforms.

Solution. We have

$$\begin{aligned}\mathcal{F}^{-1}\left[e^{i\omega\beta}F(\omega)\right] &= \int_{-\infty}^{\infty} F(\omega)e^{i\omega\beta}e^{-i\omega x}\,d\omega \\ &= \int_{-\infty}^{\infty} F(\omega)e^{-i\omega(x-\beta)}\,d\omega \\ &= f(x-\beta).\end{aligned}$$

Exercise 16.15. ✗✗

(a) Solve

$$\frac{\partial u}{\partial t} = k\frac{\partial^2 u}{\partial x^2} - \gamma u, \quad -\infty < x < \infty, \quad t > 0,$$

$$u(x,0) = f(x).$$

(b) Does your solution suggest a simplifying transformation?

Solution.

(a) If $u(x,t)$ is the solution to

$$\frac{\partial u}{\partial t} = k\frac{\partial^2 u}{\partial x^2} - \gamma u, \quad -\infty < x < \infty, \quad t > 0,$$

$$u(x,0) = f(x),$$

let

$$\widehat{u}(\omega,t) = \mathcal{F}[\,u(x,t)\,](\omega) \qquad \text{and} \qquad \widehat{u}(\omega,0) = \widehat{f}(\omega);$$

then $\widehat{u}(\omega,t)$ satisfies the initial value problem

$$\frac{d\widehat{u}}{dt} = -(k\omega^2 + \gamma)\widehat{u}, \quad t > 0,$$

$$\widehat{u}(\omega,0) = \widehat{f}(\omega),$$

with solution

$$\widehat{u}(\omega,t) = \widehat{f}(\omega)e^{-(k\omega^2+\gamma)t} = \widehat{f}(\omega)e^{-k\omega^2 t}e^{-\gamma t}.$$

The solution to the partial differential equation is

$$\begin{aligned} u(x,t) &= \mathcal{F}^{-1}\left[\,\widehat{u}(\omega,t)\,\right](x) \\ &= \mathcal{F}^{-1}\left[\widehat{f}(\omega)e^{-k\omega^2 t}e^{-\gamma t}\right](x) \\ &= e^{-\gamma t}\mathcal{F}^{-1}\left[\widehat{f}(\omega)e^{-k\omega^2 t}\right](x) \qquad \text{(since } \mathcal{F}^{-1} \text{ is linear)} \\ &= e^{-\gamma t}\frac{1}{2\pi}\, f(x) * g\,(x,t), \end{aligned}$$

where

$$g(x,t) = \sqrt{\frac{\pi}{kt}}\, e^{-x^2/4kt}.$$

Therefore,

$$u(x,t) = e^{-\gamma t}\int_{-\infty}^{\infty} f(s)\frac{e^{-(x-s)^2/4kt}}{\sqrt{4\pi kt}}\, ds$$

for $-\infty < x < \infty$ and $t > 0$.

(b) If we multiply the solution above by $e^{\gamma t}$, we find that

$$e^{\gamma t}u(x,t) = \frac{1}{2\pi}\, f(x) * g\,(x,t) = \frac{1}{\sqrt{4\pi kt}}\int_{-\infty}^{\infty} f(s)\, e^{-(x-s)^2/4kt}\, ds,$$

which looks like the solution to a homogeneous heat equation.

Indeed, if we define

$$w(x,t) = e^{\gamma t} u(x,t),$$

then

$$\begin{aligned}
\frac{\partial w}{\partial t} &= \gamma e^{\gamma t} u + e^{\gamma t} \frac{\partial u}{\partial t} \\
&= \gamma w + e^{\gamma t} \left(k \frac{\partial^2 u}{\partial x^2} - \gamma u \right) \\
&= \gamma w + k \frac{\partial^2 w}{\partial x^2} - \gamma w,
\end{aligned}$$

so that

$$\begin{aligned}
\frac{\partial w}{\partial t} &= k \frac{\partial^2 w}{\partial x^2}, \\
w(x,0) &= f(x),
\end{aligned}$$

for $-\infty < x < \infty, \quad t > 0.$

Exercise 16.16. XXX

Solve

$$\begin{aligned}
&\frac{\partial u}{\partial t} = k \frac{\partial^2 u}{\partial x^2}, \quad 0 < x < \infty, \quad t > 0, \\
&\frac{\partial u}{\partial x}(0,t) = 0, \\
&u(x,0) = f(x).
\end{aligned}$$

Solution. Since the boundary condition is a Neumann condition, we use the Fourier cosine transform. Let

$$\widetilde{u}(\omega,t) = \mathcal{C}\left[\, u(x,t) \,\right](\omega) = \frac{2}{\pi} \int_0^\infty u(x,t) \cos \omega x \, dx$$

and

$$\widetilde{f}(\omega) = \mathcal{C}\left[\, f(x) \,\right](\omega) = \frac{2}{\pi} \int_0^\infty f(x) \cos \omega x \, dx;$$

then

$$\mathcal{C}\left[\frac{\partial u}{\partial t}(x,t) \right](\omega) = \frac{\partial \widetilde{u}}{\partial t}(\omega,t)$$

and

$$\mathcal{C}\left[\frac{\partial^2 u}{\partial x^2}(x,t)\right](\omega) = -\frac{2}{\pi}\frac{\partial u}{\partial x}(0,t) - \omega^2\widetilde{u}(\omega,t),$$

and from the boundary condition, $\partial u(0,t)/\partial x = 0$, so that

$$\mathcal{C}\left[\frac{\partial^2 u}{\partial x^2}(x,t)\right](\omega) = -\omega^2\widetilde{u}(\omega,t).$$

After taking the Fourier cosine transform of both sides of the partial differential equation

$$\frac{\partial u}{\partial t} = k\frac{\partial^2 u}{\partial x^2},$$

the transform $\widetilde{u}(\omega,t)$ satisfies the initial value problem

$$\frac{d\widetilde{u}}{dt}(\omega,t) + k\omega^2\widetilde{u}(\omega,t) = 0,$$

$$\widetilde{u}(\omega,0) = \widetilde{f}(\omega),$$

with solution

$$\widetilde{u}(\omega,t) = \widetilde{u}(\omega,0)e^{-\omega^2 kt} = \widetilde{f}(\omega)e^{-\omega^2 kt}$$

for $0 < \omega < \infty$ and $t > 0$. Therefore,

$$u(x,t) = \int_0^\infty \widetilde{f}(\omega)e^{-\omega^2 kt}\cos\omega x\,d\omega$$

for $0 < x < \infty$ and $t > 0$. Note that each of the functions $\widetilde{f}(\omega)$, $e^{-\omega^2 kt}$, and $\cos\omega x$ in the integrand makes sense for all ω, and each is an even function of ω, so that

$$\int_0^\infty \widetilde{f}(\omega)e^{-\omega^2 kt}\cos\omega x\,d\omega = \frac{1}{2}\int_{-\infty}^\infty \widetilde{f}(\omega)e^{-\omega^2 kt}\cos\omega x\,d\omega.$$

Since $\sin\omega x$ is an odd function of ω, then

$$\int_{-\infty}^\infty \widetilde{f}(\omega)e^{-\omega^2 kt}\sin\omega x\,d\omega = 0,$$

and we can write the solution $u(x,t)$ as

$$\begin{aligned} u(x,t) &= \frac{1}{2}\int_{-\infty}^\infty \widetilde{f}(\omega)e^{-\omega^2 kt}(\cos\omega x - i\sin\omega x)\,d\omega \\ &= \int_{-\infty}^\infty \frac{\widetilde{f}(\omega)}{2}e^{-\omega^2 kt}e^{-i\omega x}\,d\omega; \end{aligned}$$

that is,

$$u(x,t) = \mathcal{F}^{-1}\left[\frac{\widetilde{f}(\omega)}{2}e^{-\omega^2 kt}\right](x). \qquad (16.1)$$

Let f_{even} be the even extension of $f(x)$ to $(-\infty, \infty)$; then

$$
\begin{aligned}
\frac{\widetilde{f}(\omega)}{2} &= \frac{1}{2}\frac{2}{\pi}\int_0^\infty f(x)\cos\omega x\,dx \\
&= \frac{1}{2\pi}\int_{-\infty}^\infty f_{\text{even}}(x)\cos\omega x\,dx \\
&= \frac{1}{2\pi}\int_{-\infty}^\infty f_{\text{even}}(x)(\cos\omega x + i\sin\omega x)\,dx \\
&= \frac{1}{2\pi}\int_{-\infty}^\infty f_{\text{even}}(x)e^{i\omega x}\,dx \\
&= \mathcal{F}\left[\, f_{\text{even}}(x)\,\right](\omega),
\end{aligned}
$$

so that

$$
\frac{\widetilde{f}(\omega)}{2} = \mathcal{F}\left[\, f_{\text{even}}(x)\,\right](\omega). \tag{16.2}
$$

From (16.1) and (16.2) it follows that $u(x,t)$ is the solution to the boundary value–initial value problem

$$
\begin{aligned}
&\frac{\partial u}{\partial t} = k\frac{\partial^2 u}{\partial x^2}, \qquad -\infty < x < \infty, \quad t > 0 \\
&\frac{\partial u}{\partial x}(0,t) = 0, \quad t > 0 \\
&u(x,0) = f_{\text{even}}(x), \quad -\infty < x < \infty,
\end{aligned}
$$

and therefore

$$
u(x,t) = f_{\text{even}}(x) * G(x,t),
$$

where $G(x,t)$ is the heat kernel or Gaussian kernel

$$
G(x,t) = \frac{1}{\sqrt{4\pi kt}}e^{-x^2/4kt}.
$$

The solution is then

$$
\begin{aligned}
u(x,t) &= f_{\text{even}}(x) * G(x,t) \\
&= \frac{1}{\sqrt{4\pi kt}}\int_{-\infty}^\infty f_{\text{even}}(s)e^{-(x-s)^2/4kt}\,ds \\
&= \frac{1}{\sqrt{4\pi kt}}\int_0^\infty f(s)\left(e^{-(x+s)^2/4kt} + e^{-(x-s)^2/4kt}\right)ds,
\end{aligned}
$$

so that

$$u(x,t) = \frac{1}{\sqrt{4\pi kt}} \int_0^\infty f(s) \left(e^{-(x+s)^2/4kt} + e^{-(x-s)^2/4kt} \right) ds$$

for $0 < x < \infty$, $t > 0$. Note that this is just the **method of images** we used in Chapter 9.

Exercise 16.17. XXXX

Solve the circularly symmetric diffusion equation on an infinite two-dimensional domain:

$$\frac{\partial u}{\partial t} = \frac{k}{r}\frac{\partial}{\partial r}\left(r \frac{\partial u}{\partial r} \right), \quad r > 0, \quad t > 0,$$

$$u \quad \text{bounded as } r \to 0^+,$$

$$r\, u \frac{\partial u}{\partial r} \to 0 \quad \text{as } r \to \infty,$$

$$u(r,0) = f(r).$$

Solution. We solve this problem using separation of variables. We write

$$u(r,t) = \varphi(r) \cdot T(t),$$

so that

$$\varphi(r) \cdot T'(t) = \frac{k}{r} \left(r\varphi'(r) \right)' \cdot T(t) = \left(k\varphi''(r) + \frac{k}{r}\, \varphi'(r) \right) \cdot T(t),$$

and separating variables, we have

$$\frac{T'(t)}{kT(t)} = \frac{\varphi''(r) + \frac{1}{r}\, \varphi'(r)}{\varphi(r)} = -\lambda,$$

where λ is the separation constant. Thus, we obtain the following two ordinary differential equations:

$$\varphi'' + \frac{1}{r}\, \varphi' + \lambda \varphi = 0,$$

$$T' + \lambda k T = 0.$$

Note that the boundedness conditions on φ are satisfied if φ satisfies the singular Sturm-Liouville problem

$$\varphi'' + \frac{1}{r}\varphi' + \lambda\varphi = 0, \quad 0 < r < \infty,$$

$$\varphi(r) \quad \text{bounded as } r \to 0^+,$$

$$r\varphi(r)\varphi'(r) \to 0 \quad \text{as } r \to \infty.$$

Multiplying by r, the spatial problem can be written in the form

$$r\varphi'' + \varphi' + \lambda r\varphi = 0, \quad 0 < r < \infty,$$

$$\varphi(r) \quad \text{bounded as } r \to 0^+,$$

$$r\varphi(r)\varphi'(r) \to 0 \quad \text{as } r \to \infty.$$

We solve the singular Sturm-Liouville problem for φ first. The Rayleigh quotient is

$$\lambda = \frac{-r\varphi\varphi'\Big|_0^\infty + \int_0^\infty r(\varphi')^2\,dr}{\int_0^\infty r\,\varphi^2\,dr},$$

and from the boundedness conditions,

$$-r\varphi\varphi'\Big|_0^\infty = -\lim_{r\to\infty} r\varphi(r)\varphi'(r) = 0,$$

so that $\lambda \geqslant 0$; that is, there are no negative eigenvalues.
Case (i): If $\lambda = 0$, then the differential equation is

$$(r\varphi')' = 0,$$

with general solution

$$\varphi(r) = A\log r + B.$$

Applying the boundedness condition, we have $A = 0$, and the eigenfunction is

$$\varphi_0(r) = 1$$

for $0 < r < \infty$.
Case (ii): If $\lambda > 0$, multiplying the differential equation by r, we get

$$r(r\varphi')' + \lambda r^2\varphi = 0,$$

which is Bessel's (parametric) equation of order 0, with general solution

$$\varphi(r) = AJ_0\left(\sqrt{\lambda}\,r\right) + BY_0\left(\sqrt{\lambda}\,r\right).$$

Applying the boundedness condition, we have $B = 0$, and the solution may be written

$$\varphi(r) = J_0(\mu r)$$

for $0 < r < \infty$, where $\lambda = \mu^2$. The corresponding T equation,

$$T' + \mu^2 kT = 0,$$

has solution

$$T(t) = e^{-\mu^2 kt}$$

for $t > 0$. Therefore, for each $\mu \geqslant 0$, the function

$$u(r, t, \mu) = J_0(\mu r)\, e^{-\mu^2 kt}, \quad 0 < r < \infty,\ t > 0$$

satisfies the partial differential equation and the boundedness condition, and from the superposition principle, we write

$$u(r, t) = \int_0^\infty A(\mu) J_0(\mu r)\, e^{-\mu^2 kt} \mu\, d\mu,$$

and this satisfies (formally) the diffusion equation as well as the boundedness conditions. The only thing not yet satisfied is the initial condition

$$u(r, 0) = f(r), \quad 0 < r < \infty,$$

so we want

$$f(r) = u(r, 0) = \int_0^\infty A(\mu) J_0(\mu r)\, \mu\, d\mu$$

for $0 < r < \infty$.

To determine the coefficients $A(\mu)$, we have a theorem analogous to Dirichlet's theorem, called **Hankel's integral theorem**.

Theorem. Given a function f defined on the interval $(0, \infty)$, which is piecewise continuous and of bounded variation on every finite subinterval of $(0, \infty)$, and such that

$$\int_0^\infty \sqrt{r}\, |f(r)|\, dr < \infty,$$

for each $r > 0$ we have

$$\frac{1}{2}\left[f(r^+) + f(r^-)\right] = \int_0^\infty A(\lambda) J_0(\lambda r)\, \lambda\, d\lambda,$$

where

$$A(\lambda) = \int_0^\infty f(r) J_0(\lambda r)\, r\, dr.$$

Note: The coefficient

$$A(\lambda) = \int_0^\infty f(r) J_0(\lambda r)\, r\, dr$$

is called the **Fourier-Bessel transform** of $f(r)$, or the **Hankel transform** of $f(r)$.

The solution to the circularly symmetric diffusion equation on an infinite two-dimensional domain is therefore given by

$$u(r,t) = \int_0^\infty A(\mu) J_0(\mu r)\, e^{-\mu^2 k t} \mu \, d\mu, \quad 0 < r < \infty,\ t > 0,$$

where

$$A(\mu) = \int_0^\infty f(r) J_0(\mu r)\, r \, dr$$

for $\mu > 0$.

CHAPTER 17

METHOD OF CHARACTERISTICS PROBLEMS

Exercise 17.1. ✗✗

Assume that $u(x,t)$ is the linear density of particulate matter being carried by the wind from a dump truck at the oil sands at position $x = 0$ and time t. The wind is moving in the positive x-direction with a constant speed of k meters/sec, and the particulates are condensing out of the air at a rate $ru(x,t)$, where $r > 0$ is constant. The density u satisfies the boundary value–initial value problem

$$\frac{\partial u}{\partial t}(x,t) + k\frac{\partial u}{\partial x}(x,t) = -ru(x,t), \quad 0 < x < \infty, \quad t > 0, \tag{17.1}$$

$$u(x,0) = \phi(x),$$

where $\phi(x)$ is the initial distribution of the particle density. Solve this initial value–boundary value problem using the method of characteristics.

Solution. The method of characteristics reduces the partial differential equation to a pair of ordinary differential equations, one of which is solved for the characteristic curves in the xt-plane along which the solutions to the other equation are easily found.

Partial Differential Equations: Theory and Completely Solved Problems, First Edition. By T. Hillen, I. E. Leonard, H. van Roessel

We write the partial differential equation so that the partial differential operator resembles a directional deriative or a total derivative. For example,

$$\frac{\partial u}{\partial t} + \frac{\partial u}{\partial x}\frac{dx}{dt} = -ru, \tag{17.2}$$

where

$$\frac{dx}{dt} = k.$$

The family of curves with differential equation

$$\frac{dx}{dt} = k$$

are the **characteristic curves** of the partial differential equation (17.1). If $x = x(t)$ is a characteristic curve of (17.1), then along this curve equation (17.2) becomes

$$\frac{\partial}{\partial t}\left(u(x(t),t)\right) + \frac{\partial}{\partial x}\left(u(x(t),t)\right) \cdot \frac{dx}{dt} = -ru(x(t),t);$$

that is,

$$\frac{d}{dt}\left(u(x(t),t)\right) = -ru(x(t),t),$$

which is an ordinary differential equation for $u(x(t),t)$. Letting $v(t) = u(x(t),t)$ for $t > 0$, v satisfies the ordinary differential equation

$$\frac{dv}{dt} + rv = 0,$$

that is, a first-order linear homogeneous ordinary differential equation, and to solve it we multiply by the integrating factor $M(t) = e^{rt}$, to get

$$\frac{d}{dt}\left(e^{rt}v\right) = e^{rt}\frac{dv}{dt} + re^{rt}v = e^{rt}\left(\frac{dv}{dt} + rv\right) = 0$$

for all $t > 0$. Therefore, $e^{rt}v(t)$ is a constant, so that

$$e^{rt}v(t) = e^{r0}v(0) = v(0),$$

that is,

$$e^{rt}u(x(t),t) = u(x(0),0) = \phi(x(0)),$$

so that

$$u(x(t),t) = e^{-rt}\phi(x(0)) \tag{17.3}$$

for all $t > 0$.

Given a point (x,t) in the xt-plane with $t > 0$, there is exactly one characteristic curve that passes through this point, namely,

$$x(t) = kt + x(0),$$

where

$$x = x(t) = kt + x(0),$$

and for this characteristic curve, $x(0)$ is called the **anchor point** of the characteristic through (x, t). Therefore,

$$u(x,t) = e^{-rt}\phi(x(0)) = e^{-rt}\phi(x - kt),$$

and, since (x, t) was arbitrary, the solution to the initial value problem is given by

$$u(x,t) = e^{-rt}\phi(x - kt) \qquad (17.4)$$

for $0 < x < \infty$, $t > 0$. Finally, we note that $e^{rt}u(x,t)$ is constant along the characteristic curves.

Exercise 17.2. XX
Use the method of characteristics to solve the initial value problem

$$\frac{\partial w}{\partial t} + 5\frac{\partial w}{\partial x} = e^{3t}, \quad -\infty < x < \infty, \quad t > 0,$$

$$w(x,0) = e^{-x^2}.$$

Solution. Let

$$\frac{dx}{dt} = 5;$$

then along the characteristic curve $x(t) = 5t + a$, the partial differential equation becomes

$$\frac{dw}{dt} = \frac{\partial w}{\partial t} + \frac{\partial w}{\partial x}\frac{dx}{dt} = e^{3t},$$

so that

$$w(x(t),t) = \frac{1}{3}e^{3t} + K,$$

where K is a constant, and $K = w(x(0), 0) - \frac{1}{3}$, so that

$$w(x(t),t) = \frac{1}{3}e^{3t} + w(x(0),0) - \frac{1}{3} = \frac{1}{3}e^{3t} + w(a,0) - \frac{1}{3} = \frac{1}{3}e^{3t} + e^{-a^2} - \frac{1}{3}.$$

Given the point (x, t), let $x = 5t + a$ be the unique characteristic curve passing through this point; then the anchor point is $a = x - 5t$ and the solution is

$$w(x,t) = \frac{1}{3}e^{3t} + e^{-a^2} - \frac{1}{3} = \frac{1}{3}e^{3t} + e^{-(x-5t)^2} - \frac{1}{3}$$

for $-\infty < x < \infty$, $t > 0$.

Exercise 17.3. ✗✗

Use the method of characteristics to solve the initial value problem

$$\frac{\partial w}{\partial t} - x\frac{\partial w}{\partial x} = 0, \quad -\infty < x < \infty, \quad t > 0,$$

$$w(x,0) = x^3 - 1.$$

Solution. Let

$$\frac{dx}{dt} = -x;$$

then along the characteristic curve $x(t) = x_0 e^{-t}$, the partial differential equation becomes

$$\frac{dw}{dt} = \frac{\partial w}{\partial t} + \frac{\partial w}{\partial x}\frac{dx}{dt} = 0,$$

so that

$$w(x(t),t) = K,$$

where K is a constant, and $K = w(x(0),0)$, so that

$$w(x(t),t) = w(x(0),0) = w(x_0,0) = x_0^3 - 1.$$

Given the point (x,t), let $x = x_0 e^{-t}$ be the unique characteristic curve passing through this point; then $x_0 = xe^t$ is the anchor point and the solution is

$$w(x,t) = x_0^3 - 1 = x^3 e^{3t} - 1$$

for $-\infty < x < \infty,\ t > 0$.

Exercise 17.4. ✗✗

Use the method of characteristics to solve the initial value problem

$$\frac{\partial z}{\partial t} + 3\frac{\partial z}{\partial x} = \sin 2\pi t, \quad -\infty < x < \infty, \quad t > 0,$$

$$z(x,0) = \cos x.$$

Solution. Let

$$\frac{dx}{dt} = 3;$$

then along the characteristic curve $x(t) = 3t + a$, the partial differential equation becomes

$$\frac{dz}{dt} = \frac{\partial z}{\partial t} + \frac{\partial z}{\partial x}\frac{dx}{dt} = \sin 2\pi t,$$

so that

$$z(x(t), t) = -\frac{1}{2\pi}\cos 2\pi t + K,$$

where K is a constant, and

$$z(x(0), 0) = -\frac{1}{2\pi} + K = \cos x(0) = \cos a = \cos(x(t) - 3t),$$

so that

$$K = \cos(x(t) - 3t) + \frac{1}{2\pi}.$$

Given the point (x, t), let $x = 3t + a$ be the unique characteristic curve passing through this point; then the anchor point is $a = x - 3t$ and the solution is

$$z(x, t) = -\frac{1}{2\pi}\cos 2\pi t + \cos(x - 3t) + \frac{1}{2\pi} \tag{17.5}$$

for $-\infty < x < \infty$ and $t > 0$.

As a check, we note that for

$$z(x, t) = -\frac{1}{2\pi}\cos 2\pi t + \cos(x - 3t) + \frac{1}{2\pi},$$

we have

$$\frac{\partial z}{\partial t} = \sin 2\pi t + 3\sin(x - 3t)$$

and

$$\frac{\partial z}{\partial x} = -\sin(x - 3t),$$

so that

$$\frac{\partial z}{\partial t} + 3\frac{\partial z}{\partial x} = \sin 2\pi t.$$

Also,

$$z(x, 0) = -\frac{1}{2\pi} + \cos x + \frac{1}{2\pi} = \cos x,$$

and (17.5) is a solution to the initial value problem.

Exercise 17.5. ✗✗

Solve the first-order equation

$$\frac{\partial u}{\partial t} + 3x\frac{\partial u}{\partial x} = 2t, \quad -\infty < x < \infty, \quad t > 0,$$

$$u(x, 0) = \log(1 + x^2).$$

Solution. Let

$$\frac{dx}{dt} = 3x;$$

then along the characteristic curve $x(t) = ae^{3t}$, the partial differential equation becomes

$$\frac{du}{dt} = \frac{\partial u}{\partial t} + \frac{\partial u}{\partial x}\frac{dx}{dt} = 2t,$$

so that

$$u(x(t), t) = t^2 + K,$$

where K is a constant, and $K = u(x(0), 0)$, so that

$$u(x(t), t) = t^2 + u(x(0), 0) = t^2 + u(a, 0) = t^2 + \log(1 + a^2).$$

Given the point (x, t), let $x = ae^{3t}$ be the unique characteristic curve passing through this point; then the anchor point is $a = xe^{-3t}$ and the solution is

$$u(x, t) = t^2 + \log\left(1 + x^2 e^{-6t}\right)$$

for $-\infty < x < \infty$ and $t > 0$.

Exercise 17.6. ✗✗

Using the method of characteristics, solve

$$\frac{\partial w}{\partial t} + c\frac{\partial w}{\partial x} = e^{2x}, \quad -\infty < x < \infty, \quad t > 0,$$

$$w(x, 0) = f(x).$$

Solution. Let

$$\frac{dx}{dt} = c;$$

then along the characteristic curve $x(t) = ct + a$, the partial differential equation becomes

$$\frac{dw}{dt} = \frac{\partial w}{\partial t} + \frac{\partial w}{\partial x}\frac{dx}{dt} = e^{2x(t)} = e^{2(ct+a)},$$

so that

$$w(x(t), t) = \frac{1}{2c}e^{2ct+2a} + K = \frac{1}{2c}e^{2x(t)} + K,$$

where K is a constant, and $K = w(x(0), 0) - (1/2c)e^{2x(0)}$, so that

$$w(x(t), t) = \frac{1}{2c}e^{2x(t)} + f(x(0)) - \frac{1}{2c}e^{2x(0)};$$

that is,

$$w(x(t), t) = \frac{1}{2c}e^{2x(t)} + f(x(t) - ct) - \frac{1}{2c}e^{2(x(t)-ct)}.$$

Given the point (x, t), let $x = ct + a$ be the unique characteristic curve passing through this point, then the anchor point is $a = x - ct$ and the solution is

$$w(x,t) = \frac{1}{2c} e^{2x} \left(1 - e^{-2ct}\right) + f(x - ct)$$

for $-\infty < x < \infty$ and $t > 0$.

Exercise 17.7. XX

Using the method of characteristics, solve

$$\frac{\partial w}{\partial t} + t\frac{\partial w}{\partial x} = 1, \quad -\infty < x < \infty, \quad t > 0,$$

$$w(x, 0) = f(x).$$

Solution. Let

$$\frac{dx}{dt} = t;$$

then along the characteristic curve $x(t) = (t^2/2) + a$, the partial differential equation becomes

$$\frac{dw}{dt} = \frac{\partial w}{\partial t} + \frac{\partial w}{\partial x}\frac{dx}{dt} = 1,$$

so that

$$w(x(t), t) = t + K,$$

where K is a constant, and $K = w(x(0), 0)$, so that

$$w(x(t), t) = t + w(x(0), 0) = t + f(a).$$

Given the point (x, t), let $x = (t^2/2) + a$ be the unique characteristic curve passing through this point; then the anchor point is $a = x - (t^2/2)$ and the solution is

$$w(x,t) = t + f\left(x - \frac{t^2}{2}\right)$$

for $-\infty < x < \infty$ and $t > 0$.

Exercise 17.8. XX

Consider

$$\frac{\partial u}{\partial t} + 2u\frac{\partial u}{\partial x} = 0, \quad -\infty < x < \infty, \quad t > 0,$$

$$u(x, 0) = f(x).$$

Show that the characteristics are straight lines.

Solution. Along the characteristic curve $x = x(t)$ whose differential equation is

$$\frac{dx}{dt} = 2u(x(t), t),$$

the partial differential equation becomes

$$\frac{d}{dt}\left[u(x(t), t)\right] = \frac{\partial u}{\partial t}(x(t), t) + \frac{dx}{dt} \cdot \frac{\partial u}{\partial x}(x(t), t) = 0,$$

so that $u(x(t), t) = \text{ constant } = u(x(0), 0)$, and

$$\frac{dx}{dt} = 2u(x(t), t) = 2u(x(0), 0),$$

so that

$$x(t) = 2u(x(0), 0)t + x(0) = 2f(x(0))t + x(0)$$

for $t > 0$, and the characteristic curves are the straight lines $x = 2f(x_0)t + x_0$ and intersect the x-axis at the point x_0.

Exercise 17.9. XXX

Consider

$$\frac{\partial u}{\partial t} + 2u\frac{\partial u}{\partial x} = 0, \qquad -\infty < x < \infty, \quad t > 0,$$

with

$$u(x, 0) = f(x) = \begin{cases} 1, & x < 0, \\ 1 + x/a, & 0 < x < a, \\ 2, & x > a. \end{cases}$$

(a) Determine the equations for the characteristics. Sketch the characteristics.

(b) Determine the solution $u(x, t)$. Sketch $u(x, t)$ for t fixed.

Solution.

(a) The equations for the characteristics are $x = 2f(x_0)t + x_0$, where the parameter x_0 is the intersection of the characteristic with the x-axis for $-\infty < x_0 < \infty$.

(i) For $x_0 < 0$, we have $f(x_0) = 1$, and the characteristics have the equation $x = 2t + x_0$.

(ii) For $0 < x_0 < a$, we have $f(x_0) = 1 + x_0/a$, and the characteristics have the equation $x = 2(1 + x_0/a)t + x_0$.

(iii) For $x > a$, we have $f(x_0) = 2$, and the characteristics have the equation $x = 4t + x_0$.

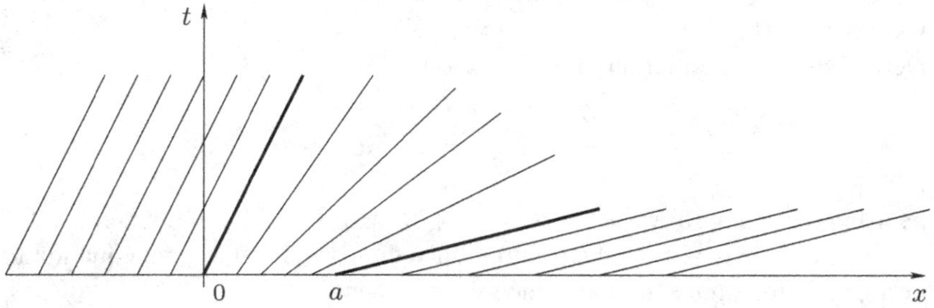

Figure 17.1 Characteristic fan.

The characteristics are sketched in Figure 17.1.

(b) The solution along the characteristic $x = 2f(x_0)t + x_0$ is given by

$$u(x,t) = f(x_0),$$

and considering the cases where

$$x_0 < 0, \qquad 0 < x_0 < a, \qquad \text{and} \qquad a < x_0,$$

we have

$$u(x,t) = \begin{cases} 1, & \text{for} \quad x < 2t, \\ \dfrac{x+a}{a+2t}, & \text{for} \quad 2t < x < 4t + a, \\ 2, & \text{for} \quad x > 4t + a. \end{cases}$$

For a fixed $t > 0$, the solution is sketched in Figure 17.2.

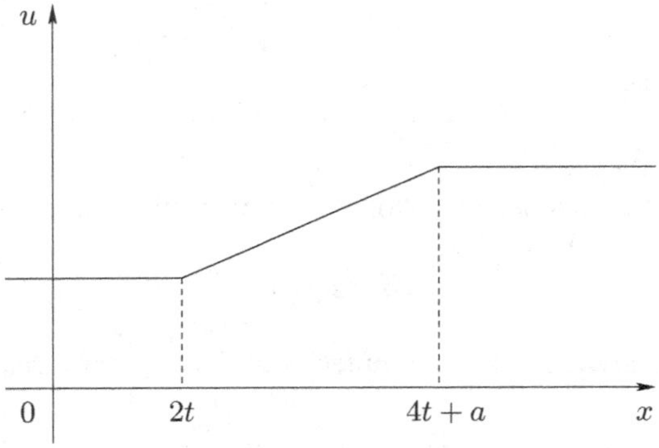

Figure 17.2 Graph of the solution.

Exercise 17.10. XX

Derive the general solution of the equation

$$a\frac{\partial u}{\partial t} + b\frac{\partial u}{\partial x} = u, \qquad a,\, b \neq 0$$

by using the change of variables $\alpha = Ax + Bt$ and $\beta = Cx + Dt$, where $A,\ B,\ C,$ and D are to be determined so as to reduce the partial differential equation to an ordinary differential equation, which we can then solve.

Solution. Let

$$\alpha = Ax + Bt \qquad \text{and} \qquad \beta = Cx + Dt;$$

from the chain rule, we have

$$\frac{\partial u}{\partial x} = A\frac{\partial u}{\partial \alpha} + C\frac{\partial u}{\partial \beta}$$
$$\frac{\partial u}{\partial t} = B\frac{\partial u}{\partial \alpha} + D\frac{\partial u}{\partial \beta},$$

and the original partial differential equation becomes

$$(aB + bA)\frac{\partial u}{\partial \alpha} + (aD + bC)\frac{\partial u}{\partial \beta} = u.$$

Now let $B = -b,\ A = a,\ C = 0,$ and $D = 1/a$; then the equation becomes

$$\frac{\partial u}{\partial \beta} - u = 0,$$

and multiplying this equation by $e^{-\beta}$, we have

$$e^{-\beta}\frac{\partial u}{\partial \beta} - e^{-\beta}u = 0;$$

that is,

$$\frac{\partial}{\partial \beta}\left(e^{-\beta}u\right) = 0,$$

and the quantity $e^{-\beta}u$ is independent of β. Therefore, the solution is

$$u(\alpha, \beta) = f(\alpha)e^{\beta},$$

where f is an arbitrary function of α. In terms of the original variables, the solution is

$$u(x, t) = f(ax - bt)e^{t/a}.$$

Exercise 17.11. XX

Solve the following partial differential equation on an infinite domain:

$$\frac{\partial w}{\partial t} + t\frac{\partial w}{\partial x} = 2t + \sin t, \quad -\infty < x < \infty, \quad t > 0,$$

$$w(x,0) = x^2 + 5.$$

At the end, you might check that your solution really satisfies this problem.

Solution. Let $dx/dt = t$, then along the characteristic curve $x(t) = (t^2/2) + a$, the partial differential equation becomes

$$\frac{dw}{dt} = \frac{\partial w}{\partial t} + \frac{\partial w}{\partial x}\frac{dx}{dt} = 2t + \sin t,$$

so that

$$w(x(t), t) = t^2 - \cos t + K,$$

where K is a constant, and $K = w(x(0), 0) + 1$, and

$$\begin{aligned} w(x(t), t) &= t^2 - \cos t + w(x(0), 0) + 1 = t^2 - \cos t + w(a, 0) + 1 \\ &= t^2 - \cos t + a^2 + 5 + 1 = t^2 - \cos t + \left(x(t) - \frac{t^2}{2}\right)^2 + 6. \end{aligned}$$

Given a point (x, t) in the xt-plane, let $x = (t^2/2) + a$ be the unique characteristic curve passing through this point; then the anchor point is $a = x - (t^2/2)$ and the solution is

$$w(x,t) = t^2 - \cos t + a^2 + 6 = t^2 - \cos t + \left(x - \frac{t^2}{2}\right)^2 + 6 \tag{17.6}$$

for $-\infty < x < \infty,\ t > 0$.

As a check, note that for this function $w(x, t)$ we have

$$\frac{\partial w}{\partial t} + t\frac{\partial w}{\partial x} = 2t + \sin t + 2\left(x - \frac{t^2}{2}\right)(-t) + 2t\left(x - \frac{t^2}{2}\right) = 2t + \sin t$$

and it satisfies the partial differential equation, while

$$w(x,0) = x^2 + 5,$$

and (17.6) is the solution to this problem.

Exercise 17.12. ✗✗✗✗

The displacement $u = u(x,t)$ of an infinitely long string is governed by the wave equation

$$\frac{\partial^2 u}{\partial t^2} = 4\frac{\partial^2 u}{\partial x^2}, \quad -\infty < x < \infty, \quad t > 0.$$

At time $t = 0$ an initial signal is given of the form

$$u(x,0) = f(x) = \begin{cases} x, & \text{for } 0 < x < 1, \\ -x+2, & \text{for } 1 < x < 2, \\ 0, & \text{otherwise,} \end{cases}$$

$$\frac{\partial u}{\partial t}(x,0) = 0, \quad -\infty < x < \infty.$$

(a) Solve this problem.

(b) Sketch the solution for times t_1, t_2, t_3, t_4, t_5, with

$$t_1 = 0, \quad 0 < t_2 < 1/4, \quad t_3 = 1/4, \quad 1/4 < t_4 < 1/2, \quad t_5 = 1/2.$$

(c) At what time does the signal reach the point $\overline{x} = 11$?

Solution.

(a) d'Alembert's solution to the wave equation is given by

$$u(x,t) = \frac{1}{2}\left[f(x+2t) + f(x-2t)\right] + \frac{1}{4}\int_{x-2t}^{x+2t} g(s)\,ds,$$

where

$$f(x+2t) = \begin{cases} x+2t, & \text{if } 0 < x+2t < 1, \\ -x-2t+2, & \text{if } 1 < x+2t < 2, \\ 0, & \text{if } x+2t \leqslant 0 \text{ or } x+2t \geqslant 2 \end{cases}$$

and

$$f(x-2t) = \begin{cases} x-2t, & \text{if } 0 < x-2t < 1, \\ -x+2t+2, & \text{if } 1 < x-2t < 2, \\ 0, & \text{if } x-2t \leqslant 0 \text{ or } x-2t \geqslant 2, \end{cases}$$

and $g(x) = 0$ for $-\infty < x < \infty$.

(b) In Figure 17.3 we sketch the solution by considering 10 regions in the xt-plane bounded by forward-facing and backward-facing characteristics. First note that $u = 0$ in regions I, IV, and VII.

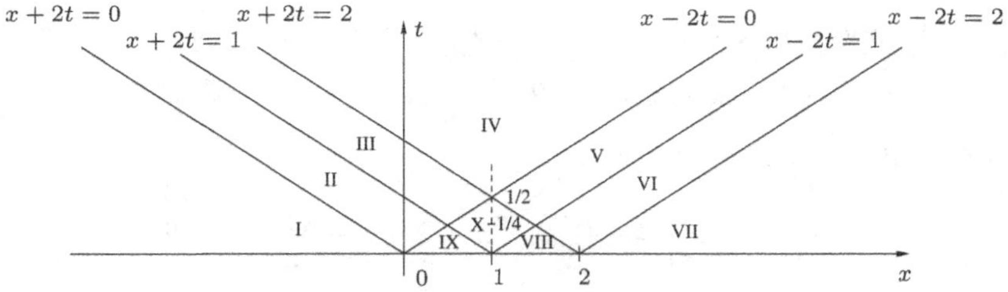

Figure 17.3

<u>Region II</u>: $0 < x + 2t < 1$ and $x - 2t < 0$; that is,

$$-2t < x < 1 - 2t \qquad \text{and} \qquad x < 2t,$$

and the solution in this region is

$$u(x, t) = \frac{1}{2}(x + 2t + 0) = \frac{1}{2}(x + 2t)\,.$$

<u>Region III</u>: $1 < x + 2t < 2$ and $x - 2t < 0$; that is,

$$1 - 2t < x < 2 - 2t \qquad \text{and} \qquad x < 2t,$$

and the solution in this region is

$$u(x, t) = \frac{1}{2}(-x - 2t + 2 + 0) = \frac{1}{2}(2 - x - 2t)\,.$$

<u>Region V</u>: $0 < x - 2t < 1$ and $x + 2t > 2$; that is,

$$2t < x < 1 + 2t \qquad \text{and} \qquad x > 2 - 2t,$$

and the solution in this region is

$$u(x, t) = \frac{1}{2}(x - 2t + 0) = \frac{1}{2}(x - 2t)\,.$$

<u>Region VI</u>: $1 < x - 2t < 2$ and $x + 2t > 2$; that is,

$$1 + 2t < x < 2 + 2t \qquad \text{and} \qquad x > 2 - 2t,$$

and the solution in this region is

$$u(x, t) = \frac{1}{2}(-x + 2t + 2 + 0) = \frac{1}{2}(2 - x + 2t)\,.$$

Region VIII: $1 < x - 2t < 2$ and $1 < x + 2t < 2$; that is,

$$1 + 2t < x < 2 + 2t \qquad \text{and} \qquad 1 - 2t < x < 2 - 2t,$$

and the solution in this region is

$$u(x,t) = \frac{1}{2}\,(-x + 2t + 2 - x - 2t + 2) = 2 - x.$$

Region IX: $0 < x - 2t < 1$ and $0 < x + 2t < 1$; that is,

$$2t < x < 1 + 2t \qquad \text{and} \qquad -2t < x < 1 - 2t,$$

and the solution in this region is

$$u(x,t) = \frac{1}{2}\,(x + 2t + x - 2t) = x.$$

Region X: $0 < x - 2t < 1$ and $1 < x + 2t < 2$; that is,

$$2t < x < 1 + 2t \qquad \text{and} \qquad 1 - 2t < x < 2 - 2t,$$

and the solution in this region is

$$u(x,t) = \frac{1}{2}\,(x - 2t - x - 2t + 2) = 1 - 2t.$$

We sketch the solution for

$$t_1 = 0, \qquad 0 < t_2 < \frac{1}{4}, \qquad t_3 = \frac{1}{4}, \qquad \frac{1}{4} < t_4 < \frac{1}{2}, \qquad t_5 = \frac{1}{2}$$

in Figure 17.4.

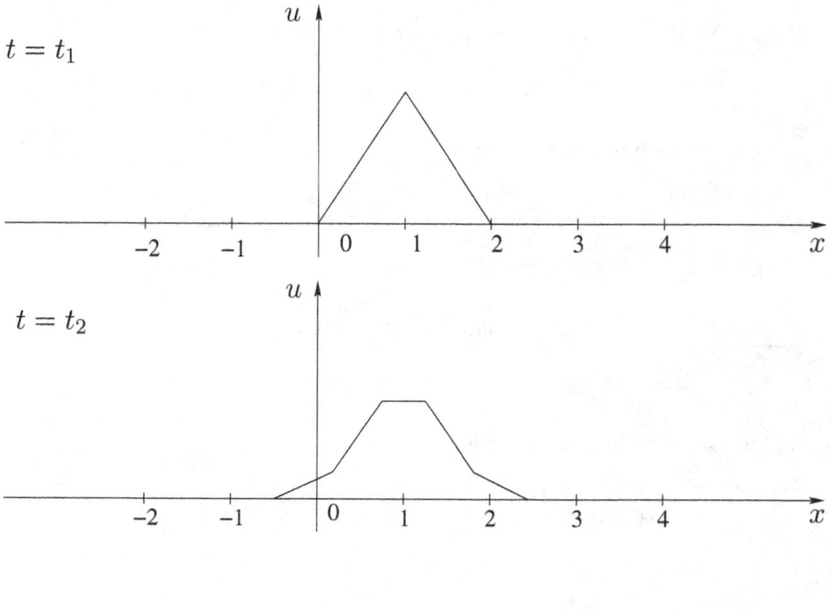

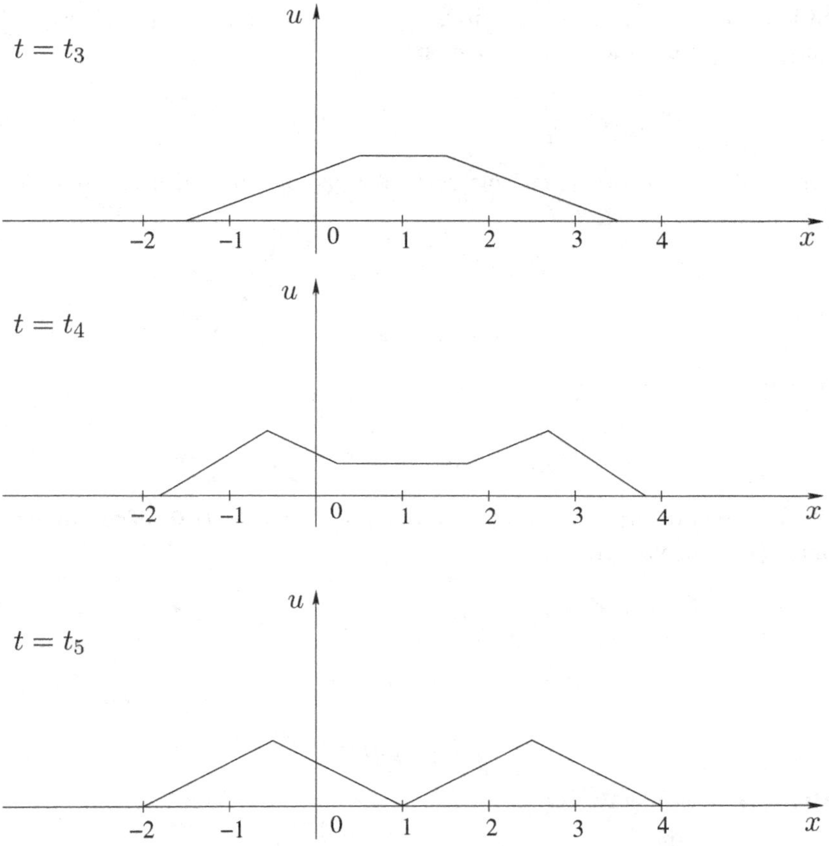

Figure 17.4

(c) Note that the right-moving signal will reach the point $\overline{x} = 11$ when $x+2t = 11$, and this characteristic hits the t-axis when $t = \frac{11}{2}$.

Exercise 17.13. ✗✗✗✗

The following hyperbolic model arises as the equation for the moment generating function $A(s,t)$ of a stochastic birth-death process.

$$\frac{\partial A}{\partial t}(s,t) - (s-1)\,(bs - \delta(t))\,\frac{\partial A}{\partial s}(s,t) = 0, \tag{17.7}$$

$$A(s,0) = s^{n_0},$$

where b denotes a constant birth rate and $\delta(t)$ a time-dependent death rate. Find $A(s,t)$.

Solution. We use the method of characteristics to solve for $A(s,t)$. Note that if $s = s(t)$ is a curve in the st-plane such that

$$\frac{ds}{dt} = -(s-1)\,(bs - \delta(t))\,, \tag{17.8}$$

then from the chain rule, along this curve the partial differential equation (17.7) becomes

$$\frac{d}{dt}\left[A(s(t),t)\right] = \frac{\partial A}{\partial t}(s(t),t) + \frac{\partial A}{\partial s}(s(t),t)\cdot\frac{ds}{dt} = 0,$$

that is,

$$A(s(t),t) = \text{ constant } = A(s(0),0) = s(0)^{n_0},$$

and we need only solve the characteristic equation

$$\frac{ds(t)}{dt} = -(s-1)\,(bs - \delta(t)) \tag{17.9}$$

for the equation of the curve passing through the point $(s(0),0)$. The differential equation (17.9) can be written as

$$\begin{aligned}\frac{ds}{dt} &= (1-s)\,(bs-\delta(t))\\ &= (1-s)\,(-b+bs+b-\delta(t))\\ &= (1-s)\,(b-\delta(t)) - b(1-s)^2,\end{aligned}$$

that is,

$$\frac{1}{(1-s)^2}\frac{ds}{dt} = \frac{b-\delta(t)}{1-s} - b,$$

and letting $y(t) = 1/[1-s(t)]$, we have

$$\frac{dy}{dt} - (b-\delta(t))y = -b. \tag{17.10}$$

This is a first-order linear differential equation for $y = y(t)$ which can be solved using an integrating factor or the method of variation of parameters. Multiplying by the integrating factor

$$\Lambda(t) = \exp\left(-\int_0^t (b-\delta(\tau))\,d\tau\right),$$

we have

$$\frac{d}{dt}\left[\Lambda(t)y(t)\right] = -b\Lambda(t).$$

Integrating and using the fact that $\Lambda(0) = 1$, we have

$$\Lambda(t)y(t) = y(0) - b\int_0^t \Lambda(\tau)\,d\tau.$$

Therefore,

$$\frac{\Lambda(t)}{1-s(t)} = \frac{1}{1-s(0)} - b\int_0^t \Lambda(\tau)\,d\tau, \tag{17.11}$$

and we can solve this equation for $s(t)$ and obtain an explicit form for the equation of the characteristic curve $s = s(t)$. However, since

$$A(s(t),t) = \text{ constant } = A(s(0),0) = s(0)^{n_0}, \tag{17.12}$$

we need to find the anchor point $(s(0),0)$. We solve Equation 17.11 for $s(0)$ to get

$$s(0) = 1 - \frac{1}{\dfrac{\Lambda(t)}{1-s(t)} + b\displaystyle\int_0^t \Lambda(\tau)\,d\tau},$$

and the solution is constant along the characteristic $s = s(t)$, so that

$$A(s(t),t) = A(s(0),0) = s(0)^{n_0} = \left[1 - \frac{1}{\dfrac{\Lambda(t)}{1-s(t)} + b\displaystyle\int_0^t \Lambda(\tau)\,d\tau}\right]^{n_0} \tag{17.13}$$

where

$$\Lambda(t) = \exp\left(-\int_0^t (b-\delta(\tau))\,d\tau\right).$$

Since the initial value problem

$$\frac{dy}{dt} - (b-\delta(t))y = -b,$$

$$y(0) = \frac{1}{1-s(0)}$$

has a unique solution, given a point (s,t) in the st-plane, there exists a unique characteristic passing through this point with $s(0) = s_0$, and

$$A(s,t) = A(s_0,0) = {s_0}^{n_0} = \left[1 - \frac{1}{\dfrac{\Lambda(t)}{1-s} + b\displaystyle\int_0^t \Lambda(\tau)\,d\tau}\right]^{n_0}, \tag{17.14}$$

where

$$\Lambda(t) = \exp\left(-\int_0^t (b-\delta(\tau))\,d\tau\right).$$

This can be simplified somewhat by substituting the expression for $\Lambda(t)$, and we find

$$A(s,t) = \left[1 - \frac{\exp\left(\displaystyle\int_0^t (b-\delta(z))\,dz\right)}{\dfrac{1}{1-s} + b\displaystyle\int_0^t \exp\left[\int_\tau^t (b-\delta(z))\,dz\right]d\tau}\right]^{n_0}. \tag{17.15}$$

Exercise 17.14. ✗✗

Show that the first-order partial differential equation

$$a\frac{\partial u}{\partial t} + b\frac{\partial u}{\partial x} = u, \quad a,\, b \neq 0 \tag{17.16}$$

has no bounded solutions other than the trivial solution.

Solution. In Exercise 17.10 we found the general solution to the partial differential equation (17.16) using the method of characteristics. Here we present a slightly different argument, originally given as the solution to Problem A3 in the 2011 Putnam competition (see *The American Mathematical Monthly*, 118, **8**, October 2011, p. 730).

Suppose that $u = u(x,t)$ is a solution to (17.16) and there is a constant $M > 0$ such that

$$|u(x,t)| \leqslant M \tag{17.17}$$

for all $x, t \in \mathbb{R}$. Let x_0 and t_0 be fixed real numbers, and let

$$x = x_0 + bs \qquad \text{and} \qquad t = t_0 + as,$$

and define

$$h(s) = u(x_0 + bs, t_0 + as)$$

for $s \in \mathbb{R}$.

From the chain rule and the partial differential equation (17.16), we have

$$\begin{aligned} h'(s) &= \frac{\partial u}{\partial t}(x_0 + bs, t_0 + as) \cdot \frac{dt}{ds} + \frac{\partial u}{\partial x}(x_0 + bs, t_0 + as) \cdot \frac{dx}{ds} \\ &= a\frac{\partial u}{\partial t}(x_0 + bs, t_0 + as) + b\frac{\partial u}{\partial x}(x_0 + bs, t_0 + as) \\ &= u(x_0 + bs, t_0 + as) \\ &= h(s) \end{aligned}$$

for $s \in \mathbb{R}$. Thus, h satisfies the first-order differential equation

$$h'(s) = h(s),$$

with solution $h(s) = h(0)\, e^s$, and from (17.17), we have

$$|h(0)| = |h(s)|\, e^{-s} \leqslant Me^{-s}$$

for all $s \in \mathbb{R}$. Therefore,

$$h(0) = u(x_0, t_0) = 0,$$

and since $(x_0, t_0) \in \mathbb{R}^2$ was arbitrary, $u(x,t) \equiv 0$, and the partial differential equation (17.16) has no nontrivial bounded solutions.

CHAPTER 18

FOUR SAMPLE MIDTERM EXAMINATIONS

These midterm examinations are actual exams held during the academic term, and each lasted 50 minutes.

18.1 MIDTERM EXAM 1

Exercise 18.1. XX

Find the values of λ^2 for which the boundary value problem

$$\frac{d^2u}{dx^2} + \lambda^2 u = 0, \quad 0 < x < \frac{\pi}{2},$$

$$u(0) = 0,$$

$$\int_0^{\pi/2} u(t)\, dt = 0$$

has nontrivial solutions.

Solution.

Case (i): $\lambda = 0$. In this case, the general solution to $d^2u/dx^2 = 0$ is given by $u(x) = Ax + B$, and $u(0) = 0$ implies that $B = 0$, so that $u(x) = Ax$. The condition $\int_0^{\pi/2} u(t)\,dt = 0$ implies that

$$\int_0^{\pi/2} A\,t\,dt = A\,\frac{t^2}{2}\Bigg|_0^{\frac{\pi}{2}} = A\,\frac{\pi^2}{8} = 0,$$

which implies that $A = 0$, and the boundary value problem has only the trivial solution in this case.

Case (ii): $\lambda \neq 0$. In this case, the general solution to $d^2u/dx^2 + \lambda^2 u = 0$ is given by

$$u(x) = A\cos\lambda x + B\sin\lambda x,$$

and $u(0) = 0$ implies that $A = 0$ so that $u(x) = B\sin\lambda x$. Now $\int_0^{\pi/2} u(t)\,dt = 0$ implies that

$$\int_0^{\pi/2} B\sin\lambda t\,dt = -\frac{B}{\lambda}\cos\lambda t\Bigg|_0^{\frac{\pi}{2}} = \frac{B}{\lambda}\left(1 - \cos\frac{\lambda\pi}{2}\right) = 0,$$

so either $B = 0$ or $\cos(\lambda\pi/2) = 1$. Therefore, a nontrivial solution exists if and only if we have $\cos(\lambda\pi/2) = 1$; that is, $\lambda\pi/2 = 2\pi n$, where $n \neq 0$ is an integer. The values of λ^2 for which the boundary value problem has nontrivial solutions are

$$\lambda_n^2 = 16n^2$$

for $n = 1, 2, 3, \ldots$.

Exercise 18.2. XX

Let $f(x) = \cos^2 x, \quad 0 \leqslant x \leqslant \pi$, and $f(x + 2\pi) = f(x)$ otherwise.

(a) Find the Fourier sine series for f on the interval $[0, \pi]$.

Hint: For $n \geqslant 1$,

$$\int \cos^2 x \sin nx\,dx = -\frac{1}{2n}\cos nx + \frac{1}{4}\int [\sin(n+2)x + \sin(n-2)x]\,dx.$$

(b) Find the Fourier cosine series for f on the interval $[0, \pi]$.

(c) For which values of x in $[0, \pi]$ do the series in (a) and (b) converge to $f(x)$?

Solution.

(a) Writing $f(x) = \cos^2 x \sim \sum_{n=1}^{\infty} b_n \sin nx$, the coefficients b_n in the Fourier sine series are computed as follows:

$$\begin{aligned}
b_n &= \frac{2}{\pi}\int_0^{\pi} \cos^2 x \sin nx\,dx = \frac{2}{\pi}\int_0^{\pi}\left(\frac{1}{2}+\frac{1}{2}\cos 2x\right)\sin nx\,dx \\
&= \frac{1}{\pi}\left(-\frac{1}{n}\cos nx\Big|_0^{\pi}\right) + \frac{1}{\pi}\int_0^{\pi}\cos 2x \sin nx\,dx \\
&= \frac{1}{n\pi}(1-(-1)^n) + \frac{1}{2\pi}\int_0^{\pi}[\sin(n-2)x + \sin(n+2)x]\,dx \\
&= \frac{1-(-1)^n}{2\pi}\left(\frac{2}{n}+\frac{1}{n-2}+\frac{1}{n+2}\right) = 0
\end{aligned}$$

if $n \neq 2$ is even, while if $n = 2$, since $\sin 2x = 2\sin x\cos x$, we have

$$\begin{aligned}
b_2 &= \frac{2}{\pi}\int_0^{\pi}\cos^2 x \sin 2x\,dx = \frac{4}{\pi}\int_0^{\pi}\sin x\cos^3 x\,dx \\
&= -\frac{1}{\pi}\cos^4 x\Big|_0^{\pi} = 0.
\end{aligned}$$

Therefore, $b_n = 0$ for all even $n \geqslant 2$. If n is odd,

$$\begin{aligned}
b_n &= \frac{2}{n\pi} + \frac{1}{\pi}\left(\frac{1}{n-2}+\frac{1}{n+2}\right) \\
&= \frac{2}{n\pi} + \frac{1}{\pi}\frac{2n}{n^2-4}.
\end{aligned}$$

The Fourier sine series for f is therefore

$$\cos^2 x \sim \frac{2}{\pi}\sum_{k=1}^{\infty}\left\{\frac{1}{2k-1}+\frac{2k-1}{(2k-1)^2-4}\right\}\sin(2k-1)x$$

for $0 < x < \pi$.

(b) Using the double-angle formula, we have

$$\cos^2 x = \frac{1}{2}+\frac{1}{2}\cos 2x,$$

which is the Fourier cosine series for f. If you integrate $\cos^2 x\cos nx$, you will find

$$a_0 = \frac{1}{2}, \quad a_2 = \frac{1}{2}, \quad \text{and} \quad a_k = 0 \quad \text{for} \quad k \neq 0,\, 2.$$

(c) From Dirichlet's theorem, the Fourier sine series in part (a) converges to $\cos^2 x$ for all $x \in (0, \pi)$ and converges to 0 for $x = 0$ and $x = \pi$. The Fourier cosine series in part (b) converges to $\cos^2 x$ for all $x \in [0, \pi]$ since the series is actually finite.

Exercise 18.3. XX

Let $v(x)$ be the steady-state solution to the boundary value–initial value problem

$$\frac{\partial^2 u}{\partial x^2} + r = \frac{1}{k}\frac{\partial u}{\partial t}, \quad 0 < x < a, \quad t > 0,$$

$$u(0, t) = T_0,$$

$$\frac{\partial u}{\partial x}(a, t) = 0,$$

where r is a constant. Find and solve the boundary value problem for the steady-state solution $v(x)$.

Solution. The steady-state solution $v(x)$ satisfies the boundary value problem

$$\frac{d^2 v}{dx^2} + r = 0, \quad 0 < x < a,$$

$$v(0) = T_0,$$

$$\frac{dv}{dx}(a) = 0,$$

and the general solution to the differential equation is

$$v(x) = -\frac{1}{2}rx^2 + Ax + B$$

and

$$\frac{dv}{dx}(x) = -rx + A.$$

Therefore,

$$v(0) = T_0 \quad \text{implies that} \quad B = T_0$$

while

$$\frac{dv}{dx}(a) = 0 \quad \text{implies that} \quad -ra + A = 0,$$

so that

$$A = ra \qquad \text{and} \qquad B = T_0.$$

The steady-state solution is therefore

$$v(x) = -\frac{1}{2}rx^2 + rax + T_0$$

for $0 \leqslant x \leqslant a$.

18.2 MIDTERM EXAM 2

Exercise 18.4. *XX*

The neutron flux u in a sphere of uranium obeys the differential equation

$$\frac{\lambda}{3}\frac{1}{r^2}\frac{d}{dr}\left(r^2\frac{du}{dr}\right)+(k-1)A\,u=0$$

for $0<r<a$, where λ is the effective distance traveled by a neutron between collisions, A is called the absorption cross section, and k is the number of neutrons produced by a collision during fission. In addition, the neutron flux at the boundary of the sphere is 0.

(a) Make the substitution

$$u=\frac{v}{r} \qquad \text{and} \qquad \mu^2=\frac{3(k-1)A}{\lambda}$$

and show that $v(r)$ satisfies $d^2v/dr^2+\mu^2\,v=0,\ 0<r<a$.

(b) Find the general solution to the differential equation in part (a) and then find $u(r)$ that satisfies the boundary condition and boundedness condition:

$$u(a)=0 \qquad \text{and} \qquad \lim_{r\to 0+}|u(r)| \text{ bounded.}$$

(c) Find the critical radius, that is, the smallest radius a for which the solution is not identically 0.

Solution.

(a) Letting $u=v/r$, then

$$\frac{du}{dr}=\frac{1}{r}\frac{dv}{dr}-\frac{1}{r^2}v \qquad \text{and} \qquad r^2\frac{du}{dr}=r\frac{dv}{dr}-v,$$

so that

$$\frac{d}{dr}\left(r^2\frac{du}{dr}\right)=r\frac{d^2v}{dr^2}+\frac{dv}{dr}-\frac{dv}{dr}=r\frac{d^2v}{dr^2},$$

and the differential equation for $v(r)$ is

$$\frac{1}{r}\frac{d^2v}{dr^2}+\frac{\mu^2}{r}v=0, \qquad \text{that is,} \qquad \frac{d^2v}{dr^2}+\mu^2\,v=0$$

for $0<r<a$.

(b) The general solution to the differential equation in part (a) is

$$v(r)=c_1\cos\mu r+c_2\sin\mu r$$

for $0 < r < a$, and the solution to the neutron flux equation is

$$u(r) = \frac{v(r)}{r} = c_1 \frac{\cos \mu r}{r} + c_2 \frac{\sin \mu r}{r}$$

for $0 < r < a$. Applying the boundedness condition, since

$$\lim_{r \to 0+} \frac{\sin \mu r}{r} = \mu \qquad \text{and} \qquad \lim_{r \to 0+} \frac{\cos \mu r}{r}$$

doesn't exist, we need $c_1 = 0$, and the solution is

$$u(r) = c_2 \frac{\sin \mu r}{r}$$

for $0 < r < a$.

(c) Applying the boundary condition

$$u(a) = \frac{c_2}{a} \sin \mu a = 0,$$

clearly, there is a nontrivial solution if and only if $\mu a = n\pi$ for some positive integer n. The critical radius is $a = \pi/\mu$.

Exercise 18.5. XXX

Show that

$$|\sin x| = \frac{2}{\pi} - \frac{2}{\pi} \sum_{n=2}^{\infty} \frac{1 + (-1)^n}{n^2 - 1} \cos nx$$

for $-\infty < x < \infty$.

Hint: Using the identity $\sin A \cos B = \frac{1}{2}\big[\sin(A+B) + \sin(A-B)\big]$, find the Fourier cosine series of the function $f(x) = \sin x$, for $0 \leqslant x \leqslant \pi$, and then use Dirichlet's theorem.

Solution. For $0 \leqslant x \leqslant \pi$, the Fourier cosine series is

$$\sin x \sim a_0 + \sum_{n=1}^{\infty} a_n \cos nx.$$

For $n = 0$,

$$a_0 = \frac{1}{\pi} \int_0^{\pi} \sin x \, dx = -\frac{1}{\pi} \cos x \Big|_0^{\pi} = -\frac{1}{\pi}(-1-1) = \frac{2}{\pi};$$

for $n = 1$,

$$\begin{aligned} a_1 &= \frac{2}{\pi} \int_0^{\pi} \sin x \cos x \, dx = \frac{1}{\pi} \int_0^{\pi} 2 \sin x \cos x \, dx \\ &= \frac{1}{\pi} \int_0^{\pi} \sin 2x \, dx = -\frac{1}{2\pi} \cos 2x \Big|_0^{\pi} \\ &= -\frac{1}{2\pi}(1-1) = 0; \end{aligned}$$

and for $n > 1$,

$$a_n = \frac{2}{\pi}\int_0^{\pi} \sin x \cos nx\, dx = \frac{1}{\pi}\int_0^{\pi} [\sin(n+1)x - \sin(n-1)x]\, dx$$

$$= -\frac{1}{\pi(n+1)}\cos(n+1)x\Big|_0^{\pi} + \frac{1}{\pi(n-1)}\cos(n-1)x\Big|_0^{\pi}$$

$$= -\frac{1}{\pi}\left[\frac{(-1)^{n+1}-1}{n+1}\right] + \frac{1}{\pi}\left[\frac{(-1)^{n-1}-1}{n-1}\right]$$

$$= -\frac{2}{\pi}\frac{1+(-1)^n}{n^2-1}.$$

The Fourier cosine series for $\sin x$ on the interval $0 \leqslant x \leqslant \pi$ is therefore

$$\sin x \sim \frac{2}{\pi} - \frac{2}{\pi}\sum_{n=2}^{\infty}\frac{1+(-1)^n}{n^2-1}\cos nx.$$

Since $\cos nx$ is an even function, this is also the Fourier series for the even extension of $\sin x$, that is, $|\sin x|$, on the interval $-\pi \leqslant x \leqslant \pi$, so that

$$|\sin x| \sim \frac{2}{\pi} - \frac{2}{\pi}\sum_{n=2}^{\infty}\frac{1+(-1)^n}{n^2-1}\cos nx,$$

and since this is a continuous piecewise smooth function on $-\pi \leqslant x \leqslant \pi$ such that $|\sin(-\pi)| = |\sin \pi| = 0$, then by Dirichlet's theorem the Fourier series converges to $|\sin x|$ for all $x \in [-\pi, \pi]$, and

$$|\sin x| = \frac{2}{\pi} - \frac{2}{\pi}\sum_{n=2}^{\infty}\frac{1+(-1)^n}{n^2-1}\cos nx$$

for all $x \in [-\pi, \pi]$. However, $|\sin x|$ is a continuous periodic function on the interval $-\infty < x < \infty$, so the Fourier series converges to $|\sin x|$ for all x, $-\infty < x < \infty$.

Exercise 18.6. ✗✗✗

Solve Laplace's equation in the square $0 < x,\ y < \pi$, with the boundary conditions

$$\frac{\partial^2 u}{\partial x^2} + \frac{\partial^2 u}{\partial y^2} = 0, \quad 0 < x < \pi, \quad 0 < y < \pi,$$

$$u(0, y) = 0,$$

$$u(\pi, y) = 0,$$

$$u(x, 0) = 0,$$

$$u(x, \pi) = 1.$$

Solution. We use separation of variables and assume a solution to Laplace's equation

$$\frac{\partial^2 u}{\partial x^2} + \frac{\partial^2 u}{\partial y^2} = 0$$

of the form $u(x, y) = X(x) \cdot Y(y)$. Separating variables, we have

$$\frac{X''(x)}{X(x)} = -\frac{Y''(t)}{Y(t)} = -\lambda,$$

and we obtain the two ordinary differential equations

$$\begin{aligned} &X'' + \lambda X = 0 \quad 0 < x < \pi, && Y'' - \lambda Y = 0, \quad 0 < y < \pi, \\ &X(0) = 0, && Y(0) = 0. \\ &X(\pi) = 0, && \end{aligned}$$

Solving the complete boundary value problem for X, the eigenvalues and eigenfunctions are given by

$$\lambda_n = n^2 \qquad \text{and} \qquad X_n(x) = \sin nx$$

for $n \geqslant 1$. The corresponding problem for Y is

$$\begin{aligned} Y'' - n^2 Y &= 0, \\ Y(0) &= 0, \end{aligned}$$

with solutions

$$Y_n(y) = \sinh ny$$

for $n \geqslant 1$.

From the superposition principle, we write

$$u(x, y) = \sum_{n=1}^{\infty} b_n \sinh ny \, \sin nx,$$

and setting $y = \pi$, we need

$$1 = u(x, \pi) = \sum_{n=1}^{\infty} b_n \sinh n\pi \, \sin nx,$$

and from the orthogonality of the eigenfunctions,

$$b_n \sinh n\pi = \frac{2}{\pi} \int_0^{\pi} \sin nx \, dx = -\frac{2}{n\pi} \cos nx \Big|_0^{\pi} = -\frac{2}{n\pi} \left[(-1)^n - 1\right],$$

so that

$$u(x, y) = \frac{2}{\pi} \sum_{n=1}^{\infty} \frac{1 - (-1)^n}{n \sinh n\pi} \sin nx \, \sinh ny$$

for $0 < x < \pi, \ 0 < y < \pi$.

18.3 MIDTERM EXAM 3

Exercise 18.7. XXX

Consider the following eigenvalue problem on the interval $(0, 1)$:

$$\begin{aligned} u''(x) + 2u'(x) - u(x) + \lambda\,(x+1)^2 e^{-2x}\, u(x) &= 0, \\ u'(0) &= 0, \\ u'(1) &= 0. \end{aligned}$$

(a) Explain the meaning of *eigenvalue problem.*

(b) Show that this eigenvalue problem is not of Sturm-Liouville type.

(c) Multiply the equation above by e^{2x} to obtain a Sturm-Liouville problem. Identify $p(x)$, $q(x)$, and $\sigma(x)$.

(d) Use the Rayleigh quotient to show that the leading eigenvalue is positive; that is, $\lambda_1 > 0$.

(e) Find an upper bound for the leading eigenvalue.

Solution.

(a) The *eigenvalue problem* consists of finding the values of λ (*eigenvalues*) for which there are nontrivial solutions (*eigenfunctions*) satisfying both the differential equation and the boundary conditions.

(b) If the eigenvalue problem

$$\begin{aligned} &u''(x) + 2u'(x) - u(x) + \lambda\,(x+1)^2 e^{-2x}\, u(x) = 0, \\ &u'(0) = 0, \\ &u'(1) = 0 \end{aligned}$$

were of Sturm-Liouville form, we would have

$$u'' + 2u' = (p\,u')',$$

that is, $p(x) = 1$ and $p'(x) = 2$, which is impossible.

(c) If we multiply the differential equation by e^{2x}, we have

$$e^{2x}u'' + 2e^{2x}u' - e^{2x}u + \lambda\,(x+1)^2 u = 0, \quad 0 < x < 1,$$

that is,

$$(e^{2x}u')' - e^{2x}u + \lambda(x+1)^2 u = 0, \quad 0 < x < 1,$$

which is of Sturm-Liouville type with

$$p(x) = e^{2x}, \quad q(x) = -e^{2x}, \quad \sigma(x) = (x+1)^2$$

for $0 < x < 1$.

(d) The eigenvalue λ and corresponding eigenfunction u are related by the Rayleigh quotient:

$$\lambda = R(u) = \frac{-p(x)u(x)u'(x)\Big|_0^1 + \displaystyle\int_0^1 \left[p(x)u'(x)^2 - q(x)u(x)^2\right] dx}{\displaystyle\int_0^1 u(x)^2\sigma(x)\, dx}$$

and

$$-p(x)u(x)u'(x)\Big|_0^1 = -e^2u(1)u'(1) + u(0)u'(0) = 0,$$

since $u'(0) = 0$ and $u'(1) = 0$, so the Rayleigh quotient becomes

$$\lambda = R(u) = \frac{\displaystyle\int_0^1 \left[e^{2x}u'(x)^2 + e^{2x}u(x)^2\right] dx}{\displaystyle\int_0^1 u(x)^2(x+1)^2\, dx} \geqslant 0,$$

and all the eigenvalues of the boundary value problem are nonnegative.

To show that $\lambda = 0$ is not an eigenvalue, we can see immediately that since u and u' are continuous on the interval $[0, 1]$, then

$$\int_0^1 \left[e^{2x}u'(x)^2 + e^{2x}u(x)^2\right] dx = 0,$$

and this implies that $u(x) = u'(x) = 0$ for all $0 < x < 1$. Hence, there is no nontrivial eigenfunction corresponding to $\lambda = 0$. Alternatively, the equation $u'' + 2u' - u = 0,\ 0 < x < 1$, has general solution

$$u(x) = Ae^{-x}e^{\sqrt{2}x} + Be^{-x}e^{-\sqrt{2}x},$$

and from the boundary conditions $u'(0) = 0$ and $u'(1) = 0$, the solution is $u(x) = 0$ for $0 < x < 1$. Therefore, the leading eigenvalue $\lambda_1 > 0$.

(e) To get an upper bound on λ_1, we try a quadratic test function v which satisfies the boundary conditions $v'(0) = 0$ and $v'(1) = 0$, say

$$v(x) = ax^2 + bx + c \qquad \text{with} \qquad v'(x) = 2ax + b;$$

then the boundary conditions imply that $a = b = 0$, so that $v(x) = c$ for $0 \leqslant x \leqslant 1$. The Rayleigh quotient for this test function is

$$R(v) = \frac{\displaystyle\int_0^1 c^2e^{2x}\, dx}{\displaystyle\int_0^1 c^2(x+1)^2\, dx} = \frac{3}{14}(e^2 - 1),$$

and since λ_1 is the minimum of $R(u)$ as u runs over all twice continuously differentiable functions that satisfy the boundary conditions,

$$0 < \lambda_1 \leqslant \frac{3}{14}(e^2 - 1) \approx 1.369.$$

Exercise 18.8. XXX

Consider Laplace's equation for the steady-state temperature distribution in a square plate of side length 1.

$$\frac{\partial^2 u}{\partial x^2} + \frac{\partial^2 u}{\partial y^2} = 0, \quad 0 < x < 1,\ 0 < y < 1,$$

$$u(0, y) = 1, \quad u(1, y) = 1,$$

$$\frac{\partial u}{\partial y}(x, 0) = 0, \quad \frac{\partial u}{\partial y}(x, 1) = 0.$$

Obviously, the solution is $u(x, y) = 1$ for $0 \leqslant x \leqslant 1,\ 0 \leqslant y \leqslant 1$. Show that this is the case by using separation of variables.

Solution.

Using separation of variables, assume that $u(x, y) = X(x) \cdot Y(y)$, and obtain the two boundary value problems

$$X''(x) - \lambda X(x) = 0, \quad 0 < x < 1, \qquad Y''(y) + \lambda Y(y) = 0, \quad 0 < y < 1,$$

$$Y'(0) = 0,$$

$$Y'(1) = 0.$$

Note that we cannot meet the boundary conditions $u(0, y) = u(1, y) = 1$ for all $0 < y < 1$ by requiring that $X(0) = 1$ and $X(1) = 1$.

We solve the complete boundary value problem for Y first since it has homogeneous boundary conditions. The eigenvalues are $\lambda_n = n^2\pi^2$ and the corresponding eigenfunctions are

$$Y_n(y) = \cos n\pi y$$

for $n = 0, 1, 2, \ldots$. The corresponding solutions to the problem for X are

$$X_0(x) = b_0 x + a_0 \qquad \text{and} \qquad X_n(x) = a_n \cosh n\pi x + b_n \sinh n\pi x$$

for $n \geqslant 1$.

From the superposition principle, we have

$$u(x,y) = b_0 x + a_0 + \sum_{n=1}^{\infty} (a_n \cosh n\pi x + b_n \sinh n\pi x) \cos n\pi y$$

for $0 < x < 1$ and $0 < y < 1$. Applying the boundary condition $u(0,y) = 1$, we get

$$1 = a_0 + \sum_{n=1}^{\infty} a_n \cos n\pi y,$$

and from the orthogonality of the eigenfunctions on the interval $[0,1]$, this implies that

$$a_0 = 1 \qquad \text{and} \qquad a_n = 0, \quad n \geqslant 1.$$

The solution is now

$$u(x,y) = 1 + b_0 x + \sum_{n=1}^{\infty} b_n \sinh n\pi x \cos n\pi y$$

for $0 < x < 1$ and $0 < y < 1$. Applying the boundary condition $u(1,y) = 1$, we get

$$1 = 1 + b_0 + \sum_{n=1}^{\infty} b_n \sinh n\pi \cos n\pi y,$$

that is,

$$0 = b_0 + \sum_{n=1}^{\infty} b_n \sinh n\pi \cos n\pi y$$

for $0 < y < 1$. Again, from the orthogonality of the eigenfunctions on the interval $[0,1]$, this implies that

$$b_0 = 0 \qquad \text{and} \qquad b_n = 0, \quad n \geqslant 1.$$

Therefore, the solution is $u(x,y) = 1$ for $0 \leqslant x \leqslant 1,\ 0 \leqslant y \leqslant 1$, as promised.

18.4 MIDTERM EXAM 4

Exercise 18.9. ✗✗

Solve the normalized wave equation

$$\frac{\partial^2 u}{\partial t^2} = \frac{\partial^2 u}{\partial x^2}, \qquad 0 < x < \pi, \qquad t > 0,$$

$$u(0,t) = 0, \quad u(\pi,t) = 0,$$

$$u(x,0) = \sin x, \qquad \frac{\partial u}{\partial t}(x,0) = \sin x.$$

Solution. Since the partial differential equation and the boundary conditions are linear and homogeneous, we can use separation of variables, and assuming a solution of the form $u(x,t) = \phi(x) \cdot G(t)$, we get two ordinary differential equations:

$$\begin{aligned}
&\phi'' + \lambda\phi = 0, \quad 0 < x < \pi, \qquad G'' + \lambda G = 0, \quad t > 0,\\
&\phi(0) = 0,\\
&\phi(\pi) = 0,
\end{aligned}$$

where λ is the separation constant. We solve the spatial problem first since it has a complete set of boundary conditions. These are homogeneous Dirichlet conditions, so the eigenvalues and eigenfunctions are given by

$$\lambda_n = n^2 \qquad \text{and} \qquad \phi_n(x) = \sin nx$$

for $n \geqslant 1$, and the corresponding solutions to the temporal equation are

$$G_n(t) = a_n \cos nt + b_n \sin nt.$$

Using the superposition principle, we write the solution as an "infinite" linear combination of $\{\phi_n \cdot G_n\}_{n\geqslant 1}$; that is,

$$u(x,t) = \sum_{n=1}^{\infty} \sin nx \left(a_n \cos nt + b_n \sin nt\right),$$

where the constants a_n and b_n are determined from the initial conditions

$$\sin x = u(x,0) = \sum_{n=1}^{\infty} a_n \sin nx$$

and

$$\sin x = \frac{\partial u}{\partial t}(x,0) = \sum_{n=1}^{\infty} n b_n \sin nx.$$

From the orthogonality of the eigenfunctions, we find

$$\begin{aligned}
a_1 &= 1, \quad a_k = 0, \quad \text{for} \quad k \neq 1,\\
b_1 &= 1, \quad b_k = 0, \quad \text{for} \quad k \neq 1,
\end{aligned}$$

and the solution is

$$u(x,t) = \sin x \cos t + \sin x \sin t$$

for $0 < x < \pi$, and $t > 0$.

Exercise 18.10. XXX

Consider the regular Sturm-Liouville problem

$$\varphi''(x) + \lambda\,\varphi(x) = 0, \quad 0 < x < 1,$$

$$\varphi(0) = 0,$$

$$\varphi(1) - h\,\varphi'(1) = 0,$$

where $h > 0$. Show that there is a single negative eigenvalue λ_0 if and only if $h < 1$. Find λ_0 and the corresponding eigenfunction $\varphi_0(x)$.

Hint: Assume that $\lambda = -\mu^2$ for some real number $\mu \neq 0$.

Solution. Following the hint, the differential equation becomes

$$\varphi''(x) - \mu^2\,\varphi(x) = 0,$$

with general solution

$$\varphi(x) = A\cosh\mu x + B\sinh\mu x$$

for $0 < x < 1$. Applying the first boundary condition, we have

$$\varphi(0) = A = 0,$$

so that

$$\varphi(x) = B\sinh\mu x \qquad \text{with} \qquad \varphi'(x) = \mu B\cosh\mu x.$$

Applying the second boundary condition, we have

$$\varphi(1) - h\,\varphi'(1) = B\,(\sinh\mu - h\,\mu\cosh\mu) = 0.$$

so that

$$B\cosh\mu\,(\tanh\mu - h\,\mu) = 0,$$

and we have a nontrivial solution if and only if

$$\tanh\mu = h\,\mu$$

for some $\mu \neq 0$. However, the graphs of $y = \tanh\mu$ and $y = h\mu$ intersect only at $\mu = 0$ if $h \geqslant 1$, and they intersect at exactly one positive value μ_0 if $0 < h < 1$. Therefore, there is exactly one negative eigenvalue for this Sturm-Liouville problem if and only if $0 < h < 1$, and the eigenvalue is

$$\lambda_0 = -\mu_0^2,$$

where μ_0 is the positive root of the equation $\tanh\mu = h\mu$. The corresponding eigenfunction is

$$\varphi_0(x) = \sinh\mu_0 x$$

for $0 \leqslant x \leqslant 1$.

Exercise 18.11. XX

Consider Laplace's equation

$$\frac{1}{r}\frac{\partial}{\partial r}\left(r\frac{\partial u}{\partial r}\right)+\frac{1}{r^2}\frac{\partial^2 u}{\partial \theta^2}=0$$

in a semicircular disk of radius a centered at the origin with boundary conditions

$$u(r,0)=0, \quad 0<r<a,$$

$$u(r,\pi)=0, \quad 0<r<a,$$

$$u(a,\theta)=\sin\theta, \quad 0<\theta<\pi,$$

$$|u(r,\theta)|<\infty \quad \text{as} \quad r\to 0^+.$$

Solve this problem using separation of variables.

Solution. Assuming a solution of the form $u(r,\theta) = R(r)\cdot\varphi(\theta)$, and separating variables, we have the following problems for R and φ :

$$\begin{aligned} &r(rR')'+\lambda R=0, \quad 0<r<a, && \varphi''+\lambda\varphi=0, \quad 0<\theta<\pi,\\ &|R(0)|<\infty, && \varphi(0)=0,\\ & && \varphi(\pi)=0. \end{aligned}$$

We solve the θ-problem first. The eigenvalues and corresponding eigenfunctions are

$$\lambda_n = n^2 \qquad \text{and} \qquad \varphi_n(x)=\sin n\theta$$

for $n \geqslant 1$. The corresponding r-equation is the Cauchy-Euler equation

$$r^2R''+rR'-n^2R=0,$$

with general solution

$$R(r)=A_nr^n+B_nr^{-n}$$

for $n \geqslant 1$. The boundedness condition $|R(0)|<\infty$ requires that $B_n=0$, so that

$$R_n(r)=A_nr^n$$

for $n \geqslant 1$. Using the superposition principle, we write

$$u(r,\theta)=\sum_{n=1}^{\infty}A_nr^n\sin n\theta$$

for $0<r<a,\ 0<\theta<\pi$, and from the boundary condition we want

$$\sin\theta=u(a,\theta)=\sum_{n=1}^{\infty}A_na^n\sin n\theta.$$

From the orthogonality of the eigenfunctions on the interval $[0, \pi]$, we have

$$aA_1 = 1 \qquad \text{and} \qquad A_n = 0,\ n \geqslant 2,$$

so the solution is

$$u(r, \theta) = \frac{r}{a} \sin \theta$$

for $0 < r < a,\ 0 < \theta < \pi$.

CHAPTER 19

FOUR SAMPLE FINAL EXAMINATIONS

These final examinations are actual exams held during the academic term, and each lasted 2 hours.

19.1 FINAL EXAM 1

Exercise 19.1. ✗

Assume that $f(x)$ is absolutely integrable and a is a given real constant. Show that

$$\mathcal{F}\left[e^{-iax}f(x)\right](\omega) = \widehat{f}(\omega - a),$$

where $\widehat{f}$ is the Fourier transform of f.

Partial Differential Equations: Theory and Completely Solved Problems, First Edition. By T. Hillen, I. E. Leonard, H. van Roessel

Solution. Since $\left|e^{-iax}\right| = 1$ and f is absolutely integrable on $(-\infty, \infty)$, $e^{-iax} f(x)$ is also absolutely integrable on $(-\infty, \infty)$ and we have

$$\begin{aligned}\mathcal{F}\left[e^{-iax} f(x)\right](\omega) &= \frac{1}{2\pi}\int_{-\infty}^{\infty} e^{-iax} f(x) e^{i\omega x}\,dx \\ &= \frac{1}{2\pi}\int_{-\infty}^{\infty} f(x) e^{i(\omega-a)x}\,dx \\ &= \widehat{f}(\omega - a)\end{aligned}$$

for all $\omega \in \mathbb{R}$.

Exercise 19.2. XXX

Hermite's differential equation reads

$$y'' - 2xy' + \lambda y = 0, \qquad -\infty < x < \infty.$$

(a) Multiply by e^{-x^2} and bring the differential equation into Sturm-Liouville form. Decide if the resulting Sturm-Liouville problem is regular or singular.

(b) Show that the Hermite polynomials

$$H_0(x) = 1, \quad H_1(x) = 2x, \quad H_2(x) = 4x^2 - 2, \quad H_3(x) = 8x^3 - 12x$$

are eigenfunctions of the Sturm-Liouville problem and find the corresponding eigenvalues.

(c) Use an appropriate weight function and show that H_1 and H_2 are orthogonal on the interval $(-\infty, \infty)$ with respect to this weight function.

Solution.

(a) Multiplying the differential equation by e^{-x^2}, we have

$$e^{-x^2} y'' - 2xe^{-x^2} y' + \lambda e^{-x^2} y = 0;$$

that is,

$$\frac{d}{dx}\left(e^{-x^2} y'\right) + \lambda e^{-x^2} y = 0.$$

This is the self-adjoint form of Hermite's equation, with $p(x) = \sigma(x) = e^{-x^2}$ and $q(x) = 0$. Even though $p,\ p',\ q,$ and σ are all continuous on the interval $(-\infty, \infty)$, this Sturm-Liouville problem is singular since the interval is infinite.

(b) The Hermite polynomial of degree n is denoted by $H_n(x)$.

- For $H_0(x) = 1$ we have

$$H_0'' - 2xH_0' + \lambda_0 H_0 = 0$$

if and only if $\lambda_0 = 0$, and the eigenvalue corresponding to the eigenfunction $H_0(x)$ is $\lambda_0 = 0$.

- For $H_1(x) = 2x$ we have

$$H_1'' - 2xH_1' + \lambda_1 H_1 = 0$$

if and only if $-4x + 2\lambda_1 x = 0$ for all x, that is, if and only if $\lambda_1 = 2$, and the eigenvalue corresponding to the eigenfunction $H_1(x)$ is $\lambda_1 = 2$.

- For $H_2(x) = 4x^2 - 2$ we have

$$\begin{aligned} H_2'' - 2xH_2' + \lambda_2 H_2 &= 8 - 2x(8x) + \lambda_2(4x^2 - 2) \\ &= -4(4x^2 - 2) + \lambda_2(4x^2 - 2) \\ &= (\lambda_2 - 4)(4x^2 - 2) \\ &= 0 \end{aligned}$$

for all x if and only if $\lambda_2 = 4$, and the eigenvalue corresponding to the eigenfunction $H_2(x)$ is $\lambda_2 = 4$.

- For $H_3(x) = 8x^3 - 12x$ we have

$$\begin{aligned} H_3'' - 2xH_3' + \lambda_3 H_3 &= 48x - 2x(24x^2 - 12) + \lambda_3(8x^3 - 12x) \\ &= -48x^3 + 72x + \lambda_3(8x^3 - 12x) \\ &= (\lambda_3 - 6)(8x^3 - 12x) \\ &= 0 \end{aligned}$$

for all x if and only if $\lambda_3 = 6$, and the eigenvalue corresponding to the eigenfunction $H_3(x)$ is $\lambda_3 = 6$.

(c) There are two ways to answer this question. The more elegant method is as follows. We can show that the Hermite polynomials H_n, for $n \geqslant 0$, are orthogonal on the interval $(-\infty, \infty)$ with respect to the weight function $r(x) = e^{-x^2}$, by noting that

$$\begin{aligned} &e^{-x^2} H_m H_n'' - 2xe^{-x^2} H_m H_n' + \lambda_n e^{-x^2} H_m H_n = 0, \quad \text{and} \\ &e^{-x^2} H_n H_m'' - 2xe^{-x^2} H_n H_m' + \lambda_m e^{-x^2} H_m H_n = 0, \end{aligned}$$

and subtracting, we have

$$\frac{d}{dx}\left[e^{-x^2}\left(H_m H_n' - H_n H_m'\right)\right] + (\lambda_n - \lambda_m)e^{-x^2} H_m H_n = 0.$$

Integrating over the real line, we have

$$(\lambda_n - \lambda_m) \int_{-\infty}^{\infty} e^{-x^2} H_m(x) H_n(x)\, dx$$

$$= \lim_{M\to\infty} e^{-x^2} \left[H_n(x)H_m'(x) - H_m(x)H_n'(x)\right]\Bigg|_{-M}^{M} = 0$$

since the exponential kills off any polynomial as $|x| \to \infty$. Therefore, if $m \neq n$, then

$$\int_{-\infty}^{\infty} e^{-x^2} H_m(x) H_n(x)\, dx = 0.$$

A more straightforward method is by integrating directly. For example, we note immediately that

$$\int_{-\infty}^{\infty} e^{-x^2} H_1(x) H_2(x)\, dx = 0,$$

since the integrand is an odd function of x and we are integrating between symmetric limits.

Exercise 19.3. ✗✗

Find all functions ϕ for which $u(x,t) = \phi(x-ct)$ is a solution of the heat equation

$$\frac{\partial^2 u}{\partial x^2} = \frac{1}{k}\frac{\partial u}{\partial t}, \qquad -\infty < x < \infty, \quad t > 0,$$

where k and c are constants.

Solution. If ϕ is a twice continuously differentiable function such that

$$u(x,t) = \phi(x - ct)$$

is a solution of the heat equation, then

$$\frac{\partial u}{\partial t} = -c\phi'(x-ct),$$
$$\frac{\partial u}{\partial x} = \phi'(x-ct),$$
$$\frac{\partial^2 u}{\partial x^2} = \phi''(x-ct),$$

and ϕ satisfies the equation

$$\phi''(x-ct) + \frac{c}{k}\phi'(x-ct) = 0$$

for all $-\infty < x < \infty$ and $t > 0$; that is,

$$\phi''(s) + \frac{c}{k}\phi'(s) = 0$$

for all $s \in \mathbb{R}$.

Therefore, the solution is given by

$$\phi(s) = A + Be^{-\frac{c}{k}s};$$

that is,

$$u(x,t) = A + Be^{-\frac{c}{k}(x-ct)},$$

where A and B are arbitrary constants.

Exercise 19.4. XXX

Consider the regular Sturm-Liouville problem

$$\begin{aligned}&\phi'' + \lambda^2\,\phi = 0, \quad 0 < x < \pi,\\ &\phi'(0) = 0,\\ &\phi(\pi) = 0.\end{aligned}$$

(a) Find the eigenvalues λ_n^2 and the corresponding eigenfunctions ϕ_n for this problem.

(b) Show directly, by integration, that eigenfunctions corresponding to distinct eigenvalues are orthogonal.

(c) Given the function $f(x) = (\pi^2 - x^2)/2, \quad 0 < x < \pi$, find the eigenfunction expansion for f.

(d) Show that

$$\frac{\pi^3}{32} = 1 - \frac{1}{3^3} + \frac{1}{5^3} - \frac{1}{7^3} + \frac{1}{9^3} - + \cdots.$$

Solution.

(a) Case(i): $\lambda = 0$. The general solution to the equation $\phi'' + \lambda^2\phi = 0$ in this case is

$$\phi(x) = c_1\,x + c_2,$$

and differentiating, $\phi'(x) = c_1$, and the condition $\phi'(0) = 0$ implies that $c_1 = 0$. The condition $\phi(\pi) = 0$ implies that $c_2 = 0$, so there are no nontrivial solutions in this case.

Case(ii): $\lambda \neq 0$. The general solution to the equation $\phi'' + \lambda^2\phi = 0$ in this case is

$$\phi(x) = c_1\cos\lambda x + c_2\sin\lambda x,$$

and differentiating, we get

$$\phi'(x) = -c_1\lambda\sin\lambda x + c_2\lambda\cos\lambda x.$$

The condition $\phi'(0) = 0$ implies that $c_2 \lambda = 0$, so $c_2 = 0$. The solution is then

$$\phi(x) = c_1 \cos \lambda x,$$

and the condition $\phi(\pi) = 0$ implies that $\cos \lambda\pi = 0$, and therefore the eigenvalues are

$$\lambda^2 = \lambda_n^2 = \left(\frac{2n-1}{2}\right)^2,$$

for $n = 1, 2, 3, \ldots$. The corresponding eigenfunctions are

$$\phi_n(x) = \cos \frac{(2n-1)x}{2}$$

for $n = 1, 2, 3, \ldots$.

(b) Let $\lambda_n = (2n-1)/2$ for $n = 1, 2, 3, \ldots$; then for $m \neq n$, we have

$$\begin{aligned}
\int_0^\pi \phi_m(x)\phi_n(x)\,dx &= \int_0^\pi \cos \lambda_m x \cos \lambda_n x\,dx \\
&= \frac{1}{2}\int_0^\pi \left\{\cos(\lambda_m + \lambda_n)x + \cos(\lambda_m - \lambda_n)x\right\} dx \\
&= \frac{1}{2(\lambda_m + \lambda_n)} \sin(\lambda_m + \lambda_n)x \Big|_0^\pi \\
&\qquad + \frac{1}{2(\lambda_m - \lambda_n)} \sin(\lambda_m - \lambda_n)x \Big|_0^\pi \\
&= \frac{1}{2(\lambda_m + \lambda_n)} \sin(\lambda_m + \lambda_n)\pi \\
&\qquad + \frac{1}{2(\lambda_m - \lambda_n)} \sin(\lambda_m - \lambda_n)\pi \\
&= 0
\end{aligned}$$

since $(\lambda_m + \lambda_n)\pi = (m+n-1)\pi$ and $(\lambda_m - \lambda_n)\pi = (m-n)\pi$. For $n = m$, we find

$$\int_0^\pi \cos^2 \lambda_n x\,dx = \frac{1}{2}\int_0^\pi \left[1 + \cos(2n-1)x\right] dx = \frac{\pi}{2}.$$

(c) Writing

$$f(x) = \frac{\pi^2 - x^2}{2} \sim \sum_{n=1}^{\infty} c_n \phi_n(x),$$

the coefficients c_n in the eigenfunction expansion are found using the orthogonality of the eigenfunctions on $[0, \pi]$:

$$\begin{aligned}
c_n &= \frac{2}{\pi}\int_0^{\pi}\left(\frac{\pi^2 - x^2}{2}\right)\cos\lambda_n x\,dx \\
&= \frac{(\pi^2 - x^2)\sin\lambda_n x}{\pi\lambda_n}\bigg|_0^{\pi} + \frac{2}{\pi\lambda_n}\int_0^{\pi} x\sin\lambda_n x\,dx \\
&= -\frac{2x\cos\lambda_n x}{\pi\lambda_n^2}\bigg|_0^{\pi} + \frac{2}{\pi\lambda_n^2}\int_0^{\pi}\cos\lambda_n x\,dx \\
&= \frac{2}{\pi}\frac{\sin\lambda_n\pi}{\lambda_n^3} \\
&= \frac{16}{\pi(2n-1)^3}\sin\frac{(2n-1)\pi}{2} \\
&= \frac{16(-1)^{n+1}}{\pi(2n-1)^3}.
\end{aligned}$$

Therefore, the eigenfunction expansion of f is given by

$$\frac{\pi^2 - x^2}{2} \sim \frac{16}{\pi}\sum_{n=1}^{\infty}\frac{(-1)^{n+1}}{(2n-1)^3}\cos\frac{(2n-1)x}{2}.$$

(d) In this particular problem, the eigenfunction expansion is actually the Fourier cosine series for f. Since the function f is piecewise smooth on the interval $[0, \pi]$ and since the even extension of f to $[-\pi, \pi]$ is continuous at $x = 0$, then by Dirichlet's theorem the series converges to $f(0) = \pi^2/2$ when $x = 0$, and therefore

$$\frac{\pi^3}{32} = 1 - \frac{1}{3^3} + \frac{1}{5^3} - \frac{1}{7^3} + \frac{1}{9^3} - + \cdots.$$

Exercise 19.5. XXX

Given the following boundary value-initial value problem for the heat equation on $[0,1]$:

$$\begin{aligned} \frac{\partial u}{\partial t} &= \frac{1}{9}\frac{\partial^2 u}{\partial x^2} - 2u, \\ u(0,t) &= 0, \\ u(1,t) &= 0, \\ u(x,0) &= 7\sin 3\pi x. \end{aligned}$$

(a) If $u(x,t)$ is the solution to this problem, find a boundary value-initial value problem satisfied by

$$w(x,t) = e^{2t}u(x,t).$$

(b) Solve the problem found in part (a) for $w(x,t)$.

(c) Find the solution $u(x,t)$ to the original problem.

(d) Find the time T_1 such that $u(x,t) < 1$ for every $x \in [0,1]$ and every $t > T_1$.

Solution.

(a) If $u(x,t)$ is the solution to the heat equation above and $w(x,t) = e^{2t}u(x,t)$, then

$$\begin{aligned} \frac{\partial w}{\partial t} &= e^{2t}\frac{\partial u}{\partial t} + 2e^{2t}u \\ &= e^{2t}\left(\frac{1}{9}\frac{\partial^2 u}{\partial x^2} - 2u\right) + 2e^{2t}u \\ &= e^{2t}\frac{1}{9}\frac{\partial^2 u}{\partial x^2}, \end{aligned}$$

so that

$$\frac{\partial w}{\partial t} = \frac{1}{9}\frac{\partial^2 w}{\partial x^2}$$

for $0 < x < 1,\ t > 0$. Therefore, $w(x,t) = e^{2t}u(x,t)$ satisfies the boundary value–initial value problem

$$\begin{aligned} \frac{\partial w}{\partial t} &= \frac{1}{9}\frac{\partial^2 w}{\partial x^2}, \quad 0 < x < 1, \quad t > 0, \\ w(0,t) &= 0, \\ w(1,t) &= 0, \\ w(x,0) &= 7\sin 3\pi x. \end{aligned}$$

(b) Assuming a solution of the form $w(x,t) = X(x) \cdot T(t)$ and separating variables, we get two ordinary differential equations,

$$X'' + \lambda X = 0 \qquad \text{and} \qquad T' + \frac{\lambda}{9} T = 0,$$

where λ is the separation constant. We can satisfy the two boundary conditions by requiring that $X(0) = X(1) = 0$, so that X satisfies the boundary value problem

$$\begin{aligned} X'' + \lambda X &= 0 \\ X(0) &= 0 \\ X(1) &= 0. \end{aligned}$$

The only nontrivial solutions occur when $\lambda > 0$, say $\lambda = \mu^2$, where $\mu \neq 0$. In this case the general solution is

$$X(x) = A \cos \mu x + B \sin \mu x$$

and from the boundary conditions, $X(0) = 0$ implies that $A = 0$, and since $X(1) = 0$, $\sin \mu = 0$. Thus, the eigenvalues are $\mu_n = n\pi$, with corresponding eigenfunctions $X_n(x) = \sin n\pi x$ for $n \geqslant 1$. For $n \geqslant 1$, the corresponding solution to

$$T' + \frac{n^2\pi^2}{9} T = 0$$

is $T_n(t) = e^{-\frac{n^2\pi^2}{9}t}$, and from the superposition principle, we write

$$w(x,t) = \sum_{n=1}^{\infty} b_n \sin n\pi x \, e^{-\frac{n^2\pi^2}{9}t}$$

for $0 < x < 1,\ t > 0$. From the initial condition, we have

$$7 \sin 3\pi x = w(x,0) = \sum_{n=1}^{\infty} b_n \sin n\pi x,$$

so that $b_n = 0$ for $n \neq 3$, while $b_3 = 7$. Therefore,

$$w(x,t) = 7 \sin 3\pi x \, e^{-\pi^2 t}$$

for $0 < x < 1,\ t > 0$.

(c) The solution to the original problem is

$$u(x,t) = e^{-2t} w(x,t) = 7 \sin 3\pi x \, e^{-(\pi^2+2)t}$$

for $0 < x < 1,\ t > 0$.

(d) Since

$$\sin 3\pi x \leqslant |\sin 3\pi x| \leqslant 1 \qquad \text{and} \qquad e^{-(\pi^2+2)t} > 0,$$

for all $x \in [0,1]$ and all $t \geqslant 0$, we can make $u(x,t) < 1$ by requiring that

$$\left|7 \sin 3\pi x \, e^{-(\pi^2+2)t}\right| < 1,$$

and this will be true if

$$7e^{-(\pi^2+2)t} < 1,$$

that is, if

$$e^{(\pi^2+2)t} > 7,$$

or equivalently, if

$$t > \frac{\log 7}{\pi^2+2},$$

so we may take

$$T_1 = \frac{\log 7}{\pi^2+2}.$$

19.2 FINAL EXAM 2

Exercise 19.6. XX

Let $0 < a < \pi$; given the function

$$f(x) = \begin{cases} \dfrac{1}{2a}, & \text{if } \ |x| < a, \\ 0, & \text{if } \ x \in [-\pi, \pi] \ \text{ and } \ |x| > a, \end{cases}$$

find the Fourier series for f and use Dirichlet's convergence theorem to show that

$$\sum_{n=1}^{\infty} \frac{\sin na}{n} = \frac{1}{2}(\pi - a)$$

for $0 < a < \pi$.

Solution. Since $f(x)$ is an even function of the interval $[-\pi, \pi]$, the Fourier series of $f(x)$ is given by

$$f(x) \sim a_0 + \sum_{n=1}^{\infty} a_n \cos nx,$$

where

$$a_0 = \frac{1}{\pi}\int_0^{\pi} f(x)\,dx = \frac{1}{\pi}\int_0^{a} \frac{1}{2a}\,dx = \frac{1}{2\pi}$$

and

$$
\begin{aligned}
a_n &= \frac{2}{\pi}\int_0^{\pi} f(x)\cos nx\,dx \\
&= \frac{2}{\pi}\int_0^{a} \frac{1}{2a}\cos nx\,dx \\
&= \frac{1}{\pi a}\int_0^{a} \cos nx\,dx \\
&= \frac{1}{\pi a}\cdot\frac{1}{n}\sin nx\Big|_0^a \\
&= \frac{1}{\pi a}\cdot\frac{\sin na}{n}
\end{aligned}
$$

for $n \geqslant 1$, and therefore

$$
f(x) \sim \frac{1}{2\pi} + \frac{1}{\pi a}\sum_{n=1}^{\infty}\frac{\sin na\cos nx}{n}
$$

for $-\pi < x < \pi$.

Since $f(x)$ is continuous on the interval $-a < x < a$, the Fourier series converges to $f(x)$ for $-a < x < a$; that is,

$$
f(x) = \frac{1}{2\pi} + \frac{1}{\pi a}\sum_{n=1}^{\infty}\frac{\sin na\cos nx}{n}
$$

for $-a < x < a$. In particular, when $x = 0$, we have

$$
\frac{1}{2a} = \frac{1}{2\pi} + \frac{1}{\pi a}\sum_{n=1}^{\infty}\frac{\sin na}{n},
$$

so that

$$
\sum_{n=1}^{\infty}\frac{\sin na}{n} = \frac{1}{2}(\pi - a)
$$

for $0 < a < \pi$.

Exercise 19.7. XX

Consider the heat equation with a steady source

$$\frac{\partial u}{\partial t} = \frac{\partial^2 u}{\partial x^2} + 7\sin 3x$$

subject to the initial and boundary conditions

$$u(0,t) = 0, \qquad u(\pi,t) = 0, \qquad \text{and} \qquad u(x,0) = 5\sin 3x.$$

Solve this problem using the method of eigenfunction expansions. Show that the solution approaches a steady-state solution as $t \to \infty$.

Solution. Since the problem already has homogeneous boundary conditions, we consider the corresponding homogeneous problem:

$$\frac{\partial u}{\partial t} = \frac{\partial^2 u}{\partial x^2}, \quad 0 < x < \pi, \quad t > 0,$$

$$u(0,t) = 0,$$

$$u(\pi,t) = 0.$$

The eigenvalues and eigenfunctions for this problem are

$$\lambda_n = n^2 \qquad \text{and} \qquad \phi_n(x) = \sin nx$$

for $n \geqslant 1$. We write the solution to the nonhomogeneous problem as an expansion in terms of these eigenfunctions:

$$u(x,t) = \sum_{n=1}^{\infty} a_n(t) \sin nx,$$

and determine the coefficients $a_n(t)$ which force this to be a solution to the nonhomogeneous problem. We will need the eigenfunction expansions for $Q(x) = 7\sin 3x$ and $f(x) = 5\sin 3x$:

$$7\sin 3x = \sum_{n=1}^{\infty} q_n \sin nx \qquad \text{with} \qquad q_n = 0 \quad \text{for} \quad n \neq 3, \quad q_3 = 7,$$

$$5\sin 3x = \sum_{n=1}^{\infty} f_n \sin nx \qquad \text{with} \qquad f_n = 0 \quad \text{for} \quad n \neq 3, \quad f_3 = 5.$$

Substituting these expansions into the nonhomogeneous equation

$$\frac{\partial u}{\partial t} = \frac{\partial^2 u}{\partial x^2} + 7\sin 3x,$$

we obtain

$$\frac{d\,a_3(t)}{dt}\sin 3x = -9\,a_3(t)\sin 3x + 7\sin 3x,$$

and the coefficient $a_3(t)$ satisfies the initial value problem

$$\frac{d\,a_3(t)}{dt} + 9\,a_3(t) = 7, \quad t > 0,$$

$$a_3(0) = 5.$$

The solution to this initial value problem is

$$a_3(t) = 5e^{-9t} + 7\int_0^t e^{-9(t-s)}\,ds;$$

that is,

$$a_3(t) = \frac{7}{9} + \left(5 - \frac{7}{9}\right)e^{-9t}, \quad t \geqslant 0.$$

Note that $\lim_{t\to\infty} a_3(t) = \frac{7}{9}$. For $n \neq 3$, we obtain

$$\frac{da_n(t)}{dt} + n^2 a_n(t) = 0, \quad t > 0,$$

$$a_n(0) = 0,$$

which implies that

$$a_n(t) = 0, \quad t \geqslant 0$$

for $n \neq 3$. The solution to the heat equation with a steady source is therefore

$$u(x,t) = \left[\frac{7}{9} + \left(5 - \frac{7}{9}\right)e^{-9t}\right]\sin 3x$$

for $0 \leqslant x \leqslant \pi$ and $t \geqslant 0$. For large values of t, this solution approaches $r(x)$, where

$$r(x) = \lim_{t\to\infty} u(x,t) = \frac{7}{9}\sin 3x$$

for $0 \leqslant x \leqslant \pi$. Differentiating this twice with respect to x, we see that

$$r''(x) = -7\sin 3x.$$

Since $r(0) = r(\pi) = 0$, the function $r(x)$ satisfies the boundary value problem

$$\frac{d^2 r}{dx^2} + 7\sin 3x = 0, \quad 0 < x < \pi,$$

$$r(0) = 0,$$
$$r(\pi) = 0,$$

which is exactly the boundary value problem for the steady-state solution; that is, $r(x)$ is the steady-state or equilibrium solution to the original heat flow problem.

Exercise 19.8. XX

(a) Using the method of characteristics, solve

$$\frac{\partial w}{\partial t} + c\frac{\partial w}{\partial x} = e^{2x}, \quad -\infty < x < \infty, \quad t \geqslant 0,$$

$$w(x,0) = \frac{1}{2}e^{2x}.$$

(b) For which values of c does this initial value problem have a time-independent solution?

Solution.

(a) Let $dx/dt = c$; then along the characteristic curve $x(t) = ct + a$, where $a = x(0)$, the partial differential equation becomes

$$\frac{dw}{dt} = \frac{\partial w}{\partial t} + \frac{\partial w}{\partial x}\frac{dx}{dt} = e^{2x(t)} = e^{2(ct+a)},$$

so that

$$w(x(t),t) = \frac{1}{2c}e^{2(ct+a)} + K = \frac{1}{2c}e^{2x(t)} + K,$$

where K is a constant, and $K = w(x(0),0) - (1/2c)e^{2x(0)}$, so that

$$w(x(t),t) = \frac{1}{2c}e^{2x(t)} + w(x(0),0) - \frac{1}{2c}e^{2x(0)};$$

that is,

$$w(x(t),t) = \frac{1}{2c}e^{2x(t)} + \frac{1}{2}e^{2(x(t)-ct)} - \frac{1}{2c}e^{2(x(t)-ct)}.$$

Given the point (x,t), let $x = ct + a$ be the unique characteristic curve passing through this point; then

$$w(x,t) = \frac{1}{2c}e^{2x} + \frac{1}{2}e^{2(x-ct)} - \frac{1}{2c}e^{2(x-ct)}$$

for $-\infty < x < \infty$ and $t > 0$.

(b) Note that if $c = 1$, the solution is

$$w(x,t) = \frac{1}{2}e^{2x}, \quad -\infty < x < \infty,$$

which is time independent.

Exercise 19.9. ✗✗

Consider torsional oscillations of a homogeneous cylindrical shaft. If $\omega(x,t)$ is the angular displacement at time t of the cross section at x, then

$$\frac{\partial^2 \omega}{\partial t^2} = a^2 \frac{\partial^2 \omega}{\partial x^2}, \qquad 0 < x < \ell, \quad t > 0,$$

where the initial conditions are

$$\omega(x,0) = f(x) \quad \text{and} \quad \frac{\partial \omega}{\partial t}(x,0) = 0,$$

and the ends of the shaft are fixed elastically:

$$\frac{\partial \omega}{\partial x}(0,t) - \alpha\,\omega(0,t) = 0 \quad \text{and} \quad \frac{\partial \omega}{\partial x}(\ell,t) + \alpha\,\omega(\ell,t) = 0,$$

with α a positive constant.

(a) Why is it possible to use separation of variables to solve this problem ?

(b) Use separation of variables and show that one of the resulting problems is a regular Sturm-Liouville problem.

(c) Show that all of the eigenvalues of this regular Sturm-Liouville problem are positive.

Note: You do not need to solve the initial value problem, just answer the questions (a), (b), and (c).

Solution.

(a) Since the partial differential equation is linear and homogeneous and the boundary conditions are linear and homogeneous, we can use separation of variables.

(b) Assuming a solution of the form

$$\omega(x,t) = \phi(x) \cdot G(t), \quad 0 < x < \ell, \quad t > 0,$$

and separating variables, we have two ordinary differential equations:

$$\begin{aligned} &\phi''(x) + \lambda\phi(x) = 0, \quad 0 < x < \ell, \qquad && G''(t) + \lambda a^2 G(t) = 0, \quad t > 0, \\ &\phi'(0) - \alpha\phi(0) = 0, && \\ &\phi'(\ell) + \alpha\phi(\ell) = 0, && \end{aligned}$$

where the ϕ-problem is a regular Sturm-Liouville problem with

$$p(x) = 1, \qquad q(x) = 0, \qquad \sigma(x) = 1$$

and

$$\alpha_1 = -\alpha, \qquad \beta_1 = 1, \qquad \alpha_2 = \alpha, \qquad \beta_2 = 1.$$

(c) We use the Rayleigh quotient to show that $\lambda > 0$ for all eigenvalues λ. Let λ be an eigenvalue of the Sturm-Liouville problem, and let $\phi(x)$ be the corresponding eigenfunction; then

$$-p(x)\phi(x)\phi'(x)\Big|_0^\ell = -\phi(\ell)\phi'(\ell) + \phi(0)\phi'(0) = \alpha(\phi(0)^2 + \phi(\ell)^2) > 0,$$

and since $q(x) = 0$ for all $0 \leqslant x \leqslant \ell$, then

$$\lambda = \frac{\alpha(\phi(0)^2 + \phi(\ell)^2) + \displaystyle\int_0^\ell \phi'(x)^2\,dx}{\displaystyle\int_0^\ell \phi(x)^2\,dx} \geqslant 0$$

since $p(x) = \sigma(x) = 1$ for $0 \leqslant x \leqslant \ell$. Note that if $\lambda = 0$, then

$$\alpha\big(\phi(0)^2 + \phi(\ell)^2\big) + \int_0^\ell \phi'(x)^2\,dx = 0$$

implies that

$$\alpha\big(\phi(0)^2 + \phi(\ell)^2\big) = 0 \qquad \text{and} \qquad \int_0^\ell \phi'(x)^2\,dx = 0.$$

Since $\alpha > 0$, this implies that $\phi(0) = 0$ and $\phi(\ell) = 0$, and since ϕ' is continuous on $[0, \ell]$, that $\phi'(x) = 0$ for $0 \leqslant x \leqslant \ell$. Therefore, $\phi(x)$ is constant on $[0, \ell]$, so that $\phi(x) = \phi(0) = 0$ for $0 \leqslant x \leqslant \ell$, and $\lambda = 0$ is not an eigenvalue. Thus, all of the eigenvalues λ of this Sturm-Liouville problem satisfy $\lambda > 0$.

19.3 FINAL EXAM 3

Exercise 19.10. ✗

Assume that $f''(t)$ is absolutely integrable and

$$\lim_{t\to\infty} f(t) = 0 \qquad \text{and} \qquad \lim_{t\to\infty} f'(t) = 0.$$

Show that

$$\mathcal{S}\,[\,f''(x)\,]\,(\omega) = -\omega^2 \mathcal{S}\,[\,f(x)\,]\,(\omega) + \frac{2}{\pi}\,\omega f(0).$$

Solution. Assuming that $\lim_{t\to\infty} f(t) = 0$, and $\lim_{t\to\infty} f'(t) = 0$, and integrating by parts, we have

$$\begin{aligned}
\mathcal{S}\,[\,f''(x)\,]\,(\omega) &= \frac{2}{\pi}\int_0^\infty f''(t)\sin\omega t\,dt \\
&= \frac{2}{\pi}\,f'(t)\sin\omega t\Big|_0^\infty - \frac{2}{\pi}\,\omega\int_0^\infty f'(t)\cos\omega t\,dt \\
&= -\frac{2}{\pi}\,\omega\int_0^\infty f'(t)\cos\omega t\,dt \\
&= -\frac{2}{\pi}\,\omega\, f(t)\cos\omega t\Big|_0^\infty + \frac{2}{\pi}\,\omega^2\int_0^\infty f(t)\sin\omega t\,dt \\
&= \frac{2}{\pi}\,\omega\, f(0) - \omega^2\mathcal{S}\,[\,f(x)\,]\,(\omega).
\end{aligned}$$

We used the fact that

$$|f'(t)\sin\omega t| \leqslant |f'(t)| \longrightarrow 0 \quad \text{as} \quad t\to\infty$$

and

$$|f(t)\cos\omega t| \leqslant |f(t)| \longrightarrow 0 \quad \text{as} \quad t\to\infty.$$

Exercise 19.11. XXX

Legendre's differential equation reads

$$(1-x^2)y'' - 2xy' + \lambda y = 0, \qquad -1 < x < 1.$$

(a) Write the differential equation in Sturm-Liouville form. Decide if the resulting Sturm-Liouville problem is regular or singular.

(b) Show that the first four Legendre polynomials,

$$P_0(x) = 1, \quad P_1(x) = x, \quad P_2(x) = \tfrac{1}{2}(3x^2-1), \quad P_3(x) = \tfrac{1}{2}(5x^3-3x),$$

are eigenfunctions of the Sturm-Liouville problem and find the corresponding eigenvalues.

(c) Use an appropriate weight function and show that P_1 and P_2 are orthogonal on the interval $(-1,1)$ with respect to this weight function.

Solution.

(a) Since

$$\frac{d}{dx}\left((1-x^2)\,\frac{dy}{dx}\right) = (1-x^2)\frac{d^2y}{dx^2} - 2x\frac{dy}{dx},$$

Legendre's equation can be written as

$$\left((1-x^2)\,y'\right)' + \lambda y = 0, \quad -1 < x < 1, \tag{19.1}$$

which is the classical Sturm-Liouville form

$$[p(x)\,y']' + [q(x) + \lambda\sigma(x)]\,y = 0, \quad a < x < b$$

with

$$p(x) = 1 - x^2, \qquad q(x) = 0, \qquad \text{and} \qquad \sigma(x) = 1$$

for $a < x < b$, where $a = -1$ and $b = 1$.

For a regular Sturm-Liouville problem we require the regularity conditions

$$p(x), \qquad p'(x), \qquad q(x), \qquad \text{and} \qquad \sigma(x)$$

are continuous on the closed interval $a \leqslant x \leqslant b$, and

$$p(x) > 0, \qquad \text{and} \qquad \sigma(x) > 0$$

for $a \leqslant x \leqslant b$. We also require the boundary conditions

$$\alpha_1 y(a) + \beta_1 y'(a) = 0 \qquad \text{and} \qquad \alpha_2 y(b) + \beta_2 y'(b) = 0,$$

where at least one of α_1 and β_1 is nonzero and at least one of α_2 and β_2 is nonzero. Thus, it is clear that (19.1) is a singular Sturm-Liouville problem (no matter what the boundary conditions are) since one of the regularity conditions is violated, namely, $p(-1) = p(1) = 0$.

(b) • For $P_0(x) = 1$, we have

$$P_0'(x) = 0 \qquad \text{and} \qquad P_0''(x) = 0$$

for $-1 < x < 1$, so that

$$(1 - x^2)P_0'' - 2xP_0' + \lambda P_0 = 0, \quad -1 < x < 1$$

is satisfied for $\lambda = 0$, and the eigenvalue corresponding to the eigenfunction $P_0(x) = 1$ is $\lambda_0 = 0$.

• For $P_1(x) = x$, we have

$$P_1'(x) = 1 \qquad \text{and} \qquad P_1''(x) = 0$$

for $-1 < x < 1$, so that

$$(1 - x^2)P_1'' - 2xP_1' + \lambda P_1 = 0, \quad -1 < x < 1$$

becomes

$$-2x \cdot 1 + \lambda x = 0, \quad -1 < x < 1,$$

which is satisfied for $\lambda = 2$, and the eigenvalue corresponding to the eigenfunction $P_1(x) = x$ is $\lambda_1 = 2$.

• For $P_2(x) = \frac{1}{2}(3x^2 - 1)$, we have

$$P_2'(x) = 3x \qquad \text{and} \qquad P_2''(x) = 3$$

for $-1 < x < 1$, so that

$$(1 - x^2)P_2'' - 2xP_2' + \lambda P_2 = 0, \quad -1 < x < 1$$

becomes

$$3(1 - x^2) - 6x^2 + \frac{\lambda}{2}(3x^2 - 1) = 0, \quad -1 < x < 1,$$

that is,

$$-3(3x^2 - 1) + \frac{\lambda}{2}(3x^2 - 1) = 0, \quad -1 < x < 1,$$

which is satisfied for $\lambda = 6$, and the eigenvalue corresponding to the eigenfunction $P_2(x) = \frac{1}{2}(3x^2 - 1)$ is $\lambda_2 = 6$.

• For $P_3(x) = \frac{1}{2}(5x^3 - 3x)$, we have

$$P_3'(x) = \frac{1}{2}(15x^2 - 3) \qquad \text{and} \qquad P_3''(x) = 15x$$

for $-1 < x < 1$, so that

$$(1 - x^2)P_3'' - 2xP_3' + \lambda P_3 = 0, \quad -1 < x < 1$$

becomes

$$15x(1 - x^2) - (15x^3 - 3x) + \frac{\lambda}{2}(5x^3 - 3x) = 0, \quad -1 < x < 1,$$

that is,

$$-6(5x^3 - 3x) + \frac{\lambda}{2}(5x^3 - 3x) = 0, \quad -1 < x < 1,$$

which is satisfied for $\lambda = 12$, and the eigenvalue corresponding to the eigenfunction $P_3(x) = \frac{1}{2}(5x^3 - 3x)$ is $\lambda_3 = 12$.

(c) Using the weight function $\sigma(x) = 1$, for $-1 < x < 1$, we have

$$\langle P_1, P_2 \rangle = \int_{-1}^{1} P_1(x) \cdot P_2(x)\, dx = 0$$

since the product $P_1(x)P_2(x)$ is an odd function integrated between symmetric limits; thus, $P_1(x)$ and $P_2(x)$ are orthogonal on the interval $-1 < x < 1$ with respect to the weight function $\sigma(x) = 1$.

Exercise 19.12. XX

Find all functions ϕ for which $u(x,t) = \phi(x+ct)$ is a solution of the heat equation

$$\frac{\partial^2 u}{\partial x^2} = \frac{1}{k}\frac{\partial u}{\partial t},$$

where k and c are constants.

Solution. If $u(x,t) = \phi(x + ct)$ is a solution to the heat equation

$$\frac{\partial^2 u}{\partial x^2} = \frac{1}{k}\frac{\partial u}{\partial t},$$

let $\xi = x + ct$; then from the chain rule we have

$$\begin{aligned}
\frac{\partial u}{\partial x} &= \frac{d\phi}{d\xi} \cdot \frac{\partial \xi}{\partial x} = \frac{d\phi}{d\xi}, \\
\frac{\partial^2 u}{\partial x^2} &= \frac{d}{d\xi}\left(\frac{d\phi}{d\xi}\right) \cdot \frac{\partial \xi}{\partial x} = \frac{d^2\phi}{d\xi^2}, \\
\frac{\partial u}{\partial t} &= \frac{d\phi}{d\xi} \cdot \frac{\partial \xi}{\partial t} = c\,\frac{d\phi}{d\xi}.
\end{aligned}$$

Therefore, ϕ satisfies the ordinary differential equation

$$\frac{d^2\phi}{d\xi^2} - \frac{c}{k}\frac{d\phi}{d\xi} = 0,$$

and the solution is given by

$$\phi(\xi) = A + B\,e^{\frac{c}{k}\xi},$$

that is,

$$u(x,t) = A + B\,e^{\frac{c}{k}(x+ct)},$$

where A and B are arbitrary constants.

Exercise 19.13. ✗✗✗

Let

$$f(x) = \begin{cases} \cos x, & |x| < \pi, \\ 0, & |x| > \pi. \end{cases}$$

(a) Find the Fourier integral of f.

(b) For which values of x does the integral converge to $f(x)$?

(c) Evaluate the integral

$$\int_0^\infty \frac{\lambda \sin \lambda\pi \cos \lambda x}{1 - \lambda^2} \, d\lambda$$

for $-\infty < x < \infty$.

Solution.

(a) The Fourier integral representation of f is given by

$$f(x) \sim \int_0^\infty [A(\lambda) \cos \lambda x + B(\lambda) \sin \lambda x] \, d\lambda, \quad -\infty < x < \infty,$$

where

$$A(\lambda) = \frac{1}{\pi} \int_{-\infty}^\infty f(t) \cos \lambda t \, dt \quad \text{and} \quad B(\lambda) = \frac{1}{\pi} \int_{-\infty}^\infty f(t) \sin \lambda t \, dt$$

for $\lambda \geqslant 0$. Since $f(t)$ is an even function on the interval $-\infty < t < \infty$, then $B(\lambda) = 0$ for all $\lambda \geqslant 0$, and

$$A(\lambda) = \frac{1}{\pi} \int_{-\infty}^\infty f(t) \cos \lambda t \, dt = \frac{2}{\pi} \int_0^\infty f(t) \cos \lambda t \, dt$$

for all $\lambda \geqslant 0$.

Now, for $\lambda \neq 1$, we have

$$\begin{aligned} A(\lambda) &= \frac{2}{\pi}\int_0^{\pi} \cos t \cos \lambda t \, dt \\ &= \frac{2}{\pi}\int_0^{\pi} \frac{1}{2}\left[\cos(1+\lambda)t + \cos(1-\lambda)t\right] dt \\ &= \frac{\sin(1+\lambda)t}{\pi(1+\lambda)}\bigg|_0^{\pi} + \frac{\sin(1-\lambda)t}{\pi(1-\lambda)}\bigg|_0^{\pi} \\ &= \frac{\sin(1+\lambda)\pi}{\pi(1+\lambda)} + \frac{\sin(1-\lambda)\pi}{\pi(1-\lambda)} \\ &= -\frac{\sin \lambda\pi}{\pi(1+\lambda)} + \frac{\sin \lambda\pi}{\pi(1-\lambda)} \\ &= \frac{2\lambda \sin \lambda\pi}{\pi(1-\lambda^2)}; \end{aligned}$$

that is,

$$A(\lambda) = \frac{2\lambda \sin \lambda\pi}{\pi(1-\lambda^2)}$$

for $\lambda \geqslant 0,\ \lambda \neq 1$. And for $\lambda = 1$, we have

$$\begin{aligned} A(1) &= \frac{2}{\pi}\int_0^{\pi} \cos^2 t \, dt = \frac{2}{\pi}\int_0^{\pi} \frac{1}{2}\left[1 + \cos 2t\right] dt \\ &= \frac{1}{\pi}\left[t + \frac{1}{2}\sin 2t\right]\bigg|_0^{\pi} = \frac{1}{\pi}\cdot \pi = 1. \end{aligned}$$

Therefore,

$$A(\lambda) = \begin{cases} \dfrac{2\lambda \sin \lambda\pi}{\pi(1-\lambda^2)}, & \text{for} \quad \lambda \geqslant 0,\ \lambda \neq 1, \\ 1, & \text{for} \quad \lambda = 1. \end{cases}$$

(b) Since $f(x)$ is continuous for all $x \neq \pm\pi$, then from Dirichlet's theorem, the Fourier integral representation converges to $f(x)$ for all such x; that is,

$$\begin{aligned} f(x) &= \int_0^{\infty} A(\lambda)\cos \lambda x \, d\lambda = \int_0^{\infty} \frac{2\lambda \sin \lambda\pi}{\pi(1-\lambda^2)} \cos \lambda x \, d\lambda \\ &= \begin{cases} \cos x, & \text{for} \quad |x| < \pi, \\ 0, & \text{for} \quad |x| > \pi \end{cases} \end{aligned}$$

for all $x \neq \pm\pi$.

When $x = \pm\pi$, from Dirichlet's theorem the Fourier integral representation converges to

$$\frac{f(\pi^+) + f(\pi^-)}{2} = \frac{0-1}{2} = -\frac{1}{2}$$

and

$$\frac{f(-\pi^+) + f(-\pi^-)}{2} = \frac{-1+0}{2} = -\frac{1}{2}.$$

(c) From part (b) we have

$$\int_0^\infty \frac{\lambda \sin \lambda\pi \cos \lambda x}{1-\lambda^2}\, d\lambda = \begin{cases} \dfrac{\pi}{2}\cos x, & \text{for } \ |x| < \pi, \\ 0, & \text{for } \ |x| > \pi, \\ -\dfrac{\pi}{4}, & \text{for } \ |x| = \pi. \end{cases}$$

19.4 FINAL EXAM 4

Exercise 19.14. XX

A fluid occupies the half-plane $y > 0$ and flows past (left to right, approximately) a plate located near the x-axis. If the x and y components of the velocity are

$$U_0 + u(x, y) \qquad \text{and} \qquad v(x, y),$$

respectively, where U_0 is the constant free-stream velocity, then under certain assumptions, the equations of motion, continuity, and state can be reduced to

$$\frac{\partial u}{\partial y} = \frac{\partial v}{\partial x}, \qquad (1 - M^2)\frac{\partial u}{\partial x} + \frac{\partial v}{\partial y} = 0, \qquad (19.2)$$

valid for all $-\infty < x < \infty$, $0 < y < \infty$. Suppose that there exists a function ϕ (called the *velocity potential*) such that

$$u = \frac{\partial \phi}{\partial x} \qquad \text{and} \qquad v = \frac{\partial \phi}{\partial y}.$$

(a) State a condition under which the first equation in (19.2) becomes an identity.

(b) Show that the second equation in (19.2) becomes (assuming that the free-stream Mach number M is a constant) a partial differential equation for ϕ which is elliptic if $M < 1$ or hyperbolic if $M > 1$.

Solution.

(a) If the velocity potential ϕ exists, then

$$\frac{\partial u}{\partial y} = \frac{\partial}{\partial y}\left(\frac{\partial \phi}{\partial x}\right) = \frac{\partial^2 \phi}{\partial y\,\partial x}$$

and

$$\frac{\partial v}{\partial x} = \frac{\partial}{\partial x}\left(\frac{\partial \phi}{\partial y}\right) = \frac{\partial^2 \phi}{\partial x\,\partial y},$$

and the mixed partial derivatives are equal at all points where they are continuous. Therefore, the first equation in (19.2) is an identity provided that

$$\frac{\partial^2 \phi}{\partial y\,\partial x} = \frac{\partial^2 \phi}{\partial x\,\partial y}$$

for all $-\infty < x < \infty$, $0 < y < \infty$. Another possible solution is then obtained by assuming that the velocity potential $\phi(x, y)$ is twice continuously differentiable.

(b) Again, assuming the existence of a velocity potential, the second equation in (19.2) becomes

$$\left(1 - M^2\right)\frac{\partial^2 \phi}{\partial x^2} + \frac{\partial^2 \phi}{\partial y^2} = 0,$$

which is elliptic if $1 - M^2 > 0$ and hyperbolic if $1 - M^2 < 0$, that is, elliptic if $M < 1$ and hyperbolic if $M > 1$.

Exercise 19.15. XX

Besides linear equations, some nonlinear equations can also result in *traveling wave solutions* of the form

$$u(x, t) = \phi(x - ct).$$

Fisher's equation, which models the spread of an advantageous gene in a population, where $u(x, t)$ is the density of the gene in the population at time t and location x, is given by

$$\frac{\partial u}{\partial t} = \frac{\partial^2 u}{\partial x^2} + u(1 - u).$$

Show that Fisher's equation has a solution of this form if ϕ satisfies the nonlinear ordinary differential equation

$$\phi'' + c\phi' + \phi(1 - \phi) = 0.$$

Solution. If $u(x,t) = \phi(x - ct)$, then

$$\begin{aligned}\frac{\partial u}{\partial x} &= \phi'(x - ct),\\ \frac{\partial^2 u}{\partial x^2} &= \phi''(x - ct),\\ \frac{\partial u}{\partial t} &= -c\phi'(x - ct),\end{aligned}$$

and Fisher's equation becomes

$$-c\phi'(x - ct) = \phi''(x - ct) + \phi(x - ct)\left(1 - \phi(x - ct)\right),$$

for all x and t, so that if ϕ satisfies the nonlinear ordinary differential equation

$$\phi''(s) + c\phi'(s) + \phi(s)(1 - \phi(s)) = 0, \quad -\infty < s < \infty,$$

then $u(x,t) = \phi(x - ct)$ is a traveling wave solution to Fisher's equation.

Exercise 19.16. XXX

Given the regular Sturm-Liouville problem

$$\begin{aligned}&\phi''(x) + \lambda^2\phi(x) = 0, \quad 0 < x < \pi\\ &\phi(0) = 0,\\ &\phi(\pi) = 0.\end{aligned}$$

(a) Find the eigenvalues λ_n^2 and corresponding eigenfunctions $\phi_n(x)$ for this problem.

(b) Show directly, by integration, that eigenfunctions corresponding to distinct eigenvalues are orthogonal on the interval $[0, \pi]$.

(c) Use the method of eigenfunction expansions to find the solution to the boundary value problem

$$\begin{aligned}&u''(x) = -x, \quad 0 < x < \pi\\ &u(0) = 0,\\ &u(\pi) = 0.\end{aligned}$$

(d) Solve the problem in (c) by direct integration and use this result to show that

$$\frac{x(\pi^2 - x^2)}{6} = 2\sum_{n=1}^{\infty}(-1)^{n+1}\frac{\sin nx}{n^3}$$

for $-\pi \leqslant x \leqslant \pi$.

Solution.

(a) Since the parameter $\lambda^2 \geqslant 0$, we need consider only two cases:

(i) If $\lambda = 0$, the differential equation is $\phi'' = 0$ and has general solution $\phi(x) = Ax + B$. The boundary condition $\phi(0) = 0$ implies that $B = 0$, while the boundary condition $\phi(\pi) = 0$ implies that $A = 0$, and there are no nontrivial solutions in this case.

(ii) If $\lambda \neq 0$, the differential equation $\phi'' + \lambda^2 \phi = 0$ has general solution $\phi(x) = A\cos\lambda x + B\sin\lambda x$. The boundary condition $\phi(0) = 0$ implies that $A = 0$, while the boundary condition $\phi(\pi) = 0$ implies that $\lambda\pi = n\pi$ for some positive integer n. Therefore, the eigenvalues and corresponding eigenfunctions are

$$\lambda_n^2 = n^2 \qquad \text{and} \qquad \phi_n(x) = \sin nx$$

for $n \geqslant 1$.

(b) If m and n are distinct positive integers, then

$$\begin{aligned}\int_0^\pi \sin mx \sin nx\,dx &= \int_0^\pi \frac{1}{2}\left[\cos(m-n)x - \cos(m+n)x\right]dx \\ &= \frac{\sin(m-n)x}{2(m-n)}\bigg|_0^\pi - \frac{\sin(m+n)x}{2(m+n)}\bigg|_0^\pi \\ &= 0\end{aligned}$$

and eigenfunctions corresponding to distinct eigenvalue are orthogonal.

(c) Let $u(x)$ be a solution to the specified boundary value problem on the interval $[0, \pi]$. Expanding $u(x)$ in terms of the eigenfunctions of the Sturm-Liouville problem, we have

$$u(x) = \sum_{n=1}^{\infty} c_n \sin nx$$

for $0 \leqslant x \leqslant \pi$. Differentiating this twice, we get

$$u''(x) = \sum_{n=1}^{\infty} (-c_n n^2) \sin nx;$$

that is,

$$x = \sum_{n=1}^{\infty} c_n n^2 \sin nx$$

for $0 \leqslant x \leqslant \pi$. Multiplying by $\sin mx$ and integrating, we have

$$\int_0^\pi x \sin mx\,dx = c_m m^2 \int_0^\pi \sin^2 mx\,dx.$$

From the orthogonality conditions, we find that

$$\int_0^\pi x\sin mx\,dx = -\frac{x\cos mx}{m}\bigg|_0^\pi + \frac{1}{m}\int_0^\pi \cos mx\,dx$$

$$= -\pi\frac{\cos m\pi}{m} = \frac{\pi(-1)^{m+1}}{m}$$

and

$$\int_0^\pi \sin^2 mx\,dx = \frac{\pi}{2},$$

so that

$$c_m = \frac{\displaystyle\int_0^\pi x\sin mx\,dx}{\displaystyle m^2\int_0^\pi \sin^2 mx\,dx} = \frac{2(-1)^{m+1}}{m^3}$$

for $m \geqslant 1$. Therefore,

$$u(x) = 2\sum_{n=1}^{\infty}(-1)^{n+1}\frac{\sin nx}{n^3}$$

for $0 \leqslant x \leqslant \pi$.

(d) The general solution to the differential equation $u''(x) = -x$ is

$$u(x) = -\frac{x^3}{6} + Ax + B,$$

where A and B are constants. Applying the boundary conditions, $u(0) = 0$ implies that $B = 0$, while $u(\pi) = 0$ implies that $A = \pi^2/6$, and the solution is

$$u(x) = \frac{x\left(\pi^2 - x^2\right)}{6}$$

for $0 \leqslant x \leqslant \pi$. From part (c) we have

$$\frac{x\left(\pi^2 - x^2\right)}{6} = 2\sum_{n=1}^{\infty}(-1)^{n+1}\frac{\sin nx}{n^3} \tag{19.3}$$

for $0 \leqslant x \leqslant \pi$. The series on the right-hand side is the Fourier sine series for the odd function

$$\frac{x\left(\pi^2 - x^2\right)}{6}$$

on the interval $[0, \pi]$, and since the odd extension is continuous on $[-\pi, \pi]$, Dirichlet's theorem says that (19.3) holds for all x with $-\pi \leqslant x \leqslant \pi$.

Exercise 19.17. XXX

Find the solution to Laplace's equation on the rectangle:

$$\frac{\partial^2 u}{\partial x^2} + \frac{\partial^2 u}{\partial y^2} = 0, \quad 0 < x < a, \quad 0 < y < b,$$

$$u(0, y) = 1,$$

$$u(a, y) = 1,$$

$$\frac{\partial u}{\partial y}(x, 0) = 0,$$

$$\frac{\partial u}{\partial y}(x, b) = 0$$

using the method of separation of variables. Is your solution what you expected?

Solution. Writing $u(x, y) = X(x) \cdot Y(y)$, we obtain

$$\frac{X''}{X} = -\frac{Y''}{Y} = \lambda^2,$$

where λ is the separation constant, and hence we get the two ordinary differential equations

$$X'' - \lambda^2 X = 0 \quad 0 < x < a, \qquad Y'' + \lambda^2 Y = 0, \quad 0 < y < b,$$

$$Y'(0) = 0,$$

$$Y'(b) = 0.$$

Solving the regular Sturm-Liouville problem for Y, for the eigenvalue $\lambda_0^2 = 0$ the corresponding eigenfunction is

$$Y_0(y) = 1,$$

and the corresponding solution to the first equation is

$$X_0(x) = b_0 x + a_0.$$

For the eigenvalues $\lambda_n^2 = (n\pi/b)^2$, the corresponding eigenfunctions are

$$Y_n(y) = \cos \lambda_n y,$$

and the corresponding solutions to the first equation are

$$X_n(x) = a_n \cosh \lambda_n x + b_n \sinh \lambda_n x$$

for $n = 1, 2, 3, \ldots$. Using the superposition principle, we write

$$u(x, y) = b_0 x + a_0 + \sum_{n=1}^{\infty} \big(a_n \cosh \lambda_n x + b_n \sinh \lambda_n x\big) \cos \lambda_n y.$$

From the boundary condition $u(0, y) = 1$, we have

$$1 = a_0 + \sum_{n=1}^{\infty} a_n \cos \lambda_n y,$$

so that

$$a_0 = \frac{1}{b} \int_0^b 1 \, dy = 1,$$

while

$$a_n = \frac{2}{b} \int_0^b \cos \lambda_n y \, dy = \frac{2}{n\pi} \sin \lambda_n y \Big|_0^b = 0$$

for $n = 1, 2, 3, \ldots$. From the boundary condition $u(a, y) = 1$, we have

$$1 = b_0 \, a + 1 + \sum_{n=1}^{\infty} b_n \sinh \lambda_n a \, \cos \lambda_n y$$

and integrating this equation from 0 to b we get $b_0 \, a \, b = 0$, and therefore $b_0 = 0$, so that

$$0 = \sum_{n=1}^{\infty} b_n \sinh \lambda_n a \, \cos \lambda_n y.$$

To evaluate the b_n's, we multiply this equation by $\cos(m\pi/b)y$ and integrate from 0 to b, to obtain $b_m \, \sinh(m\pi/b)a = 0$; that is, $b_m = 0$ for $m = 1, 2, 3, \ldots$. Therefore, the solution is $u(x, y) \equiv 1$, which is not totally unexpected. The solution is unique and it is clear from the statement of the problem that $u(x, y) = 1$ satisfies Laplace's equation on the rectangle and also satisfies all of the boundary conditions.

Exercise 19.18. XX

Solve the following initial value problem for the damped wave equation:

$$\frac{\partial^2 u}{\partial t^2} + 2\frac{\partial u}{\partial t} + u = \frac{\partial^2 u}{\partial x^2}, \qquad -\infty < x < \infty, \quad t > 0,$$

$$u(x, 0) = \frac{1}{1 + x^2},$$

$$\frac{\partial u}{\partial t}(x, 0) = 1.$$

Hint: Do not use separation of variables; instead, solve the boundary value-initial value problem satisfied by $w(x, t) = e^t \cdot u(x, t)$.

Solution. Note that $u(x, t) = e^{-t} \cdot w(x, t)$, so that

$$\frac{\partial^2 u}{\partial x^2} = e^{-t} \frac{\partial^2 w}{\partial x^2}$$

and

$$\frac{\partial u}{\partial t} = -e^{-t}w + e^{-t}\frac{\partial w}{\partial t},$$

so

$$\frac{\partial^2 u}{\partial t^2} = e^{-t}w - 2e^{-t}\frac{\partial w}{\partial t} + e^{-t}\frac{\partial^2 w}{\partial t^2}.$$

Therefore,

$$\frac{\partial^2 u}{\partial t^2} + 2\frac{\partial u}{\partial t} + u = e^{-t}\frac{\partial^2 w}{\partial t^2},$$

so that

$$e^{-t}\frac{\partial^2 w}{\partial x^2} = e^{-t}\frac{\partial^2 w}{\partial t^2}.$$

Thus, if u is a solution to the original partial differential equation, since $e^{-t} \neq 0$, w satisfies the initial value problem

$$\begin{aligned}
&\frac{\partial^2 w}{\partial t^2} = \frac{\partial^2 w}{\partial x^2}, \qquad -\infty < x < \infty, \quad t > 0,\\
&w(x,0) = \frac{1}{1+x^2},\\
&\frac{\partial w}{\partial t}(x,0) = 1 + \frac{1}{1+x^2}.
\end{aligned}$$

From d'Alembert's solution to the wave equation, we have (since $c = 1$)

$$w(x,t) = \frac{1}{2}\left[\frac{1}{1+(x+t)^2} + \frac{1}{1+(x-t)^2}\right] + \frac{1}{2}\int_{x-t}^{x+t}\left(1 + \frac{1}{1+s^2}\right)ds,$$

so that

$$\begin{aligned}
u(x,t) = \frac{e^{-t}}{2}&\left[\frac{1}{1+(x+t)^2} + \frac{1}{1+(x-t)^2}\right]\\
&+ \frac{e^{-t}}{2}\left[2t + \tan^{-1}(x+t) - \tan^{-1}(x-t)\right]
\end{aligned}$$

for $-\infty < x < \infty, \quad t > 0.$

APPENDIX A

GAMMA FUNCTION

We have used the gamma function $\Gamma(x)$ in the text without ever having defined it. Here we give the definition and a few of the more elementary properties of $\Gamma(x)$.

Definition A.1. The **gamma function** is defined for all $x > 0$ as

$$\Gamma(x) = \int_0^\infty t^{x-1}e^{-t}\,dt.$$

However, we have to make sure that this definition makes sense, that is, that the improper integral converges for all $x > 0$.

Theorem A.2. For each positive real number x, the integral

$$\int_0^\infty t^{x-1}e^{-t}\,dt$$

converges absolutely.

Proof. Let x be a fixed real number, and let $f(t) = t^{x-1}e^{-t}$ for $t \geqslant 1$; then

$$\lim_{t\to\infty} t^{x-1}e^{-t/2} = 0.$$

Partial Differential Equations: Theory and Completely Solved Problems, First Edition. By T. Hillen, I. E. Leonard, H. van Roessel

Hence, there exists a constant $M > 0$ such that

$$t^{x-1}e^{-t/2} \leqslant M$$

for all $t \geqslant 1$, so

$$t^{x-1}e^{-t} \leqslant M\,e^{-t/2}$$

for all $t \geqslant 1$. Therefore, for $b \geqslant 1$ we have

$$\int_1^b t^{x-1}e^{-t}\,dt \leqslant M\int_1^b e^{-t/2}\,dt \leqslant M\int_0^b e^{-t/2}\,dt = 2M\big(1 - e^{-b/2}\big) < 2M,$$

and letting $b \to \infty$,

$$\int_1^\infty t^{x-1}e^{-t}\,dt \leqslant 2M.$$

Hence, the improper integral converges absolutely for all real numbers x.

On the other hand, for $x > 0$, if $\epsilon > 0$ we have

$$\int_\epsilon^1 t^{x-1}e^{-t}\,dt = \int_\epsilon^1 \frac{e^{-t}}{t^{1-x}}\,dt \leqslant \int_\epsilon^1 \frac{dt}{t^{1-x}} = \frac{1}{x}(1 - \epsilon^x),$$

and in this case

$$\lim_{\epsilon \to 0+} (1 - \epsilon^x) = 1,$$

so that

$$\int_0^1 t^{x-1}e^{-t}\,dt = \lim_{\epsilon \to 0+} \int_\epsilon^1 t^{x-1}e^{-t}\,dt \leqslant \frac{1}{x},$$

and the integral converges for all $x > 0$. We leave it as an exercise to show that this integral diverges for $x \leqslant 0$. Therefore,

$$\int_0^\infty t^{x-1}e^{-t}\,dt = \int_0^1 t^{x-1}e^{-t}\,dt + \int_1^\infty t^{x-1}e^{-t}\,dt$$

converges for all $x > 0$.

□

We have the following characteristic property or functional equation for the gamma function:

Theorem A.3. For $x > 0$,

$$\Gamma(x+1) = x\Gamma(x).$$

Proof. For $x > 0$, integrating by parts, we have

$$\Gamma(x+1) = \int_0^\infty t^x e^{-t}\,dt = -t^x e^{-t}\Big|_0^\infty + x\int_0^\infty t^{x-1}e^{-t}\,dt = x\Gamma(x).$$

□

Note: For $x = 1$, we have

$$\Gamma(1) = \int_0^\infty e^{-t}\,dt = -e^{-t}\Big|_0^\infty = 1,$$

and if $x = n$ is a positive integer, then

$$\Gamma(n+1) = n\Gamma(n) = n(n-1)\Gamma(n-1) = n(n-1)\cdots 3\cdot 2\cdot 1\cdot \Gamma(1) = n!.$$

In fact, if $x > 0$ is not necessarily an integer, then from the characteristic property of the gamma function, if n is any positive integer,

$$\Gamma(x+n) = (x+n-1)(x+n-2)\cdots x\cdot\Gamma(x);$$

thus,

$$\Gamma(x) = \frac{\Gamma(x+n)}{x(x+1)(x+2)\cdots(x+n-1)} \tag{A.1}$$

for all $n \geqslant 1$. The expression on the right side of (A.1) makes sense for any real number x such that $x+n > 0$; that is, $x > -n$. This allows us to extend the definition of the gamma function for negative x:

Definition A.4. The **gamma function** is defined by

$$\Gamma(x) = \begin{cases} \displaystyle\int_0^\infty t^{x-1}e^{-t}\,dt, & \text{for } \quad x > 0 \\ \displaystyle\frac{\Gamma(x+n)}{x(x+1)(x+2)\cdots(x+n-1)}, & \text{for } \quad -n < x < -n+1, \\ & \qquad n = 1, 2, 3, \ldots. \end{cases}$$

Thus, $\Gamma(x)$ is now defined for all $x \neq -n$ where n is a nonnegative integer. The graph of $\Gamma(x)$ is shown in Figure A.1.

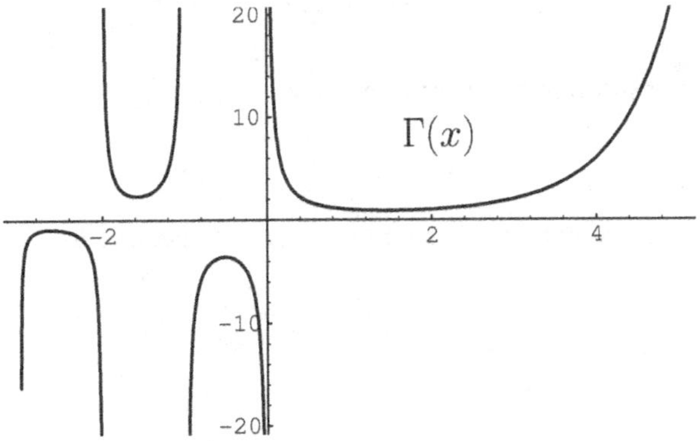

Figure A.1 Plot of $y = \Gamma(x)$.

We give a few more useful results.

Theorem A.5. If $\alpha \in \mathbb{R}$, with $0 < \alpha < 1$, then

$$\Gamma(\alpha)\,\Gamma(1-\alpha) = \frac{\pi}{\sin \pi\alpha},$$

and, in particular,

$$\Gamma\left(\frac{1}{2}\right) = \sqrt{\pi}.$$

Proof. For $0 < \alpha < 1$, we have

$$\Gamma(\alpha) = \int_0^\infty e^{-t} t^{\alpha-1}\, dt$$

and making the substitution $t = y^2$,

$$\Gamma(\alpha) = \int_0^\infty e^{-y^2} (y^2)^{\alpha-1} 2y\, dy$$

$$= 2\int_0^\infty e^{-y^2} y^{2\alpha-1}\, dy.$$

Therefore,

$$\Gamma(\alpha)\,\Gamma(1-\alpha) = \left(2\int_0^\infty e^{-y^2} y^{2\alpha-1}\, dy\right)\left(2\int_0^\infty e^{-x^2} x^{1-2\alpha}\, dx\right)$$

$$= 4\int_0^\infty \int_0^\infty \left(\frac{y}{x}\right)^{2\alpha-1} e^{-(x^2+y^2)}\, dx\, dy$$

$$= 4\int_0^{\frac{\pi}{2}} \int_0^\infty (\tan\theta)^{2\alpha-1} e^{-r^2}\, r\, dr\, d\theta$$

$$= 2\int_0^{\frac{\pi}{2}} (\tan\theta)^{2\alpha-1}\, d\theta,$$

and making the substitution $t = \tan^2\theta$, we have

$$\Gamma(\alpha)\,\Gamma(1-\alpha) = \int_0^\infty \frac{t^{\alpha-1}}{1+t}\, dt,$$

This last integral can be evaluated using the theory of residues, as

$$\int_0^\infty \frac{t^{\alpha-1}}{1+t}\, dt = \frac{\pi}{\sin \pi\alpha},$$

so that

$$\Gamma(\alpha)\,\Gamma(1-\alpha) = \frac{\pi}{\sin \pi\alpha} \tag{A.2}$$

for $0 < \alpha < 1$. Although this is proven only for $0 < \alpha < 1$, it can be extended to all nonintegral values of α by using the recurrence relation for the gamma function.

Evaluating (A.2) at $\alpha = 1/2$, we have

$$\Gamma\left(\frac{1}{2}\right)\Gamma\left(\frac{1}{2}\right) = \frac{\pi}{\sin(\pi/2)} = \pi,$$

so that

$$\Gamma\left(\frac{1}{2}\right) = \sqrt{\pi}.$$

□

One last formula is the following:

Theorem A.6. If $(1+\mu)/\mu$ is not a negative integer or 0, then

$$\int_0^\infty e^{-t^\mu}\,dt = \Gamma\left(\frac{1+\mu}{\mu}\right).$$

Proof. We have

$$\int_0^\infty e^{-t^\mu}\,dt = \frac{1}{\mu}\int_0^\infty e^{-s}s^{\frac{1}{\mu}-1}\,ds = \frac{1}{\mu}\Gamma\left(\frac{1}{\mu}\right) = \Gamma\left(\frac{1+\mu}{\mu}\right)$$

from the characteristic property of the gamma function. □

The main properties of the gamma function are summarized in

Theorem A.7. The gamma function satisfies the following:

(a) $\Gamma(x+1) = x\Gamma(x)$.

(b) $\Gamma(1) = 1$.

(c) $\Gamma(n+1) = n\Gamma(n) = n!$ for $n = 1, 2, 3, \ldots$.

(d) $\Gamma(\alpha)\Gamma(1-\alpha) = \pi/\sin \pi\alpha$, provided that α is not an integer.

BIBLIOGRAPHY

1. Abramowitz, M., and I.A. Stegun, eds., *Handbook of Mathematical Functions with Formulas, Graphs, and Mathematical Tables*, Dover Publications, Inc., New York, NY, 1972; also available at `http://dlmf.nist.gov/`.
2. Allen, L.J.S., *An Introduction to Mathematical Biology*, Pearson Education, Inc., Upper Saddle River, NJ, 2007.
3. Andrews, L.C., *Elementary Partial Differential Equations with Boundary Value Problems*, Academic Press, Inc., Orlando, FL, 1986.
4. Apostol, T.M., *Mathematical Analysis*, 2nd ed., Addison-Wesley Publishing Company, Inc., Reading, MA, 1974.
5. Asmar, N.H., *Partial Differential Equations with Fourier Series and Boundary Value Problems*, 2nd ed., Pearson Education, Inc., Upper Saddle River, NJ, 2005.
6. Berg, P.W., and J.L. McGregor, *Elementary Partial Differential Equations*, Holden-Day, Inc., San Francisco, CA, 1964.
7. Boas, M.L., *Mathematical Methods in the Physical Sciences*, John Wiley & Sons, Inc., New York, NY, 1966.
8. Boyce, W.E., and R.C. DiPrima, *Elementary Differential Equations and Boundary Value Problems*, 8th ed., John Wiley & Sons, Inc., Hoboken, NJ, 2005.
9. Brauer, F., and J.A. Nohel, *Introduction to Differential Equations with Applications*, Harper & Row, Publishers, New York, NY, 1986.
10. Bromwich, T.J., *Theory of Infinite Series,* 2nd ed., Blackie & Sons, Glasgow, 1926.

Partial Differential Equations: Theory and Completely Solved Problems, First Edition. By T. Hillen,
I. E. Leonard, H. van Roessel

11. Brown, J.W., and R.V. Churchill, *Fourier Series and Boundary Value Problems*, 7th ed., McGraw-Hill Higher Education, Boston, MA, 2008.

12. Carslaw, H.S., *An Introduction to the Theory of Fourier's Series and Integrals*, 3rd ed., Dover Publications, Inc., New York, NY, 1950.

13. Carslaw, H.S., and J.C. Jaeger, *Conduction of Heat in Solids*, 2nd ed., Oxford University Press, Oxford, England, 2000.

14. Chester, C.R., *Techniques in Partial Differential Equations*, McGraw-Hill Book Company, New York, NY, 1971.

15. Copson, E.T., *Partial Differential Equations*, Cambridge University Press, Cambridge, England, 1975.

16. Courant, R., and D. Hilbert, *Methods of Mathematical Physics*, vols. 1 and 2, John Wiley & Sons, Inc., New York, NY, 1996.

17. Dennemeyer, R., *Introduction to Partial Differential Equations and Boundary Value Problems*, McGraw-Hill Book Company, New York, NY, 1968.

18. Dettman, J.W., *Mathematical Methods in Physics and Engineering*, McGraw-Hill Book Company, Inc., New York, NY, 1962.

19. de Vries, G., T. Hillen, M. Lewis, J. Müller, and B. Schönfisch, *A Course in Mathematical Biology: Quantitative Modeling with Mathematical and Computational Methods*, SIAM, Philadelphia, PA, 2006.

20. Dym, H., and H.P. McKean, *Fourier Series and Integrals*, Academic Press, New York, NY, 1972.

21. Erdélyi, A., *Operational Calculus and Generalized Functions*, Holt, Rinehart and Winston, New York, NY, 1962.

22. Folland, G.B., *Fourier Analysis and Its Applications*, Brooks/Cole Publishing Company, Pacific Grove, CA, 1992.

23. Friedman, B., *Principles and Techniques of Applied Mathematics*, John Wiley & Sons, Inc., New York, NY, 1965.

24. Guenther, R.B., and J.W. Lee, *Partial Differential Equations of Mathematical Physics and Integral Equations*, Prentice-Hall, Englewood Cliffs, NJ, 1988.

25. Haberman, R., *Applied Partial Differential Equations with Fourier Series and Boundary Value Problems*, 4th ed., Pearson Education, Inc., Upper Saddle River, NJ, 2004.

26. Hardy, G.H., *A Course of Pure Mathematics*, 10th ed., Cambridge University Press, Cambridge, England, 1967.

27. Hellwig, G., *Partial Differential Equations*, B.G. Teubner, Stuttgart, Germany, 1977.

28. Hildebrand, F.B., *Advanced Calculus for Applications, Fourier Series and Boundary Value Problems*, 2nd ed., Prentice-Hall, Inc., Upper Saddle River, NJ, 1976.

29. Hille, E., *Lectures on Ordinary Differential Equations*, Addison-Wesley Publishing Company, Inc., Reading, MA, 1969.

30. John, F., *Partial Differential Equations*, 3rd ed., Springer-Verlag, New York, NY, 1978.

31. Keane, M.K., *A Very Applied First Course in Partial Differential Equations*, Prentice-Hall, Inc., Upper Saddle River, NJ, 2002.

32. Kreider, D.L., R.G. Kuller, and D.R. Ostberg, *Elementary Differential Equations*, Addison-Wesley Publishing Company, Inc., Reading, MA, 1968.
33. Kreider, D.L., R.G. Kuller, D.R. Ostberg, and F.W. Perkins, *An Introduction to Linear Analysis*, Addison-Wesley Publishing Company, Inc., Reading, MA, 1966.
34. Lax, P.D., *Linear Algebra and Its Applications*, 2nd ed., John Wiley & Sons, Hoboken, NJ, 2007.
35. Mathews, J., and R.L. Walker, *Mathematical Methods of Physics*, 2nd ed., W. A. Benjamin, Inc., Menlo Park, CA, 1964.
36. Miles, J., *Integral Transforms in Applied Mathematics*, Cambridge University Press, Cambridge, England, 1971.
37. Myint-U, T., and L. Debnath, *Linear Partial Differential Equations for Scientists and Engineers*, 4th ed., Birkhäuser, Boston, MA, 2007.
38. O'Neil, P.V., *Advanced Engineering Mathematics*, 2nd ed., Wadsworth Publishing Company, Belmont, CA, 1987.
39. Pinsky, M.A., *Partial Differential Equations and Boundary-Value Problems with Applications*, 3rd ed., American Mathematical Society, Providence, RI, 2011.
40. Polya, G., and G. Szegö, *Problems and Theorems in Analysis*, vols. I and II, Springer-Verlag, New York, NY, 1972.
41. Powers, D.L., *Boundary Value Problems*, 5th ed., Elsevier Science and Technology Books, San Diego, CA, 2006.
42. Randolf, J.F., *Basic Real and Abstract Analysis*, Academic Press, New York, NY, 1968.
43. Seeley, R., *An Introduction to Fourier Series and Integrals*, W. A. Benjamin, Inc., New York, NY, 1966.
44. Simmons, G.F., *Differential Equations with Applications and Historical Notes*, McGraw-Hill Book Company, New York, NY, 1972.
45. Sneddon, I.N., *Elements of Partial Differential Equations*, Dover Publications, Inc., Mineola, NY, 2006.
46. Sneddon, I.N., *Fourier Transforms*, Dover Publications, Inc., Mineola, NY, 1995.
47. Stade, E., *Fourier Analysis*, John Wiley & Sons, Inc., Hoboken, NJ, 2005.
48. Strauss, W.A., *Partial Differential Equations: An Introduction*, John Wiley & Sons, Inc., Hoboken, NJ, 2008.
49. Stromberg, K.A., *An Introduction to Classical Real Analysis*, Wadsworth, Inc., Belmont, CA, 1981.
50. Tikhonov, A.N., and A.A. Samarskii, *Equations of Mathematical Physics*, Pergamon Press Ltd., Oxford, England, 1963.
51. Weinberger, H.F., *A First Course in Partial Differential Equations with Complex Variables and Transform Methods*, Blaisdell Publishing Company, Waltham, MA, 1965.
52. Wylie, C.R., *Advanced Engineering Mathematics*, 4th ed., McGraw-Hill Book Company, New York, NY, 1975.
53. Young, N., *An Introduction to Hilbert Space*, Cambridge University Press, Cambridge, England, 1988.
54. Zauderer, E., *Partial Differential Equations of Applied Mathematics*, John Wiley & Sons, Inc., New York, NY, 1989.

INDEX

A

Abel's theorem, 381
Absolutely integrable, 269
Absorption cross section, 619
Acceleration due to gravity, 518
Advection equation, 6
Airy's equation, 193
Analytic, 191, 228
Anchor point, 331, 333, 599
Angular problem, 182, 189
Arc length, 173
Associated Legendre equation, 227, 249, 524
    Bounded solutions, 251
Associated Legendre functions of the first kind
    Degree n, order m, 251, 530
    Properties, 251
Associated Legendre functions of the second kind
    Degree n, order m, 251, 530
Auxiliary conditions, 10
Average of the jump, 58
Azimuthal angle, 222
Azimuthal angular problem, 476, 529

B

Bacterial flux, 24
Basis vectors, 44
Bessel function
    First kind of order $3m$, 517
    First kind of order μ, 195, 198
    First kind of order n, 468
    First kind of order p, 507
    Second kind of order $3m$, 517
    Second kind of order n, 201, 468
Bessel function J_0, 453, 504
Bessel functions, 195
    Asymptotic behavior, 201
    Derivatives, 202
    Differentiation formulas, 204
    Generating function, 206
    Integral representation, 204
    Integration formulas, 204
    Orthogonality relation, 210
    Oscillation, 202
    Properties, 201
    Recurrence relations
        $J_n(x)$, 203
        $J_n'(x)$, 203
    Spacing between zeros, 202
Bessel's equation, 119, 179, 188, 191
    Nonparametric form, 215
    Order μ, 194

Order $n + \frac{1}{2}$, 252
Bessel's equation of order $3m$, 517
Bessel's equation of order n, 468
Bessel's equation of order zero, 418, 453, 505, 513, 521
Bessel's inequality, 66, 69
Black-Scholes formula, 3
Boundary conditions, 4, 10, 89
Dirichlet conditions, 11
Homogeneous Dirichlet conditions, 11
Homogeneous Neumann conditions, 11
Mixed, 415
Neumann conditions, 11, 84
Nonhomogeneous, 95
Replaced by boundedness conditions, 116
Robin conditions, 11
Sturm-Liouville type, 117, 415
Boundary value problem, 117
Boundary value–initial value problem, 108, 188
Bounded spectrum, 125
Boundedness conditions, 116, 181, 319
Burger's equation, 14

C

Cauchy-Euler equation, 123, 190, 240, 256
Characteristic equation, 184, 191
Cauchy-Schwarz inequality, 69, 244
Cell density, 24
Change of coordinates, 9
Characteristic curves, 331, 337, 597
Characteristic equation, 91
Characteristic equations, 337, 339
Characteristic ODEs, 332
Circular harmonics, 467
Circular plate
Steady-state temperature distribution, 180
Coefficient matrix, 402
Coefficient matrix of a PDE, 8
Complete problem for a PDE, 11
Complete well-posed problem, 163
Complex form of Fourier series, 364
Complex Fourier series, 80
Conservation law
Energy, 17
Mass, 17
Conservation of energy, 28
Vibrating string, 496
Constant coefficient equation, 91
Continuous, 39
Convective velocity, 442
Converges absolutely, 661
Convolution, 567
Definition, 289
Gauss-Weierstrass kernel, 328
Properties, 289, 574
Sine function, 290
Theorem, 290
Convolution product, 289
Convolution theorem, 290, 567
Critical radius, 619
Curve of anchor points, 336

D

d'Alembert's general solution, 158
d'Alembert's solution, 156, 157, 300, 329, 344, 490, 532
Damped wave equation, 659
Density of substance, 27
Dependent variable, 5
Derivatives of Fourier transforms, 282
Diffusion equation, 24
Diffusion through a cell membrane, 16
Dimension of a PDE, 6
Directional derivative, 173
Dirichlet conditions, 11, 25, 32
Dirichlet's discontinuous factor, 265
Dirichlet's integral, 271
Dirichlet's theorem, 128, 134, 213, 269, 382, 595
see also Fourier's theorem, 57
Discontinuity of the first kind, 39
Jump discontinuity, 40
Removable discontinuity, 40
Discontinuity of the second kind, 40
Discontinuous, 39
Divergence theorem, 29
Dot product, 43
Driving force, 397
Dump truck, 597

E

Eigenfunction expansion, 127, 132
Eigenfunction-eigenvalue pair, 116
Eigenfunctions, 86, 108, 116, 132, 151
Complete, 411
Nonunique, 184
Orthogonal, 175, 411
Two dimensions, 172
Eigenpair, 116, 417
Eigenvalue problem, 116
Periodic, 404
Stable, 134
Unstable, 134
Eigenvalues, 86, 108, 116, 132, 151
Leading, 134
Nonnegativity, 176

Ordered, 125
Elliptic PDE, 8
Energy method, 396
Equidimensional equation, 123, 190
Equilibrium solution, 16, 98, 425
Equilibrium temperature distribution, 425
Error function, 308, 329, 586
Euclidean basis, 43
Euler-Mascheroni constant, 201
Euler's formula, 78
Even extension, 42, 53, 134, 268, 285
Even function, 41, 52, 268
Even periodic extension, 42
Exact differential, 415
Exterior Dirichlet problem for a disk, 552

F

Fick's first law, 24
Fick's second law, 25
First-order linear ODE, 116
Fisher's equation, 654
Floor function, 356
Flux vector, 29
Forcing term, 29
Fourier coefficients, 50
Complex, 80
Fourier cosine integral, 268
Fourier cosine series, 53, 134, 152, 374
Term-by-term differentiation, 71
Fourier cosine transform, 285, 287, 578, 590
Of derivatives, 288
Operational properties, 288
Fourier integral, 262
Fourier integral coefficients, 264
Fourier integral formula, 262, 568
Fourier integral representation, 264, 568
Fourier integral theorem, 264, 276
Proof, 276
Fourier series, 48, 50
Bessel's inequality, 66
Complex, 79, 364
Linearity, 64, 370
Mean square error, 75
Parseval's identity, 78
Representation, 50
Term-by-term differentiation, 70
Term-by-term integration, 72
Uniqueness, 63
Fourier sine integral, 269
Fourier sine series, 53, 152, 374
Term-by-term differentiation, 71
Fourier sine transform, 284, 287, 580
Of derivatives, 288
Operational properties, 288
Fourier transform, 279, 567
Derivatives of, 283
Duality, 298
ω-world, 298
x-world, 298
Gaussian function, 296
Linearity, 281, 587
Methods in PDEs, 299
Of derivatives, 282
Of integrals, 283
Operational properties, 281
Reciprocity relation, 573
Rectangular pulse, 264
Shift theorems, 281, 588
Fourier transform of a Gaussian function, 296
Fourier transform pair, 279
Fourier transform worlds
ω-world, 299
x-world, 299
Fourier's law, 27
Fourier's theorem, 57, 382
Fourier-Bessel expansion, 510
Fourier-Bessel expansion theorem, 213
Fourier-Bessel series, 210, 213, 454, 470, 508
Fourier-Bessel transform, 595
Fourier-Legendre expansion, 246
Fourier-Legendre expansion theorem, 245
Fourier-Legendre series, 245, 561
Free-stream Mach number, 653
Free-stream velocity, 653
Frequencies of oscillation, 85, 88
Frequency domain, 298
Frequency of oscillation, 500
Frictional force, 31
Frobenius series, 193
Fundamental solution to heat equation, 306

G

Gamma function, 198, 661
Definition, 663
Functional equation, 662
Graph, 663
Properties, 665
Gauss-Weierstrass kernel, 308, 328
General linear homogeneous PDE, 89
General second-order linear homogeneous ODE
Irregular singular point, 192
Ordinary point, 191
Regular singular point, 192
Singular point, 191
General solution, 85, 91
Homogeneous equation, 397
Nonhomogeneous equation, 397
Generalized Fourier coefficients, 128

Generalized Fourier series, 109, 128
Geometric series, 380
Gibbs' phenomenon, 61, 81
Gradient, 5
Gravitational acceleration g, 519
Green's first identity, 173
Green's identities, 172
Green's identity
First, 173
Second, 173
Green's second identity, 174
Green's theorem, 172
Gregory's series for $\tan^{-1}x$, 381

H

Hanging chain problem, 518
Hankel transform, 595
Hankel's integral theorem, 595
Harmonic
At a point P, 180
In a region R, 180
Harmonic functions, 163, 180, 319, 533
Heat capacity, 27
Heat equation, 3, 6, 24, 26, 305, 311
Dirichlet conditions, 427
Fundamental solution, 306, 308
Gauss-Weierstrass kernel, 308
Homogeneous, 427
Neumann conditions, 432
Nonhomogeneous, 95, 107
Rectangle plate, 150
Steady source, 448
Thin two-dimensional plate, 447
Three-dimensional, 224
Two-dimensional, 150
Heat flow
Circular disk, 451, 465
Time-independent source, 451
Hemisphere, 473
Hemispherical solid, 558
Homogeneous bar, 456
Poorly insulated sides, 456
Robin conditions, 458
Infinite rod, 305
Method of images, 558
Quarter-plane, 554
Rectangular plate, 150
Semi-infinite plate, 463
Semi-infinite rod, 311
Dirichlet condition, 311
Neumann condition, 317
Spherical shell, 471
Steady-state, 558
Heat flux vector, 27
Heat kernel, 313
Heat transport
Nonhomogeneous rod, 445
Heat transport with convection, 442
Helmholtz equation, 172, 175
Homogeneous, 224
Surface, 226
Hemispherical solid, 526
Hermite polynomials, 632
Hermite's equation, 632
Homogeneous cylindrical shaft
Torsional oscillations, 645
Homogeneous PDE, 5
Homogeneous second-order linear ODE, 115
Hooke's law, 29
Hyperbolic PDE, 8

I

Imperfect insulation, 149
Independent variables, 5
Indicial equation, 197
Indicial exponent, 193, 196
Infinite discrete set, 119, 172
Infinite domain, 261
Infinite domains, 299
Infinite spectrum, 125
Initial conditions, 4, 10, 90
Initial curve, 336
Initial displacement, 32, 87
Initial value problem, 110
Initial velocity, 32, 88, 488
Inner product on $PWC(a, b)$, 47
Inner product, 47
Norm, 47
Null function, 47
Orthogonality, 47
Weight function, 47
Inner product space, 46
Input function, 2π-periodic, 397
Insulation
Imperfect, 149
Perfect, 149
Integral curves, 337
Integrating factor, 110, 115
Interior Dirichlet problem, 35
Disk, 179, 188
Interior Dirichlet problem for a sphere, 253
Internal heat generation, 27, 145, 150
Inverse Fourier cosine transform, 287, 578
Inverse Fourier sine transform, 287, 580
Inverse Fourier transform, 279, 567
Isotherms, 323

J

Jacobian determinant, 9, 180, 222

Jump discontinuity, 40

K

KdV equation, 6
Kelvin's transformation, 562
Kinetic energy, 496
Kingdoms of PDEs, 7
Kronecker delta, 43

L

Lagrange's trigonometric identity, 366
Laguerre polynomials, 403
Laplace's equation, 3, 6, 34, 163, 239, 533
  Circular annulus, 548
  Disk, 179
  Half-plane, 319
  Kelvin's transformation, 562
  Maximum principle, 167
  Quarter-plane, 554
  Rectangle, 163, 538, 539, 658
  Semicircular disk, 629
  Semi-infinite strip, 324
  Spherical coordinates, 254
  Spherical harmonics, 240
  Square, 543
  Unit square, 545
   Temperature at center, 547
Laplacian, 5
  Cylindrical coordinates, 534
  Polar coordinates, 180, 534
  Spherical coordinates, 222, 531
Leading eigenvalue, 134
  Upper and lower estimates, 134
Legendre functions, 227, 229
Legendre functions of the second kind, 248
Legendre polynomial of degree n, 476
Legendre polynomials, 118, 228, 231, 232, 559, 647
  Boundedness, 243
  Formula for, 231
  Generating function, 238
  L^2 norm, 237
  Orthogonality, 232
  Properties, 244
  Recurrence relations, 234, 241
  Rodrigues' formula, 233
Legendre's equation, 118, 227, 228, 472, 476, 559, 647
  Bounded solutions, 249
Leibniz's formula, 24
  For $\frac{\pi}{4}$, 381
Leibniz's rule
  Differentiating a product, 250
  Differentiating an integral, 24
Length of a vector, 45
Linear combination, 92
Linear density, 247
Linear differential operator, 5
  Fourth order, 415
Linear homogeneous PDE, 14
Linear operator, 13
Linear PDE, 5, 12
  Homogeneous, 13
  Nonhomogeneous, 13
Linear sink, 334
Linear source, 334
Lommel's integrals, 417

M

Maclaurin series, 199
Maximum principle
  Laplace's equation, 167, 186
  Poisson's equation, 167
Maxwell's equations, 3
Mean of the jump, 58
Mean square error, 75
Method of characteristics, 331, 597
Method of eigenfunction expansions, 100, 108, 448
Method of Frobenius, 191, 193
Method of images, 311, 558, 593
  Odd extension of boundary data, 558
Method of separation of variables, 93
Method of undetermined coefficients, 398
Method of variation of parameters, 109
Mixed boundary conditions, 415
Moment generating function, 611

N

Natural frequency of vibration, 398, 515
Neumann conditions, 11, 12, 32
Neumann function, 201
Neumann problem
  Circular annulus, 551
  Compatibility condition, 552
Neutron flux, 619
Newton's law of cooling, 11
Newton's second law, 519
No internal heat generation, 465
Nonhomogeneities, 95
Nonhomogeneous boundary conditions, 107
Nonhomogeneous equations, 95
Nonhomogeneous PDE, 5, 15
Nonnegativity of eigenvalues, 176
Nonparametric Bessel's equation
  Order n, 190
Nonphysical boundaries, 181
Nonsingular transformation, 9

Nontrivial solutions, 86, 90, 116, 132
Norm of a vector, 45
Normal mode of vibration, 398
Normal vector to a surface, 336
Normalization constant, 424, 461
Nuclear fission, 619

O

Odd extension, 42, 53, 268, 284
Odd function, 41, 52, 268
Odd periodic extension, 42, 93
Oil sands, 597
One-dimensional wave equation, 29, 153, 532
  d'Alembert's solution, 300, 344
  Damped, 89
  Dirichlet boundary conditions, 481
One-sided limits, 39
Order of a PDE, 6
Ordered eigenvalues, 125
Ordinary differential equation, 85, 90, 109
Ordinary point, 191, 228
Orthogonal basis, 45
Orthogonal functions, 43
Orthogonal set of eigenfunctions, 228
Orthogonal set of functions on (a, b), 48
Orthogonality, 87, 109, 132
  Cosines, 357
  Sines, 358
Orthogonality of eigenfunctions, 175
Outward-pointing unit normal, 173

P

Parabola $f(x) = x^2,\ -a \leqslant x \leqslant a$
  Differentiating Fourier series, 363
  Fourier series, 362
Parabolic PDE, 8
Parseval's identity, 78
Partial derivatives, 4
Partial differential equation, 3, 5
  Coefficient matrix, 8
  Elliptic, 3, 8
  Homogeneous, 5
  Hyperbolic, 3, 8
  Linear, 4, 5
  Nonhomogeneous, 5, 95
  Parabolic, 3, 8
  Principal part, 7
Partial sum
  Fourier series, 60
Particular solution, 15, 397
Perfect insulation, 149
Perfectly elastic string, 29
Period of oscillation, 500
Periodic boundary conditions, 36
Periodic extension, 42
Periodic function, 41
Periodicity conditions, 36, 181, 254, 400, 415
Phase shift, 156
Phase velocity, 156
Pie-shaped membrane, 515
Piecewise continuous, 279
  On every finite interval, 279
Piecewise continuous (PWC), 40
Piecewise smooth, 134, 279
  On every finite interval, 279
Piecewise smooth (PWS), 41
Poisson integral formula
  Disk, 186, 188, 466
  Half-plane, 321, 564
Poisson's equation, 34
Polar angle, 222
Polar angular problem, 529
Polar coordinate transformation, 180
Polar coordinates, 180
Pole O, 180
Population dynamics, 24
Potential
  Disk, 180
  Rectangle, 163
Potential energy, 496
Potential equation, 34, 163
Potential function, 239
Potential functions, 180
Potential in a sphere, 253
Potential in a spherical domain, 239
Principal part, 7
Principle of superposition, 14
Product solution, 90
Putnam competition, 614

Q

Quadratic test function, 138
Quasilinear PDEs, 348

R

Radial problem, 182, 189, 476, 528
Rayleigh quotient, 126, 134, 135, 190, 225, 411, 419, 446
Reaction-diffusion equations, 3
Real analytic, 191
Rectangular pulse, 264, 293
Recurrence relation, 197, 228
Regular Sturm-Liouville problem, 116, 132
  Eigenfunctions, 119
    Orthogonal, 119
  Eigenvalues, 119
    Distinct, 119
    Real, 121

Properties, 119
Spectrum, 119
Removable discontinuity, 40
Restoring force, 31
Riemann sum, 263
Ripping membrane, 507
Robin conditions, 11, 12, 32
Rule of thumb for side conditions, 12, 181

S

Sawtooth function, 50
Schauder basis, 413
Schrödinger's equation, 3
Second-order linear PDEs, 7
Sectionally continuous, 40
Self-adjoint form, 115
Semi-infinite strip, 324
Separated solution, 19, 83
Separation constant, 20, 90, 151
Separation of variables, 4, 19, 83, 93, 150
Series solutions of ODEs, 191
Side conditions, 4, 10
Auxiliary conditions, 10
Boundary conditions, 10
Initial conditions, 10
Simple discontinuity, 39
Singular Sturm-Liouville problem, 116
Spatial equation, 85
Specific heat, 442
Spectral decomposition, 141
Spectral theorem
Sturm-Liouville problems, 411
Spectrum, 116
Sphere of uranium, 619
Spherical Bessel equation, 226, 252
Order n, 530
Spherical Bessel functions, 226, 252
Properties, 253
Spherical Bessel functions of the first kind
Order $2m+1$, 477
Order m, 530
Order n, 252
Spherical Bessel functions of the second kind
Order $2m+1$, 477
Order n, 252
Spherical Bessel's equation
Order $2m+1$, 477
Spherical coordinate transformation, 222
Spherical coordinates, 221
Spherical harmonics, 226, 240
Spherical waves, 531
Spring constant, 33
Steady-state solution, 16, 98, 425, 448
Semi-infinite plate, 463
Heat loss at faces, 463
Stochastic birth-death process
Moment generating function, 611
Sturm-Liouville form, 115, 117, 190
Sturm-Liouville problem, 116
Singular, 116, 228
Spectral theorem, 141
Sturm-Liouville spectral theorem, 125
Superposition principle, 12, 87, 92, 152
Surface harmonics, 226
Symbol of a PDE, 8

T

Taylor series, 191, 380
Taylor series expansion, 228
Telegrapher's equation, 31
Temporal problem, 189, 529
Tension, 31
In chain, 519
Term-by-term differentiation, 70
Fourier cosine series, 71
Fourier sine series, 71
Term-by-term integration
Fourier series, 72
Test function, 137, 138
Linear, 138
Quadratic, 138
Thermal conductivity, 27, 246
Thermal diffusivity, 29, 437, 447
Time equation, 85, 151
Time problem, 529
Time-dependent source, 107
Torsional oscillations, 645
Total energy, 496
Transcendental equation, 129, 130, 148, 421, 460
Transform, solve, invert, 300
Transformation
Jacobian determinant, 9
Nonsingular, 9
Transient solution, 98
Transient temperature, 475
Translation of initial data, 334
Traveling wave, 156, 334
Moving to the left, 156, 157
Moving to the right, 156, 157
Traveling wave solutions, 654
Triangular wave, $2a$-periodic, 356
Trigonometric identity, 357, 358
Lagrange's, 366
Trigonometric polynomial, 74
Trivial solutions, 20, 85, 116
Two-dimensional heat equation, 150
Two-dimensional wave equation, 168

Rectangle, 169
Type of a PDE, 8

U

Uniform convergence
Fourier series, 68
Sequences, 68
Uniform material, 31
Uniform vibrating string, 496
Unit tangent vector, 173
Upper half-plane, 319

V

Variation of parameters, 109
Vector field, 336
Velocity potential, 653
Vertical displacement, 154, 160
Finite string, 154, 496
Membrane, 169
Semi-infinite string, 160
Vibrating circular membrane, 188, 507, 511
Radially symmetric, 504
Solution, 214
Vibrating membrane
Square, 501

W

Wave equation, 3, 6, 29
Damped, 659
Damping term, 89
Disk, 512
Hanging chain, 518
Hemisphere, 526
Homogeneous Neumann problem, 84
Infinite domain, 300, 608
Pie-shaped membrane, 515
Spherical waves, 531
Three-dimensional, 224
Two-dimensional, 168, 501
Wave operator, 345
Weber function, 201
Weierstrass M-test, 69
Weight function, 47, 128
Weight of chain, 519

Z

Zero matrix, 402
Zonal harmonics of order n, 258

Printed in the USA
J092507SCI010615 01S29053000000000229